FANUC Certified Edu

CNC Machining Center Programming, Setup, and Operation

A Guide to Mastering the Use of CNC Machining Centers

SECOND EDITION
Now stresses model 0iF and shows examples using the Metric System

MIKE LYNCH

Developed jointly with FANUC America

© 2017 CNC Concepts, Inc.
44 Little Cahill Road • Cary, IL 60013
Phone 847.639.8847 • Fax 847.639.8857
Internet: www.cncci.com
Email: lynch@cncci.com

5/24/17

Notice
The information in this book is meant to supplement, not replace, proper CNC machine usage training. In all cases, information contained in the machine tool builder's manual will supersede material presented in this book. Like any industrial equipment that requires skill to master, CNC machine tools pose some risk. The publisher advises readers to take full responsibility for their safety and know their limits, as well as the limits of the CNC machine tool being used. Before practicing the skills and techniques described in this book, read all safety advice presented in the machine tool builder's manuals, be sure that your equipment is well maintained, and do not take risks beyond your level of experience, aptitude, training, and comfort level.

Copyright 2017, CNC Concepts, Inc.

All rights reserved. No part of this publication may be reproduced or transmitted in any form or by any means, electronic or mechanical, including photocopying, recording, or any other information storage and retrieval system, without the written permission of the publisher.

Library of Congress Cataloging-in-Publication Data

Lynch, Mike

FANUC CNC Machining Center Programming, Setup, and Operation - 2nd Edition   p.   cm.

Includes index.

ISBN-13: 978-1546946731

ISBN-10: 154694673X

Distributed to the book trade by CNC Concepts, Inc.

Table of Contents

Preface 12
   Prerequisites 12
      Basic Machining Practice Experience 12
      Math 13
      Motivation 13
   Controls Covered 13
      Controls Other Than FANUC 14
   Limitations 14
   Scope 14
   Key Concepts Approach 15
      Lesson Structure 15
      Practice, Practice, Practice! 15
      Key Concepts and Lessons 15
   New in this Edition 16
   About Measurement System Examples 16
   Enjoy! 17

KNOW THE MACHINING CENTER FROM A PROGRAMMER'S VIEWPOINT 18
   Experienced with Conventional (non-CNC) Machine Tools? 18

Lesson 1.1 - Machine Configurations 20
   Objectives 20
   Introduction 20
   Vertical Machining Centers 20
      C-frame Style 20
         Directions of Motion (axes) for a C-Frame-Style Vertical Machining Center 21
         Axis Polarity 21
      Knee-Style Vertical CNC Milling Machines 22
      Bridge-Style Vertical Machining Center (also called gantry-style) 23
   Horizontal Machining Centers 24
      Directions of Motion (axes) for a Horizontal Machining Center 25
         Axis Polarity 25
   Programmable Functions of a Machining Center 26
      Selecting the Measurement System 26
      Spindle 27
         Spindle Speed 27
         Spindle Activation and Direction 28
         Spindle Range 28
      Feedrate 29
         Feed-Per-Minute 29
         Feed-Per-Revolution 30
      Coolant 30
      Automatic Tool Changer 30
   What Else Might be Programmable? 31
   Key Points for Lesson 1.1: 32
   Quiz 32

Lesson 1.1a – CNC Certification Cart Machine Configuration 34
   Objectives 34
   CNC Certification Cart Vertical Machining Center 34
      Directions of Motion (axes) for the CNC Certification Cart 35
   Basic Programmable Functions of the CNC Certification Cart 35
      Spindle 35
         Spindle Speed 35
         Spindle Activation and Direction 35
         Spindle Range 35
      Feedrate 35
      Coolant 36
      Automatic Tool Changer 36
      Measurement System Mode (inch or metric) 37
   Key Points for Lesson 1.1a: 37
   Quiz 37

Lesson 1.2 – CNC Job Workflow 38
   Objectives 38
   Companies that Use CNC Machining Centers 38
      What Will You be Doing? 39
   The CNC Job Workflow 40
      Study the Workpiece Drawing 40
      Decide which CNC Machining Center to Use 40
      Determine the Sequence of Operations and Tooling 40
      Order and Check Tooling 40
      Create the Part Program 40
      Create Setup and Production Documentation 40
      Load the Program Into the CNC's Memory 41
      Setup the Machine 41
      Verify the Program 41
      First Part Inspection 41
      Production 41
      Store the Corrected Version of the Program 41
   Key Points for Lesson 1.2: 41

Lesson 1.3 – Visualizing the Execution of a Program 44
   Objectives 44
   Introduction 44
   Program Structure 45
   Order of Program Execution 45
      An Example of Program Execution 45
         Manual Milling Machine Procedure: 46
         CNC Program 46
   Sequence Numbers 47
   Decimal Point Programming 48

Modal Words 48
Initialized Words 48
   Word Order in a Block 48
Common Beginner Mistakes 48
   Omitting Decimal Points 48
      Decimal Point Tip 49
   Other Mistakes of Omission 49
   Letter O or Number Zero? 49
Key Points for Lesson 1.3: 50
Quiz 50

Lesson 1.4 – Understanding the Workpiece Coordinate System 52
Objectives 52
Introduction 52
A Graph Analogy 52
   What About the Z-Axis? 54
Understanding Polarity 54
Wisely Choosing the Workpiece Coordinate System Zero Location 57
   Selecting the X and Y Workpiece Coordinate System Zero 57
      A Reminder About Axis Movement 58
   Selecting the Z-Axis Workpiece Coordinate System Zero 59
Absolute Versus Incremental Modes 59
A Decimal Point Reminder 61
Key Points for Lesson 1.4: 61
Quiz 62

Lesson 1.5 – Determining Workpiece Coordinate System Offsets 64
Objectives 64
Introduction 64
Determining the Workpiece Coordinate System Offset Values 65
   What is Machine Zero Position? 67
How to Determine the Workpiece Coordinate System Offset Values 67
   How Predictable is the Work Holding? 67
   What is a Qualified Work Holding Setup? 68
Machine, Workpiece, and Relative Coordinate Systems 68
   Using the Relative Coordinate System and an Edge-Finder for Measurement 69
Manually Measuring the Workpiece Coordinate System Offset Values 71
   Measuring the X and Y Axes for a Rectangular-Shaped Workpiece 72
      Step 1: Load workpiece and edge-finder 72
      Step 2: Touch off the workpiece edge in X 72
      Step 3: Set the relative position for the X-axis to zero 73
      Step 4: Back off in Z, move over 5.0 millimeters in X 73
      Step 5: Zero the X-axis relative position again 74
      Step 6: Touch off the workpiece edge in Y 74
      Step 7: Set the relative position for the Y-axis to zero 75
      Step 8: Back off in Z, move over -5.0 millimeters in Y 75
      Step 9: Reset the Y-axis relative position again 76
      Step 10: Send the X and Y axes to the zero return (reference) position 76
   Measuring the X and Y Axes for a Circular-Shaped Workpiece 76
      Step 1: Load the workpiece and dial indicator 77
      Step 2: Indicate the center of the workpiece. 77
      Step 3: Zero the X and Y relative positions 77
      Step 4: Send the X and Y axes to the zero return position 78
   Measuring the Z-axis 78
      Step 1: Make the work holding setup and load a workpiece 78
      Step 2: Touch the spindle nose to the gauge block 79
      Step 3: Preset the Z-axis relative position to the height of the gauge block 79
      Step 4: Send the Z-axis to the zero return position 80
Measuring the Workpiece Coordinate System Offsets with a Probe 80
How the Workpiece Coordinate System Zero Selection Affects the Setup Person 81
Key Points for Lesson 1.5: 83
Quiz 83

Lesson 1.6 – Setting Workpiece Coordinate System Offsets 84
Objectives 84
Introduction 84
Understanding Workpiece Coordinate System Offsets 84
A Full Example of Using the Workpiece Coordinate System 86
   The Program 86
      What is the G91 Doing in Block N090? 87
   The Setup 87
   What is the External/Common Work Coordinate System Offset? 88
Key Points for Lesson 1.6: 89

Lesson 1.7 – Introduction to Programming Words 92
Objectives 92

Introduction 92
   Decimal Points in Words 93
O-Word 93
N-Word 93
G-Word 93
X-Word 94
Y-Word 94
Z-Word 94
A-Word 94
B-Word 94
C-Word 94
      Note About Rotary Axis Designators and Indexer Activators 95
R-Word 95
I, J, K-Words 95
Q-Word 95
P-Word 95
L-Word 95
F-Word 95
M-Word 96
S-Word 96
T-Word 96
H-Word 96
D-Word 96
EOB (end of block character) 96
/ (slash code) 97
G-codes and M-codes 97
   G-codes 97
      G-code Limitations: 97
      Optional G-Codes 97
      What Does Initialized Mean? 97
      What Does Modal Mean? 97
      Commonly Used G-Codes 98
   0i-F G-codes 98
   M-codes 100
      Common M-codes Used on a CNC Machining Center 100
Key Points for Lesson 1.7: 101

YOU MUST PREPARE TO WRITE PROGRAMS 102
Preparation and Time 102
Preparation and Safety 103
Typical Mistakes 104
   Syntax Mistakes 104
   Motion Mistakes 104
   Mistakes of Omission 104
   Process Mistakes 104

Lesson 2.1–Preparation for Programming 106
Objectives 106
Introduction 106
Plan the Sequence of Machining Operations 106
Develop the Cutting Conditions 108
   An Example 109
   Cutting Conditions Can Be Subjective 110
Do the Math and Mark-up the Print 111
   Marking up the Workpiece Engineering Drawing 113
   Doing the Math 113
   What About Contour Milling Operations? 115
      Planning a Milling Cutter's Centerline Tool Path 116
Check the Required Tooling 116
Plan the Work Holding Set-up 117
Other Documentation Needed for the Job 118
   Production Run Documentation 118
   Program Listing 119
Is it all Worth It? 119
Key Points for Lesson 2.1: 119

UNDERSTANDING MOTION TYPES 122
Understanding Interpolation 122

Lesson 3.1–Programming the Three Basic Motion Types 126
Objectives 126
Introduction 126
   Motion Type Commonalities 126
Understanding the Point Programmed for Each Cutting Tool 127
   Hole-Making Tools 127
      Center Drill 127
      Spot Drill (not shown above) 127
      Drill 127
      Reamer 127
      Tap 128
      Boring Bar 128
   Milling Cutting Tools 128
G00 Rapid Motion (also called positioning) 129
   How Many Axes Can Be Included in a Rapid Motion Command? 131
   About Dog-Leg Motion... 131
   When Do You Use Rapid Motion? 131
   What Is a Safe Approach Distance? 132
G01 Linear Interpolation (straight-line motion) 133
   Using G01 for a Fast-Feed Approach 135
   A Milling Example 136
   Drill Holes with G01? 137
G02 and G03–Circular Interpolation (circular motion) 138
   Which Positions to Program 138
   Specifying Arc Size with the R-word 138
      The R-Word is Not Modal 141

More About Clockwise Versus Counterclockwise Circular Interpolation 141
Specifying Circular Interpolation with Directional Vectors (I, J, and K) 142
Arc Limitations 143
    Simultaneously Specifying R with I, J or K 143
    Specifying a Semicircle 143
    Specifying Arcs Greater than 180° Using the R-word 143
    Programming a Full Circle 143
        Example Using I, J and K 144
        Example Using R-word 145
Planning Your Own Tool Paths 147
Key Points for Lesson 3.1: 147
Exercise: Work On Your First CNC Program 148

KNOW THE COMPENSATION TYPES 152

Lesson 4.1–Introduction to Compensation 154
Objectives 154
What are Compensations and Why are They Needed? 154
More On Tolerances 155
    The Initial Setting for Compensation 155
    When is Trial Machining Required? 155
    What Happens as Tools Begin to Wear? 156
    What do You Shoot For? 156
    Why do Programmers Have to Know About Compensations? 156
Understanding Offsets 156
    Offset Organization 157
        Offsets Related to Cutting Tools 157
        Offsets Related to Work Coordinate System Zero 158
    How Are Offsets Instated? 159
Key Points for Lesson 4.1: 159

Lesson 4.2–Tool Length Compensation 160
Objectives 160
Introduction 160
Reasons Why Tool Length Compensation is Needed 160
    No Two Tools Will Have Exactly the Same Length 160
    Tool's Length Will Vary Each Time it is Assembled 161
    Tool Data is Entered Separately From the Program 161
    Sizing and Trial Machining Must Often be Done 161
        What About Interference and Reach? 161
    Programming Tool Length Compensation 161
        Choosing the Offset Number to Be Used with Each Tool 162

An Example Program 162
The Setup Person's Responsibilities with Tool Length Compensation 165
    Using the Tool's Length as the Tool Length Compensation (offset) Value 165
        Determining Tool Length Compensation Values 166
        Measuring Tool Lengths Off Line–with a Tool Length Measuring Gauge 167
        Entering Tool Length Compensation Offsets 167
    Using the Distance from the Tool Tip to Workpiece Coordinate System Zero as the Tool Length Compensation (offset) Value 169
        Measuring tool length compensation values with this second method 169
Reasons for Using the Recommended Method 170
Typical Mistakes with Tool Length Compensation 170
    Forgetting to Instate Tool Length Compensation 170
    Forgetting to Enter the Tool Length Compensation Value 171
    Mismatching Offsets 171
Trial Machining with Tool Length Compensation 171
    When Trial Machining is Not Required 172
    Sizing with Tool Length Compensation 172
    A Tip for Remembering Which Way to Adjust the Offset 172
    What If I Use the Second Method Shown for Tool Length Compensation? 172
    Do I Have to Make All These Calculations When Adjusting Offsets? 172
    Why Can't I Just Change the Z Coordinate/s In the Program to Make Sizing Adjustments? 172
Key points for lesson 4.2: 173
Exercise: Practice Programming Tool Length Compensation 173

Lesson 4.3–Cutter Radius Compensation 178
Objectives 178
Introduction 178
    Will You Need to Learn this Feature? 178
Reasons Why Cutter Radius Compensation is Required 178
    Calculations Are Simplified for Manual Programmers 178
    Range of Cutter Sizes 180
        Do You Use Re-Sharpened (re-ground) Cutters? 181
    Trial Machining and Sizing 181
    Rough and Finish Milling with the Same Coordinates 182

Table Of Contents

Do You Have a CAM System? 182
How Cutter Radius Compensation Works 182
Steps to Programming Cutter Radius Compensation 183
   Step 1: Instate Cutter Radius Compensation 184
      The XY Motion To the Prior Position 184
      The Z Motions To the Z-Axis Work Surface 185
      Instate Cutter Radius Compensation and Position the Cutting Tool To the First Surface To Mill 186
      G41 or G42? 186
      The Offset Used with Cutter Radius Compensation 187
      The Motion to the First Work Surface 188
   Step 2: Program the Tool Path to Be Machined 189
   Step 3: Cancel Cutter Radius Compensation 192
      Don't Forget To Cancel! 193
      What If I Have More Than One Contour To Mill? 194
Examples 194
   Example 1 194
   Example 2 196
What if I Use a CAM System to Prepare Programs? 197
Setup Person's Cutter Radius Compensation Responsibilities 198
Rough and Finish Milling with the Same Set of Tool Path Coordinates 198
   A Warning 201
Trial Machining with Cutter Radius Compensation 201
   What If the Milling Cutter Machines On Both Sides of the Workpiece? 202
   A Tip for Remembering Which Way to Adjust the Offset 203
   How Important Is It To Make Your First Workpiece a Good One? 203
   When Trial Machining Is Not Required 203
Sizing with Cutter Radius Compensation 203
   What Do You Shoot For When Making Sizing Adjustments? 204
   What If I Use a CAM System and the Offset Value Is the Deviation From the Planned Cutter Size? 204
   Do I Have to Make All These Calculations When Adjusting Offsets? 204
   Why Can't I Just Change Programmed Coordinate/s In the Program To Make Sizing Adjustments? 204
Key Points for Lesson 4.3: 204
Exercise: Practice Programming Cutter Radius Compensation 205

Lesson 4.4–Workpiece Coordinate System Offsets 210

Objectives 210
Introduction 210
   Do You Need to Learn More about Workpiece Coordinate System Offsets? 210
Assigning Multiple Workpiece Coordinate System Zero Points 211
   Programming with Multiple Workpiece Coordinate System Zero Points 213
      The Potential Trade-Off with this Method 216
      Reminder About Tool Length Compensation Values 216
Shifting the Workpiece Coordinate System Offsets Reference Point 216
Programming Workpiece Coordinate System Offset Entries 218
Some Other Applications for the External Work Offset 219
   Allowing for Variations in Pallet Changers 219
   Allowing for Variations after a Mishap 219
   Differences in Spindle Gap from One Machine to Another 219
   To Enhance Safety During Dry-Runs 219
Key Points for Lesson 4.4: 220

YOU MUST PROVIDE STRUCTURE TO YOUR CNC PROGRAMS 222

Lesson 5.1–Introduction to Program Structure 224

Objectives 224
Introduction 224
Objectives of Your Chosen Program Structure 224
Reasons for Structuring Programs with Consistent Format 225
   Familiarization 225
   Consistency 225
   Re-running Tools in the Program 225
   Efficiency Limitations 226
Machine Variations That Affect Program Structure 226
   M-Code Differences 226
   Automatic Tool Changer Variations 227
      T-Word Brings a Tool To the Ready Station, M06 Commands the Tool Change 227
      Single-Arm or Umbrella Tool Changers 229
      Certification Cart Tool Changer 230
      T-Word Does Everything 230
      Tool Change at Beginning Or End? 231
      Does the Machine Even Have an Automatic Tool Changer? 231
   Understanding the G28 Command 232
      What About G53? 232
A Possible Problem with Initialized Modes 233

Key Points for Lesson 5.1: 233

Lesson 5.2–Structured Program Format 234
Objectives 234
Introduction 234
Structured Program Format for Vertical Machining Centers 235
   Program Start-Up Structure: 235
   Tool Start-Up Structure: 235
   Tool Ending Structure: 235
   Program Ending Structure: 235
   The Percent Signs That Begin and End the Program (%) 235
   Important Note about Safety Commands 236
   A Note About Documentation Comments 237
Example Program for Vertical Machining Centers 238
   A Few Questions About the Program: 241
   More On the Optional Stop Word (M01) 241
   Where Is the Restart Command For Each Tool? 241
Format for Horizontal Machining Centers 243
   Program-Startup Structure: 243
   Tool-Startup Structure: 243
   Tool-End Structure: 243
   Program-End Structure: 243
Key Points for Lesson 5.2: 245
Exercise: Write Your First Program By Yourself 246

FEATURES THAT HELP SIMPLIFY PROGRAMMING 248

Lesson 6.1–Canned Cycles For Drilling 250
Objectives 250
Introduction 250
   What Does "Canned" Mean? 250
   Invoking Canned Cycles: 250
Canned Cycle Commonalities 251
Description of Each Canned Cycle 251
   G80 – Cancel the Canned Cycle Mode 251
   G81 – Drilling Cycle 251
   G73 – High-Speed Peck Drilling Cycle 251
   G83 – Peck Drilling Cycle (full retract between pecks) 252
   G84 – Right-Hand Tapping Cycle (also used for rigid tapping) 253
      Rigid Tapping 253
      Feedrate for Tapping 253
      Rapid Plane for Tapping 253
      Tapping Can Be a Little Scary 253
      Coolant for Tapping? 254
      When To Tap 254
   G74 – Left-Hand Tapping Cycle 254
   G82 – Counter-Boring Cycle 254
   G86 – Boring Cycle (leaves drag line witness mark) 255
   G89 – Boring Cycle with Dwell (leaves drag line witness mark) 255
   G76 – Fine Boring Cycle (leaves no witness mark) 255
      Controlling Move-Over At Hole-Bottom 255
      A Tip for Boring Bar Tip Pointing 256
   G85 – Boring Cycle (Feeds in and feeds out) 256
   G87 and G88 – Manual Cycles (not recommended) 257
Words Used In Canned Cycles 257
A Simple Example 257
   Understanding G98 and G99 259
Canned Cycles and the Z-Axis 262
Extended Example Showing Canned Cycle Usage 264
Using Canned Cycles In the Incremental Positioning Mode 271
Key Points for Lesson 6.1: 273
Exercise: Practice Programming Canned Cycles 274

Lesson 6.2–Subprogram Techniques 279
Objectives 279
Introduction 279
The Difference Between a Main Program and a Subprogram 279
Loading the Main Program and Subprograms 281
Words Used with Subprograms 281
Nesting Subprograms 282
Machining Multiple Identical Pockets 282
   Understanding G52–Local Coordinate System 283
Multiple Hole-machining Operations on a Series of Holes 284
   Want to Include All of the Hole-Locations in the Subprogram? 285
Rough and Finish Side Milling 287
Utility Applications for Subprograms 288
   Control Programs 288
What is Parametric Programming (Custom Macro)? 289
   Part Families 289
   User Defined Canned Cycles 289
   Utilities 289
   Complex Motions and Shapes 290
   Driving Accessory Devices 290
Key Points for Lesson 6.2: 290

Lesson 6.3–Other Special Programming Features 291
Objectives 291

Table Of Contents

Introduction 291
Block Delete (also called optional block skip) 291
   Applications for Block Delete 292
      Another Optional Stop 292
      Trial Machining 292
      Trial Boring 293
      A Warning About Block Delete Applications 295
Sequence Number (N-word) Techniques 295
   Eliminating Sequence Numbers 295
   Using Special Sequence Numbers In Program Restart Commands 295
   Using Sequence Numbers As Statement Labels 296
      Using Block Delete To Exit a Series of Commands 296
Other G-codes of Interest 297
   Thread Milling, G02 & G03 (helical interpolation) 297
   G04 - Dwell Command 299
   G09 and G61 - Exact Stop Check 300
   G10 - Offset Setting By Programmed Command 301
      Applications for G10 301
      Programming Tool Length and Cutter Radius Compensation Offset Entries 301
      Handling Differences Among Pallets 303
   Polar Coordinates (G15 and G16) 303
   Plane Selection Commands (G17, G18, and G19) 304
   Inch/metric Mode Selection G20 and G21 305
   Secondary Reference Position, G30 306
   Scaling Commands, G50 and G51 306
   G50.1 and G51.1 - Mirror Image Commands 306
      Applications for Mirror Image 306
      The Two Ways To Activate Mirror Image 307
   Motion Relative to Zero Return Position (machine coordinate system), G53 308
   Single Direction Positioning Mode, G60 308
   Coordinate Rotation G68 and G69 309
      Adjusting For Work Holding Devices That Are Not Square with the Axes 311
Key Points for Lesson 6.3: 312

Lesson 6.4–Programming Rotary Devices 313
Objectives 313
Introduction 313
The Difference Between an Indexer and a Rotary Axis 313
A Note to Horizontal Machining Center Programmers 313
Benefits of Rotary Devices 314
Indexers 314
   Programming Indexer Rotation 314
      90 Degree and 45 Degree Indexers 314

      Five Degree Indexers 315
      One Degree Indexer 315
Rotary Axes 315
   How To Program a Rotary Axis Departure 316
   Comparison To Other Axes 316
      Zero Return Position 316
      Polarity 317
      Designation of Workpiece Coordinate System Zero 317
      Absolute Positioning Mode 318
      Incremental Positioning Mode 321
      Clamping the Rotary Axis for Machining After Rotation 323
      Rapid and Straight Line (linear) Motion 323
      Canned Cycle Usage 324
      Measurement system mode comment 324
Approaching Rotary Device Applications 324
   Workpiece Coordinate System Zero Selection 324
      Assigning One Workpiece Coordinate System Zero Point Per Side 326
      Using One Central Workpiece Coordinate System Zero Point 326
      Calculating Coordinates When Center of Rotation Is the Workpiece Coordinate System Zero Point 327
Example Program Using a Rotary Device 328
Key Points for Lesson 6.4: 335

KNOW YOUR MACHINE FROM AN OPERATOR'S POINT OF VIEW 337
Are You Only Interested In Setup and Operation? 337
   From Lesson 1.1: Machine configurations 337
   From Lesson 1.2: CNC job work flow 337
   From Lesson 1.4: Understanding the workpiece coordinate system 337
   From Lesson 1.5: Determining workpiece coordinate system values 337
   From Lesson 1.6: Setting workpiece coordinate system offset values 338
   From Lesson 4.1: Introduction to compensation 338
   From Lesson 4.2: Tool length compensation 338
   From Lesson 4.3: Cutter radius compensation 338
   From Lesson 4.4: Workpiece coordinate system setting offsets 338
   From Lesson 5.1: Introduction to program structure 338
The Need for Hands-On Experience 338
Key Concept Number Seven 339
   More About Axis Polarity 339
Procedures You Must Know 341
   Machine Power-Up 341

Table Of Contents

Sending the Machine To Its Zero Return (reference) Position 341
Manually Moving Each Axis 341
   An Example of Manual Axis Movement 341
Manually Starting the Spindle 342
Manually Making Tool Changes 342
Manipulating the Display Screen 342
   The Position Display Screen 342
   The Program Display Screen 343
   The Offset/Setting Display Screen 343
   The Program-Check Display Screen 343
Loading Programs 343

Lesson 7.1–Tasks Related to Setup and Running Production 345
Objectives 345
Introduction 345
   Setup-Related Tasks 345
   Running Production-Related Tasks 345
A CNC Job From Start to Finish 347
Setup Documentation 349
   Tear Down the Previous Setup and Put Everything Away 350
   Gather the Components Needed To Make the Setup 350
   Make the Work Holding Setup 351
   Assign the Workpiece Coordinate System Zero Point 352
      Measure Workpiece Coordinate System Zero Assignment Values 352
      Enter Workpiece Coordinate System Zero Assignment Values Into Workpiece Coordinate System Offset Number One 353
   Assemble the Cutting Tools Needed for the Job 353
   Measure Tool Length- and Cutter-Radius-Compensation Values 355
      The Static Nature of Cutting Tool Measurements 355
   Enter Tool Length and Cutter Radius Compensation Values Into Offsets 355
   Load Cutting Tools Into the Machine's Automatic Tool Changer Magazine 356
   Load the CNC Program 356
   The Physical Tasks Related To Setup are Now Completed 357
   Verify the Correctness of a New or Modified Program 357
   Verify the Correctness of the Setup 358
      A Tip That Will Save a Crash Some Day 358
      Dry Running Our Example Program 358
   Cautiously Run the First Workpiece 360
      The Most Dangerous Time 360
      Making Sure the First Workpiece Is a Good One 361

   Machining the First Workpiece In Our Example Job 361
   If Necessary, Optimize the Program for Better Efficiency 363
   If Changes Have Been Made To the Program, Save the Corrected Version of the Program 363
Production Run Documentation 363
   A Note To Programmers: 364
   Remove the Previous Workpiece 365
   Load the Next Workpiece 365
   Activate the Cycle 365
   Monitor the Cycle 366
   Clean and De-Burr the Workpiece 366
   Perform Specified Measurements 366
   Make Offset Adjustments To Maintain Size for Critical Dimensions (sizing) 367
   Replace Worn Tools 367
   Clean the Machine 368
   Preventive Maintenance 368
Key Points for Lesson 7.1: 368

Lesson 7.2–Operation Panels 371
Objectives 371
Introduction 371
The Two Most Important Operation Panels 371
The Control Panel Buttons and Switches 372
   Display Screen Mode Keys 373
      Position Display Pages 374
      Program Display Pages 374
      Offset Display Pages 375
      Graph Display Pages 376
      Other Display Screen Modes 376
   The Keyboard 376
      Letter Keys 377
      The Slash Key (/) 377
      Number Keys 377
      Decimal Point Key 377
      CAN Key 377
      EOB Key 377
      The INPUT Key 377
      Cursor Control Keys 377
      Program Editing Keys 377
      RESET Key 377
The Machine Panel 378
   Power Buttons 378
   MODE switch 378
   CYCLE START Button 379
   FEED HOLD Button 379
   FEEDRATE OVERRIDE Switch 379
   RAPID OVERRIDE Switch 379
   EMERGENCY STOP Button 380
   Conditional Switches 380
      DRY RUN On/Off Switch 380
      SINGLE BLOCK On/Off Switch 380

Table Of Contents

BLOCK DELETE On/Off Switch (also called optional block skip) 380
OPTIONAL STOP On/Off Switch 380
Buttons and Switches for Manual Functions 380
    Axis Jogging Controls 381
    Hand-Wheel Controls 381
    Spindle Control 381
    Automatic Tool Changer Control 381
Indicator Lights and Meters 381
    Spindle RPM and Horsepower Meters 382
    Axis Drive-Motor Horsepower Meter 382
    Cycle Indicator Lights 382
    Zero Return Position Indicator Lights 382
    Optional Stop Indicator Light 382
Other Buttons and Switches on the Machine Panel 382
Other Operation Panels on Your Machining Center 382
Key points for lesson 7.2: 382

KNOW THE BASIC MODES OF OPERATION 383

Lesson 8.1–Operation Modes 385

Objectives 385
Introduction 385
The Manual Mode Switch Positions 385
The Manual Data Input (MDI) Mode Switch Position 386
    Commanding an MDI Zero Return 386
    The Procedure to Give an MDI Command 386
    Commanding a Tool Change 387
    Commanding Spindle Activation with MDI 387
    Other Times When MDI is Used 387
    Can You Make Motion Commands with MDI? 387
The Edit Mode Switch Position 388
    To Make a Program in Memory the Active Program (to call up a program) 388
    To Enter a New Program 389
The Program Operation Mode 389
    To Run the Active Program from the Beginning 389
Key Points for Lesson 8.1: 390

THE IMPORTANCE OF PROCEDURES 391

Lesson 9.1–The Key Operation Procedures 393

Objectives 393
Introduction 393
Develop Your Own Operation Handbooks 393
    Manual Procedures: 393
    MDI Procedures 393
    Program Manipulation Procedures 394
    Setup Procedures 394
    Program Running Procedures 394
    Blank Procedure Form 394
Sample Operation Handbook: Levil Certification Cart (mill) with 0iF 396
    Manual Procedures 396
        To Power-Up the Machine 396
        To Do a Reference Return (send each axis to its reference position) 396
        To Start the Spindle 396
        To jog axes (using continuous jog) 396
        To Jog Axes (using incremental jog) 396
        To Use the Hand-Wheel 396
        To Set Axis Displays 397
        To Enter Tool Offsets (GEOM(H), WEAR(H), GEOM(D), WEAR(D)) 397
        To Enter Workpiece Coordinate System Offsets 397
    MDI Procedures 398
        To Execute an MDI Command 398
    Program Manipulation Procedures 398
        To Get Ready to Edit Programs 398
        To Show a Directory of Programs 398
        To Call Up a Program from Within the CNC Memory (make it the active/main program) 398
        To Load a Program 398
        To Delete a Program 399
        To Search Within a Program 399
        To Alter, Insert, & Delete 399
        To Save Programs to an External Device 400
        To Use Background Edit 400
    Setup Procedures 400
        To Mount the Work Holding Device on the Table 400
        To Measure the Workpiece Coordinate System Zero Assignment Values in the X and Y Axes 400
        To Measure the Workpiece Coordinate System Zero Assignment Value in the Z-Axis 401
        To Measure and Enter Tool Length Compensation Values 401
    Program Running Procedures 401
        To Run the Program (in normal production - no verification techniques) 401
        To Rerun Tools 402
        To Do a Free Flowing Dry Run 402
        To Do a Normal Air Cutting Run 402
        To Run the First Workpiece 402
        To Cancel a Cycle 403
        To Clear an Alarm 403
Key Points for Lesson 9.1: 403

KNOW HOW TO SAFELY VERIFY PROGRAMS 405

Safety Priorities 405
  Operator Safety 405
  Machine Tool Safety 406
  Workpiece Safety 406

Lesson 10.1–Program Verification 407
  Objectives 407
  Introduction 407
  Two More Procedures 407
    Canceling the CNC Cycle 407
    To Re-Run a Tool 408
  Verifying a Job that Contains Mistakes 410
    Loading the Program 414
    The Dry Run to Check for Setup Mistakes 415
    Cautiously Running the First Workpiece 417
  Key Points for Lesson 10.1: 418
  Index 419

Preface

CNC machining centers are among the most popular forms of CNC metal-cutting machine tools. Just about every company that uses metal-cutting CNC machines has at least one machining center.

This course introduces students to programming, setup, and operation techniques for CNC machining centers with FANUC CNC. This text is part of the approved curriculum for FANUC Certified Education CNC Training.

We begin with basic concepts, ensuring that newcomers to CNC will be able to follow the presentations. We use a building block approach–so as you get deeper into the material–we will be adding to what you already know. When you are finished, you will have a thorough understanding of what it takes to program, setup, and run a CNC machining center that has a FANUC CNC control.

We use a Key Concepts approach. The Key Concepts allow us to minimize the number of major topics you must master in order to become proficient with a CNC machining center. In this text, there are ten Key Concepts related to CNC machining center programming, setup, and operation.

Key Concepts 1 through 4 are the most important building blocks, giving you a way to organize your thoughts. Starting with Key Concept 5 (program formatting), we'll begin putting it all together. Students should concentrate most on understanding the points made early in each Key Concept. It is as important to know why you are doing things as it is to know how to do them. Concentrate first on the whys. It will be impossible for beginners to totally memorize and comprehend every technique used with machining centers the very first time it is presented.

Rest assured that if you can understand the basic reasoning behind why each CNC feature is required, it will be much easier to master the use of the feature. Once this basic reasoning is understood, it will be easy to review material to extract specific details of how each CNC feature is used–so you can start putting a CNC machining center to good use.

If you have previous CNC machining center experience using other types of CNCs, you will find it relatively easy to adapt what you already know to FANUC CNCs, the most popular CNCs in the industry. If you have had programming experience with other types of machine tools, such as turning centers or wire EDM machines, this text will help you adapt what you know to CNC machining centers.

There are many ways to utilize CNC equipment. This course will show you one or two safe ways to accomplish tasks. You can use your own common sense and experience to develop your own style.

Prerequisites

This text will cover CNC machining centers from the ground up. We will assume that you have absolutely no previous experience with CNC programming, setup and operation. However, there are certain things we do assume about students taking this course.

Basic Machining Practice Experience

We will assume that you have at least some experience with basic machining practice and machining operations. We will assume you possess basic knowledge of operations like face milling, end milling, drilling, reaming, and tapping. We will also assume that you understand tooling and related cutting conditions required to perform operations safely and efficiently.

If you have experience with any form of conventional milling or drilling equipment, like a milling machine or a drill press, we think you will find it remarkably easy to learn how to program, setup, and operate a CNC machining center. Think of it this way: You already know what you want the machine to do. It will be a relatively easy task to learn how to tell the CNC machining center to perform the machining operation. This is why machinists make good CNC programmers.

On the other hand, if you have no previous basic machining experience, or worse, no shop experience, your challenge will be much greater. You not only need to learn CNC programming, setup, and operation, you also need to learn basic machining practice. If you have no previous machining practice experience, we

Preface

strongly recommend that you enroll in a course related to basic machining practice in conjunction with taking this course.

Math

The word numerical in computer numerical control implies that numbers are highly involved with CNC. Indeed, every CNC command includes numbers, and almost every CNC command requires an arithmetic calculation to be made. However, most of the calculations are quite simple. The types of arithmetic calculations required for the typical CNC machining center program include addition, subtraction, multiplication, and division. For more complex workpieces, some right-angle trigonometry may also be required.

This text will require very little in the way of math (though your instructor may wish to include more math). For the most part, we will assume that you can add, subtract, multiply, and divide. We will be teaching CNC machining center programing, setup and operation, not mathematics. For this reason, our examples and practice exercises will be quite simple regarding the math you need to know.

We do not wish to understate the importance of the math required to prepare real-world programs for CNC machining centers. In real life, the CNC programmer must be prepared to perform rather complex calculations involving trigonometry in order to come up with the coordinates needed in the CNC program. A computer aided manufacturing (CAM) system or conversational programming system can be used to reduce the amount of math required for programming.

Motivation

This should go without saying. We assume that you are motivated to learn. If you are highly motivated to learn about CNC machining centers, it will make your task much easier. Your motivation will help you overcome any problems you may have with learning the material in this course. With motivation, you'll stick to it until you understand and can apply your knowledge.

Controls Covered

Since this course is being used as part of a FANUC Certified Education CNC Training curriculum, examples are given in FANUC format. Keep in mind, however, that the Key Concepts approach we use throughout this text will make it possible for you to learn techniques that can be applied to just about any CNC. FANUC provides the most popular CNCs available so many CNC manufacturers make their programming FANUC compatible. If you understand the basic concepts, and if you understand how specific techniques are applied to one particular CNC type, it will be relatively easy to adapt what you know to just about any CNC machining center being used today.

Several specific FANUC control models are covered by this course. The primary examples in this course are provided in the format of the FANUC Series 0i-F, the 0i-MD and the FANUC Series 30i/31i/32i CNCs. However, FANUC maintains basic program compatibility with each new generation of their CNCs, so much of this course is applicable to all FANUC CNCs. In many cases, we will point out some differences for older FANUC CNCs.

Prior to the FANUC Series 30i/31i/32i, there was always a CNC model for lathe and a different model for machining centers. For example, the FANUC Series 0i-TD is for turning and the 0i-MD is for machining centers.

Due to the fact that mill-turn or turn-mill machines are becoming increasingly popular, the Series 30i/31i/32i CNCs can support turning, milling or both turning and milling.

Without getting into too much history of how FANUC controls have evolved, you should also know that FANUC has always enhanced each control model within its lifetime. Generally speaking, FANUC adds a letter after the control model name to determine the actual control version (such as A, B, or C). The first release of a new control is the A version, and as new versions are introduced, the letter is changed (to B and C). For example, the 0-M control began as a 0-MA, and evolved into the 0-MB, and 0-MC as FANUC continued to improve its hardware and software capabilities.

FANUC's older conversational machining center controls are an exception to this naming convention rule. FANUC labeled some controls with an "F" designation to indicate the conversational version. The 0-MF, 6-MF, 11-MF, and 15-MF are examples of conversational machining center controls. These days, the controls support conversational programming without a special designation. FANUC's conversational programming, MANUAL GUIDE i, is just another feature in the CNC. Though CNCs can be programmed conversationally, keep in mind that all FANUC conversational controls can also be programmed by conventional methods using G-code. This course will not cover the conversational features of FANUC CNCs.

Note that in the past, FANUC CNCs were also sold under the GE Fanuc and General Numerics brands in the United States.

Controls Other Than FANUC

Though the techniques given during this text are specific to FANUC controlled CNC machining centers, keep in mind that most CNC machining center controls are programmed with very similar techniques. In fact, since FANUC is so popular and holds the largest market share of CNC controls in the world, many CNC machining center manufacturers make sure that their CNCs are FANUC compatible. This means that programs written for a FANUC CNC will run in a CNC that is FANUC compatible with only minor modifications.

Limitations

Our primary goal will be to acquaint you with the programming, setup and operation of 3-axis machining centers (X, Y, and Z). For the majority of this course, only these three linear axes will be discussed. In Key Concept 6, we will introduce rotary devices, including indexers and rotary axes that are almost always supplied with horizontal machining centers and newer vertical machining centers. These devices allow you to expose more than one side of the workpiece to the spindle for machining during the CNC cycle. While programming for rotary devices is not especially difficult, programs utilizing them tend to be quite lengthy due to the additional surfaces of the workpiece they can expose to tooling for machining.

We will be discussing both vertical as well as horizontal machining centers throughout this text. Since vertical machining centers are a little easier to work with and visualize, and since vertical machining centers are much more popular than horizontals, most of the example programs given will be for vertical machining centers.

This text will not specifically address 5-axis machining centers. Though many of the same principles apply, we will not discuss 5-axis machines during this introductory course.

Scope

As the name of this course implies, we address all three tasks a person must master in order to utilize a CNC machining center, including programming, setup, and operation.

Programming is the act of preparing a series of commands that tell the CNC machining center how to machine a workpiece. It involves determining machining processes, selecting cutting tools, designing a work holding setup, and actually creating the CNC program. The method of programming we'll be using in this course is called manual programming (also called G-code level programming). While there are other ways to create programs (using a computer aided manufacturing, or CAM, system, or conversational programming for example), it is with G-code level, manual programming techniques that you can be the most intimate with a CNC machining center – commanding every function of the machine. Every CNC person must understand this form of programming in order to make modifications to the CNC program at the machine when a job is run.

Setup is the act of preparing the CNC machining center to run a series of workpieces (called a job or production run). Tasks involve (among other things) making the work holding setup, assembling, measuring, and loading cutting tools, entering certain offsets, loading the program, and verifying that the program is correct. We'll be discussing the related tasks in the approximate order setups are made.

Operation actually involves two things. First, you must be comfortable with the general operation of a CNC machining center. This involves knowing the various components on the machine, the buttons and switches, and how to perform several important procedures. Second, you must be able to complete a production run once the job is setup. Tasks involve (among others) workpiece load and unload, cycle activation, measuring completed workpieces and making adjustments if necessary, and the replacement of dull or broken tools.

Preface

Key Concepts Approach

This effective presentation method will allow you to organize your thoughts as you take this course. The course includes ten Key Concepts (six for programming and four for setup and operation). There are several benefits to this presentation method.

1) Any good training program should put a light at the end of the tunnel. All students want to know where they stand throughout any training course. With our Key Concepts approach, you will always have a clear understanding of your progress throughout the course.

2) During each Key Concept, we will first present the main idea behind the concept. As stated earlier, we say it is at least as important to understand why you are doing things as it is to understand how to do them. Think of these early presentations for each Key Concept as the why. From there, we will present the specific techniques that are related to each concept.

3) The Key Concepts allow us to use a building blocks approach and present information in a tutorial manner. We will be constantly building on previously presented information.

4) The Key Concepts approach allows us to limit the number of new ideas you must understand in order to grasp information presented within the text. Think of it this way: If you can understand but ten basic ideas, you will be well on your way to becoming proficient with CNC machining centers!

Lesson Structure

We further divide the ten Key Concepts into lessons. This makes it possible to organize the most important topics related to machining center usage.

Practice, Practice, Practice!

We have included lots of practice within the course to help you confirm your understanding of the presented material. Answers are provided right in the text, close to the exercise so you can check your own work.

There are also separate exercises that accompany this text. After many lessons, you do an exercise. Many of the exercises include programming activities. Many instructors who are using this course in their CNC curriculums use the exercises as homework assignments–to confirm your understanding of the material.

Key Concepts and Lessons

Here is a list of the ten Key Concepts and the associated lessons that comprise this text. Again, Key Concepts 1 through 6 are related to programming – Key Concepts 7 through 10 are related to setup and operation.

1: Know your machine from a programmer's viewpoint
   1.1: Machine configurations
   1.1a: CNC Certification Cart machine configuration
   1.2: General flow of the CNC process
   1.3: Visualizing program execution
   1.4: Program zero and the rectangular coordinate system
   1.5: Determining program zero assignment values
   1.6: Assigning program zero
   1.7: Introduction to programming words

2: You must prepare to write programs
   2.1: Preparation steps for programming

3: Understand the motion types
   3.1: Programming the three most basic motion types

4: Know the compensation types
   4.1: Introduction to compensation
   4.2: Tool length compensation
   4.3: Cutter radius compensation
   4.4: Workpiece coordinate system setting offsets

5: You must provide structure to your CNC programs

5.1: Introduction to program structure
5.2: Four types of program format

6: Special features that help with programming
6.1: Canned cycles for drilling
6.2: Working with subprograms
6.3: Other special programming features
6.4: Programming rotary devices

7: Know your machine from a setup person or operator's viewpoint
7.1: Tasks related to setup and running production
7.2: Buttons and switches on the operation panels

8: Know the three basic modes of operation
8.1: The three modes of operation

9: Understand the importance of procedures
9.1: The key operation procedures

10: You must know how to safely verify programs
10.1: Program verification

Note once again that we do present the programming discussions first. While you may be more interested in setup and operation, there are many programming-related topics that setup people and operators must understand. And it is during a presentation about programming that many CNC features are best introduced.

When you complete the programming Key Concepts, you may be surprised at how many setup and operation topics you already understand.

Some readers may not be at all interested in programming. You can skip the programming-related lessons and jump right to the setup- and operation-related lessons. At the beginning of Key Concept 7, we do provide a list of important setup and operation related topics that are presented during programming. You can review just the listed presentations in the programming lessons.

New in this Edition
This edition includes two major enhancements:

1. The FANUC CNC model shown in all illustrations will be the FANUC model 0iF (the previous edition shows the 0iD). Other current model FANUC CNCs, like the 30iA and 30iB, are remarkably similar to the 0iF.
2. Examples are shown using the Metric measurement system (in addition to the Imperial/Inch system).

About Measurement System Examples
There are two measurement systems in use with modern CNCs: the Metric measurement system and the Imperial measurement system. Generally speaking, a company will choose one or the other and stick with it. It is rare to see a company that will switch measurement systems from day to day or job to job. When a CNC machine is first installed in a company, it will be set to power up using the company's measurement system mode of choice.

With the Metric measurement system, the most basic unit of measurement for CNC application is a millimeter. With the Imperial measurement system, it is an inch. We refer to measurement system modes throughout this text as <u>metric mode</u> and <u>inch mode</u> - and we show both modes for almost all discussions and examples that involve units of measurement. Generally, we will show the metric mode value or example first. Often we will place inch mode values in parentheses for further clarification.

If you know which measurement system you will be using, you can avoid confusion by studying only the examples for that measurement system and ignore those for the other measurement system. If you do not know which measurement system you will be using, we recommend that you study the metric-mode examples, since it is the more popular measurement system globally.

Drawings shown will often use nominal dimensions in each measurements system mode, not necessarily exact conversions. This means drawings will not be perfectly to scale for one of the examples. For example,

Preface

a dimension of 12.0 millimeters for the metric-mode example may be shown as 0.5 inches for the inch-mode example. Of course 12.0 millimeters is not exactly 0.5 inches. Since we want to show realistic examples, we place the priority on clarity over precise measurement system conversions.

Getting Familiar with Measurement Systems

Size comparison

A millimeter is about the thickness of a U.S. quarter dollar. An inch is about the thickness of two decks of playing cards stacked together. The exact relationship is that one inch is equal to 25.4 millimeters.

- Inches = millimeters divided by 25.4
- Millimeters = inches times 25.4

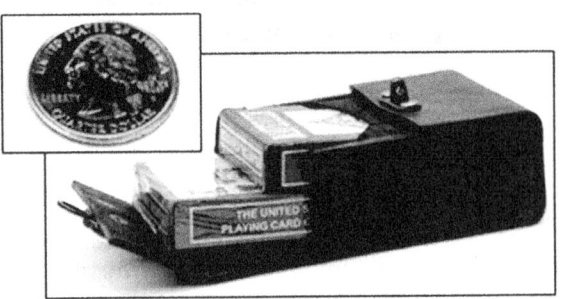

Tiny increments

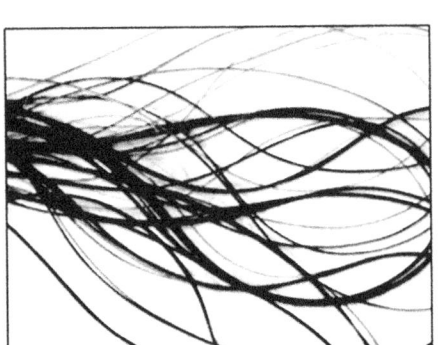

You will be working with some pretty small values in a machine shop. The smallest value typically used with CNC machining centers is 0.001 millimeters when working in metric mode. In inch mode, the smallest increment is 0.0001 inches.

- A human hair is about 0.050 millimeters (or 0.002 inches) in diameter

Pronouncing values in a machine shop

Metric mode:

- 1.000: Pronounced "one millimeter"
- 0.100: Pronounced "one hundred microns"
- 0.010: Pronounced "ten microns"
- 0.001: Pronounced "one micron"

Note: 0.1 is the same as 0.100 and 0.01 is the same as 0.010

Examples:

- 3.0: "Three millimeters"
- 0.2: "Two hundred microns"
- 0.37: "Three hundred seventy microns"
- 0.226: "Two hundred twenty-six microns"

Inch mode:

- 1.0000: "One inch"
- 0.1000: "One hundred thousandths"
- 0.0100: "Ten thousandths"
- 0.0010: "One thousandth"
- 0.0001: "One ten-thousandth" (often shortened to simply "one tenth")

Examples:

- 3.0: "Three inches"
- 0.2: "Two hundred thousandths"
- 0.37: "Three hundred seventy thousandths"
- 0.226: "Two hundred twenty-six thousandths"
- 0.3857: "Three hundred eighty-five thousandths and seven tenths"

Enjoy!

We at CNC Concepts, Inc. and FANUC America wish you the best of luck with this course. We hope you find it easy to understand our written presentations and the presentations of your instructor (live and/or recorded). Once completed, we hope this course makes your introduction to FANUC controlled CNC machining centers as easy and enjoyable as possible.

Key Concept 1

Know the Machining Center from a Programmer's Viewpoint

You must come to know a CNC machining center from two distinctly different perspectives. In Key Concept 1, we look at the machine from a programmer's viewpoint. Later, during Key Concept 7, we will look at the machine from a setup person's or operator's viewpoint.

Key Concept 1 is the longest of the Key Concepts. It contains several lessons:

    1.1: Machine configurations

    1.1a: CNC Milling Certification Cart machine configuration

    1.2: General flow of programming

    1.3: Visualizing program execution

    1.4: Understanding the workpiece coordinate system

    1.5: Determining Workpiece Coordinate Offsets

    1.6: How to Enter Workpiece Coordinate Offsets

    1.7: Introduction to programming words

A CNC programmer need not be as intimate with a specific CNC machining center as a setup person or operator–but they must understand enough about the machine to:

- ✓ Create programs
- ✓ Instruct setup people and operators
- ✓ Create related setup and production documentation.

First and foremost, a CNC programmer must understand what the CNC machining center is designed to do. That is, they must:

- ✓ Understand the machining operations a machining center can perform
- ✓ Be able to develop a workable process (sequence of machining operations)
- ✓ Select appropriate cutting tools for each machining operation
- ✓ Determine cutting conditions (feeds and speeds) for each cutting tool
- ✓ Design a work holding setup.

All of these skills, of course, are related to basic machining practice–which as we state in the preface–are beyond the scope of this course. For the most part, we'll be assuming you already possess these important skills. That said, we do include some important information about the machining operations that can be performed on CNC machining centers throughout this course. For example, we provide a description of hole-machining and contour milling operations. We provide suggestions about how machining operations can be programmed in the appropriate lessons. This information should be adequate to help you understand enough about machining operations to begin working with CNC machining centers.

Experienced with Conventional (non-CNC) Machine Tools?

A CNC machining center can be compared to a conventional milling machine. Many of the same operations performed on a milling machine are performed on a CNC machining center. If you have experience with manually operated milling machines, you already have a good foundation on which to build your knowledge of CNC machining centers.

This is why machinists make good CNC programmers. With a good understanding of basic machining practice, you can easily learn to program CNC equipment. You already know <u>what</u> you want the CNC machine to do. It is a relatively simple matter of learning <u>how to tell</u> the CNC machine to do it.

If you have experience with machining operations like face milling, end milling, drilling, tapping, boring, and reaming, and if you understand the processing of workpieces to be machined, you are well on your way to understanding how to program a CNC machining center. Your previous experience has prepared you for learning to program a CNC machining center.

We can also compare the importance of knowing basic machining practice in order to write CNC programs to how important it is for a speaker to be well versed with the topic they will be presenting. If not well versed with their topic, the speaker will not make much sense during the presentation. In the same way, a CNC machining center programmer who is not well versed in basic machining practice will not be able to prepare a program that makes any sense to experienced machinists.

Lesson 1.1 - Machine Configurations

As a programmer, you must understand the characteristics of a CNC machining center. You must be able to identify its basic components —you must understand the moving components of the machine (called axes)–and you must know the various functions of your machine that are programmable.

Objectives
After completing this lesson, students should be able to:

- ✓ Describe several types of machining centers
- ✓ Describe the basic components of a machining center
- ✓ Describe the primary moving axes and their polarity and the best way to view polarity from a programmer's viewpoint
- ✓ Describe the basic programmable functions of the spindle
- ✓ Calculate speeds (S-word) and feeds (F-word) for a machining center
- ✓ Describe the basic functions of an automatic tool changer
- ✓ Describe the basic programmable modes of inch and metric

Introduction
Beginners can be a little intimidated when they see a machining center in operation for the first time. Admittedly, there will be a number of new functions to learn. The first point to make is that you must not let the machine intimidate you. As you go along in this course, you will find that a machining center is very logical and easy to understand with proper instruction.

You can think of any CNC machine as being nothing more than the standard manual machine it is replacing with motion control and other machine functions automated. Instead of moving the axes manually using hand-wheels and selecting or activating functions using switches and levers, you will be preparing a program that tells the machine what to do. Virtually anything that needs to be done on a true machining center can be controlled in a program–meaning anything you need the machine to do can be commanded in a program.

There are two basic types of machining centers that we will be addressing in this course. They are the vertical machining center (VMC) and horizontal machining center (HMC). Let's start by describing the most common features of each.

Vertical Machining Centers
A vertical machining center has its spindle oriented vertically. The spindle, and therefore the cutting tool, point downwards toward the machine's table and the part. Because of this spindle/tool orientation, chips will tend to collect and build up on the workpiece, and may eventually interfere with machining operations. However, this is a very popular type of CNC machining center because it closely resembles the knee-mill–a popular type of conventional machine. For anyone with experience using a knee-mill, a vertical machining center should be quite familiar.

C-frame Style
A common type of vertical machining center is called a C-frame-style machining center because the headstock, column, and bed, when viewed from the left-hand side, form the letter "C". An automatic tool changer is mounted to the machine (usually on the left side) to allow tools to be loaded into the spindle without operator intervention. The table has a series of tee-slots and/or location/clamping holes to allow work holding devices (like a table vise) to be mounted on the table.

Key Concept 1: Know Your Machine From A Programmer's Viewpoint

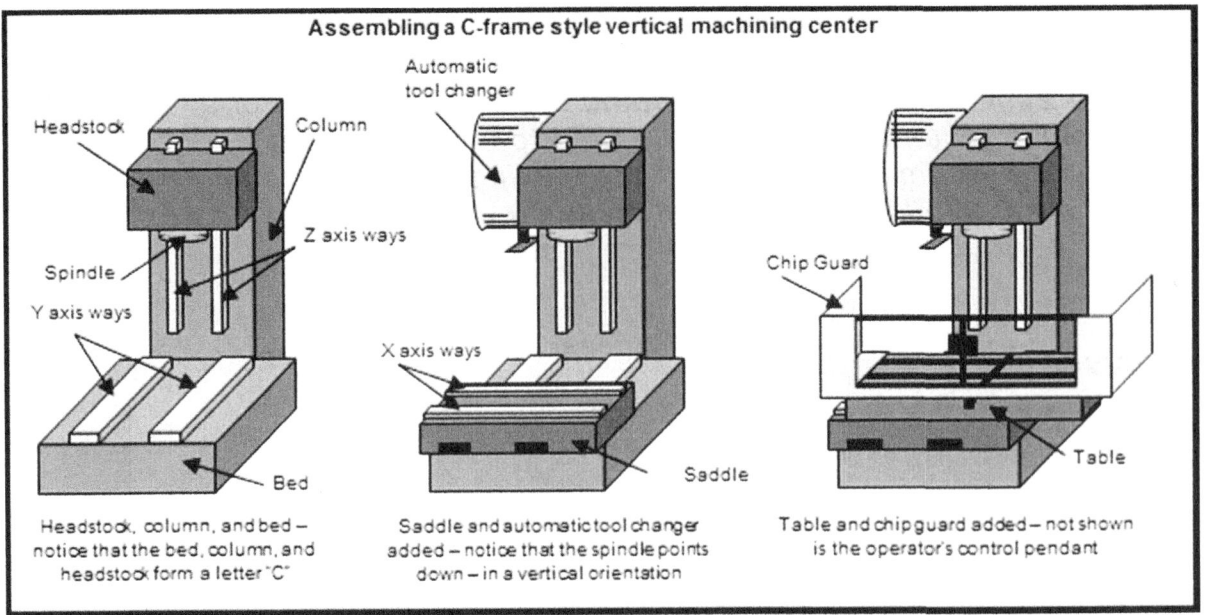

Figure 1.1: Primary components of a C-frame-style vertical machining center

Directions of Motion (axes) for a C-Frame-Style Vertical Machining Center

Basic vertical machining centers allow three directions of motion, or *axes*. These three basic motions are linear axes—allowing motion along a straight line. With a C-frame-style vertical machining center, the table can move left/right (the X-axis)—the table can move fore/aft (the Y-axis)—and the headstock or spindle can move up/down (the Z-axis). Figure 1.2 shows the axes of a C-frame-style vertical machining center.

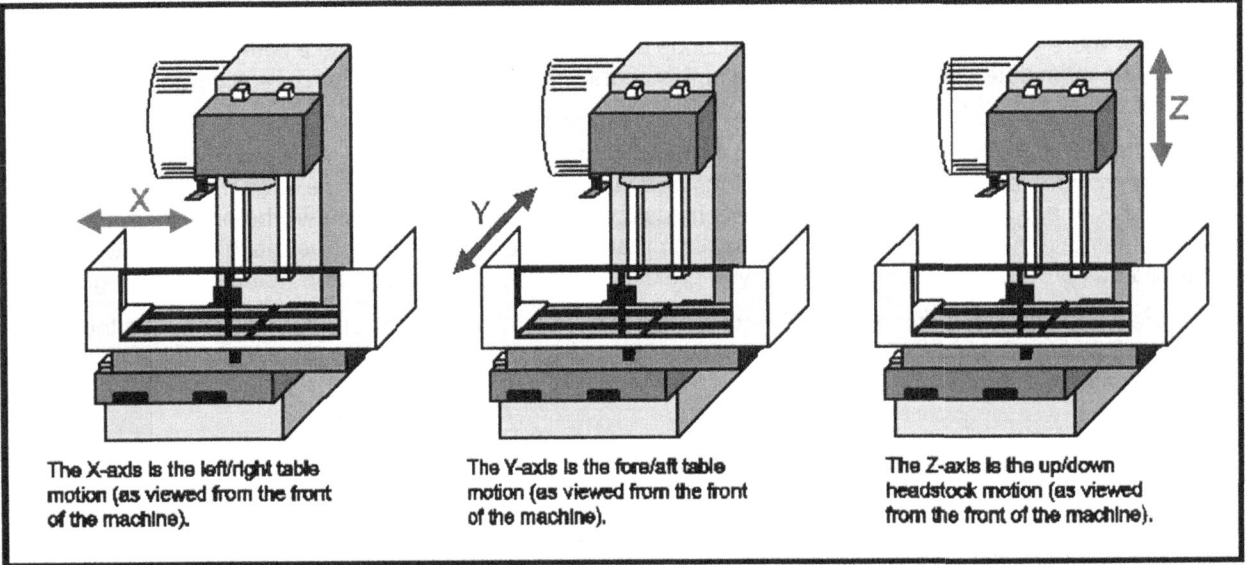

Figure 1.2: Directions of motion (axes) of a C-frame-style vertical machining center

With this kind of machine, notice that the cutting tool does not move in the X and Y-axis. The table and therefore the part moves in X and Y in relation to the tool. The tool only actually moves in the Z-axis.

Axis Polarity

Though not depicted in figure 1.2, each axis has a polarity (plus and minus direction). As the table moves to the left, it is moving in the X-plus direction. As it moves to the right, it is moving in the X-minus direction. As the table moves toward you, it is moving in the Y-plus direction. As it moves away from you, it is moving in the Y-minus direction. As the headstock/cutting tool moves up, it is moving in the Z-plus direction. As it moves down, it is moving in the Z-minus direction.

Since the cutting tool does not move in the X and Y axes, it can be a little confusing (especially for programmers) to understand polarity by looking at table motion. From a programmer's viewpoint, it is much easier to understand polarity if you imagine that the cutting tool is moving in all axis. Figure 1.3 shows how to visualize polarity with this method.

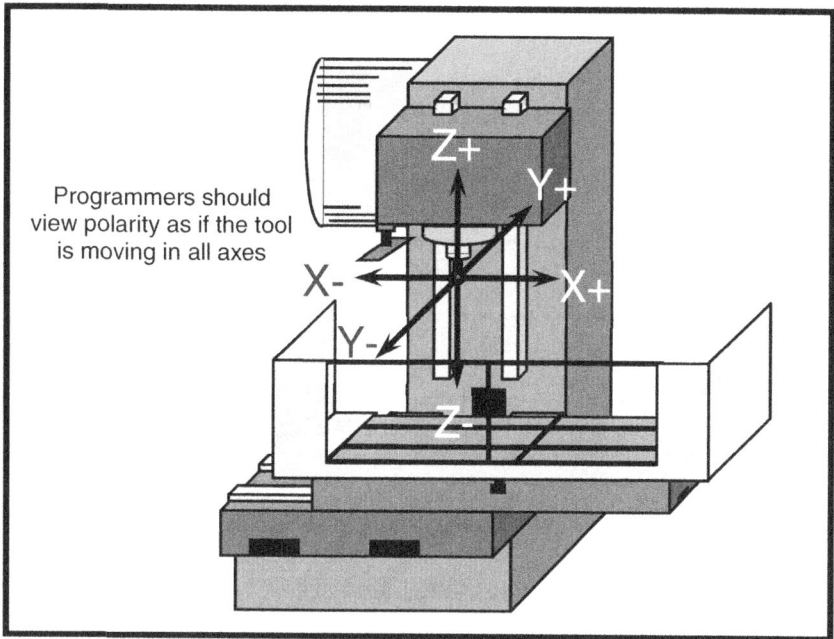

Figure 1.3: Viewing polarity as if the tool is moving in each axis

If you imagine that the cutting tool is moving in X and Y, determining polarity will be easier (especially when we introduce the workpiece coordinate system used with CNC programming in lesson 1.4). As the cutting tool moves to the right, it is moving in the X-plus direction. (But remember, the cutting tool does not really move to the right in the X-axis–it is the relative motion of the tool and part as the table moves to the left–which again, is the X-plus direction.) As the cutting tool moves to the left, it is moving in the X-minus direction. As the tool moves away from you, it is moving in the Y-plus direction. As it moves toward you, it is moving in the Y-minus direction. In Z, of course, the tool is really moving with the axis, so polarity is much easier to understand–up is Z-plus, down is Z-minus.

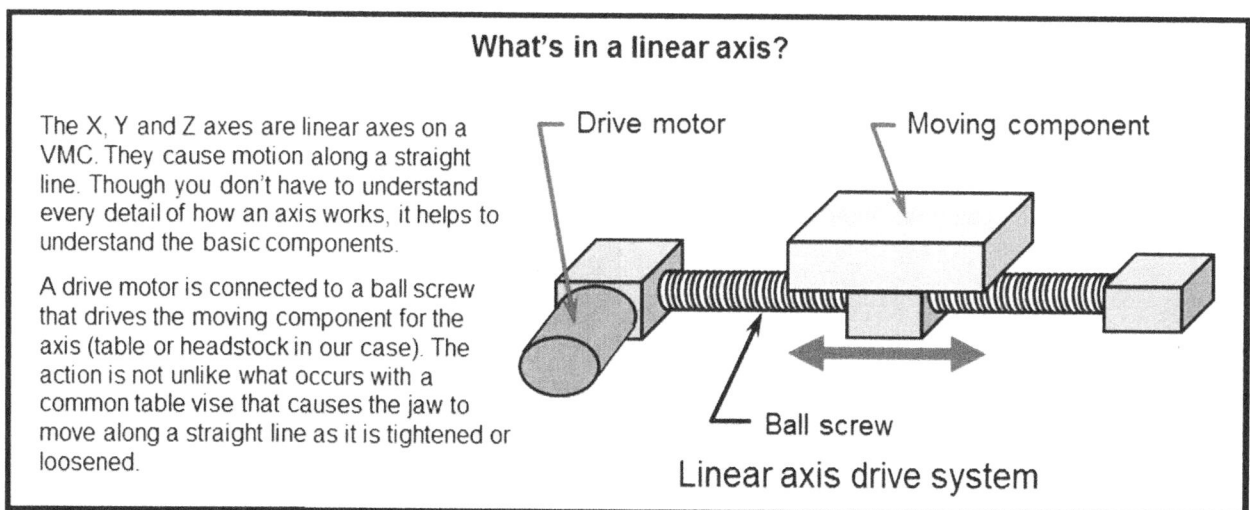

Knee-Style Vertical CNC Milling Machines

While the C-frame style of vertical machining center is the most popular type of production vertical machining center, it is but one of several types. Another common type is called the knee-style vertical CNC milling machine. It is identical to a conventional knee-style milling machine with the addition of CNC motion control. Many knee-style vertical milling machines do not have a tool changer–which is why they are typically called CNC milling machines instead of CNC machining centers. If the machine does not have an automatic tool changer, cutting tools must be manually loaded during the CNC cycle. The figure below shows a typical knee-style CNC milling machine.

Key Concept 1: Know Your Machine From A Programmer's Viewpoint

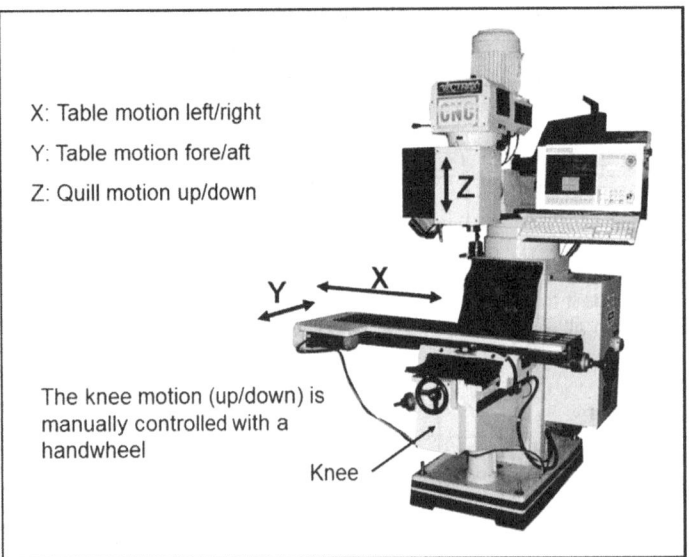

Figure 1.4: Knee-style CNC milling machine with axes directions shown

Notice how similar the axes of this machine are to the C-frame-style CNC machining center. The only difference is that the Z-axis of a knee-style CNC milling machine has a quill that moves within the headstock–as opposed to the entire moving headstock of the C-frame style machining center.

The polarity for each axis is also the same. X-plus is table motion to the left. X-minus is table motion to the right. Y-plus is table motion toward you. Y-minus is table motion away from you. Z-plus is quill motion up. Z-minus is quill motion down. And since the tool does not move along with the X and Y axes, we recommend that you view polarity as if the tool is moving in X and Y–just as we suggested with a C-frame-style machining center.

Bridge-Style Vertical Machining Center (also called gantry-style)

Yet another type of vertical machining center is called the bridge-type machining center. The table remains stationary on this kind of machine, which is necessary when machining a very large or heavy workpiece–like larger aircraft and energy generation components.

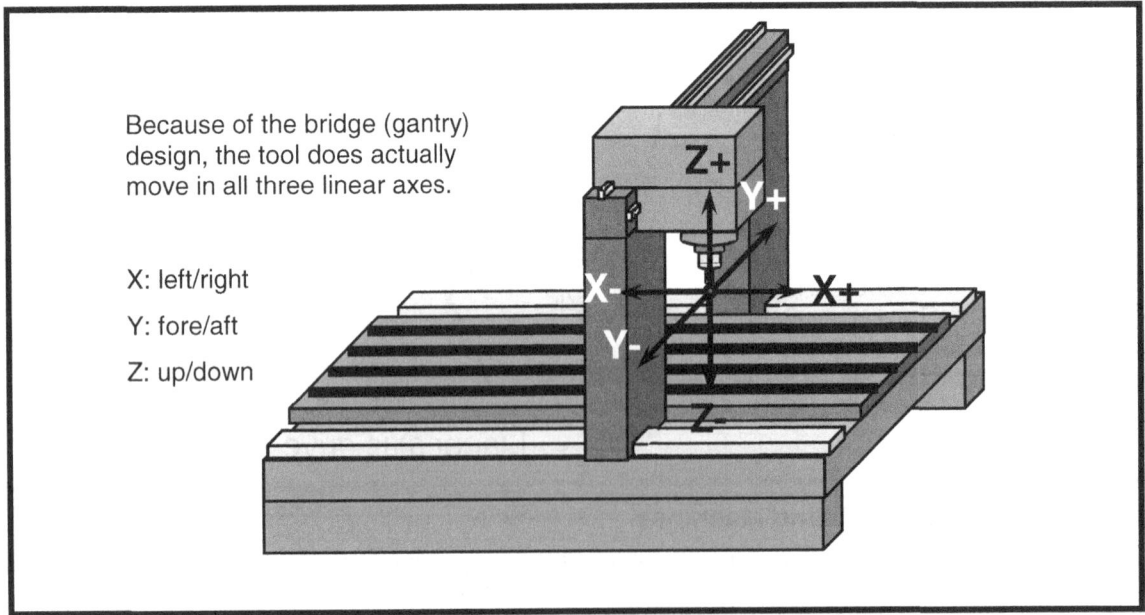

Figure 1.5: Bridge type vertical machining center axes

With the bridge-style vertical machining center, the tool <u>does</u> move along with each axis. This makes understanding polarity much easier. X-plus is tool movement to the right. X-minus is tool movement to the left. Y-plus is tool movement away from the operator. Y-minus is tool movement toward the operator. Z-plus is tool movement up. Z-minus is tool movement down.

Horizontal Machining Centers

A horizontal machining center has its spindle oriented horizontally. This provides one advantage over a vertical machining center. With a vertical machining center, chips tend to gather on the workpiece right around the cutting tool during machining and if they are not removed by some means (manually by the operator or flushed away by coolant or air blast), they will eventually interfere with the machining operation. By comparison, with a horizontal machining center, chips fall away from the machining operation under gravity.

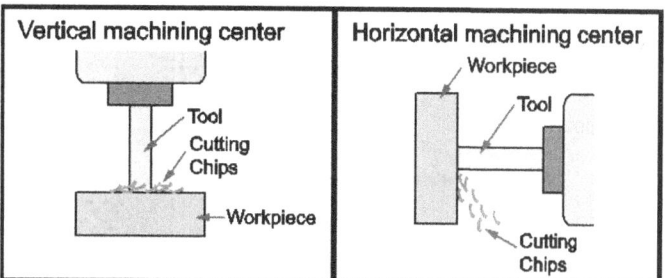

Figure 1.6: Chip evacuation on vertical machining center versus horizontal machining center

Another advantage of a horizontal machining center is that it is easier to incorporate a rotary device that allows the workpiece to be rotated during the machining cycle. This allows several surfaces of the workpiece to be exposed to the spindle for machining in a single setup. The most common form of rotary device for a horizontal machining center is a rotary axis, which is called the B-axis. Figure 1.7 shows the primary components of a horizontal machining center.

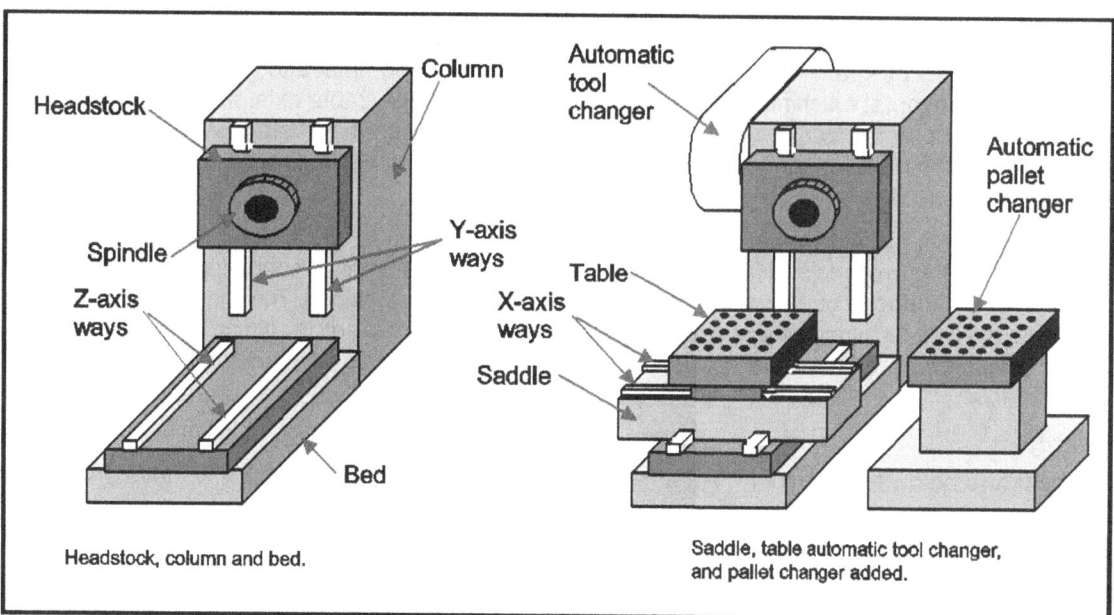

Figure 1.7: Components of a horizontal machining center

Most horizontal machining centers also incorporate an automatic pallet changer (shown in the right-side illustration). This device allows the operator to be loading a workpiece on the external pallet while a workpiece on the active pallet is being machined. While it is not shown in the illustration above, most horizontal machining centers have guarding that completely surrounds the active pallet during machining, which keeps the operator safe while loading workpieces on the external pallet.

Key Concept 1: Know Your Machine From A Programmer's Viewpoint

Directions of Motion (axes) for a Horizontal Machining Center

At first glance, the axes of a horizontal machining center (X, Y, and Z) appear to be radically different from a vertical machining center. But if you look more closely, you'll see that a horizontal machining center is similar to a vertical machining center placed on its back. In fact, a program written for a vertical machining center will run properly in a horizontal machining center, at least with regard to axis movements.

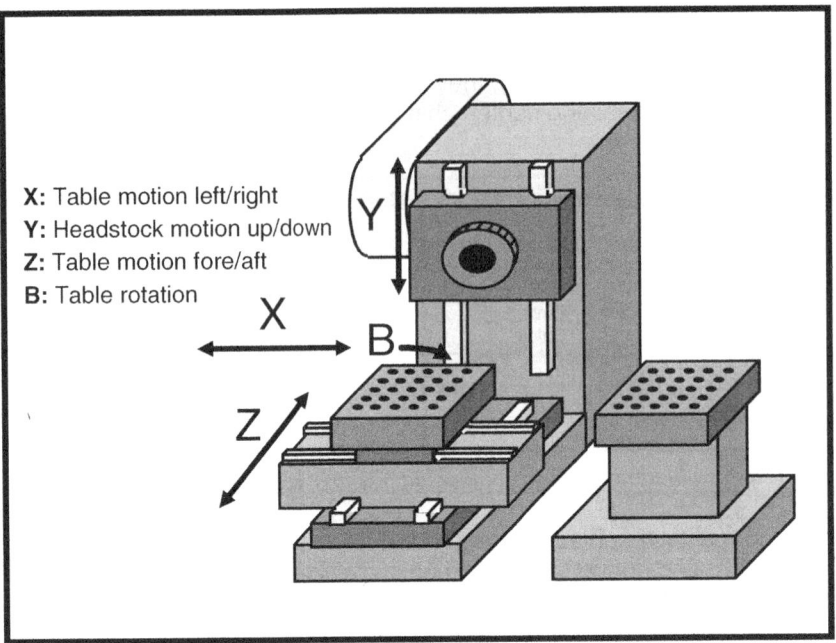

Figure 1.8: Horizontal machining center axes

Also notice the addition of the rotary axis (B-axis) within the table. This allows the table to be rotated during the execution of a CNC program—exposing different surfaces of the workpiece to the spindle and tool for machining. Some horizontal machining centers do not have a full rotary table axis. Instead, they have a simple indexing device. A rotary axis allows machining to occur as the table rotates, whereas with an indexer, machining can only occur after the table has finished rotating. We discuss both types of rotary devices in Key Concept 6.

There is a misconception that horizontal machining centers are harder to work with than vertical machining centers. In practice, the only major difference is that programs for horizontal machining centers tend to be longer since several surfaces of the workpiece are exposed to the spindle and more cutting tools will be required. While this will require a longer program, the program will not be more difficult to write. In fact, a vertical machining center will require several setups and programs to machine the same number of workpiece surfaces.

Axis Polarity

Notice in the figure below that the tool will not move along with the X and Z axes (it will move only with the Y-axis). And as we said during the discussion of C-frame vertical machining centers, this tends to cause some confusion when it comes to determining the polarity of each axis. Again, we recommend that you view polarity as if the tool is actually moving along with each axis.

When it comes to polarity for the B-axis, most machine tool builders will make clockwise table rotation (as viewed from above the table) the plus direction. But not all machine tool builders adhere to this standard.

> **Note:** Since vertical machining centers are more popular than horizontal machining centers, most example programs in this text are for a vertical machining center. During our discussion of rotary devices in Key Concept 6, we discuss special programming considerations for horizontal machining centers.

Lesson 1.1: Machine Configurations

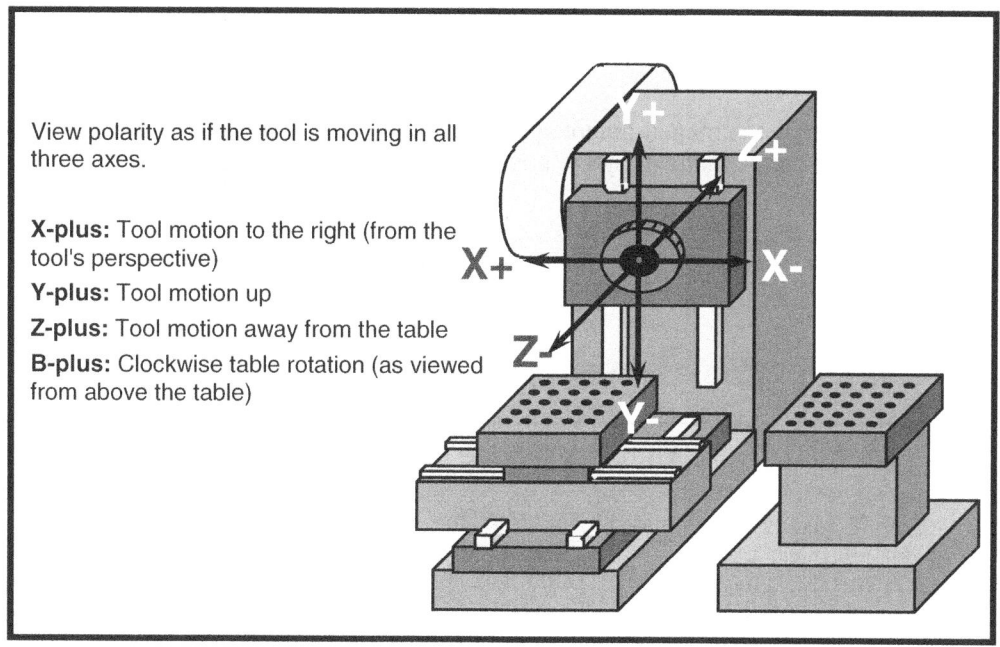

Figure 1.9: Horizontal machining center axes polarity

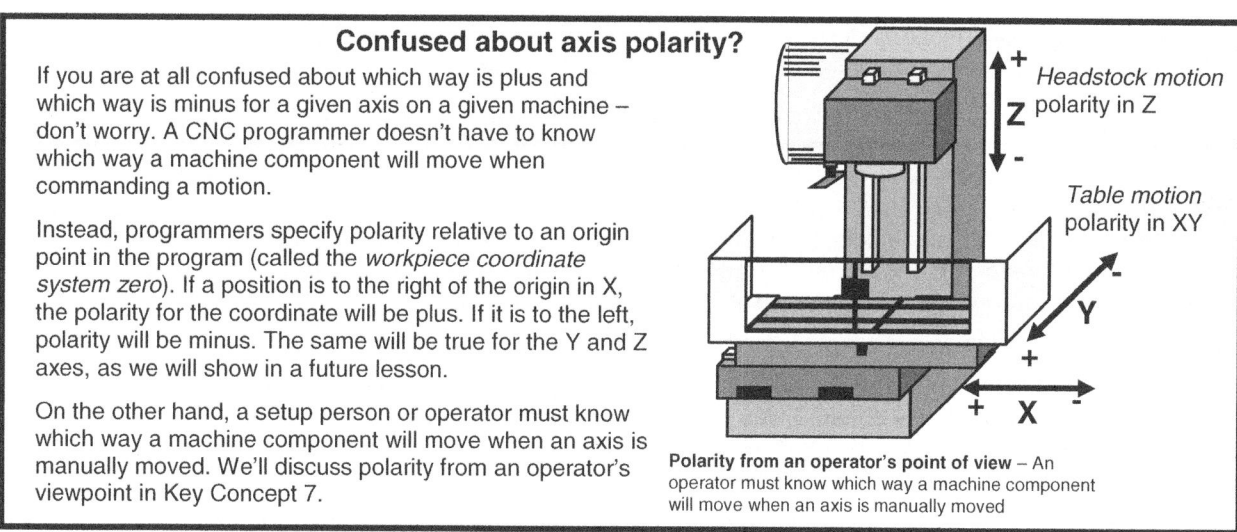

Programmable Functions of a Machining Center

A true CNC machining center will allow you to control just about all of its functions in a program. There should be very little or no operator intervention during a CNC machining cycle. Below are some common functions that can be programmed on most machining centers. While we do show the related CNC words used to command these functions, it is not our intention to teach programming commands in this lesson. We are simply making you aware of the kinds of things a programmer can control in a program.

Selecting the Measurement System

All FANUC CNCs allow the use of either the Metric measurement system or the Imperial measurement system. The most basic CNC-related unit for the Metric measurement system is a millimeter. The most basic unit in the Imperial measurement system is an inch.

When you power up a FANUC CNC, it will automatically select a measurement system. This *initialized* measurement system can be changed, meaning the company you work for can set the CNC to use the measurement system that is most often used by your company. While we show how to select the measurement system from within a program, we assume that your CNC has been set to initialize using your company's measurement system of choice.

Key Concept 1: Know Your Machine From A Programmer's Viewpoint

For CNC use and throughout this text, we refer to the two measurement systems as <u>inch mode</u> or <u>metric mode</u>. There are two G-codes are used to select the measurement system. G20 is used to select inch mode and G21 is used to select metric mode.

<u>Many</u> CNC functions are affected by the measurement system choice. In inch mode, these functions are commonly specified in inches. In metric mode, they are specified in millimeters.

For example, the command

    `G20`

selects inch mode. Any coordinate value following G20 will be interpreted in inches. The command

    `X1.0 Y1.0`

specifies a value of one inch in each axis (as opposed to one millimeter).

What is a G code?

G codes are called *preparatory functions*. They prepare the machine for what is coming up – in the current command and possibly in up-coming commands. In many cases, they set *modes*, meaning once a G code is *instated* it will remain in effect until the mode is changed or cancelled.

Here we list a few common G codes, but don't worry if they don't make much sense yet. Upcoming discussions will clarify.

Common G codes:

G00: Rapid motion	G42: Cutter comp. right
G01: Straight line motion	G43: Tool length comp.
G02: Circular motion (CW)	G81: Drilling cycle
G03: Circular motion (CCW)	G83: Peck drilling cycle
G04: Dwell	G84: Tapping cycle
G20: Inch mode selection	G90: Absolute mode
G21: Metric mode selection	G91: Incremental mode
G28: Zero return command	G98: Initial plane
G40: Cancel cutter radius comp.	G99: R plane
G41: Cutter comp. left	

Look in your control manufacturer's manual for a full list of G codes.

Spindle

The spindle of all machining centers can be programmed in at least three ways, activation (start/stop), direction (forward/reverse), and speed (in revolutions-per-minute or rpm). Some machining centers also provide multiple power or gear ranges (like the transmission of an automobile).

Spindle Speed

You can precisely control how fast the spindle of a machining center rotates in one rpm increments. An S-word is used to specify spindle speed. If you want the spindle to rotate at 350 rpm, program S350. Since spindle speed is specified in whole numbers, you must not include a decimal point in the S-word. Also, the S-word by itself does <u>not</u> actually start or activate the spindle.

Most cutting tool manufacturers provide spindle speed recommendations for a particular material and tool in *surface <u>meters</u> per minute* (smpm) or *surface <u>feet</u> per minute* (sfm), depending upon your measurement system choice - metric mode or inch mode. This allows them to use a single recommendation for a wide range of tools. A calculation is required to determine the spindle speed for the program in rpm.

Here is the rpm formula when you are using <u>metric mode</u>:

    rpm = 318.3 times smpm divided by cutting tool diameter in millimeters

For example, consider drilling a 12.0-mm diameter hole in mild steel with a high-speed-steel twist drill. The cutting tool manufacturer recommends drilling mild steel at 25 smpm.

    rpm = 318.3 * 25 smpm / 12-mm = 663.1 rpm

The S-word does not allow decimal point programming, so we must round the result to the closest whole number. The word S663 will be used to specify the spindle speed.

Here is the rpm formula if you using <u>inch mode</u>:

    rpm = 3.82 times sfm divided by cutting tool diameter in inches

For example, consider drilling a 0.5-in diameter hole in mild steel with a high-speed-steel twist drill. The cutting tool manufacturer recommends drilling mild steel at 70 sfm.

    rpm = 3.82 * 70 sfm / 0.5-in = 534.8 rpm

Because the S-word does not allow decimal point programming, we round the result to the closest whole number. The word S535 will be used to specify the spindle speed.

Spindle Activation and Direction

You can control which direction the spindle rotates–forward or reverse–and stop the spindle using M-codes. The forward direction is used for right-hand tooling. It will appear as counter-clockwise when viewed from in front of (below) the spindle. The reverse direction is used for left-hand tooling and will appear as clockwise when viewed from in front of the spindle.

Three M-codes control spindle activation. M03 turns the spindle on in the forward direction. M04 turns the spindle on in a reverse direction. M05 turns the spindle off.

What is an M code?

M codes control many of the programmable functions on a machining center. In many cases, you can think of them as being like programmable on/off switches. "M" stands for "miscellaneous" or "machine" function.

M codes are created by the machine tool builder, and will often vary from one machining center to another. Here we show some common M codes, but you must look in your machine tool builder's programming manual to find the complete list of M codes for a given machining center.

These M codes don't vary:
M00: Program stop
M01: Optional stop
M03: Spindle on (forward)
M04: Spindle on (reverse)
M05: Spindle off
M06: Activate tool changer
M08: Flood coolant on
M09: Coolant off
M30: End of program
M60: Activate pallet changer

These M codes vary:
M___: Automatic door open
M___: Automatic door close
M___: Chip conveyor on
M___: Chip conveyor off
M___: _____
M___: _____
M___: _____
M___: _____
M___: _____

Spindle Range

Some, especially larger machining centers, have two or more spindle ranges. Spindle ranges are like the gears in an automobile transmission. Lower ranges are used for power–higher ranges are used for speed. With most modern machining centers, spindle range selection is automatic and transparent. The spindle range will be automatically selected when you specify a spindle speed (S-word). For this reason, some programmers don't even know the machine that they are programming has two or more spindle ranges! For older machines, you may have to change gears in the program using M-codes–check the machine tool builder's programming manual.

Figure 1.10 shows the power curve chart for a machining center that has two spindle ranges. If your machining center has more than one spindle range, you can find this kind of chart in your machine tool builder's programming manual.

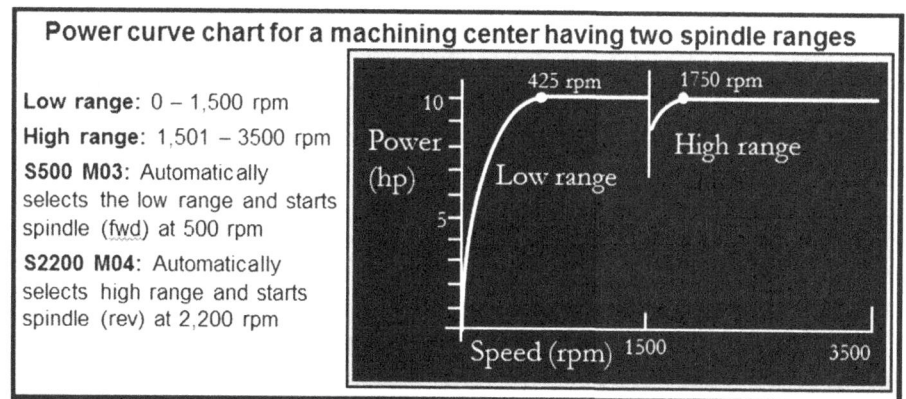

Figure 1.10: Typical power curve chart for a machining center having two spindle ranges.

For the example machining center in figure 1.10, full power is achieved in the low spindle range at 425 rpm and continues up to 1,500 rpm. In the high spindle range, full power is achieved at 1,750 rpm. The start of the high range is at 1,500 rpm. This means there is a gap between 1,501 and 1,749 rpm where full power is not achievable. This is important information, especially when you are programming machining operations that require more power.

Key Concept 1: Know Your Machine From A Programmer's Viewpoint

For example, consider a 50-mm (1.968-in) face mill for a rough machining operation where the tooling manufacturer recommends using a cutting speed of 250 surface meters per minute (smpm), which is 820 surface feet per minute (sfm). The calculated rpm is:

Using metric mode: rpm = 318.3 * 250 smpm / 50-mm = 1,591.5 rpm

or

Using inch mode: rpm = 3.82 * 820 sfm / 1.968-in = 1,591.6 rpm

If S1591 is specified as the spindle speed, the machine will automatically select the high spindle range, and there may not be enough power to perform the powerful machining operation–the spindle may stall. In this case, it will be necessary to compromise spindle speed and program to S1499 to force the machine to select the low spindle range in order to achieve the power required.

Feedrate

As you know, a machining center has three linear axes, X, Y, and Z. You must be able to control how quickly these axes move, especially during machining. Feedrate is the rate at which the cutting tool will move during a machining operation. It is a programmable function for all machining centers. Feedrate is specified with an F-word (F for feedrate).

Feed-Per-Minute

For machining centers, feedrate is commonly specified as the distance moved in a minute, either inches-per-minute or millimeters-per-minute. Most cutting tool manufacturers provide feedrate recommendations in distance-per-revolution. That is in inches-per-rev (ipr) or millimeters-per-rev (mmpr). To determine the per-minute feedrate, you must multiply the (recommended) per-revolution value by the previously calculated spindle speed in rpm.

If using metric mode:

Again, consider drilling a 12.0-mm diameter hole in mild steel where the tooling manufacturer recommends a speed of 25 surface meters per minute (smpm). The calculated spindle speed is 663 rpm (again, 318.3 * 25 smpm / 12-mm). The cutting tool manufacturer also recommends a feedrate of 0.25 millimeters-per-revolution (mmpr). In this case, you need to calculate the rate in millimeters-per-minute (mmpm):

mmpm = mmpr * rpm = 0.25 mmpr * 663 rpm = 165.75 mmpm

You are allowed to include a decimal point with the F-word. Program a feedrate of F165.75.

If using inch mode:

Consider drilling a 0.5-in diameter hole in mild steel where the tooling manufacturer recommends a speed of 70 surface feet per minute (sfm). The calculated spindle speed is 535 rpm (3.82 * 70 sfm / 0.5-in). The cutting tool manufacturer also recommends a feedrate of 0.008 inches-per-revolution (ipr). In this case, you need to calculate the rate in inches-per-minute (ipm):

ipm = ipr * rpm = 0.007 * 535 = 3.745 ipm

Since the F-word allows the use of a decimal point, program a feedrate of F3.745.

Cutting tool manufacturers sometimes recommend feedrate with feed-per-tooth values. This is common for milling cutters. To determine the distance-per-revolution feedrate amount, multiply the feed-per-tooth value by the number of teeth or flutes on the cutter and then multiply the result by the rpm.

If using metric mode:

Consider side milling with a 25-mm solid carbide cutter. The manufacturers recommended feedrate is 0.1 millimeters per tooth (mmpt). The tool has 4 flutes. The recommended cutting speed is 160 surface meters per minute (smpm).

rpm = 318.3 * 160 smpm / 25-mm = 2,037 rpm

mmpm = 0.1 mmpt * 4 (teeth) * 2,037 rpm = 814.8 mmpm (F814.8)

If using inch mode:

Consider side milling with a 1.0-in solid carbide cutter. The manufacturers recommended feedrate is 0.005-ipt (inch-per-tooth). The tool has 4 flutes. The recommended cutting speed is 490 sfm (surface-feet-per-minute).

rpm = 3.82 * 490 sfm / 1.0-in = 1872 rpm

ipm = 0.005 ipt * 4 (teeth) * 2521 rpm = 37.44 ipm (F37.44)

Feed-Per-Revolution

Modern CNCs allow feedrate to be specified directly in distance-per-revolution (inches-per-revolution or millimeters-per-revolution) - which minimizes calculations. If both feedrate specification methods are allowed (inches-per-minute and inches-per-revolution), G-codes are used to specify the feedrate type.

G94 is used to select feed-per-minute mode and G95 is used to select feed-per-revolution mode. Any F-word following a G94 will be considered as a feed-per-minute feedrate. Any F-word following a G95 will be considered as a feed-per-revolution feedrate.

To make sure programs will run on older as well as newer machines, it is recommended that G94 feed-per-minute be used for all normal machining; after all feed-per-minute involves but a simple calculation. The one application that will benefit from using feed-per-revolution programming is tapping, where the feedrate must be equal to the thread's pitch.

> Check with an experienced person in your school to find out whether or not your machining center allows feedrate specification in feed-per-revolution. The CNC Certification Cart does provide both feed-per-minute and feed-per-revolution.

Coolant

Coolant is the fluid used to flush chips away from the cutting area. It also cools and lubricates the cutting operation. All machining centers provide flood coolant capability. Flood coolant is turned on and off with an M-code—M08 coolant on; M09 coolant off.

Some machine tool builders provide additional coolant capabilities, including high-pressure coolant, through-the-tool coolant, and air-blast (to clear chips from the work area). These special coolant functions are almost always programmed with M-codes, but the specific M-code numbers may vary from one machine tool builder to another. You must look in your machine tool builder's programming manual to see if your machine has any special coolant capabilities, as well as to determine the related M-codes.

Automatic Tool Changer

All true machining centers have automatic tool changers (ATC) that load and unload tools into the spindle during the program cycle. This allows several machining operations to be performed on a workpiece with one program cycle and without operator intervention. Machines vary when it comes to how many cutting tools they can hold. They range from under ten to over one-hundred.

Automatic tool changer designs vary from machine to machine, but programming methods remain similar, falling into one of two basic programming styles. We'll discuss the specific differences during Key Concept 5.

For now, we'll just mention that a T-word is used to select the next tool. This may rotate an automatic tool changer mechanism. With most machines, the T-word does not actually cause a tool change to occur. Again, it just rotates the magazine to its ready position (also called the waiting position) or specifies the next tool to be used.

ATC magazine stations are numbered. A machine having a tool changer magazine that can hold twenty tools will commonly have T-words ranging from T01 to T20. The T-word is used to specify the desired tool station number that will be brought to the ready position.

An M06 M-code is used to actually make the tool change. For most machining centers, M06 will cause the tool that is in the spindle to be placed back into the magazine—and the tool in the ready position will then be placed into the spindle. For example, the command

```
T05 M06
```

will cause tool station number 5 to rotate to the ready station, then the tool in station 5 is placed into the spindle, and whatever tool was previously in the spindle will be placed back into the magazine at the correct position. Figure 1.11 illustrates the activation of a double-arm-style automatic tool changer.

Key Concept 1: Know Your Machine From A Programmer's Viewpoint

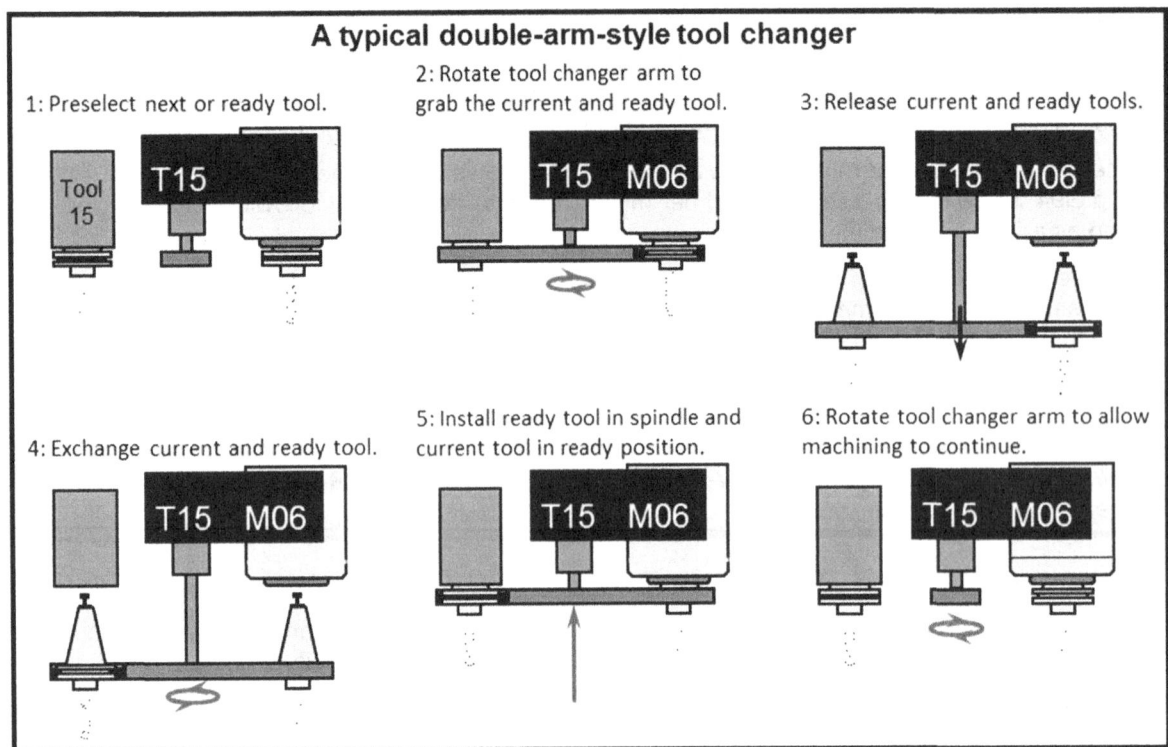

Figure 1.11: Steps of a double arm automatic tool changer

In step 1 of Figure 1.11, the T15 word will place the tool in station number fifteen in the ready position. Starting from step 2, the M06 will make the tool change. We're showing a double-arm automatic tool changer, which is the most efficient style of tool changer. Notice how the tool in the spindle is replaced with the tool in the ready station.

> The programming for automatic tool changers will vary from one machine builder to another. With the FANUC Certification Cart, for example, it is important that the M06 word precedes the T-word. To command tool number 3 be placed in the spindle, you must specify "M06 T03" not "T03 M06".

We will discuss automatic tool changers in more detail during Key Concept 5.

What Else Might be Programmable?

While this text will acquaint you with the most common programmable functions of machining centers, you must be prepared for more. Other programmable devices that may be equipped on your machining center include pallet changer, programmable chip conveyer, spindle probe, tool length measuring probe, and a variety of other application-based features. If you have any of these features, you must reference your machine tool builder's programming manual to learn how these special features are programmed.

> Check with an instructor in your school to find out what other programmable features are available on the machining center you will be working with.

Key Points for Lesson 1.1:
- ✓ There are two types of machining centers–vertical and horizontal.
- ✓ The C-frame-style vertical machining center is the most popular type of production machining center.
- ✓ You should be able to recognize the basic machine components.
- ✓ Every machining center has three linear axes, X, Y, and Z.
- ✓ Some machining centers have one or more rotary axes (B-axis on a horizontal machining center).
- ✓ Each axis has polarity (plus and minus direction).
- ✓ For linear axes, it is best to view polarity as if the tool is moving rather than the table.
- ✓ You must understand the functions of your machining center that can be programmed.
- ✓ Spindle can be controlled in at least three ways (activation, direction, and speed).
- ✓ Feedrate specifies the rate of motion for machining operations.
- ✓ Coolant can be activated to remove chips and cool and lubricate the cutting operation.
- ✓ All true machining centers have an automatic tool changer.
- ✓ Machines in the U.S. allow the use of inch or metric mode.
- ✓ You must determine what else is programmable on your machining center(s).

Quiz

1. Which style of machining center is the most popular?
   a. C-frame style vertical
   b. Knee-style vertical
   c. Bridge-style vertical
   d. Horizontal

2. Specify the correct axis for each moving component below (C-frame style vertical machining center).
   a. _____Table motion in/out
   b. _____Headstock motion up/down
   c. _____Table motion left/right

3. Specify the correct axis and polarity for each tool motion below (C-frame style vertical machining center).
   a. _____Tool motion left
   b. _____Tool motion away from you
   c. _____Tool motion down

4. Provide the CNC word or command needed to activate the following:
   a. Start spindle (fwd) at 300 rpm: _____
   b. Start spindle (reverse) at 2,000 rpm: _____
   c. Stop spindle: _____
   d. Turn on the flood coolant: _____
   e. Turn off the coolant: _____
   f. Place tool station seven in spindle: _____
   g. Select inch mode: _____

5. You plan to use a high speed steel 8-mm milling tool to machine aluminum. The tooling manufacturer recommends 80 smpm and 0.08 mmpr. Calculate:
   a. Spindle speed in rpm S_____
   b. Feedrate in mmpm F_____

6. You plan to use a high speed steel 0.25-in milling tool to machine aluminum. The tooling manufacturer recommends 250 sfm and 0.003 ipr. Calculate:
   a. Spindle speed in rpm S_____
   b. Feedrate in ipm F_____

7. Review the machine tool builder's manual for the machines in your company or school.
   a. Do your machining centers have more than one spindle range?
   b. If so, what are the cut-off points for each range?
   c. At what rpm does the spindle achieve maximum horsepower in each range?
   d. Do any of your machines have high pressure coolant systems? If so, what is the related M-code?
   e. How many cutting tools can your machining centers hold?

Answers: 1: a, 2a: Y, 2b: Z, 2c: X, 3a: X-, 3b: Y+, 3c: Z-, 4a: S300 M03, 4b: S2000 M04, 4c: M05, 4d: M08, 4e: M09, 4f: T07 M06, 4g: G20, 5a: S3183, 5b: F254.64, 6a: S3820, 6b: F11.46, 7: no answer can be provided

Lesson 1.1a – CNC Certification Cart Machine Configuration

You must understand the characteristics of the CNC Certification Cart CNC machining center to be able to program it. You must be able to identify its basic components – you must understand the moving components of the machine axes – and you must know the various functions of the machine that are programmable. This lesson builds on lesson 1.1, so read it before reading this lesson.

Objectives

After completing this lesson, students should be able to:

- ✓ Apply the information learned in lesson 1.1 to the CNC Certification Cart
- ✓ Describe the basic components of the CNC Certification Cart VMC
- ✓ Describe the primary moving axes and their polarity from a programmer's viewpoint
- ✓ Describe the basic programmable functions of the spindle
- ✓ Describe the basic operation of the automatic tool changer

CNC Certification Cart Vertical Machining Center

The Levil CNC Certification Cart is a vertical machining center. The CNC Certification Cart is very similar to a bridge-style vertical machining center, except it only has one side of the bridge.

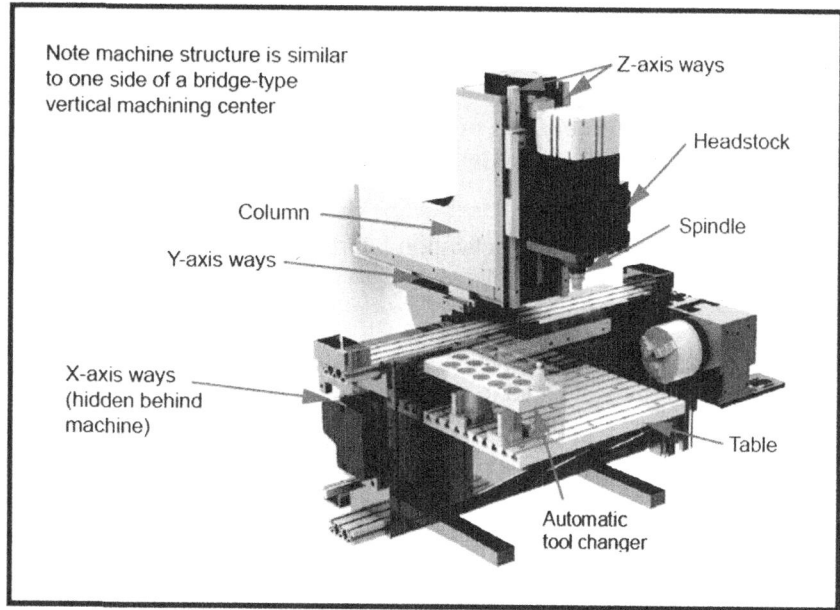

Figure 1.1 CNC Certification Cart vertical machining center with axes shown

Key Concept 1: Know Your Machine From A Programmer's Viewpoint

Directions of Motion (axes) for the CNC Certification Cart

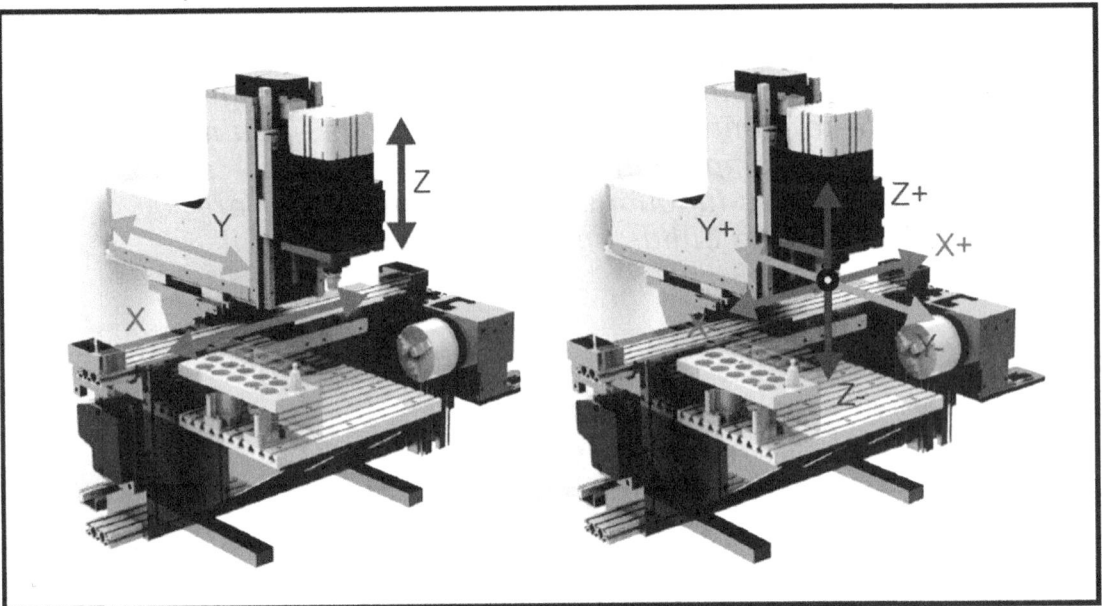

Figure 1.2 CNC Certification Cart axes motion and polarity

Like the bridge-style vertical machining center, the table is completely stationary and the tool moves with each axis. This makes understanding polarity much easier. X-plus is tool movement to the right. X-minus is tool movement to the left. Y-plus is tool movement away from the operator. Y-minus is tool movement toward the operator. Z-plus is tool movement up. Z-minus is tool movement down.

Basic Programmable Functions of the CNC Certification Cart

The CNC Certification Cart is a true CNC machining center that allows you to control just about all of its functions in a program. There should be very little or no operator intervention during a CNC machining cycle. Below are some basic functions that can be programmed.

Spindle

The spindle can be programmed in three ways, activation (start/stop), direction (forward/reverse), and speed (in revolutions-per-minute or rpm). The CNC Certification Cart has a single power range.

Spindle Speed

You can control how fast the spindle of a machining center rotates in one rpm increments. An S-word is used to specify spindle speed. If you want the spindle to rotate at 350 rpm, program S350. Since spindle speed is specified in whole numbers, you must not include a decimal point with the S-word. Also, the S-word by itself does not actually start or activate the spindle.

Spindle Activation and Direction

You can control which direction the spindle rotates – forward or reverse – and stop the spindle using M-codes. The forward direction is used for right-hand tooling. It will appear as counter-clockwise when viewed from in front of (below) the spindle. The reverse direction is used for left-hand tooling and will appear as clockwise when viewed from in front of the spindle.

Three M-codes control spindle activation. M03 turns the spindle on in the forward direction. M04 turns the spindle on in a reverse direction. M05 turns the spindle off.

Spindle Range

The CNC Certification cart has a single spindle range.

Feedrate

The CNC Certification Cart vertical machining center has three linear axes, X, Y, and Z. Feedrate is specified with an F-word (F for feedrate). The CNC Certification Cart allows feedrate to be specified in the distance-per-minute (inches-per-minute or millimeters-per-minute) or directly in inches- or millimeters-per-revolution.

G94 is used to select feed-per-minute mode and G95 is used to select feed-per-revolution mode. Any F-word following a G94 will be considered as a feed-per-minute feedrate. Any F-word following a G95 will be considered as a feed-per-revolution feedrate.

Coolant

Coolant is the fluid used to flush chips away from the cutting area. It also cools and lubricates the cutting operation. The CNC Certification Cart provides flood coolant capability. Flood coolant is turned on and off with an M-code–M08 coolant on; M09 coolant off.

There are no other special coolant capabilities.

Automatic Tool Changer

The CNC Certification Cart has a unique automatic tool changer (ATC) that loads and unloads tools into the spindle during the program cycle. This allows several machining operations to be performed on a workpiece with one program cycle and without operator intervention.

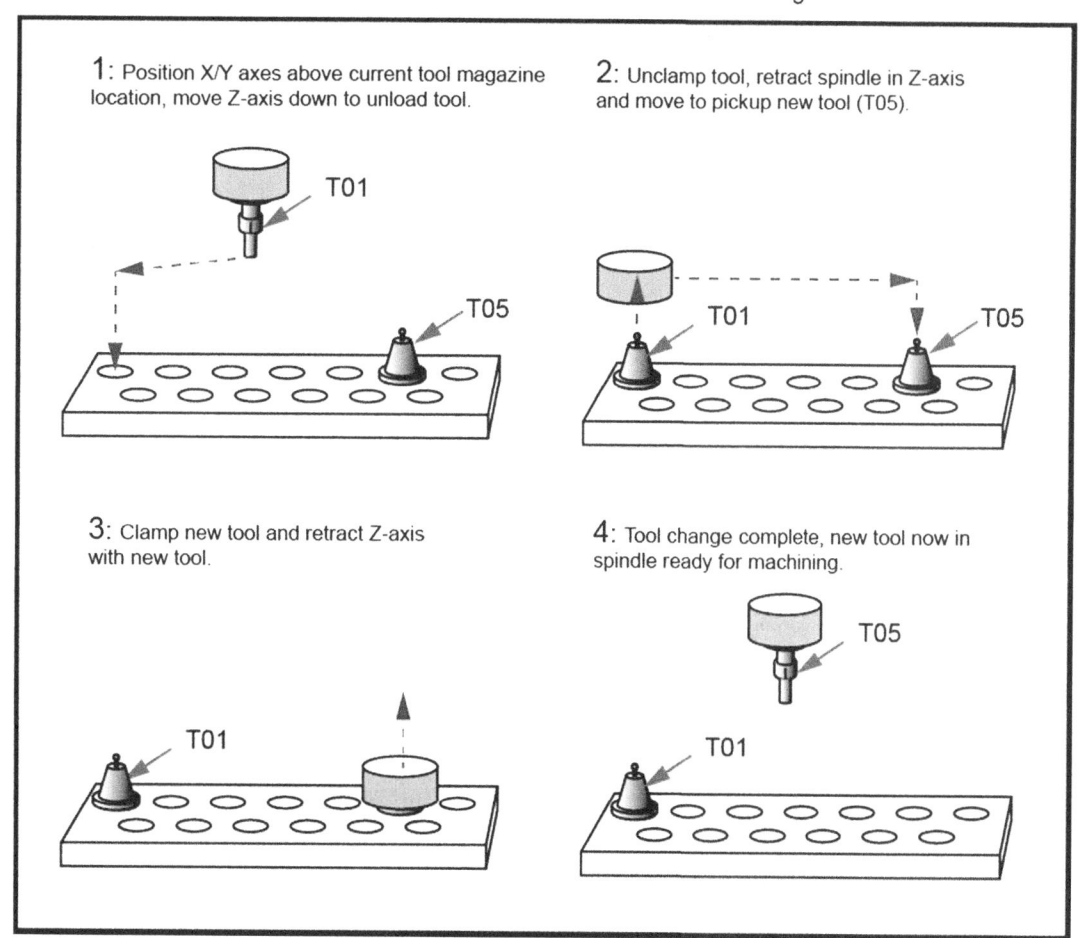

Programming the CNC Certification Cart automatic tool changer is similar to programming other machining center ATCs. We will discuss the specific differences during Key Concept 5.

The number of tools that the CNC Certification Cart can hold is eleven (11). The T-word is used to select the tool. ATC magazine/rack stations are numbered 1 through 11. A programmed T-word ranging from T01 to T11 is used to specify the desired tool station number that will be placed in the spindle. As with most machines, the T-word does not actually cause a tool change to occur. An M06 M-code is used to start the tool change. An M06 will cause the tool that is in the spindle to be placed back into the tool rack–and the programmed tool will then be placed into the spindle. The M06 must precede the T-code in the tool change command. For example, the command

```
M06 T05
```

will cause the tool in the spindle to be placed back in the tool rack, then the tool in station 5 is placed into the spindle. Figure 1.3 illustrates the CNC Certification Cart automatic tool changer.

Figure 1.3 – Steps of the CNC Certification Cart automatic tool changer

Key Concept 1: Know Your Machine From A Programmer's Viewpoint

In step 1 of Figure 1.3, the M06 will first cause the tool in the spindle to be placed back in the tool rack. The T05 makes ATC select the tool in magazine position 5 and inserts it in the spindle.

Measurement System Mode (inch or metric)

The CNC Certification Cart can use either the Imperial (inch) system or the Metric system. We refer to the two measurement systems as inch mode or metric mode. G20 is used to select the inch mode and G21 is used to select the metric mode. The default measurement system can also be selected on the CNC Handy Settings page or in parameters.

Many CNC functions are affected by your measurement system of choice. In the inch mode, these values are specified in inches. In the metric mode, they are specified in millimeters.

For example, the command

    G20

selects the inch mode. Any coordinate or feedrate following G20 will be interpreted in inches. The command

    X1.0 Y1.0

specifies a value of one inch in each axis (as opposed to one millimeter).

Feedrates specified after a G20 will be taken in inches per minute (ipm). Feedrates after a G21 will be taken in millimeters-per-minute (mmpm). (Do remember that the Certification Cart does allow feed per revolution specification. You can specify feedrate in inches-per-revolution or millimeters-per-revolution.)

Key Points for Lesson 1.1a:
- ✓ The CNC Certification Cart is a vertical machining center.
- ✓ It has three linear axes, X, Y, and Z.
- ✓ Because the tool and spindle move in all axes, it is easy to understand the polarity.
- ✓ You must understand machine functions that can be programmed.
- ✓ Spindle can be controlled in three ways (activation, direction, and speed).
- ✓ Feedrate specifies the rate of motion for machining operations.
- ✓ Coolant can be activated to remove chips and cool and lubricate the cutting operation.
- ✓ It has an 11 station automatic tool changer.
- ✓ You can program inch or metric mode.

Quiz

1. Specify the correct axis for each moving component of the CNC Certification Cart.
   a. _____ Headstock motion up/down
   b. _____ Table motion in/out
   c. _____ Table motion left/right

2. Specify the correct axis and polarity for each tool motion of the CNC Certification Cart.
   a. _____ Tool motion right
   b. _____ Tool motion away from you
   c. _____ Tool motion down

3. Provide the CNC word or command needed to activate the following:
   a. Start spindle (forward) at 2,000 rpm: _____
   b. Stop spindle: _____
   c. Turn on the flood coolant: _____
   d. Turn off the coolant: _____
   e. Place tool station eight in spindle: _____

4. Review the CNC Certification Cart Operator's Manual.
   a. Does the machine have more than one spindle range? _____
   b. How many cutting tools can the machining center hold? _____

Answers: 1a: Z, 1b: Y, 1c: X, 2a: X+, 2b: Y+, 2c: Z-, 3a: S2000 M03, 3b: M05, 3c: M08, 3d: M09, 3e: M06 T08, 4a: No, 4b: 11

Lesson 1.2 – CNC Job Workflow

Programming, setup, and operation are but three of the tasks that must be completed in order to manufacture a part on a CNC. It helps to understand how these tasks fit into the complete CNC job workflow and how company type may affect employee roles.

Objectives
After completing this lesson, students should be able to:

- ✓ Describe the four types of companies that use CNC machining centers
- ✓ Describe how the type of company may impact job roles for CNC people
- ✓ Describe some of the titles or roles of people working with CNC machining centers
- ✓ Describe the basic steps or tasks in the CNC job workflow

CNC machine tools are used by all sorts of companies, from large fortune 500 companies with hundreds or thousands of employees to small job shops that operate with just a handful of people. For some companies, CNC machining is one of their core competencies. For example, automobile powertrain, aircraft engines, and agricultural machines are all machining centric companies. Some companies specialize in making CNC machined parts for other companies under contract, and that is all they do. For other companies, CNC machining is a much smaller part of their operations. For example, a company making plastic injection molded products may only use CNC machining to manufacture tools and dies used in the injection molding machines.

In general, if a company manufacturers anything, it's likely that they're using at least some CNC machine tools, even if indirectly. With the diversity of company, size, industry application, and corporate culture, there is also diversity in what is expected of employees that are involved with CNC machines. Understanding how a particular company fits into this diverse group may help you understand what will be expected of you.

Companies that Use CNC Machining Centers

There are many factors that contribute to how a CNC-using company applies its CNC machining centers. These factors include (among others) lot sizes, lead times, percentage of new jobs, closeness of tolerances held, materials machined, and company type. The most important of these factors is company type.

When it comes right down to it, there are only four types of companies that use CNC machine tools:

Product producing companies – get revenue from the sale of a product. They are also sometimes called Original Equipment Manufacturers or OEMs.

Workpiece producing companies – get revenue from the sale of production workpieces to the product producing companies. They are also call job-shops or contract manufacturers depending on their size and product or industry focus.

Tooling producing companies – get revenue from the sale of manufacturing tooling (fixtures, jigs, molds, dies, gauges, cutting tools, etc.) to product producing and workpiece producing companies. They are also sometimes called tool and die shops.

Prototype producing companies – get revenue from the sale of prototypes to product producing companies.

Not all companies fall neatly into one category. For example, some product producing companies have their own tool producing capabilities or tool room in which CNC machine tools are used. Others may have a research and development department that produces prototypes internally. Some workpiece producing or tooling producing companies also have products of their own, it is just not their primary focus.

While there will be exceptions to the guidelines presented here, some generalizations can be made about the roles of CNC people based upon just the company-type.

Product producing companies tend to be larger and have more resources than workpiece producing, tooling producing, and prototype producing companies. Since global competition typically influences the selling price of many commodity products, the company's profitability depends on their efficiency, or their ability to make products at the lowest cost.

Since product making companies tend to make lots of the same product or product family, they engineer and document all facets of the manufacturing process. They also break up the tasks in the CNC job workflow and have people specialize in performing one of the tasks as efficiently as possible. In many companies, all employees participate in continuous improvement programs in the never ending journey to improve efficiency and customer satisfaction.

One measure of efficiency at product producing companies is maximizing machine tool utilization, ensuring that machines are making parts at as high a percentage of time as possible. Many CNC job workflow tasks are performed in parallel to production. So while a CNC machine operator is running one job on the CNC machine, other people are programming, assembling and checking cutting tools, and gathering material and fixtures for the jobs that will run on the machine next. This minimizes the amount of time that a CNC machine sits idle between production runs.

Some larger workpiece producing companies will operate much the same as product producing companies, especially if they specialize such that they get repeat business and the associated economy of scale. They are being paid a fixed price for producing a subcomponent of a product, and efficiency is a key driver of their profitability.

In contrast, smaller workpiece producing, tooling producing, and prototype producing companies most often charge an hourly rate for machine time. Often one of the key metrics in these companies is lead-time – the product making companies want their tools and prototypes as soon as possible so they can get their products to market faster. In these companies resources are more limited and each person may be responsible for multiple tasks in the CNC job workflow. They do care about efficiency but will sacrifice machine tool utilization to some extent in order to run jobs quickly with a smaller number of people. This means machines may sit idle while the machine operator performs some or all of the tasks in the CNC job workflow.

What Will You be Doing?

This course includes the information to allow you to master programming, setup and operating a CNC machining center, and you'll probably want to learn it all. If you work for a product producing or larger workpiece producing company, it's likely that you will specialize on one or two of the tasks. If you work for a smaller workpiece producing, tooling producing, or prototype producing company, you will probably have to master several if not all of the tasks.

Common job titles or roles in the CNC machining environment include:

Process engineer – This role determines which machine will be used to produce a workpiece and develops the sequence of machining operations, or process.

Tooling engineer – This role designs any special work holding devices (fixtures) needed for a job and determines which cutting tools will be used to machine the workpiece

CNC programmer – This role creates the CNC program and in larger companies may develop the setup and production run documentation used by the machine operator.

CNC setup person – This role sets up any new tooling and work holding on the CNC machine that is required to run the next job. They also produce the first part for inspection.

Inspector – This role checks the first workpiece to confirm that it is within specifications before the production runs. It may also involve some sampling of the production components to ensure the machine continues to make parts within specification.

CNC operator – Once the first workpiece passes inspection, this role runs the CNC machine until the required quantity of parts are completed.

CNC helper – This role does as much as possible to help the other people in the CNC shop. Tasks may include gathering components (fixtures, cutting tools, inserts, material blanks etc.), cleaning machines and performing basic preventive maintenance.

Tool crib attendant – This role gathers and assembles tooling components for upcoming jobs and supplies perishable tooling (like inserts) to CNC operators currently running jobs.

> Ask an experienced person in your company what your CNC-related responsibilities will be.

The CNC Job Workflow

This is an example of how a typical shop may handle the CNC job workflow. It will not be the same for all shops. Smaller workpiece producing, tooling producing and prototype companies tend to have fewer people handling several steps in the workflow. Product producing and larger workpiece producing companies will usually break processes into smaller tasks, each to be handled by a different specialist. We present the CNC job workflow tasks in the approximate order that they must happen.

Study the Workpiece Drawing

All questions about how a workpiece must be machined are answered by the workpiece drawing. Everyone involved with a job must get acquainted with the workpiece to be produced. Most engineering drawings describe a part from several directions (front, side, top, sections) or views and you must be able to use these to visualize the finished part in three dimensions. Some of the key information on the engineering drawings are the material blank type and size, dimensions and geometry of the part features, and the manufacturing accuracy or tolerance requirements of each part feature.

Decide which CNC Machining Center to Use

If a company has more than one CNC machining center, someone must decide which machine will be used to produce a given workpiece. Deciding factors include, machine availability, operator skill and availability, workpiece geometry, size, weight and lot size, the material to be machined, and the accuracy and the surface finish quality required.

Determine the Sequence of Operations and Tooling

Most parts require more than one machining operation to complete. The best sequence of machining operations must be determined. Some are obvious, drill a hole before tapping it, perform roughing machining before finishing machining, and others are based on good machining practice. For example, some hole-making processes produce holes with a higher tolerance than others—reaming versus drilling. The CNC part program will follow this machining sequence of operations. During this step, as each machining process is considered, the cutting tools needed for the job will also be determined and cutting conditions (speeds, feeds, and depths-of-cut) are calculated.

Order and Check Tooling

Based on the sequence of operations and the tooling selected, the required tooling must be obtained. If the tooling is already available, it must be checked. If not, it must be purchased. Tooling includes work holding tooling as well as cutting tools.

Create the Part Program

In this step, the programmer codes the sequence of operations into a language that the CNC machine can understand. This step is one of the major topics of this course. Six Key Concepts are included to help you understand this step. A G-code program can be created right on the CNC, but this is rather cumbersome for anything but the shortest program. The keyboard on a CNC is fine for program editing, but since it is typically mounted vertically it is not optimal for entering large sections of code, especially when compared to the layout of a PC keyboard. Theoretically, a program could be entered while the machine is running using the background editor. Most G-code programs are developed offline in an office environment using a PC with a simple text editor.

Create Setup and Production Documentation

Part of the programmer's responsibility is to make it clear how the machine must be setup to machine the part and how the production run is to be completed. Drawings are commonly made to describe the work holding setup. A cutting tool list is made, specifying the components needed for each tool, the offsets to be used for each tool, and the automatic tool changer magazine station number to be used for each tool. Production documentation commonly includes workpiece loading instructions and instructions related to holding critical tolerances on the workpiece during the production run. Anything that helps a setup person or operator should be documented.

Key Concept 1: Know Your Machine From A Programmer's Viewpoint

Load the Program Into the CNC's Memory

Once the program is created, typically offline using a PC, there are several ways to load it into the CNC memory.

Newer FANUC CNCs support Ethernet-based communications, USB memory sticks, PCMCIA flash memory interfaces and RS-232 serial port communications. Older FANUC CNCs may be limited to RS-232 serial port communications or even paper tape with holes. Transferring programs using serial communications used to be called distributive numerical control or direct numerical control or DNC for short. Some very large part programs may be too large to fit into the CNCs memory. In this case, there are several solutions to "drip feed" (also referred to as DNC) sections of the program into the CNCs memory during program execution.

Setup the Machine

Before a program can be run on the machine, the setup person must complete several setup tasks using the setup instructions prepared by the programmer. Setup tasks commonly include the work holding setup, assembling cutting tools and loading them into the proper magazine tool stations, measuring the position of program zero and entering the workpiece coordinate offsets into the CNC, and measuring the length of each tool and entering the tool length and tool radius geometry offsets into the CNC.

Verify the Program

It is very rare that a new program requires no modification at the machine. Even if the programmer does a good job programming the required cutting tool motions and activating all the machine functions appropriately, there will almost always be some optimizing that is necessary to minimize program execution time (especially for larger lots). Sometimes the exact tooling planned when the program is created is not available at production time, requiring the adjustment of feeds and speeds. If the program was originally created for one machine but the program is setup on a different machine, the tool magazine pocket assignments may need to be modified.

First Part Inspection

If the production quantity required is more than one, it is common for the first part to be produced by the setup person. The first part may be inspected by the setup person or by an inspector. Measurement of the part may require the setup person to fine tune the setup using the tool wear offsets.

Production

At this point the machine is turned over to the CNC operator to complete the production run. While the programmer's and setup person's job could be considered finished at this point, there may be some long term problems that do not present themselves until several workpieces are run. For example, cutting tool wear may be excessive. Tools may have to be replaced more often than the company would like. In this case, the speeds and feeds in the program may have to be adjusted. If the production run is long, the operator may have to adjust tool wear offsets to keep the parts produced within specification. They may also have to replace tooling and reset the tool wear offsets back to the nominal values.

Store the Corrected Version of the Program

Many workpieces that run on CNC machines will be repeated, especially in product producing companies. At some future date it will be necessary to run the workpiece again. If changes have been made during a new program's verification, it will be necessary to transmit the corrected CNC program back to the program storage device. If this step is not done, the program will have to be verified again the next time the job is run.

Key Points for Lesson 1.2:
- ✓ There are four types of companies that use CNC machine tools.
- ✓ The type of company may determine what is expected of CNC people.
- ✓ You must know what you will be expected to do in the CNC environment.
- ✓ It helps to understand the bigger picture of how a job is processed through a CNC using company.

> **Talk to experienced people in your company...**
>
> ... to learn more about how your company operates. If you work for a CNC-using company, here are questions you'll want to get answered.
>
> 1) What kind of company do you work for: product producing, workpiece producing, tooling producing, or prototype producing?
>
> 2) Who are the programmers? Who does setups? Is there a tool crib? If so, who is the attendant? Do setup people inspect their own workpieces or is there a special inspector that inspects the first workpiece? What will *you* be doing?
>
> 3) Who develops the machining process for jobs run on CNC machines? Is it the CNC programmer?
>
> 4) Do programmers write programs manually or do they use some kind of computer aided manufacturing (CAM) system?
>
> 5) Does the company have a distributive numerical control (DNC) system? If so, where is it and how is it used?
>
> 6) How often are new programs run? Does the setup person verify programs alone, or does the programmer come to the machine to help?

Lesson 1.3 – Visualizing the Execution of a Program

A CNC programmer must possess the ability to visualize the movements a CNC machine will make as it executes a program. The better a person can visualize what the machining center must do in order to machine a part, the easier it will be to create a workable CNC program.

Objectives

After completing this lesson, students should be able to:

- ✓ Describe the importance of being able to visualize a program
- ✓ Describe the basic structure of a simple program
- ✓ Describe the order of execution of a simple program
- ✓ Describe some of the basic mistakes that new programmers make and how to avoid them

Introduction

In order to be able to write a part program for a CNC machining center, you must be able to visualize (see in your mind) the movements of the machine axes required to machine the part geometry. You must also be able to visualize the activation of the various machine functions required including spindle start and stop, tool selection, and coolant flow. Experience machining parts with a conventional milling machine may be useful when visualizing a CNC machining center executing a part program.

When a machinist prepares to machine a workpiece on a conventional milling machine, they have the advantage that everything they need for the job is right in front of them. The machine, cutting tools, work holding setup, and workpiece engineering drawing are all at hand to be used or referenced. Because of this, it is unlikely that the machinist will make a basic mistake like forgetting to start the spindle before trying to machine the workpiece.

When a CNC machining center programmer writes a program, they only have the workpiece engineering drawing to reference. The machine, tooling, work holding setup, and material blank are not physically available. For this reason, a programmer must be able to visualize what will happen during the execution of the program. A beginner programmer will be prone to forget certain things–sometimes very basic things like starting the spindle prior to machining the workpiece.

In this lesson, we will acquaint you with the things a programmer must be able to visualize. We will also show you an elementary program to demonstrate the importance of visualization.

Visualization is necessary to develop any kind of instructions

Consider the visualization it takes to write a set of travel instructions to get someone from the airport to your company.

Before you can write the instructions, you must visualize the path from the airport to your company. If you cannot visualize the path, you cannot write the instructions. Worse, if you think you can visualize the path but you are wrong, you will write incorrect instructions and the person following them will get lost.

In similar fashion, if you cannot visualize the path of a cutting tool and the auxiliary operations, such as selecting the tool and starting the spindle, you cannot write an effective program for the CNC that will drive the cutter to make the right part.

1) Take airport exit to Highland Dr. Turn left.
2) Take Highland Dr. 4 mi to Elm St. Turn right.
3) Take Elm St. 1 mi To March Ave. Turn left.
4) Take March Ave 2 mi to Lance Dr. Turn right.
5) Take Lance Dr. 1/2 mi to company (on right).

Key Concept 1: Know Your Machine From A Programmer's Viewpoint

Program Structure

Like the sentences that make up a set of instructions, a CNC program is made up of blocks. Each block is made up of words. Each word is made up of a letter address (N, X, Z, T, etc.) and a numerical value. The figure below shows the beginning of a CNC program that illustrates this basic program structure.

Following instructions

CNC programs are executed sequentially. First the CNC read, interpets and executes the first block in the program. Then it moves on to the second block. Read, interpret, execute. The CNC continues in this manner until it reaches the end of the program command.

Compare this to how you follow any step-by-step instructions

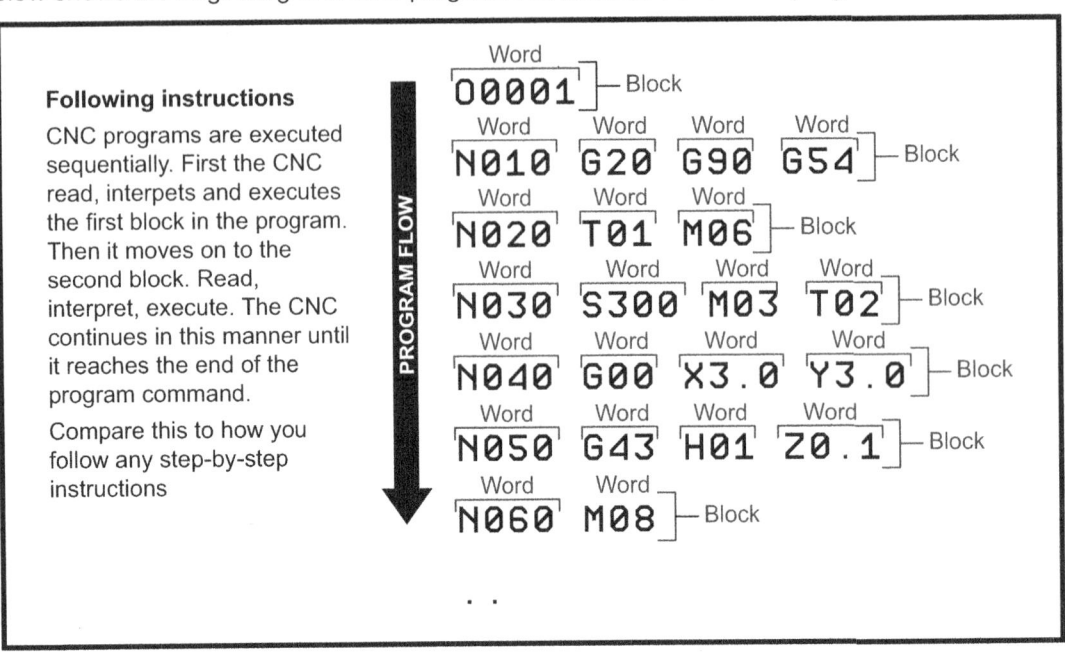

```
O0001
N010 G20 G90 G54
N020 T01 M06
N030 S300 M03 T02
N040 G00 X3.0 Y3.0
N050 G43 H01 Z0.1
N060 M08
. .
```

Figure 1.12: The flow of a CNC program

Order of Program Execution

You can compare writing a CNC program to writing a set of step-by-step instructions. For example, say you have just purchased a bookcase that requires assembly. The instructions you receive will be in sequential order. You perform step number one before proceeding to step number two. Each step will include an explanation of what it is you are supposed to do to complete the step. As you complete the procedure in each step, you are one step closer to finishing the complete assembly.

A CNC program is also executed in sequential order. The CNC will read, interpret, and execute the first block in the program. It will then go on to the next block. Read, interpret, execute. The CNC continues reading, interpreting and executing blocks until the end of the program command is reached.

The CNC will make no assumptions and executes each command in the program explicitly. Compare this to the set of instructions for assembling the bookcase. The manufacturer may assume that you have a screwdriver and that you know how to use it. Poorly written assembly instructions may be rather vague in describing exactly what you are supposed to do in a given step so that different people may interpret the instructions slightly differently. In contrast, each CNC command will have only one resultant machine action or set of actions.

An Example of Program Execution

To stress the sequential order of execution, and the visualization that is necessary to write programs, let's look at a very simple machining center example. We will first show the steps a machinist will perform to machine a very simple workpiece on a conventional milling machine. Then we will show the equivalent CNC program that will perform the same machining operation on a CNC machining center. In each case, we assume that the workpiece is already mounted in a vise on the table.

The figure below shows the print for this machining operation. In this case, we are simply drilling a 0.500-in diameter hole to 1.0-in deep.

This is a very simple example to illustrate the sequential order by which a machinist will machine a workpiece and to visualize the steps necessary to write a CNC program to perform the same machining operation.

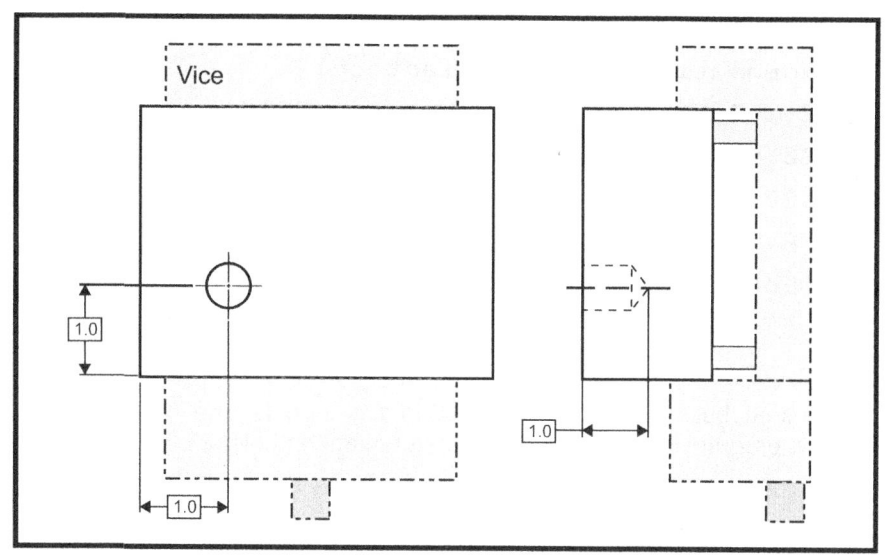

Figure 1.13: Drawing for example illustrating program execution flow

Manual Milling Machine Procedure:

First let's look at the steps a machinist will take on a conventional milling machine or drill press.

1. Mount the tool (drill) in the spindle.
2. Turn spindle on in the forward direction and set the spindle speed to 600 RPM.
3. Move the table slides to position the workpiece under the drill.
4. Advance the headstock quill to move the drill close to the surface of the part.
5. Spray coolant on the tip of the tool
6. Advance the quill at the desired feedrate to drill the 0.500-in hole, adding more coolant as required.
7. Retract the drill from the hole.
8. Move the tool away and turn off the spindle.

CNC Program

Though the CNC commands in this program will not make much sense to you yet, the comments within parentheses should clarify what is happening in each program block. By the way, you can include comments within parentheses in your own programs. Aside from the preamble in blocks N010 and N020 to set inch mode, absolute mode and the program zero, the program closely follows the steps that the machinist makes on a conventional mill, it is just that each function is automated under CNC program control.

The percent signs (%) in the programs below are called *file delimiters*. They are required in the program's computer file (commonly a txt file) for program loading purposes. A percent sign must be placed before the program's beginning and after its end.

Inch-mode example:
Now here is a CNC program to drill the 0.500-in diameter hole in the workpiece on a CNC machining center.

```
%
O0001 (Program number)
N010 G20 G90 (Select inch & absolute programming modes)
N020 G54 (Set the workpiece coordinate system zero)
N030 T01 M06 (Load the drill into spindle)
N040 S600 M03 (turn spindle on CW at 600 RPM)
N050 G00 X1.0 Y1.0 (Move tool above the hole in X and Y)
```

Key Concept 1: Know Your Machine From A Programmer's Viewpoint

```
N060 G43 H01 Z0.1 M08 (Rapid to workpiece surface, instate tool
length compensation, turn coolant on)
N070 G01 Z-1.0 F3.5 (Drill hole at 3.5IPM)
N080 G00 Z0.1 M09 (Retract drill, coolant off)
N090 G91 G28 Z0. M05 (Rapid to Z-ref position, spindle off)
N100 G28 X0. Y0. (Rapid to X/Y reference positions)
N110 M30 (End of program)
%
```

Metric-mode example:
The setup remains the same, but instead of drilling a 0.5-in dimeter hole, one inch deep and one inch from the lower-left corner, this program machines a 12 millimeter diameter hole, 25.0-mm deep and 25.0-mm from the lower left corner.

```
%
O0001 (Program number)
N010 G21 G90 (Select metric & absolute programming modes)
N020 G54 (Set the workpiece coordinate system zero)
N030 T01 M06 (Load the drill into spindle)
N040 S600 M03 (turn spindle on CW at 600 RPM)
N050 G00 X25.0 Y25.0 (Move tool above the hole in X and Y)
N060 G43 H01 Z2.5 M08 (Rapid to workpiece surface, instate tool
length compensation, turn coolant on)
N070 G01 Z-25.0 F85.0 (Drill hole at 85 mmpm)
N080 G00 Z2.5 M09 (Retract drill, coolant off)
N090 G91 G28 Z0. M05 (Rapid to Z-ref position, spindle off)
N100 G28 X0. Y0. (Rapid to X/Y reference positions)
N110 M30 (End of program)
%
```

It is not our intention to teach you the programming words being used (yet). We just want to stress two things:

1. The sequential order in which the program will be executed: The CNC will execute the line that begins with N010 before moving on to block number N020. Then block number N030. And so on–until the entire program is executed.
2. The visualization that is necessary in order to write programs: For this example, you must be able to visualize the drill machining the workpiece before you can write the program. If you cannot visualize how you will want a tool to move as it performs a machining operation, you will not be able to write the related CNC commands. For those machining operations with which you are not familiar, you will need the help of an experienced machinist.

Sequence Numbers

Look at the example program again. Notice that each command in the program begins with an N-word or sequence number. In this example, sequence numbers are simply block identification labels to help keep track of individual blocks in the program. For the most part, the CNC control ignores them. Sequence numbers are used extensively in the examples in this course to allow us to point out things about a specific program block–they confirm which block you should be looking at that is related to the discussion in the text.

For longer programs, sequence numbers help you confirm that you are truly looking at the correct block as many commands may be repeated and may look alike. This is extremely important if you are going to make modifications to the program.

Since the CNC control ignores sequence numbers, they can be left out of the program if you prefer. Some programmers minimize using sequence numbers, especially if the CNC's memory capacity is small, only adding them at key places in the program that identify the beginning of an operation they wish to search for later.

We recommend that you include them in your programs during this course–and place them in a logical order. It will be a useful communication tool between you and your instructor.

Throughout this course every block of every program will have a sequence number. And you'll notice that we skip ten numbers between sequence numbers. That is, we'll use N010, then N020, then N030, and so on. We recommend using this technique so you can insert additional blocks to your programs, say between blocks N020 and N030, perhaps labeling it N015, without having to renumber every block that follows.

Decimal Point Programming

You may have noticed that certain CNC words in the example above use a decimal point in their numerical values and others do not. All current CNC controls allow a decimal point to be placed in CNC words which are real numbers (numbers that require a portion of a whole number). Most CNCs do not allow a decimal point in words that always require integer values (whole numbers). All axis words (like X, Y, and Z) often require real numbers, so a decimal point is allowed and recommended. Spindle speed is specified with an integer value, so you are not allowed to include a decimal point in an S-word.

Letter address words that typically use the decimal point include X, Y, and Z for axis positions, I, J, and K, for circular motion commands, R for radius, and F for feedrate. Words that do not typically allow a decimal point and must be programmed as integer values including N, G, H, D, L, M, S, T, O, and P. The latest FANUC controls do use a decimal point in some G-words.

Modal Words

Many CNC words are modal. This means that the CNC value remains in effect until changed or canceled. In the previous example program, notice the G00 command in block N050. This happens to be the command that activates rapid motion. The axis motion that results from block N050 will move at the rapid traverse rate. The very next block (N060) commands the Z-axis to move down to just above the workpiece surface, but there is no motion type explicitly programmed in this block. This movement will also occur at the machine's rapid rate because G00 is modal. Modal words do not have to be repeated in every block.

Initialized Words

Certain CNC words are initialized. This means the CNC will automatically instate these words at power-up. For example, most machines used in the United States will be initialized in the inch mode (again, the instating word is G20). If a company will be using the inch mode exclusively, they can depend upon the CNC to be in this state at all times (they'll never select the metric mode), meaning there is no need to actually include a G20 in the program. However, it is still good practice to include initialized codes in the header of a program to document what is expected. It is a good communication tool between the programmer and the setup person and machine operator.

Word Order in a Block

With very few exceptions, the order in which CNC words appear in a program block has no bearing upon how the command will be executed. For example, the command

    N050 G00 X15.0 Y12.5 M08

will be executed in exactly the same way as

    N050 M08 X15.0 G00 Y12.5

Common Beginner Mistakes

Omitting Decimal Points

Beginning programmers have the tendency to leave out decimal points. You must remember to include a decimal point within each word that allows a decimal point. If you do not, some very strange things can happen.

Key Concept 1: Know Your Machine From A Programmer's Viewpoint

Older NC or numerical controls (before about 1975) did not allow a decimal point in any programmed word. These NCs used a fixed format for all real numbers in the program. Trailing zeros were used to imply the position of the decimal point. For example, for an NC that assumed four decimal places in the inch mode, an X-word 5.0 inches was specified as "X50000" (note the four trailing zeros). The decimal point position was implied by placing it four places from the right most digit–X50000 becomes X5.0000.

Today's CNCs may offer backward compatibility so that older programs can still run on newer machine tools. The backward compatibility may be determined by CNC system parameters. If a current CNC control sees a word that allows a decimal point, but the decimal point is not specified in the word, it may assume the decimal point position automatically and place it four places to the left of the right-most digit (in the inch mode), just like the older NCs.

Here's an example that illustrates the kind of mistake you can make. Say you intend to specify an X-word of 3.0 inches or 75.0 millimeters. The correct way to designate this movement is X3.0 or X75.0. What happens if you specify "X3" or "X75", leaving out the decimal point and the trailing zero? The CNC may interpret the X-word as 0.0003 or 0.075", automatically placing the decimal point four - or three - places to the left of the right-most digit (most CNCs use a three-placed fixed format when using metric mode). In this case, "X3" will be interpreted as X0.0003, not X3.0 (inches). "X75" will be interpreted as X0.075, not X75.0 (millimeters).

Concentrate on overcoming this tendency for making mistakes of decimal point omission. Get in the habit of including a decimal point within those words that allow it.

Decimal Point Tip

When programming numbers in CNC words that allow a decimal point, be sure to carry the value out to the first zero after the decimal point. For example, write

> X4.0

instead of

> X4.

This will force you to write/type the decimal point.

In a similar fashion, when specifying values under one, begin the value with the zero to the left of the decimal point. For example, write

> X0.375

instead of

> X.375

Again, this will force you to write/type the decimal point.

Other Mistakes of Omission

Knowing the mistakes beginners are prone to make may help you avoid them. Beginners tend to forget things in their programs. As stated above, beginners may forget to program a decimal point. They forget to turn on the spindle before machining a surface. They forget to turn on the coolant. They forget to drill a hole before tapping it. You will have to concentrate very hard on avoiding these mistakes. We hope that if you know about this tendency to make mistakes of omission, that it will help you avoid them.

Letter O or Number Zero?

As you review the example program above, you see several zeroes. The only letter O in this program is the very first character (which happens to be the letter address that specifies a program number). All other characters that look like the letter O are actually zeroes.

For example, notice the spindle-on-forward M-code in block N040 (M03). This is specified as M-zero-three, not M-oh-three. Again, the character in the middle is a number zero, not the letter O. This is very important. The control will not behave properly if you use a letter O here. One very common beginner's mistake is to use the upper-case letter O instead of a number zero.

When you view a program on a FANUC, the mistake should be more obvious. Zeros displayed on the FANUC CNC have a line through their centers.

Key Points for Lesson 1.3:

- ✓ You must be able to visualize the motions cutting tools will make as they machine workpieces.
- ✓ CNC programs are executed in sequential order, command by command.
- ✓ Commands are made up of words.
- ✓ Words are comprised of a letter address and a numerical value.
- ✓ You've seen your first complete CNC program. Though it was short, you've seen the sequential order by which CNC programs are executed–and you should understand the importance of being able to visualize a program's execution.
- ✓ We've introduced some program-structure-related points, including decimal point programming, modal words, and initialized words.
- ✓ You've seen some of the mistakes beginners are prone to making – like mistakes of omission and using an upper case letter O instead of a number zero.

Quiz

1. Basic machining practice experience helps you visualize cutting tool motions.
   a. True
   b. False

2. A CNC program is executed
   a. randomly
   b. sequentially
   c. right to left
   d. none of the above

3. A decimal point should be included in all CNC words.
   ☐ True
   ☐ False

4. A modal word
   a. is not allowed
   b. is automatically instated at power-up
   c. remains in effect until changed or cancelled
   d. only takes effect in the block in which it is included

5. An initialized word
   a. is not allowed
   b. is automatically instated at power-up
   c. remains in effect until changed or cancelled
   d. only takes effect in the block in which it is included

Answers: 1: b True, 2: b 3: False, 4: c 5: b

Lesson 1.4 – Understanding the Workpiece Coordinate System

A programmer must be able to specify positions to which cutting tools will move as they machine a workpiece. The easiest way to do this is to specify each position relative to a common origin point called the workpiece coordinate system zero or program zero.

Objectives

After completing this lesson, students should be able to:

- ✓ Describe the workpiece coordinate system for the X, Y and Z axes
- ✓ Describe axis polarity in the workpiece coordinate system
- ✓ Describe how to wisely choose a workpiece coordinate system zero point
- ✓ Describe the difference between the absolute and incremental programming modes

Introduction

You know from a previous lesson that machining centers have three linear axes–X, Y, and Z. You also know these axes move and that they have a polarity (plus and minus directions). In order to machine a workpiece in the desired manner, each axis must be moved in a controlled manner. One of the ways you must be able to control each axis is with precise positioning within the workpiece coordinate system.

The workpiece coordinate system has an origin point that is called the workpiece coordinate system zero. It allows you to specify all positions or coordinates from this central location. As a programmer, you will be choosing the location for the workpiece coordinate system zero–and if you choose it wisely, many of the coordinates you will use in the program will come directly from your workpiece engineering drawing, meaning the number and difficulty of the calculations required to create a program can be reduced.

A Graph Analogy

A simple graph helps understanding of the CNCs workpiece coordinate system. Since everyone has had to interpret a graph at one time or another, we can relate what you already know to CNC coordinates. The figure below is a graph showing a company's productivity over a year.

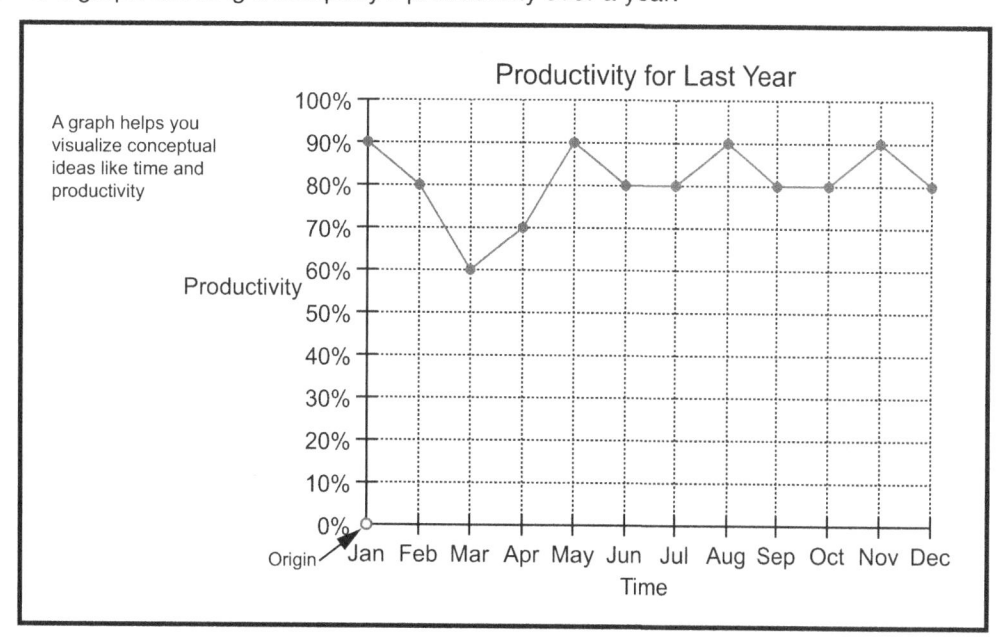

Figure 1.14: Graph example used to illustrate a coordinate system

Key Concept 1: Know Your Machine From A Programmer's Viewpoint

In the figure above, the horizontal base line represents time. The increment of the time baseline is specified in months. One whole year is the range January through December. The vertical baseline represents productivity. The increment for this baseline is specified in 10% increments and ranges from 0% to 100% productivity. The point at which the horizontal baseline and the vertical baseline cross is called the origin.

In order to make this graph, a person must have the productivity data for the year. To plot the point for January, they locate January on the horizontal baseline and then move up vertically until they are parallel with the value of 90% on the vertical baseline. To plot the point for February, they locate February on the horizontal baseline and then move up vertically until they are parallel with the value of 80% on the vertical baseline. This procedure is repeated for every month of the year. Once all of the points are plotted, a line or curve can be passed through each point to show anyone at a glance how the company did last year.

A graph is very similar to the workpiece coordinate system used with CNC. Look at the figure below.

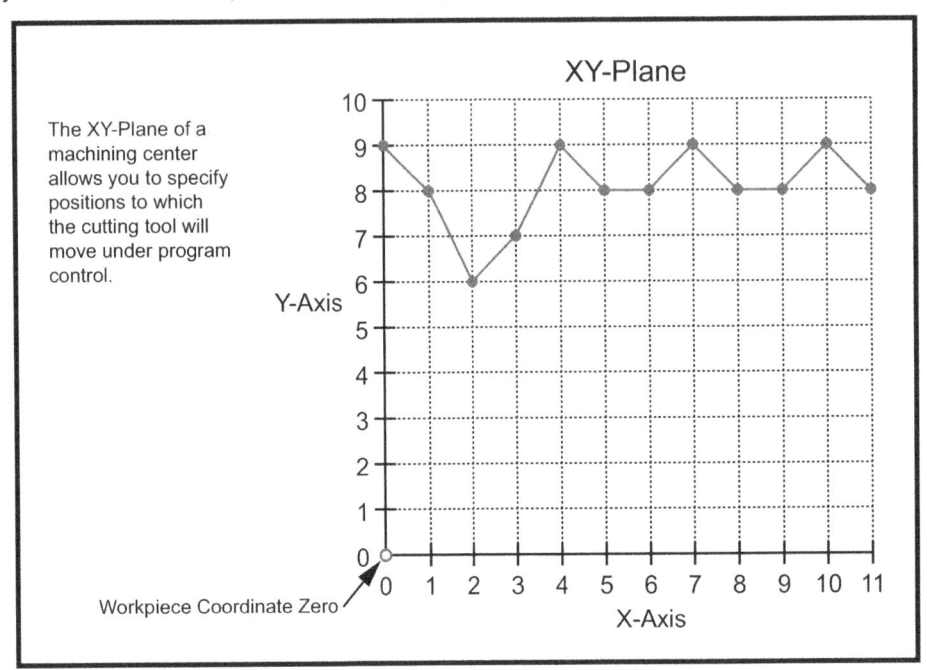

Figure 1.15: The coordinate system of a machining center (XY plane)

For the workpiece coordinate system used with CNC machining centers, the horizontal baseline represents the positions or coordinates of the X-axis. The vertical baseline represents the positions or coordinates of the Y-axis. (The Z-axis is at a right-angle to the page, toward and away from you. For now, let's concentrate on the X and Y axes.)

The increment of each base line is given in linear units. Working in inch mode, each increment is given in smallest increment programmable on the CNC. For many CNCs, the smallest programmable increment (using inch mode) is 0.0001 inches, meaning each CNC axis has a very fine grid. In metric mode, the increment will be in millimeters. In metric mode, the smallest programmable increment is typically 0.001 mm.

FANUC CNCs typically support several programmable increment systems (called IS-A, IS-B and IS-C). The increment system used is typically controlled by a CNC system parameter and is set by the machine tool builder.

Increment system	Smallest increment
IS-A (rather coarse for a CNC machining center)	0.01 mm 0.001 in 0.01 deg
IS-B (most common for CNC machining centers)	**0.001 mm** **0.0001 in** **0.001 deg**
IS-C (used for jig-grinding or jig-boring machines)	0.0001 mm 0.00001 in 0.0001 deg

The programmable increment should not be confused with the positional accuracy (resolution) of CNC axis control. The accuracy of a machine depends on several factors, including mechanical and dynamic issues, and not all are controllable by the CNC. The CNC may control axis positions to a much greater accuracy–the latest FANUC CNC, for example, use nanometer resolution (0.000000001mm).

The range for each axis is the amount of travel in the axis (from one over-travel limit to the other).

What About the Z-Axis?

The figure above describes only two of the machining center's axes, X and Y. The Z-axis behaves in exactly the same manner as X and Y. There is a zero point, plus and minus polarity, the same programmable increment is used, and the range is the maximum travel for the Z-axis. When taken all together, the X, Y, and Z axes provide you with a three dimensional grid. It is within this grid that you will be specifying positions (coordinates) that your tools move to, as we show in the figure below.

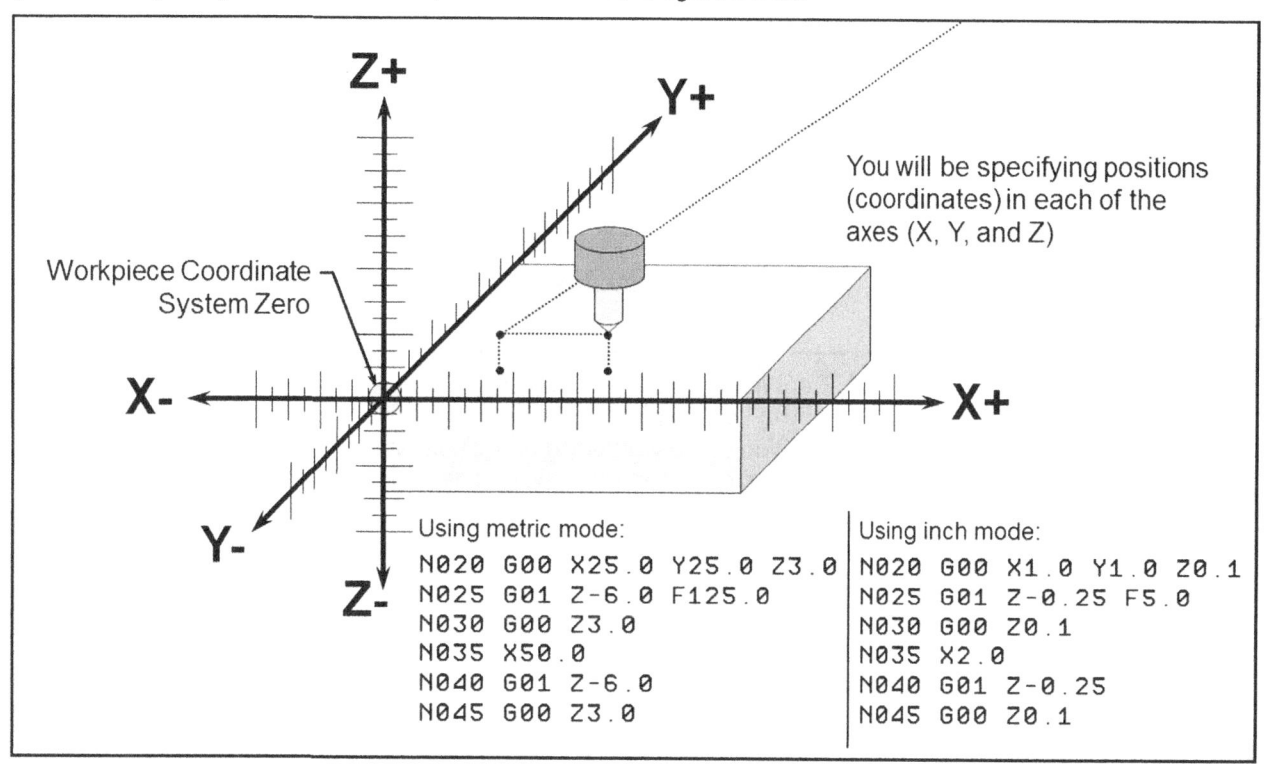

Using metric mode:
```
N020 G00 X25.0 Y25.0 Z3.0
N025 G01 Z-6.0 F125.0
N030 G00 Z3.0
N035 X50.0
N040 G01 Z-6.0
N045 G00 Z3.0
```

Using inch mode:
```
N020 G00 X1.0 Y1.0 Z0.1
N025 G01 Z-0.25 F5.0
N030 G00 Z0.1
N035 X2.0
N040 G01 Z-0.25
N045 G00 Z0.1
```

Figure 1.16: Three dimensional coordinate system of a machining center

Understanding Polarity

In the graph example previously illustrated, all the points plotted are above and to the right of the origin point. The area above and to the right of the two baselines is called a quadrant. This particular quadrant is quadrant number one where the coordinates in both axes are positive. The person creating the productivity graph intentionally planned for coordinates to fall in quadrant number one in order to make the graph easy to read. It is not uncommon on CNC machines that end points within the program fall in other quadrants. When this happens, at least one of the coordinates must be specified as negative or minus value.

You know from a prior lesson that each CNC axis has a polarity. You also know that since sometimes the cutting tool moves and sometime the table moves, it can be confusing to remember which way is plus and which way is minus. (Consider the X and Y axes of a C-frame style vertical machining center, for example. As the table moves to the left, it is moving in the X plus direction.) To make things easier, we asked you to view polarity as if the tool is actually moving in each axis.

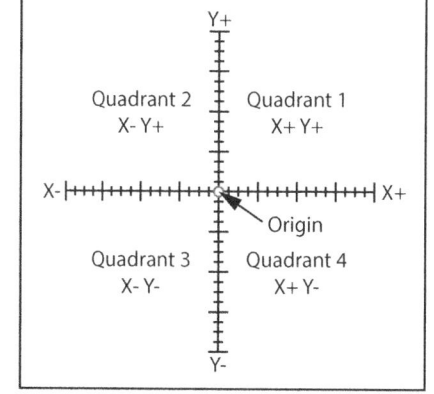

Key Concept 1: Know Your Machine From A Programmer's Viewpoint

The workpiece coordinate system makes determining the polarity of coordinates used in a program very simple. The figure below shows the polarity for the X-axis in the workpiece coordinate system.

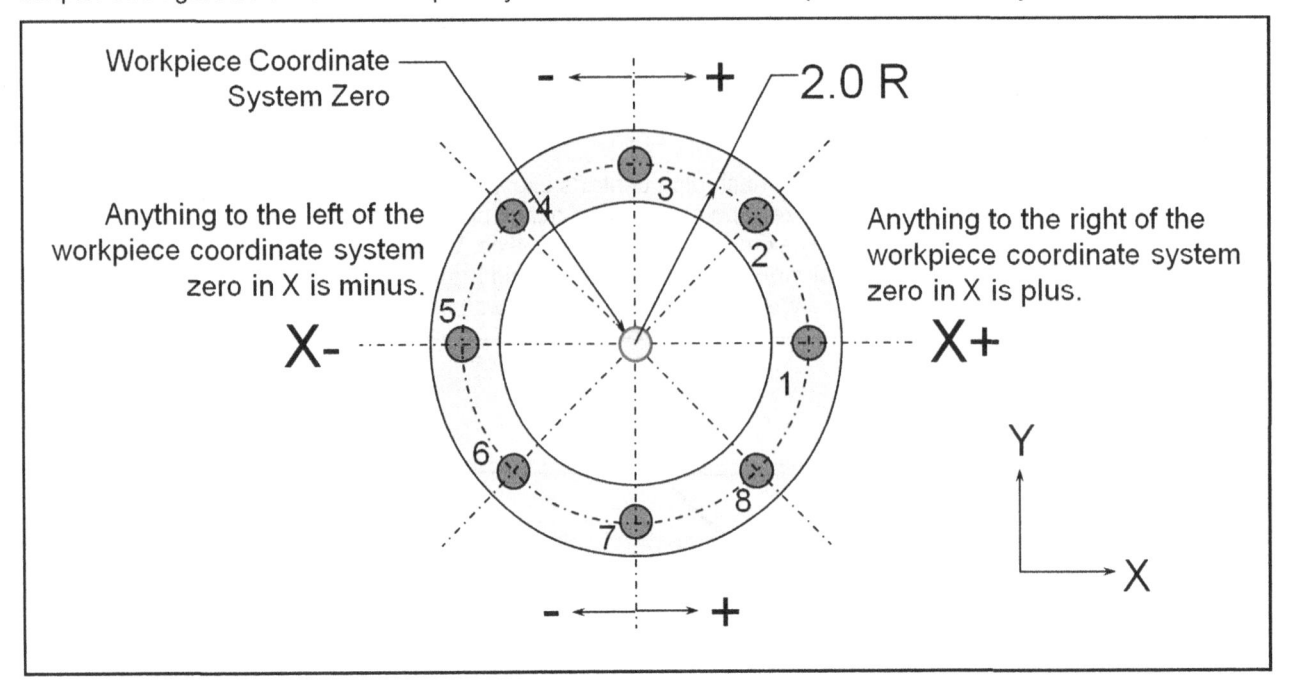

Figure 1.17: X-axis polarity

Notice that polarity is based upon the location of the workpiece coordinate system zero–as it will be for all axes. For X, anything to the right of the workpiece coordinate system zero is positive (plus). Anything to the left of workpiece coordinate system zero is negative (minus).

The figure below shows polarity for the Y-axis.

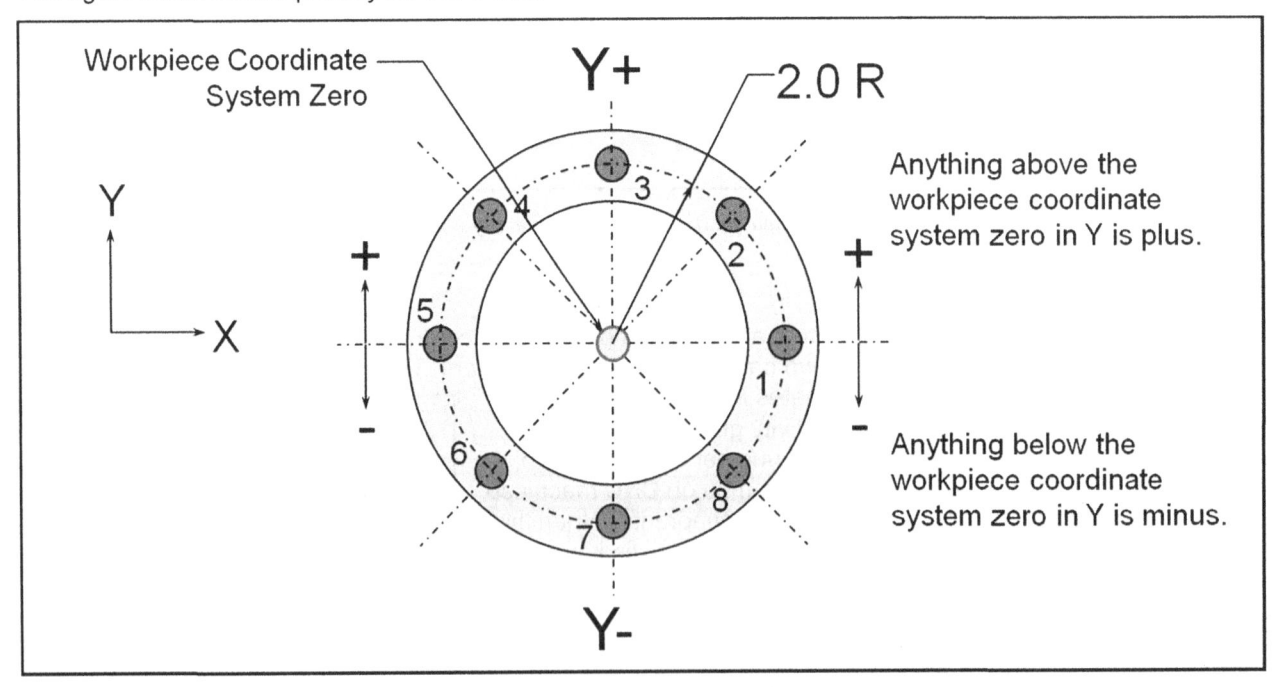

Figure 1.18: Y-axis polarity

Anything above the workpiece coordinate system zero in Y is positive (plus). Anything below the workpiece coordinate system zero in Y is negative (minus).

Lesson 1.4: Understanding the Workpiece Coordinate System

The figure below shows the coordinates for all of the holes. Note that holes are in all four quadrants.

Notice the polarity of each coordinate. The plus sign is assumed. Only the minus sign needs to be programmed.

Anything to the right of and in front of the workpiece coordinate system zero is plus. Anything to the left of and below the zero is minus.

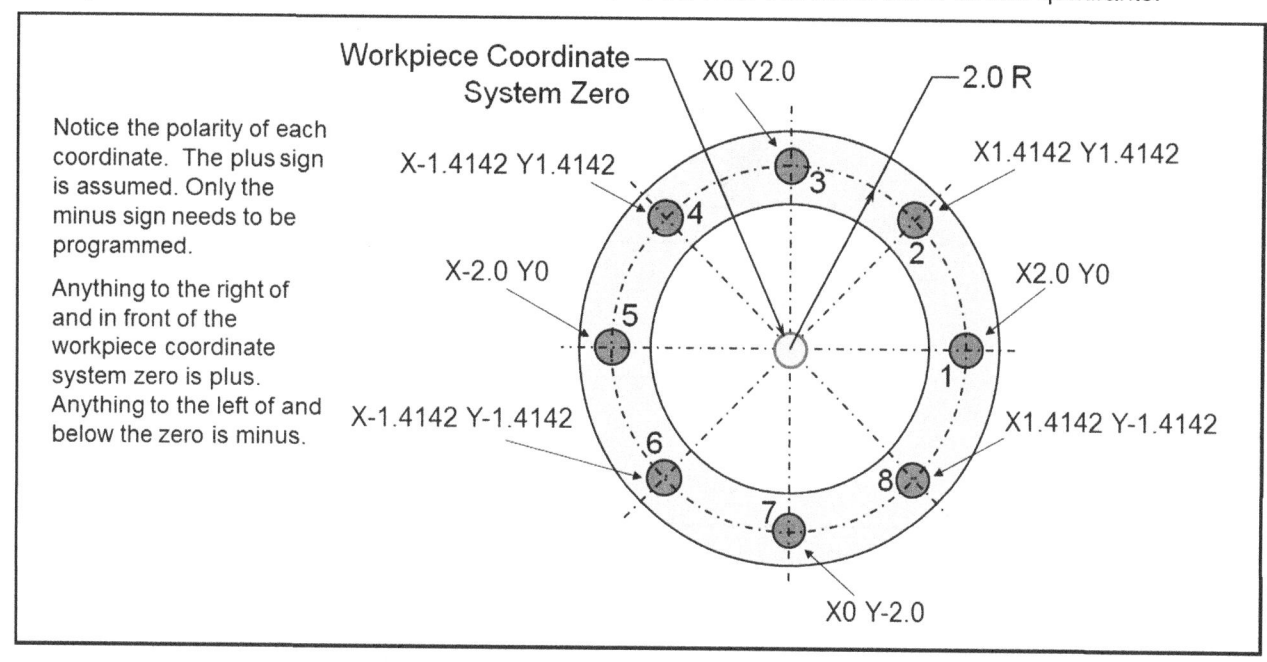

Figure 1.19: Example of polarity for XY-plane (shown using inch mode)

The figure above shows a series of holes on a bolt pattern that are easily specified relative to the center of the bolt pattern, which has been selected as the workpiece coordinate system zero. As you can see, any hole to the left of program zero requires a negative X coordinate. Any hole below program zero requires a negative Y coordinate.

A CNC will assume that a coordinate is positive (plus) unless a minus sign is specified. The CNC word X2.0, for example, specifies a position along the X-axis of positive 2.0-inches. Some CNCs may generate an alarm if the plus sign is included within the word, meaning you must let the control assume positive values. It is common practice to only include a polarity sign for negative (-) values.

The workpiece coordinate system zero must also be specified in the Z-axis. The figure below shows the XZ plane (looking at a vertical machining center from the front).

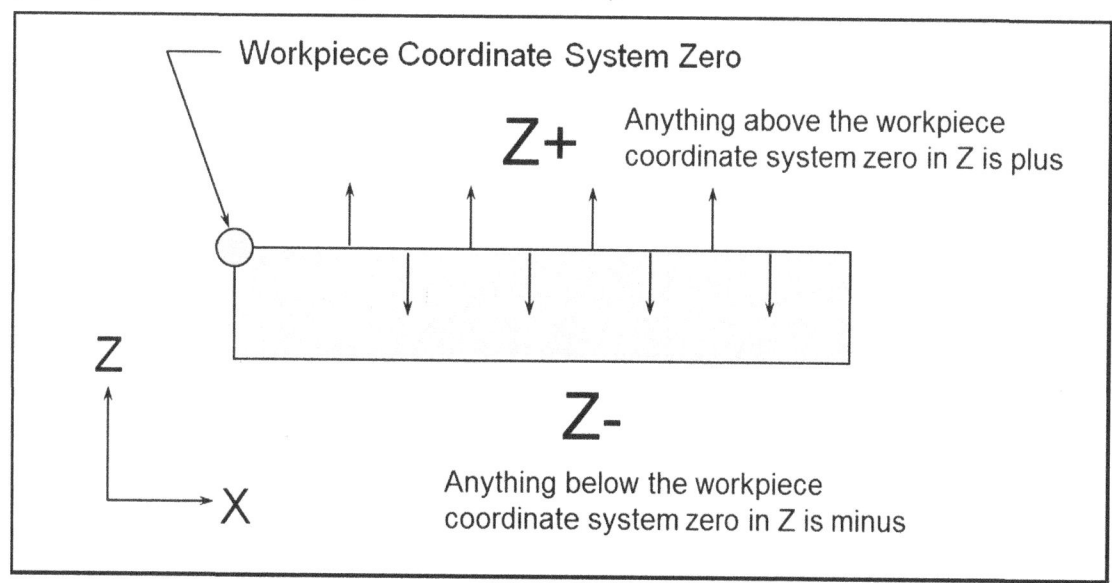

Figure 1.20: Polarity for the Z-axis

Anything above workpiece coordinate system zero in Z is positive (plus). Anything below workpiece coordinate system zero point is negative (minus).

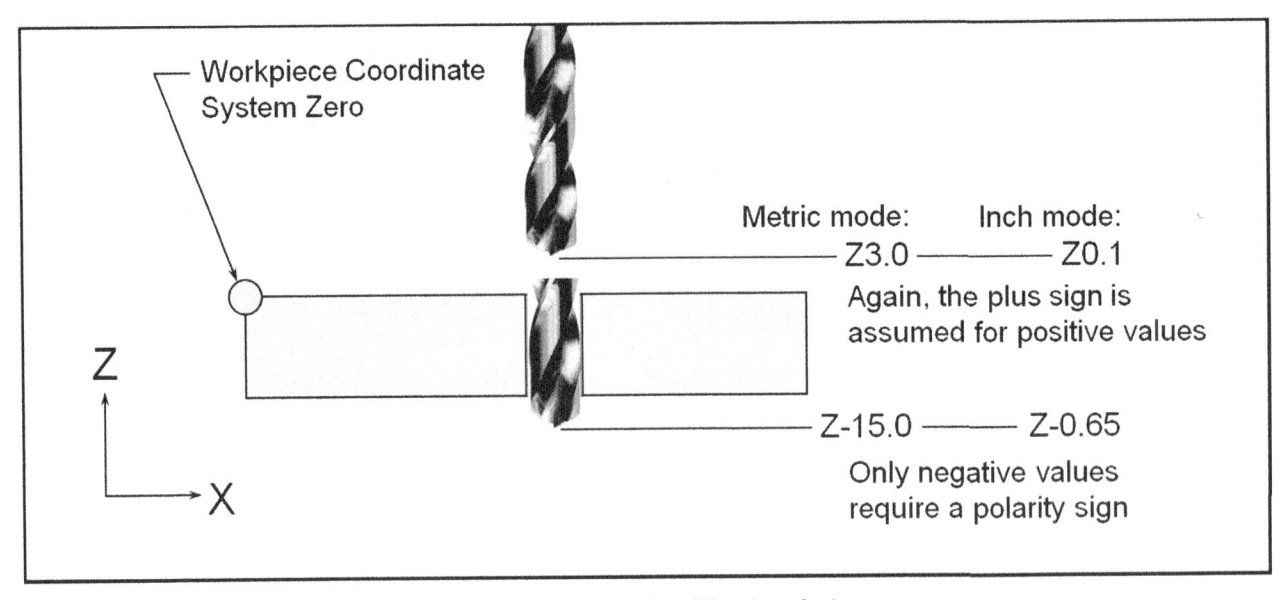

Key Concept 1: Know Your Machine From A Programmer's Viewpoint

Figure 1.21: Example of Z-axis polarity

Notice that we have selected the top of the workpiece as the workpiece coordinate system zero in Z (which is a common workpiece coordinate system zero for vertical machining center applications). Any tool position above the top of the workpiece is positive (plus) in Z. Any position below the top of the workpiece is negative (minus) in Z.

Wisely Choosing the Workpiece Coordinate System Zero Location

As the programmer, you determine the workpiece coordinate system zero point location for every program you write. Theoretically, the workpiece coordinate system zero could be placed at any location. As long as all the coordinates used in your program are specified from that workpiece coordinate system zero point, the program will function correctly. However, a wise selection of the workpiece coordinate system zero point will make programming much easier. It may also make it easier for the setup person, as we will see in the next lesson.

Selecting the X and Y Workpiece Coordinate System Zero

A good rule-of-thumb for selecting the workpiece coordinate system zero point location is to base your decision on how the workpiece engineering drawing is dimensioned. Look for datum points or workpiece surfaces from which dimensions begin. Though design engineers vary with how obvious they make it, you should be able to find one surface in each axis from which most dimensions for an axis start. Look at the figure below, for example.

With this drawing, the workpiece coordinate system zero selection is pretty obvious.

Since the lower-left corner is the datum surface for each direction, it makes the perfect workpiece coordinate system zero.

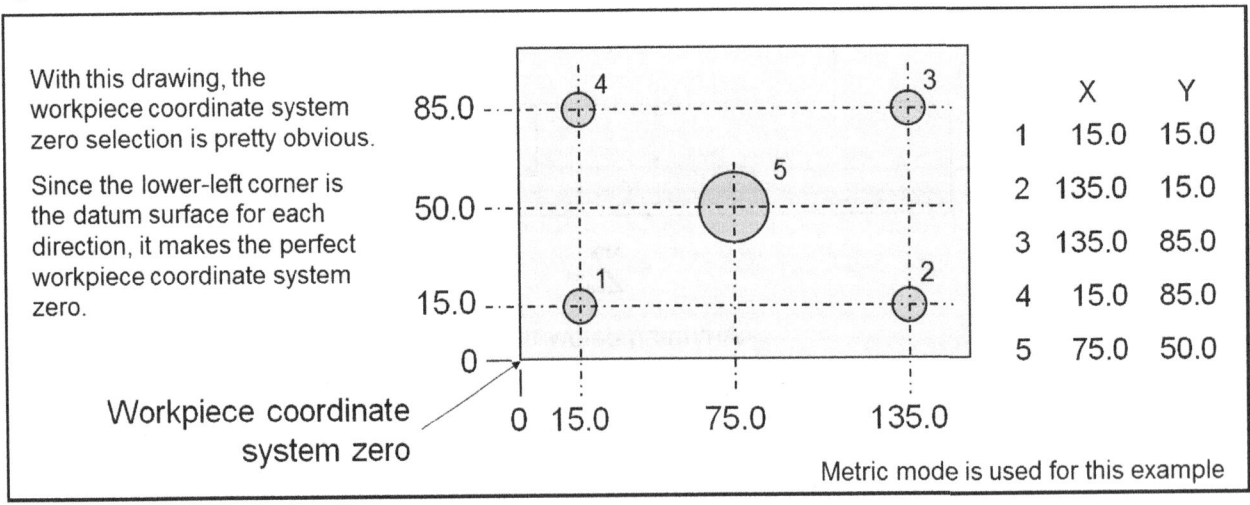

	X	Y
1	15.0	15.0
2	135.0	15.0
3	135.0	85.0
4	15.0	85.0
5	75.0	50.0

Metric mode is used for this example

Figure 1.22: Example of how to select the workpiece coordinate system zero point – Datum surface dimensioning

Lesson 1.4: Understanding the Workpiece Coordinate System

In this workpiece engineering drawing, all the horizontal dimensions start at the left side of the part. All the vertical dimensions start from the bottom of the part. Therefore, it is logical to place the workpiece coordinate system zero in the bottom left of the part. The coordinates of hole #1 will be X15.0 Y15.0. With a wise selection of a workpiece coordinate system zero, we can often take the coordinates directly from the workpiece engineering drawing.

Design engineers don't always make it so obvious. Look at the figure below.

With this drawing, the wokrkpiece coordinate system zero selection is not so obvious.

The upper-left hand corner is the best place for the workpiece coordinate system zero.

Note that since all the X coordinates are to the right of the zero, they are plus. All the Y coordinates are below the zero, so they are minus values.

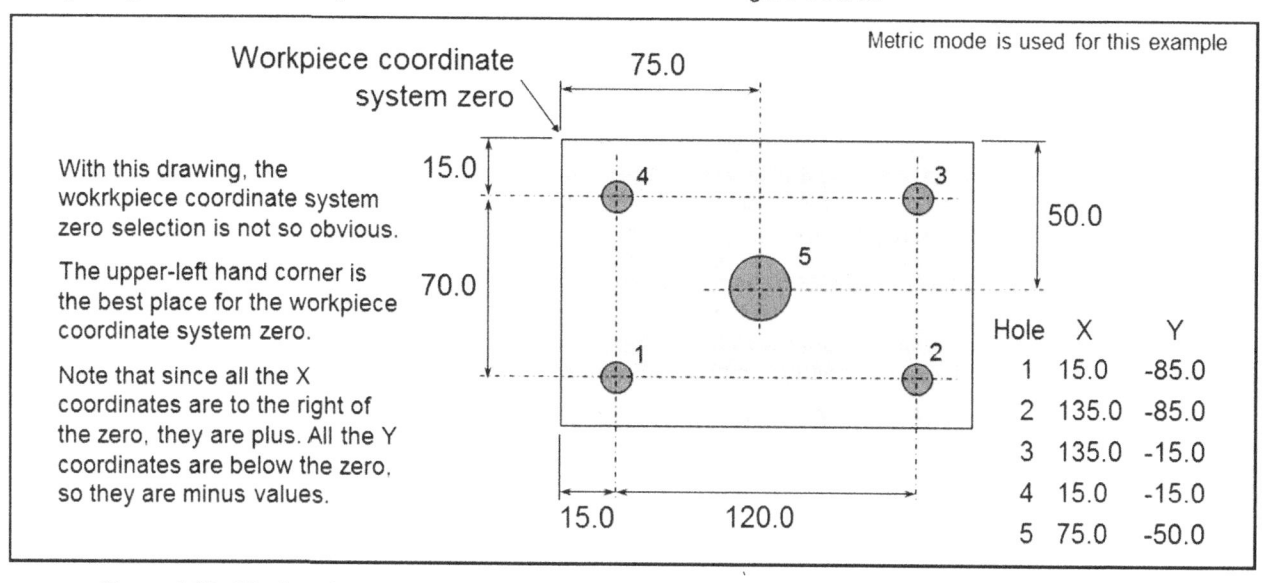

Metric mode is used for this example

Hole	X	Y
1	15.0	-85.0
2	135.0	-85.0
3	135.0	-15.0
4	15.0	-15.0
5	75.0	-50.0

Figure 1.23: The best location for workpiece coordinate system zero may be more difficult to determine

Initially, it appears that there are dimensions given from several different places. However, careful study will reveal that they actually all come from the top-left hand corner of the workpiece.

With the location of the workpiece coordinate system zero selected, all coordinates must be calculated from this point. With datum surface dimensioning (Figure 1.22), it is very easy. Coordinates used in your program will almost always match the dimensions on the engineering drawing. But if the design engineer doesn't use datum surface dimensioning techniques (Figure 1.23), you have to make some calculations.

Remember that you must determine how far it is from the workpiece coordinate system zero point to each coordinate you need in your program. As we show for the X and Y coordinates of point number two in the figure above, you'll often have to do some arithmetic.

> For the X value for the center of hole #2, you must add 15.0 plus 120.0, totaling X135.0

> For the Y value for the center of hole #2, you must add 15.0 plus 70.0, totaling Y-85.0 (note that the result is negative, since it is below workpiece coordinate system zero)

Remember, anything to the right of workpiece coordinate system zero in X is plus. Anything to the left is minus. Anything above workpiece coordinate system zero in Y is plus. Anything below is minus.

A Reminder About Axis Movement

Though we don't want to confuse you, we need to bring this up one more time. Remember, with a C-frame style vertical machining center, the tool does not move with the X and Y axes. Instead, the table moves in these axes.

Consider once again the workpiece in Figure 22 with the workpiece coordinate system zero in the lower left corner in X and Y. When you want the tool to move to hole #5, what coordinates will you specify? You should easily agree that you'll send the tool to the coordinates

```
X75.0 Y50.0
```

Notice that these coordinates are both specified as positive values (plus sign is assumed). At the completion of the motion, the tool will be resting over the center of hole #5. Here is the key question: Which way (plus or minus) does the table have to move in each axis to bring the tool to this position? The answer is: *"Who cares?"*

Key Concept 1: Know Your Machine From A Programmer's Viewpoint

This is the reason why we ask you to view polarity as if the tool is moving in all axes. You will always specify polarity in your program relative to the workpiece coordinate system zero point (with one exception that we'll show shortly). You will never have to concern yourself about which way the table (or an axis) must move in order to get the tool to the desired position.

Say the tool is now resting at hole #5 (in Figure 1.22). If you want to move to hole #1 (lower-left hole). What coordinates will you specify? Hopefully, you agree that you will specify:

```
X15.0 Y15.0
```

Again, you specify all coordinates relative to the workpiece coordinate system zero. Which way (plus or minus) will the machine move during this motion? Though the answer again is: *"Who cares?"*, it just so happens that the tool will be moving in the negative direction in both axes during this motion–moving from a large coordinate value in each axis (X75.0 Y50.0) to a smaller one (X15.0 Y15.0). And by the way, remember from the first lesson that the negative direction for X is table motion on a vertical machining center is to the right. The negative Y direction is table motion away from you.

Hopefully, you are beginning to understand how much the workpiece coordinate system is doing for you. It makes it very simple to specify positions within your CNC program regardless of which way the tool, axis, or moving component of the machine must move to get there. You need only concern yourself with where you want the tool to move–and always specify this position relative to the program zero point.

Selecting the Z-Axis Workpiece Coordinate System Zero

As with XY, you must specify a program zero point in the Z-axis. Though the location of Z-axis workpiece coordinate system zero can vary, most programmers will make the workpiece coordinate system zero surface in Z the top surface of the workpiece no matter how the part is dimensioned. See the figure below. If this is done, any machining that occurs into the workpiece (below the top surface) will require a negative (-) Z-axis coordinate. Any positioning motions above the part will be a positive value. This is the Z-axis workpiece coordinate system zero location used for all examples shown in this course.

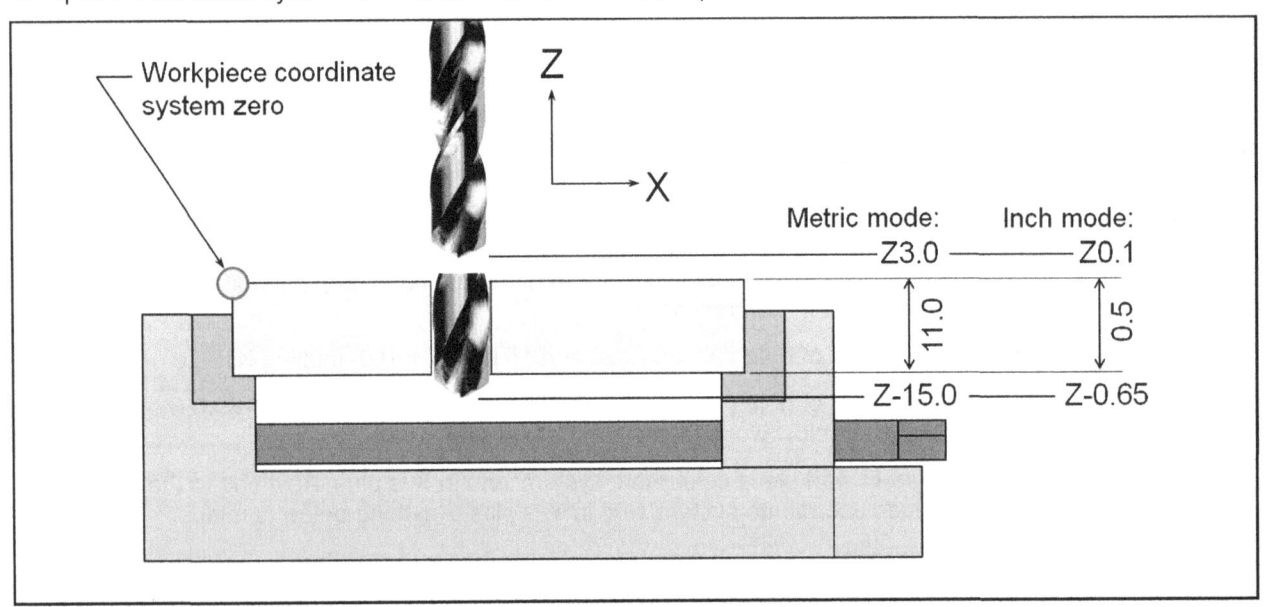

Figure 1.24: Most programmers make program zero in Z the top of the workpiece for vertical machining center programs

Absolute Versus Incremental Modes

Though we have not actually said so yet, when you specify coordinates from the workpiece coordinate system zero point, it is called the absolute positioning mode. The absolute positioning mode is specified using a G90 word. Once a G90 is specified, all coordinates are taken to be from the workpiece coordinate system zero since G90 is modal.

Everything introduced to this point has been related to the absolute positioning mode. And again, the point of reference for absolute positioning mode is the workpiece coordinate system zero point.

Lesson 1.4: Understanding the Workpiece Coordinate System

There is another method of axis positioning called the incremental positioning mode. G91 is used to specify incremental mode. Unlike absolute mode, the point of reference for all specified positions in the incremental mode is the tool's current position–the location of the tool at the beginning of the motion.

In the incremental mode, each movement is specified as a distance and direction from the tool's current position to the next position. At first glance, it may seem easier to work in the incremental mode than in the absolute mode. But you will soon find that programming with incremental positioning is quite difficult and error-prone. And by the way, if you make a mistake in a series of incrementally specified motions, every movement from the mistake on will also be incorrect.

While there are some excellent applications for incremental mode (we'll show them in Key Concept 6), beginning programmers should work exclusively in the absolute mode. All examples in this course (with the exception of some we show in Key Concept 6) use the absolute mode.

Any series of motions can be commanded in either the absolute or incremental mode. Look at the figure below.

Metric mode is used for this example

Workpiece Coordinate System Zero

Positioning to the first hole is done with absolute positioning in both examples.

The polarity of coordinates with incremental positioning is based upon the cutting tool's starting position.

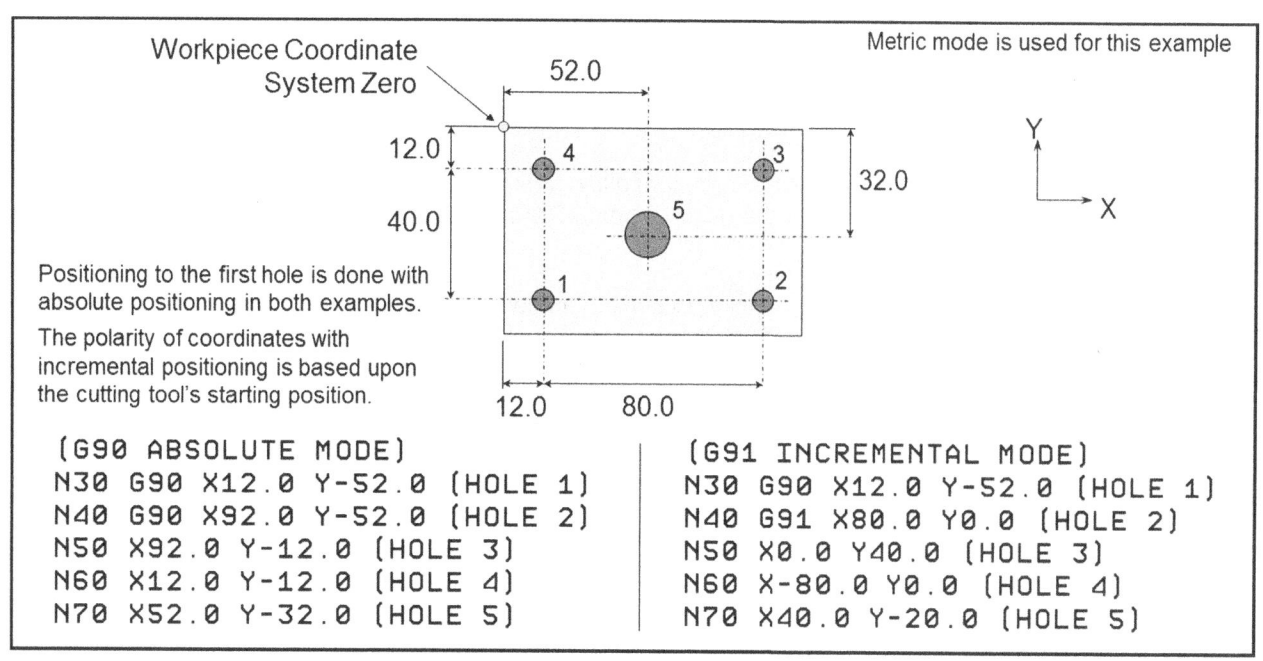

```
(G90 ABSOLUTE MODE)              (G91 INCREMENTAL MODE)
N30 G90 X12.0 Y-52.0 (HOLE 1)    N30 G90 X12.0 Y-52.0 (HOLE 1)
N40 G90 X92.0 Y-52.0 (HOLE 2)    N40 G91 X80.0 Y0.0 (HOLE 2)
N50 X92.0 Y-12.0 (HOLE 3)        N50 X0.0 Y40.0 (HOLE 3)
N60 X12.0 Y-12.0 (HOLE 4)        N60 X-80.0 Y0.0 (HOLE 4)
N70 X52.0 Y-32.0 (HOLE 5)        N70 X40.0 Y-20.0 (HOLE 5)
```

Figure 1.25: Movements can be specified in both the absolute and incremental positioning mode.

As you can see, absolute positioning makes more sense. Coordinates often match print dimensions – but even when they don't–the point of reference for each position is the same–the workpiece coordinate system zero. Incremental positioning doesn't make much sense. Positions are nothing more than a whole series of disjointed movements, each taken from the tool's previous position.

What about tolerances?

Very few dimensions shown in this course specify a tolerance. But design engineers specify a tolerance for every dimension on a workpiece engineering drawing.

Which value do you program?
Specify the mean value for every dimension (coordinate) you use in your program. The mean value is right in the middle of the tolerance band. With some tolerances, it is more difficult to determine the mean value.

Plus/minus tolerance: 4.0 +/- 0.002
The high limit is 4.002 and the low limit is 3.998. The mean value is specified right in the dimension. 4.0 is the mean value of the plus/minus tolerance.

Uneven tolerance: 4.0 +0.003 -0.001
The high limit is 4.003 and the low limit is 3.999. The mean value must be calculated. Add the high and low limit values together and divide the result by two. (4.003 + 3.999) / 2 = 4.001.

High and low limits specified: 4.004 / 4.001
The high and low limits are specified and the mean value must be calculated as shown above. (4.004 + 4.001) / 2 = 4.0025

Inch mode is used for these examples

Key Concept 1: Know Your Machine From A Programmer's Viewpoint

A Decimal Point Reminder

As stated in a previous lesson, coordinates (X, Y, and Z) require real numbers. That is, values that often include a portion of a whole number after a decimal point. You must remember to include a decimal point with coordinates, even when you are specifying a whole number value.

With metric mode:
If you want to specify an X coordinate of twelve millimeters, you should specify

    `X12.0` not `X12`

If you specify "X12", the control may use the fixed format for the X-word. It will automatically place the decimal point in a position three places to the left of the right-most digit (in metric mode). The value "X12" will be taken as X0.012, and not X12.0.

With inch mode:
If you want to specify an X coordinate of three-inches, you should specify

    `X3.0` not `X3`

If you specify "X3", the control may use the fixed format for the X-word. It will automatically place the decimal point in a position four places to the left of the right-most digit (in inch mode). The value "X3" will be taken as X0.0003, and not X3.0.

When programming whole numbers in CNC words that allow a decimal point (real number values), be sure to carry the value out to the first zero after the decimal point. For example, use

    `X4.0` instead of `X4`.

This will force you to write/type the decimal point. In similar fashion, when specifying values smaller than 1, begin the value with the zero to the left of the decimal point. For example, use

    `X0.375` instead of `X.375`

Again, this will force you to write/type the decimal point.

Most programmers prefer to use decimal point programming for obvious reasons. It doesn't make much sense to program using the fixed format. But be ready for some computer aided manufacturing (CAM) systems that automatically output G-code CNC programs in fixed format—without the decimal point. They can be a little tough to interpret.

Key Points for Lesson 1.4:

- ✓ You must be able to specify positions (coordinates) within CNC programs.
- ✓ You know that the workpiece coordinate system of a CNC machine is very similar to the coordinate system used to plot a graph.
- ✓ The origin for the workpiece coordinate system is called the workpiece coordinate system zero, or simply the program zero.
- ✓ From a programmer's viewpoint, polarity for each motion is based upon the commanded position's relationship to workpiece coordinate system zero.
- ✓ The workpiece coordinate system zero location is determined based upon how the workpiece engineering drawing is dimensioned.
- ✓ When you specify coordinates relative to the workpiece coordinate system zero, you're working in the absolute mode.

Quiz

Mark the best location for the workpiece coordinate system zero based on how the workpiece engineering drawing is dimensioned. Then fill in the coordinate worksheet, entering the X and Y coordinates of the center of each of the holes 1 through 9. We provide two exercises, first for metric mode, second for inch mode.

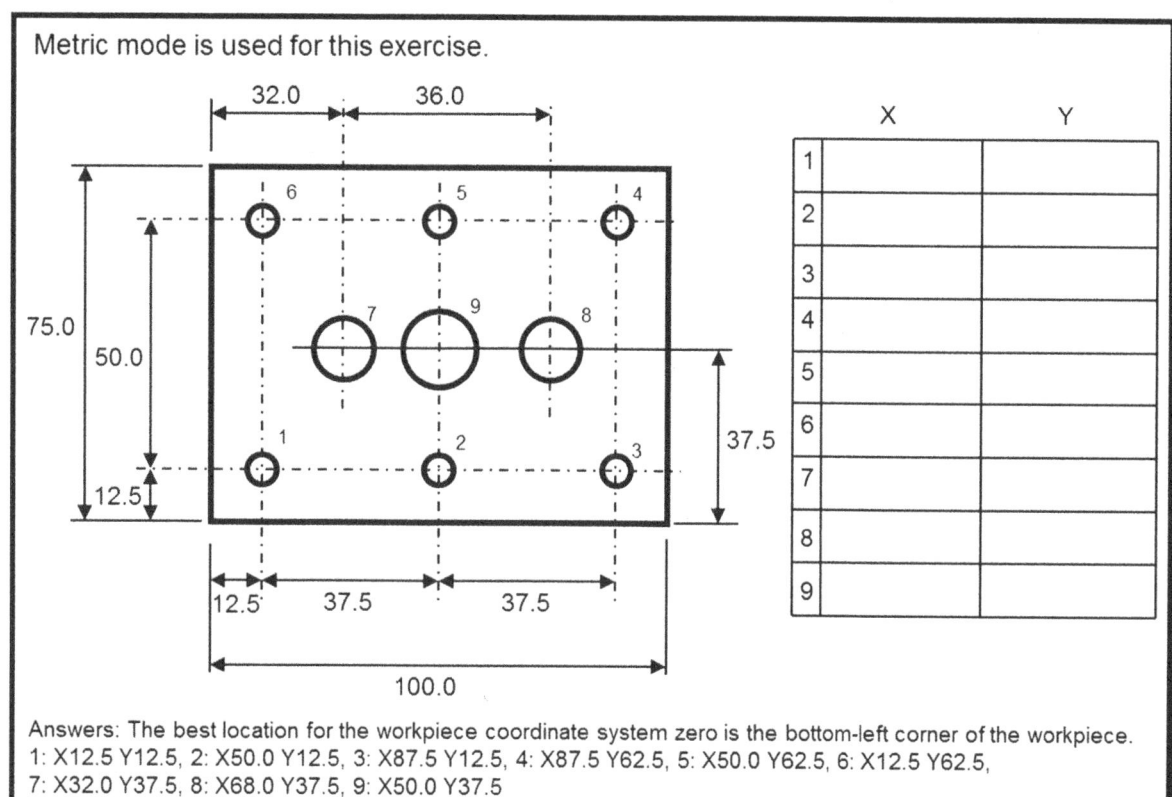

Metric mode is used for this exercise.

Answers: The best location for the workpiece coordinate system zero is the bottom-left corner of the workpiece.
1: X12.5 Y12.5, 2: X50.0 Y12.5, 3: X87.5 Y12.5, 4: X87.5 Y62.5, 5: X50.0 Y62.5, 6: X12.5 Y62.5, 7: X32.0 Y37.5, 8: X68.0 Y37.5, 9: X50.0 Y37.5

Exercise using the inch mode:

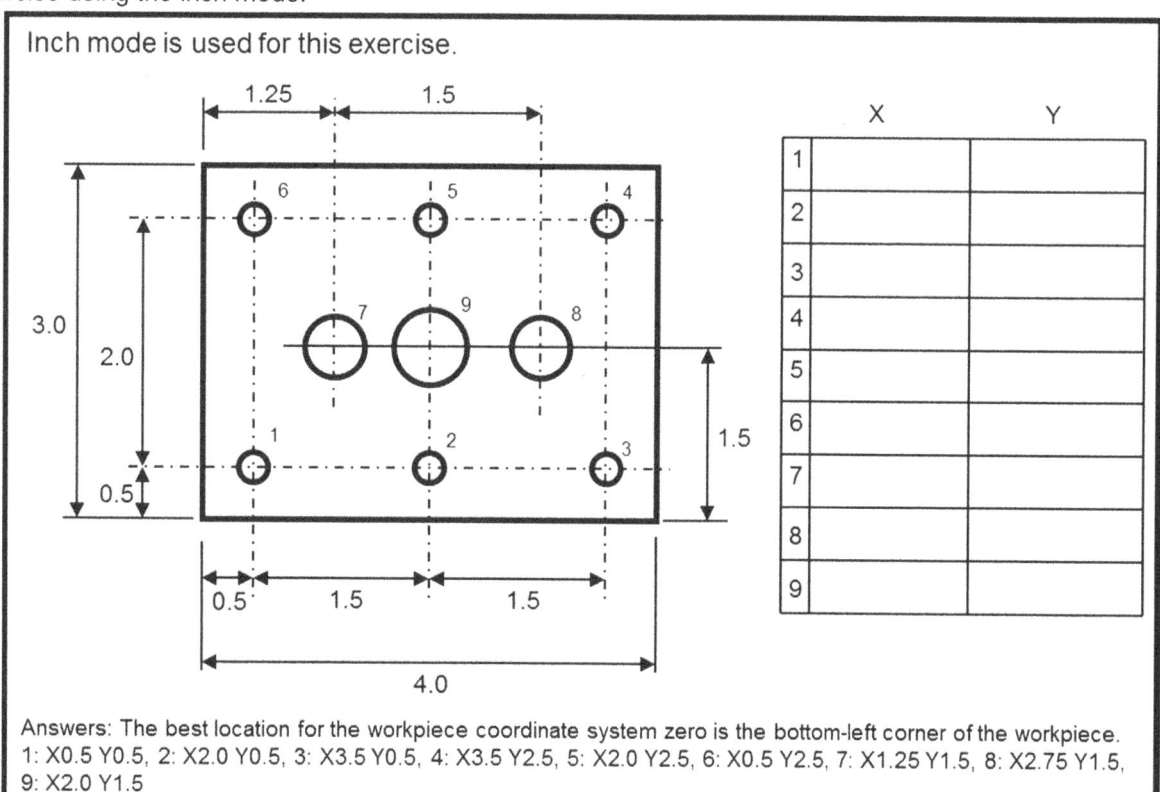

Inch mode is used for this exercise.

Answers: The best location for the workpiece coordinate system zero is the bottom-left corner of the workpiece.
1: X0.5 Y0.5, 2: X2.0 Y0.5, 3: X3.5 Y0.5, 4: X3.5 Y2.5, 5: X2.0 Y2.5, 6: X0.5 Y2.5, 7: X1.25 Y1.5, 8: X2.75 Y1.5, 9: X2.0 Y1.5

Lesson 1.5 – Determining Workpiece Coordinate System Offsets

The programmer chooses the workpiece coordinate system zero point location. But the CNC machining center must also be told where program zero is located so it can move cutting tools accordingly.

Objectives

After completing this lesson, students should be able to:

- ✓ Describe workpiece coordinate system offsets
- ✓ Describe when workpiece coordinate system offsets must be measured
- ✓ Describe the differences between the machine, relative and workpiece coordinate systems
- ✓ Describe an overview of a method to manually measure the X, Y and Z workpiece coordinate system offset values for both a rectangular-shaped and circular-shaped workpieces
- ✓ Describe the advantages of using a spindle probe to measure workpiece coordinate system offset values
- ✓ Describe how selecting the workpiece coordinate system zero point can affect the setup persons job

Introduction

You know from the previous lesson that the workpiece coordinate system zero or workpiece coordinate system zero is the origin for the coordinates in your program. In the absolute positioning mode (G90), all coordinates specified in your program are taken from the workpiece coordinate system zero. You also know that the workpiece coordinate system point is selected based upon how the engineering drawing is dimensioned. The workpiece coordinate system zero is typically placed at the drawing datum for each axis, the place from which the majority of the dimensions begin.

It is important to realize that just because you select the workpiece coordinate system zero to be at a given location doesn't mean that the CNC is automatically going to know where it is located. Explicit action must be taken by the operator to assign values to the workpiece coordinate system offsets so the CNC does know where you assigned the zero position. Otherwise the CNC will assume that all coordinates are dimensioned from the machine coordinate system zero. Entering values in the workpiece coordinate system offsets marries the part program to the work holding setup on the machine and compensates for the difference between the workpiece coordinate system zero and the machine coordinate system zero.

Much of what is presented in the next two lessons is more related to setup than it is to programming. However, a CNC programmer must be able to instruct and direct setup people, providing instructions for how to setup the machine to make a specific part. This means they must understand almost as much about setups as setup people–and this includes an understanding of how the values in the workpiece coordinate system offsets are measured and set in the CNC.

Key Concept 1: Know Your Machine From A Programmer's Viewpoint

More About Measurement System Modes (metric mode / inch mode)

FANUC CNCs can be adjusted to power up in your company's measurement system mode of choice. One way to select the measurement system is to use the SETTING (HANDY) display page. If the INPUT UNIT register is set to zero (0), the machine will power up in the metric mode. If this register is set 1, the machine will power up in the inch mode.

You will be learning in this lesson to take measurements involving the position display page. This page shows values based upon which measurement system mode your company uses. In metric mode, values are shown in millimeters. In inch mode, they are shown in inches.

Which Mode is Selected?

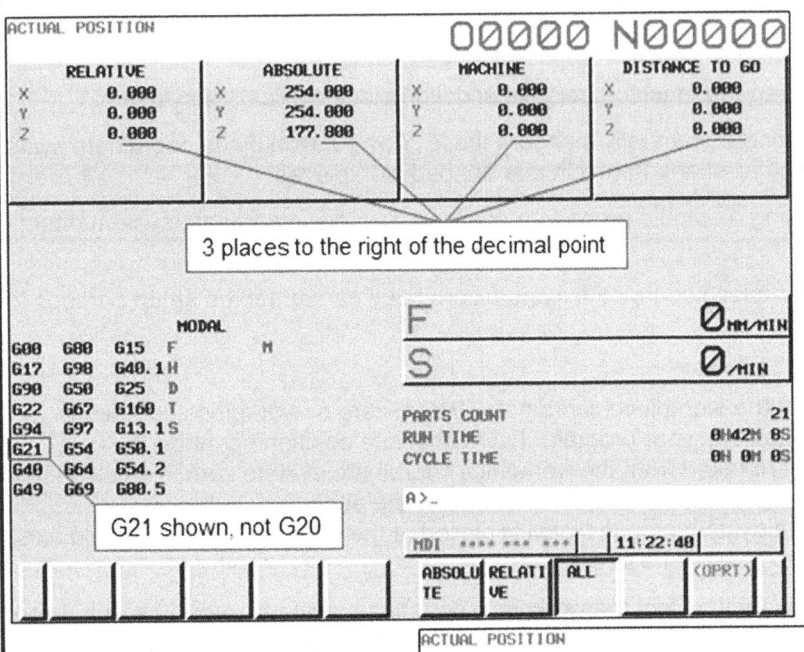

Metric mode

You can easily tell when metric mode is selected by looking at the position display page. With almost all CNC machining centers, you will see three places to the right of the decimal point for coordinate values when metric mode is selected. To be absolutely sure, confirm that there is a G21 (not G20) displayed in the list of currently instated G-codes.

Inch mode

In similar fashion, look at decimal point placement to determine if inch mode is selected. The vast majority of CNC machining centers will display four places to the right of the decimal point for coordinate values if the inch mode is selected. And you can be absolutely sure if you see a G20 (not G21) displayed in the list of currently instated G-codes.

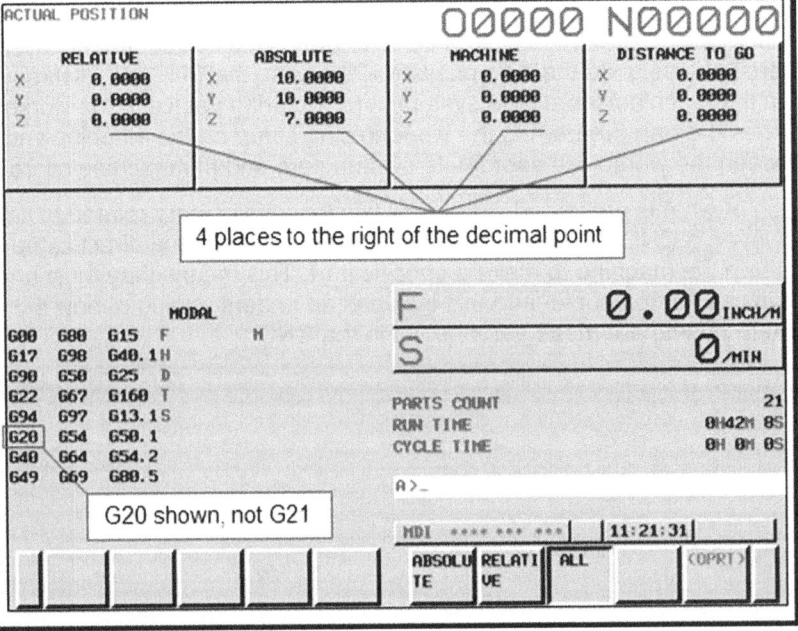

Determining the Workpiece Coordinate System Offset Values

Before you can set the workpiece coordinate system offset values, you must measure the distance between the workpiece coordinate system zero on the physical part as it is mounted in the work holding device and the machine coordinate system zero for each axis. For the X and Y axes, the reference position for the

Lesson 1.5: Determining Workpiece Coordinate System Offsets

machine coordinate system zero is the center of the spindle when it is sitting at the machine zero or home position. The workpiece coordinate system offset values in X and Y are the distances between the workpiece coordinate system zero on the part and the centerline of the spindle when they are resting at the machine zero position. The figure below illustrates the workpiece coordinate system zero offset values needed for the X and Y axes. The machine happens to be in inch mode, but the concept is exactly the same in metric mode.

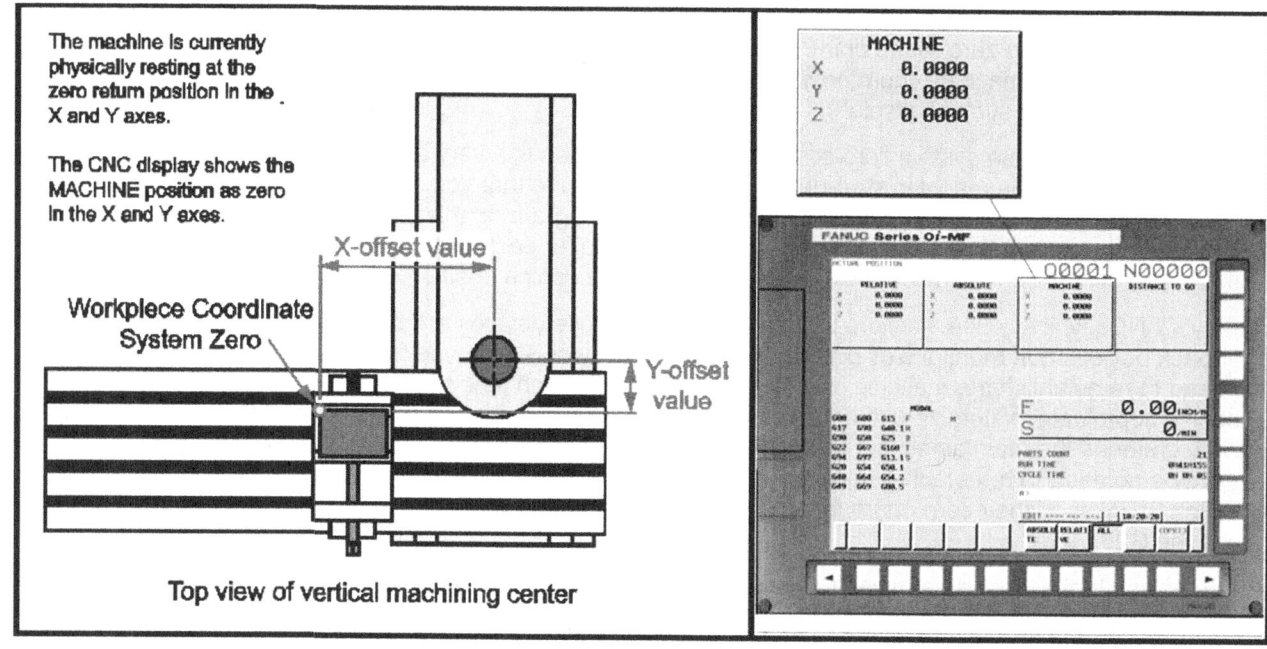

Figure 1.26: You must determine the program zero assignment values for the X and Y axes

There are two primary methods for establishing the Z-axis workpiece coordinate system offset value. The method used depends on how tool length compensation is applied. We don't introduce tool length compensation until a later lesson, so for now, we will only show the recommended method. Using the recommended method, the workpiece coordinate system offset value for the Z-axis is the distance between the workpiece coordinate system zero on the part (commonly the top surface of the workpiece) and the spindle nose (the bottom surface of the spindle itself) when it rests at the machine coordinate system zero, the zero return position. The figure below illustrates the workpiece coordinate system offset value needed for the Z-axis. The machine in the illustration happens to be in inch mode.

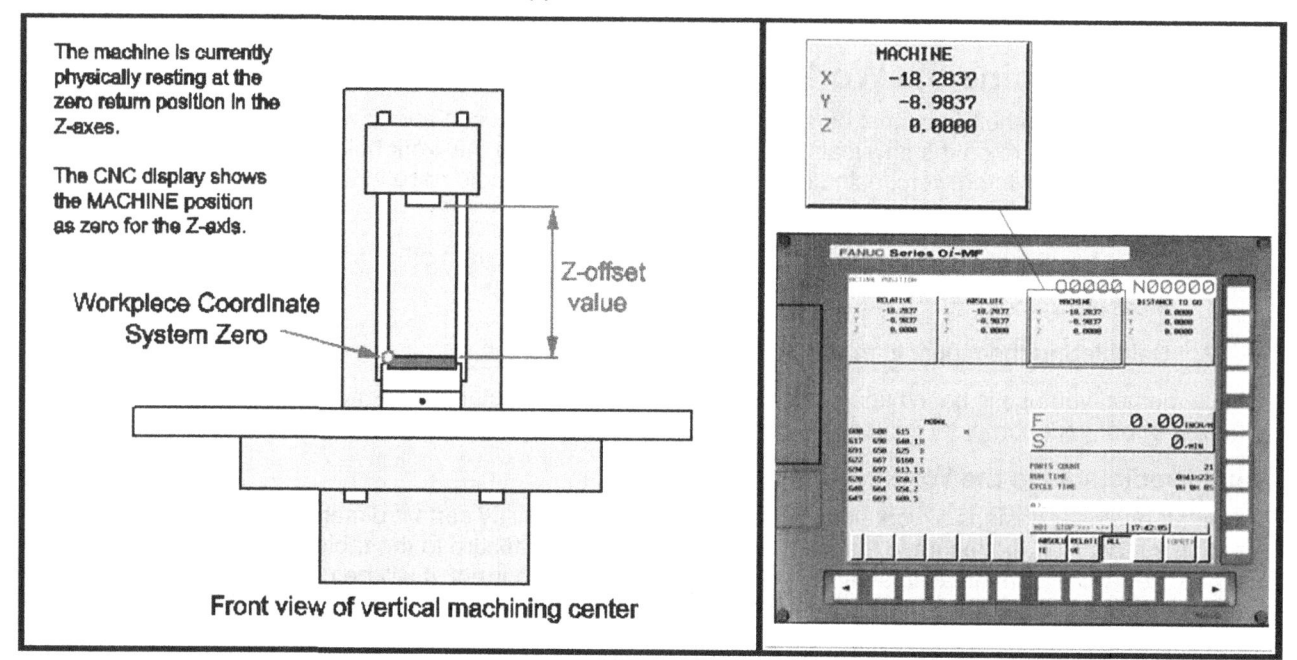

Figure 1.27: You must determine the program zero assignment value for the Z-axis

Key Concept 1: Know Your Machine From A Programmer's Viewpoint

What is Machine Zero Position?

In the two figures above, the measurements required to determine the X, Y and Z workpiece coordinate system zero offset values describe the machine axes resting at something called the machine zero position. Some people also call this position the home position or reference return zero. So there are actually two coordinate system zero points that are of extreme importance to you–the workpiece coordinate system zero position which is on the workpiece as specified in the part program, and the machine coordinate system zero which is the machine's zero return point. The difference or the distance between the workpiece coordinate system zero and the machine coordinate system zero, in each axis, is the value that must be entered in the workpiece coordinate system offsets.

The machine coordinate system zero or the zero return position is a very accurate location along each axis. On older machines, the machine start-up procedure may require that you send each axis to its zero return position before you are allowed to move the axes using a program. In this case, when the machine is sent to its zero return position, three axis origin lights come on (one for each axis) to indicate that the machine is truly resting at its zero return position. At this position, the machine position display shows zero in each axis.

Today's CNCs typically use absolute feedback devices. These devices are battery-backed, intelligent feedback devices that monitor axis positions even when power is removed from the CNC. Therefore, there is no need to re-establish the machine coordinate system zero each time the machine is powered on. The machine coordinate system zero is established when the machine is manufactured. As long as the feedback device batteries are maintained and the batteries are only replaced when the CNC is under power, there should be no need to re-establish the machine coordinate system zero. However, the CNC still supports sending the axes to their zero return position manually, using the reference cycle, and under program control using the G28 cycle.

The machine tool builder determines the location of the machine coordinate system zero or zero return position.

> **Vertical machining centers**–the zero return position is commonly placed at the extreme plus limit of each axis. When a vertical machining center is resting at zero return position, the table is all the way to the left in X, all the way forward in Y (toward the operator), and the headstock of the machine is all the way up in Z. The two figures above show this configuration.
>
> **Horizontal machining centers**–the zero return position for Y is usually when the headstock of the machine is all the way up. For Z it is when the table is as far away from the spindle nose as possible. The Y and Z axes are at their extreme plus limits. But with the X-axis, machine tool builders vary. Some place it at the center of the X-axis travel (center of the table aligned with center of the spindle) while others place it at the extreme plus end of the X-axis (table all the way to the left as viewed from the spindle).

How to Determine the Workpiece Coordinate System Offset Values

By one means or another, you must be able to determine the distance in each axis between the workpiece coordinate system zero on the physical workpiece when mounted in the work holding device and the machine coordinate system zero, spindle center in X and Y, the spindle nose in Z, when they are at the zero return position.

There are two basic ways to determine the workpiece coordinate system offset values:

1. **Measuring** them during setup
2. **Calculating** them prior to making the setup

Which method you use is based upon whether or not you use predictable work holding tools (fixtures) and whether or not your company makes qualified work holding setups.

How Predictable is the Work Holding?

When an engineer designs a work holding tool or fixture, they specify certain dimensions from workpiece location surfaces on the fixture to the components that locate the fixture to the table (commonly keys or pins). If the fixture is made accurately and these dimensions are correct, it will be possible to calculate the workpiece coordinate system offset values before the setup is ever made. This also requires you to know some important distances on the machine itself–like the distances from the machine coordinate system zero in each axis to key-slots or location holes on the machine table. In practice, not many programmers know the related dimensions or make the effort to find or measure them–so very few programmers calculate program zero assignment values.

What is a Qualified Work Holding Setup?

By qualified work holding setup, we mean one that can be replaced on the machine table in exactly the same location–over and over again. Product producing companies can often justify the higher costs to make qualified work holding setups. The fixture discussed above, for example, is qualified. It will be located in machine table tee-slots or pin holes. If a work holding setup is truly qualified, the workpiece coordinate system offset values will be exactly the same every time the setup is made. While it may not be possible to calculate workpiece coordinate system offset values the first time the job is setup, it should not be necessary to re-measure them after the setup is made for the first time. Many companies run the same jobs over and over (lots of repeat business), and for them, qualifying work holding setups can save a lot of setup time.

Machine, Workpiece, and Relative Coordinate Systems

Shortly we will introduce using the relative coordinate system to measure the workpiece coordinate system offset values. The CNC has three coordinate systems to keep track of the machines movement. When the machine moves a certain distance, say 1-inch, the position display for each of these coordinate systems will change by the same amount, 1-inch. Right after the CNC is turned on and if there are no values in the active workpiece coordinate system offset, they would be completely in synchronization. In this case, if the axes are at the zero return position, the relative, absolute, and machine positions all read zero, as you can see below.

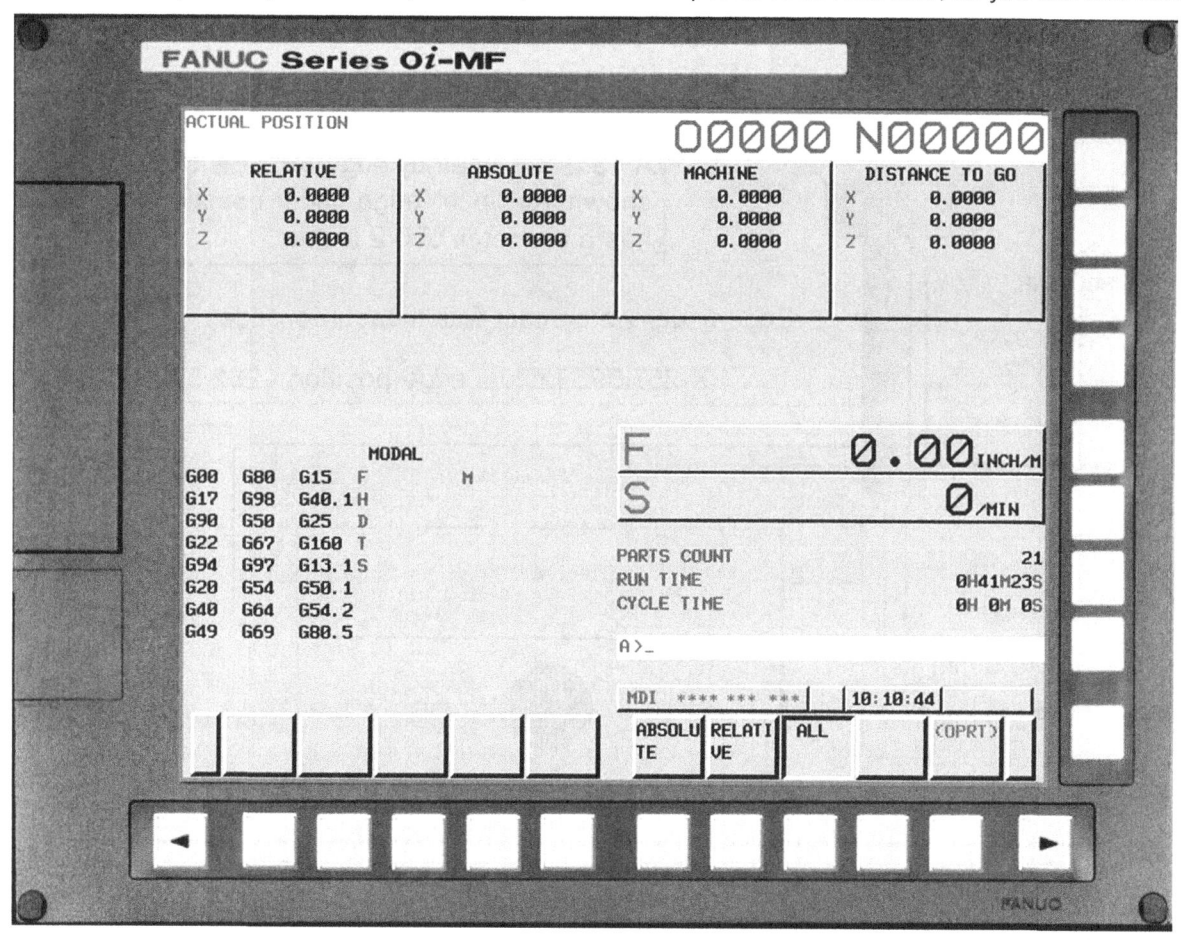

Figure 1.28: Relative, absolute and machine positions all at zero

The only difference between the machine coordinate system, the workpiece coordinate system, and the relative coordinate system is their zero positions. The machine coordinate system zero is always at the machine's zero return position. When we set the workpiece coordinate system offset values appropriately, the workpiece coordinate system zero will be located at the position you designate on the workpiece. The relative coordinate system zero is available for you to set anywhere and is primarily used to take measurements on the machine.

Key Concept 1: Know Your Machine From A Programmer's Viewpoint

Each current position of each of these coordinate systems can be displayed on the CNC screen.

Coordinate System	Zero Location	Zero Setting	Display Name
Machine Coordinate System	Zero return position for each axis.	By the machine tool builder and return to zero cycle.	MACHINE
Workpiece Coordinate System	The program zero position YOU select on the part.	By setting the workpiece coordinate system offset values.	ABSOLUTE
Relative Coordinate System	Anywhere YOU like	By pressing the [ORIGIN] or [PRESET] soft keys on the RELATIVE CNC screen.	RELATIVE

Using the Relative Coordinate System and an Edge-Finder for Measurement

In the procedures that follow, we will use the relative coordinate system position display or registers to measure the workpiece coordinate system offset values. At first you might find it confusing that we are using the relative coordinate system to measure the difference between the workpiece coordinate system and the machine coordinate system zero position. But measurement is the primary purpose for having the relative coordinate system in the CNC. It is the flexibility we have in zeroing or presetting the position of the relative coordinate system that makes it so useful for taking measurements.

The basic idea is that we use a tool called an edge-finder to move the CNC axis being measured a known distance (the radius of the edge-finder) away from a datum surface.

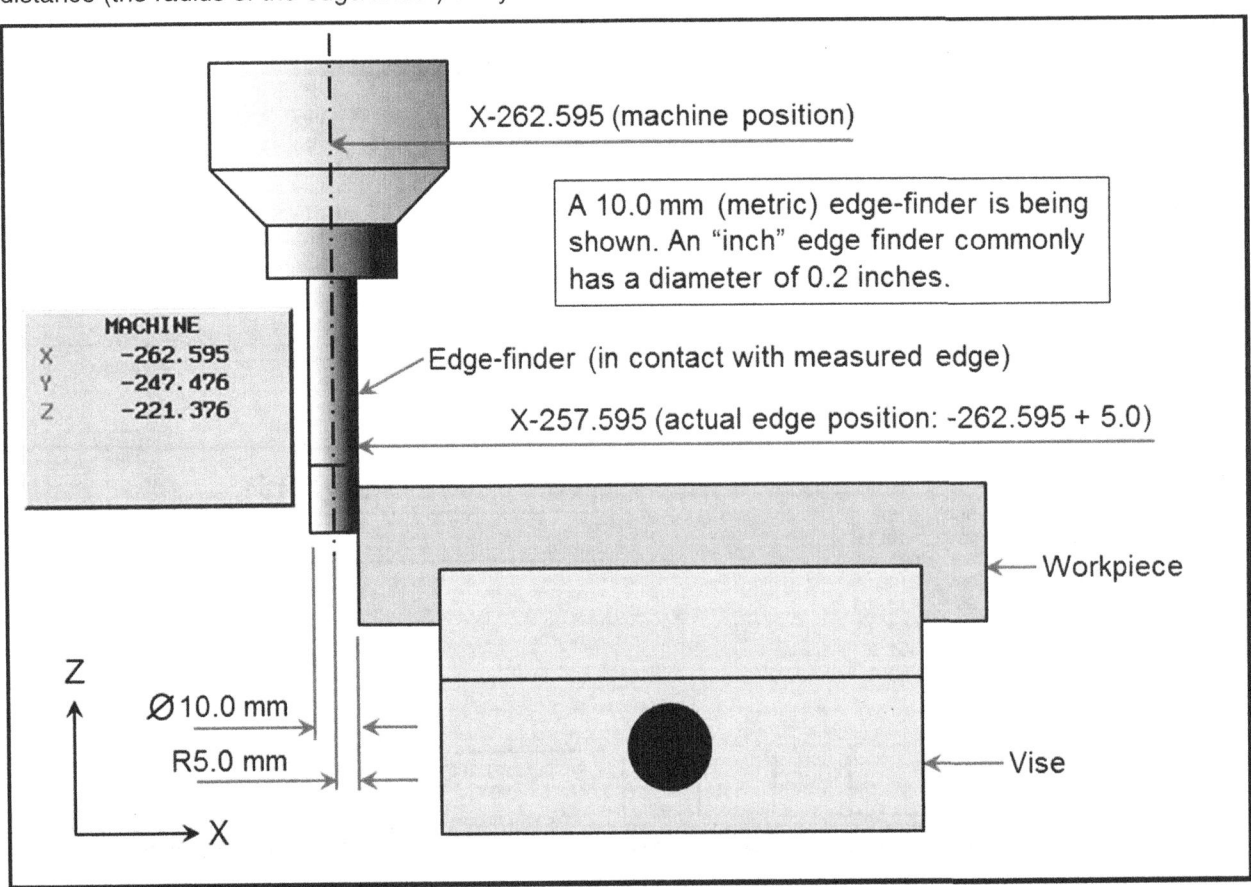

Figure 1.29: Detecting the position of the left-hand edge of a part with an edge-finder (shown using metric mode)

In the figure above, the edge-finder has been positioned so it is touching the left-hand side of the part. The machine position for the X-axis is currently -262.595. Since we are on the left-hand edge of the part, the workpiece edge is actually half the diameter of the edge-finder in the positive direction. Since the diameter of the edge-finder is 10.0 millimeters, the edge is plus 5.0 millimeters more in the X-axis than -262.595, or at the position -257.595. Another way to think of it is that the current axis position is -5.0 millimeters away from the datum edge being measured.

The goal is to set the relative position for the X-axis to zero when the center line of the spindle is directly over the edge being detected, but we are currently -5.0 millimeters away from that position.

Lesson 1.5: Determining Workpiece Coordinate System Offsets

There are two methods to allow for the -5.0 millimeters, depending on your preference and the features available on the CNC.

1. If the CNC has the relative position [PRESET] soft key, you can simply preset the relative coordinate system position.

   a. Select the CNC relative screen.

   b. Press the [OPRT)) operations key until the [PRESET] soft key is displayed.

   c. Enter the axis letter address followed by the current position in the desired workpiece coordinate system. Since we want the edge to be zero in the workpiece coordinate system, we are currently at X-5.0. So enter X-5.0 and press the [PRESET] soft key.

If you find this confusing, just consider that if the current position was X-5.0 and you programmed a move to X0.0, the axis would move 5.0-millimeters in the positive direction and place the center of the spindle directly over the edge, which is what we are trying to do.

2. In this method, we will actually move the X-axis above the edge and set the relative position to zero.

   a. Select the CNC relative screen.

   b. Press the [OPRT)) operations key until the [ORIGIN] soft key is displayed.

   c. Enter the letter address X into the key input buffer and then press the [ORIGIN] soft key. The relative position for the X-axis is set to zero.

   d. Retract the Z-axis so the edge-finder is above the workpiece.

   e. Use incremental jog or the electronic handle wheel with the smallest increment setting to move the axis precisely the radius of the edge-finder so the center of the spindle is precisely above the edge being measured. In the example above, with a 10.0 millimeter diameter edge-finder, the relative position display will be X5.0 when the center of the spindle is directly above the edge.

   f. Enter the letter address X into the key input buffer and then press the [ORIGIN] soft key again. The relative position for the X-axis is set to zero and the center of the spindle is directly above the edge.

After using either of these methods, return the X axis to the zero return (reference) position. The machine position display will be zero and the relative display for the axis measured will have the value that must be entered in the workpiece coordinate system offset.

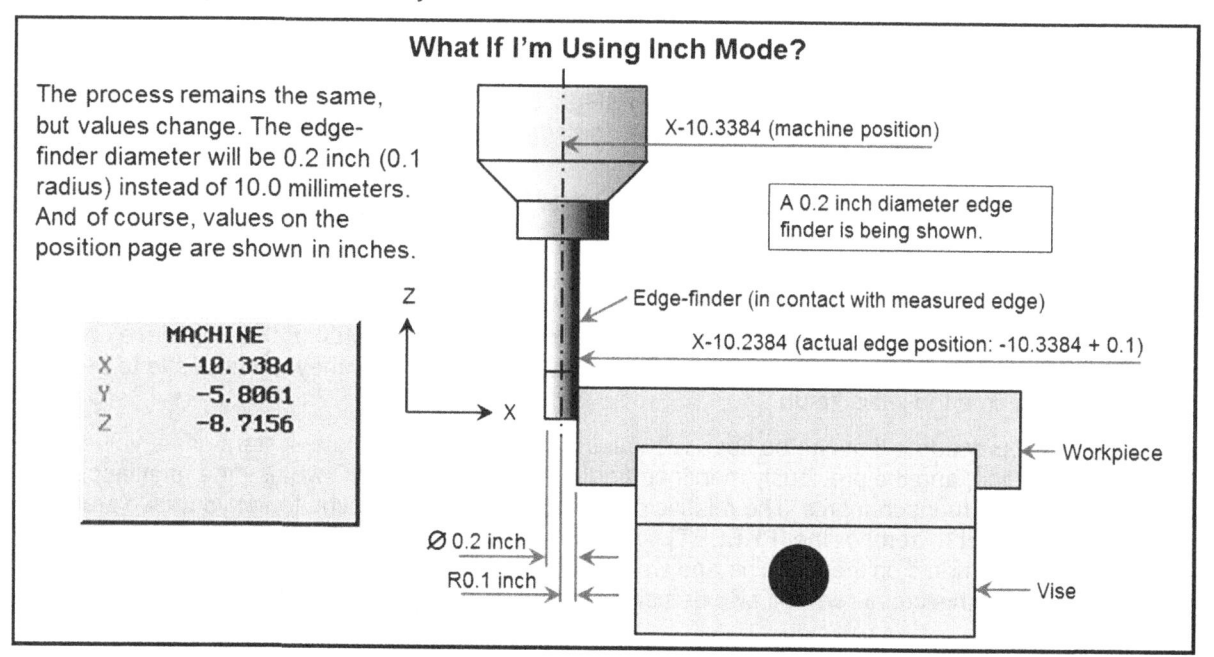

What If I'm Using Inch Mode?

The process remains the same, but values change. The edge-finder diameter will be 0.2 inch (0.1 radius) instead of 10.0 millimeters. And of course, values on the position page are shown in inches.

MACHINE
X   -10.3384
Y   -5.8061
Z   -8.7156

X-10.3384 (machine position)

A 0.2 inch diameter edge finder is being shown.

Edge-finder (in contact with measured edge)

X-10.2384 (actual edge position: -10.3384 + 0.1)

Ø 0.2 inch
R0.1 inch

Workpiece

Vise

It is important to realize that in method 1, you do not always add the radius of the edge-finder. You may have to subtract it. This depends which edge you are measuring. Consider the situation when the workpiece coordinate zero is on the right-hand side of the workpiece.

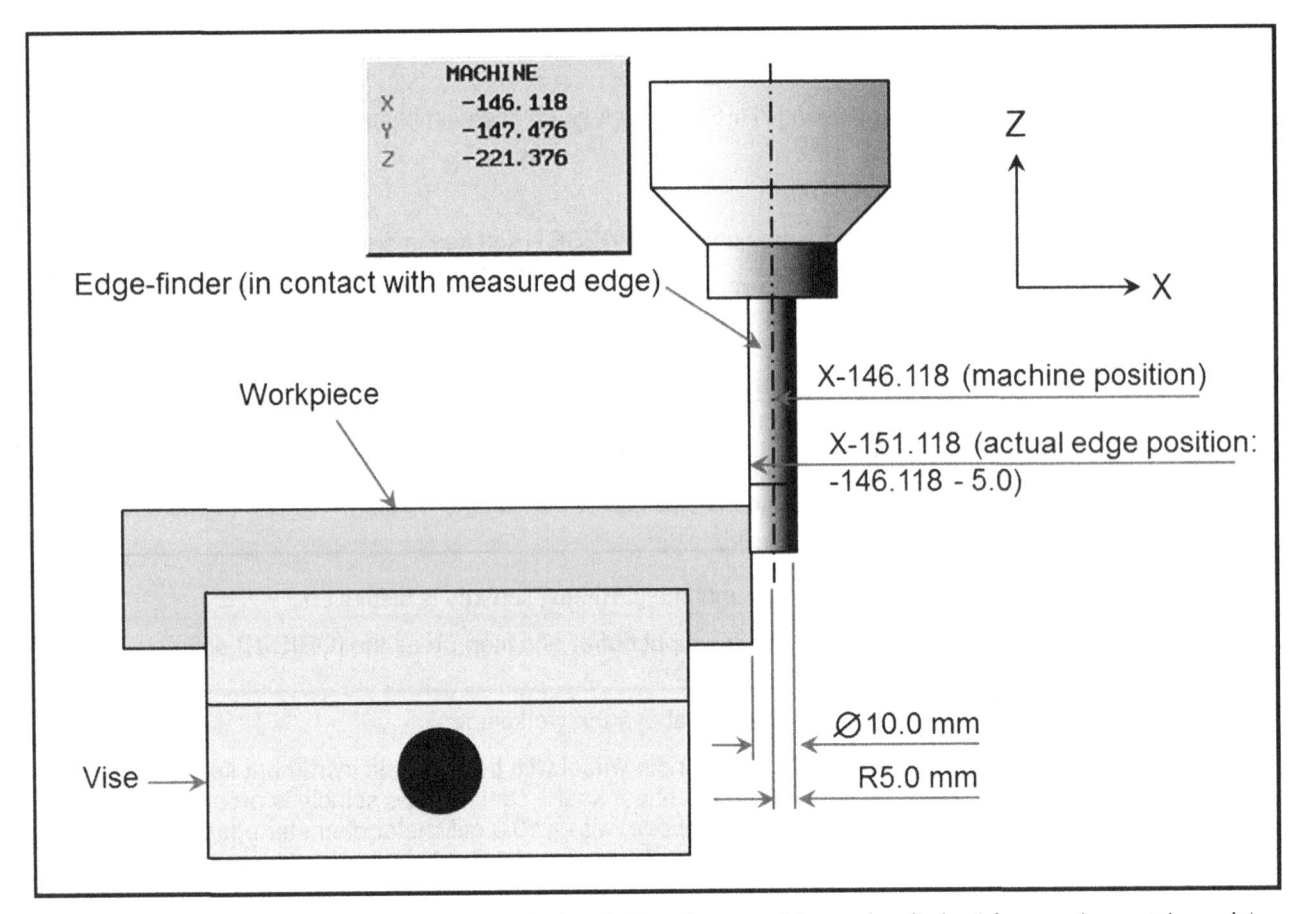

Key Concept 1: Know Your Machine From A Programmer's Viewpoint

Figure 1.30: Detecting the position of the right-hand edge of a part with an edge-finder (shown using metric mode)

Now when we contact the edge, the center of the spindle is a half a diameter of the edge-finder in the positive direction. In method 1 above, we must set the relative position to X5.0 in step c. The axis would then have to move in the negative direction by 0.5.0 millimeters to place the center line of the spindle on the edge. In method 2, the axis is moved precisely 5.0 millimeters in the negative direction in step e until the relative position display is at X-5.0.

Method 1 is quicker, but you must be sure to enter the correct polarity when presetting the relative position to half the radius of the edge-finder. Method 2 has more steps, but it automatically sets the adjustment for the edge-finder as you visually move the edge-finder over the edge being measured.

Manually Measuring the Workpiece Coordinate System Offset Values

Work holding devices are seldom predictable enough to allow the calculating of program zero assignment values. This means another method of determining program zero assignment values must be used–actually measuring them at the machine during setup. If your company uses qualified work holding, this measurement should only be necessary the very first time each new job is setup. If the workpiece coordinate system zero offset values are documented (and if the setup is truly qualified), they will not have to be re-measured the next time the job is setup.

Below is a general procedure that can be used to measure workpiece coordinate system offset values using the CNC, the machine, and the previously mentioned edge-finder. On the CNC we use the relative position display screen to take measurements. The relative coordinate system allows you to set your own zero point using the [ORIGIN] (set to zero) or the [PRESET] soft keys. The relative coordinate system is commonly used to take measurements on the machine and we can use it to measure the workpiece coordinate system offset values. Later in this course we will also be using it to measure tool length compensation offsets.

The procedures we describe below assume that you know how to run a machine–at least to be able to move the axes with continuous jog or the hand-wheel, start and stop the spindle, and select the relative position screen on the CNC and set the relative coordinate system zero using soft keys. While we haven't presented this information to you yet, hopefully you have sufficient general machine background to understand this overview of how to measure workpiece coordinate system offset values. The procedures presented in this lesson are intentionally brief and general. More detailed and specific procedures and instructions are provided in the setup and operation part of this course–starting in Key Concept 7.

Measuring the X and Y Axes for a Rectangular-Shaped Workpiece

This procedure involves using an edge-finder (also called a wiggler) and techniques similar to those used on manual milling machines when picking up an edge. As with many setup related tasks, this requires some basic machining practice experience. While you may be able to easily understand this section, if you do have questions, ask an experienced person in your company or school to demonstrate how an edge-finder is used.

The procedure uses an edge-finder that has a 10.0-mm diameter (5.0-mm radius). When the edge-finder is flush with the edge of the workpiece, the centerline of the spindle will be precisely 5.0 millimeters away from the surface being touched by the edge-finder.

Step 1: Load workpiece and edge-finder

Begin by setting up the work holding device and load a workpiece. Place the edge-finder in the spindle and start the spindle at about 500 rpm or the spindle speed recommended by the edge-finder manufacturer. If necessary, push on the edge-finder stylus to make it wobble or run out (it is now ready to pick up a surface in X or Y).

Step 2: Touch off the workpiece edge in X

Using continues jog, incremental jog or the hand-wheel, move the table so that the edge-finder comes in gentle contact with the edge of the workpiece in the X-axis–left side of our example workpiece as shown in the drawing below. If the X-axis relative position has not already been set or preset since power-up it will be the same as the machine position for the X-axis and it will be the distance from the zero return position.

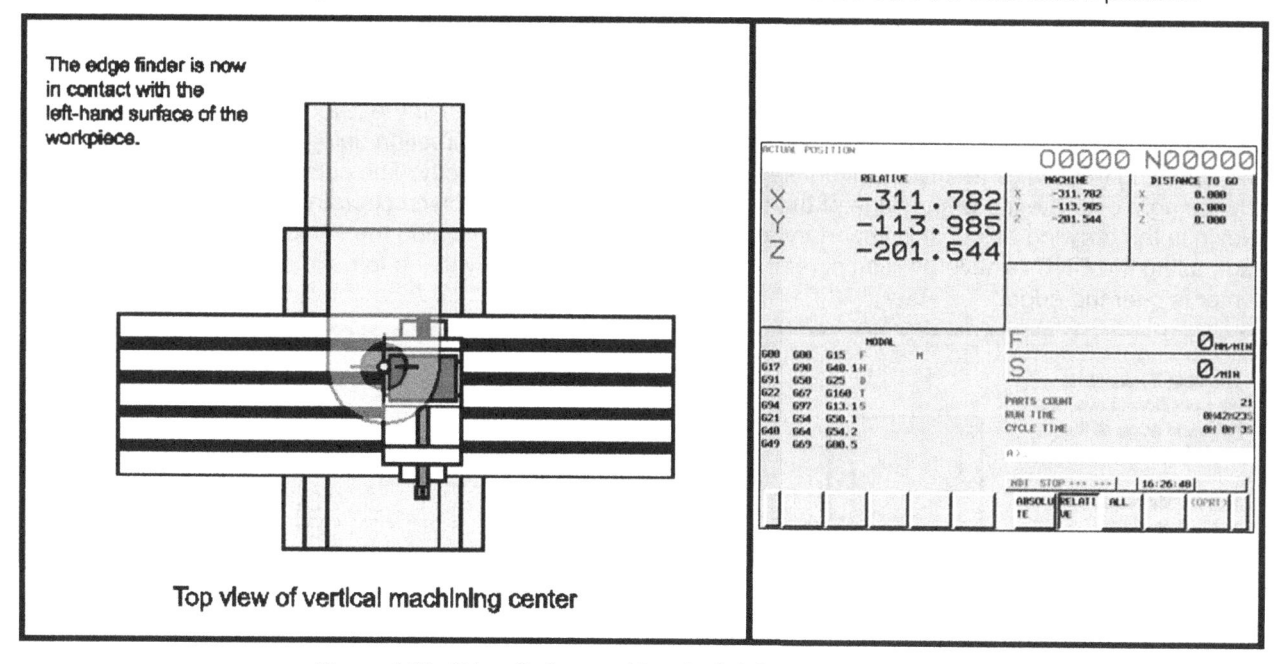

Figure 1.31: Edge-finder touching the left-hand edge of the part

Key Concept 1: Know Your Machine From A Programmer's Viewpoint

Step 3: Set the relative position for the X-axis to zero

With the X-axis at this position, select the relative position page. Press the X-axis letter address and then the [ORIGIN] soft key to set the X-axis relative position. The X-axis relative position will now be zero.

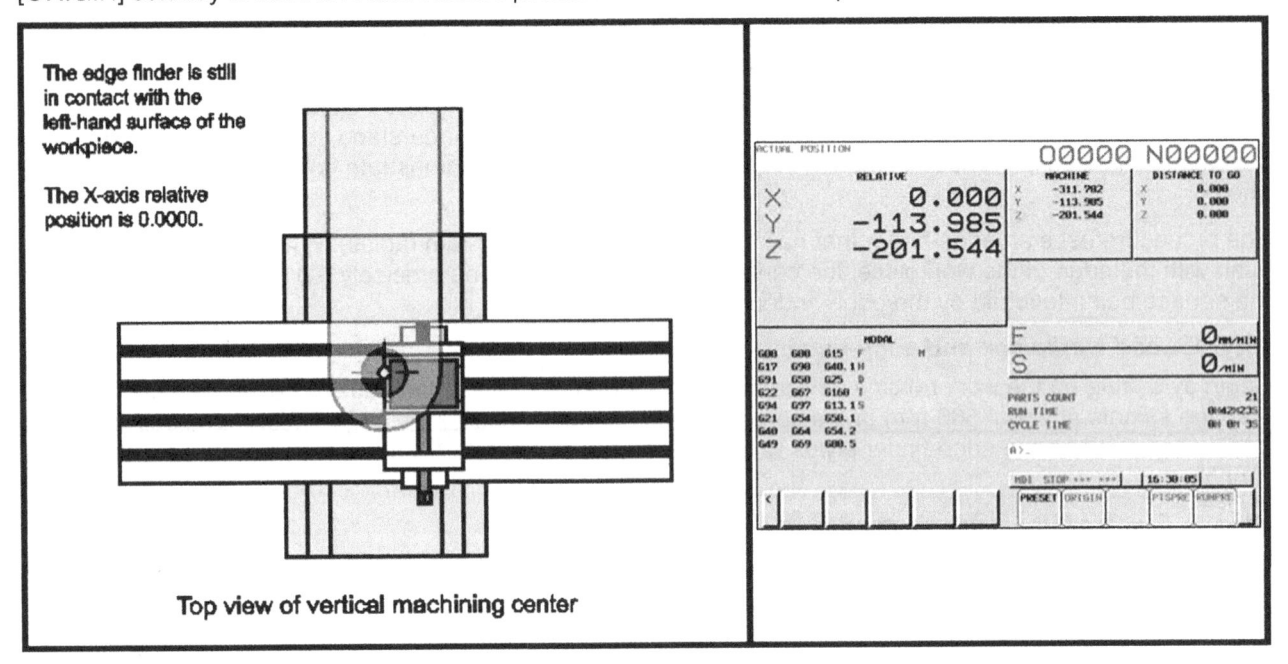

Figure 1.32: Edge-finder touching the left-hand edge of the part and the X-axis relative position zeroed

Step 4: Back off in Z, move over 5.0 millimeters in X

Using incremental jog or the hand-wheel, move the edge-finder up (plus) in the Z-axis to clear the top of the workpiece. Then move the X-axis precisely 5.0 millimeters in the plus direction using a suitable increment and watching the X-axis relative position display until it shows X5.0 exactly. The centerline of the spindle will now be right over the top of the edge of the part in X, which is the workpiece coordinate system zero in X as shown in the drawing below. It is important to realize that you are positioning the X-axis exactly over the edge using the CNC relative position screen for the X-axis, not visually trying to estimate when the spindle center is over the edge.

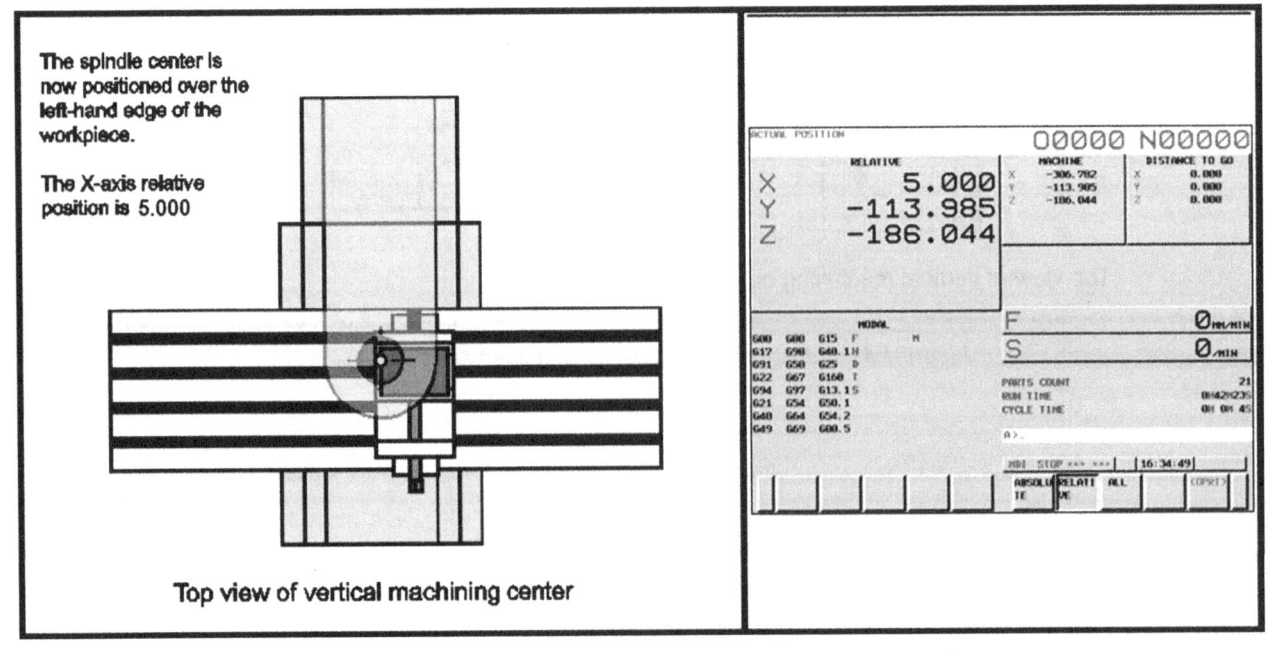

Figure 1.33: Spindle center positioned over the left-hand edge of the part

Lesson 1.5: Determining Workpiece Coordinate System Offsets

Step 5: Zero the X-axis relative position again

Now press the X-axis letter address and the [ORIGIN] soft key again to set the relative axis position to zero.

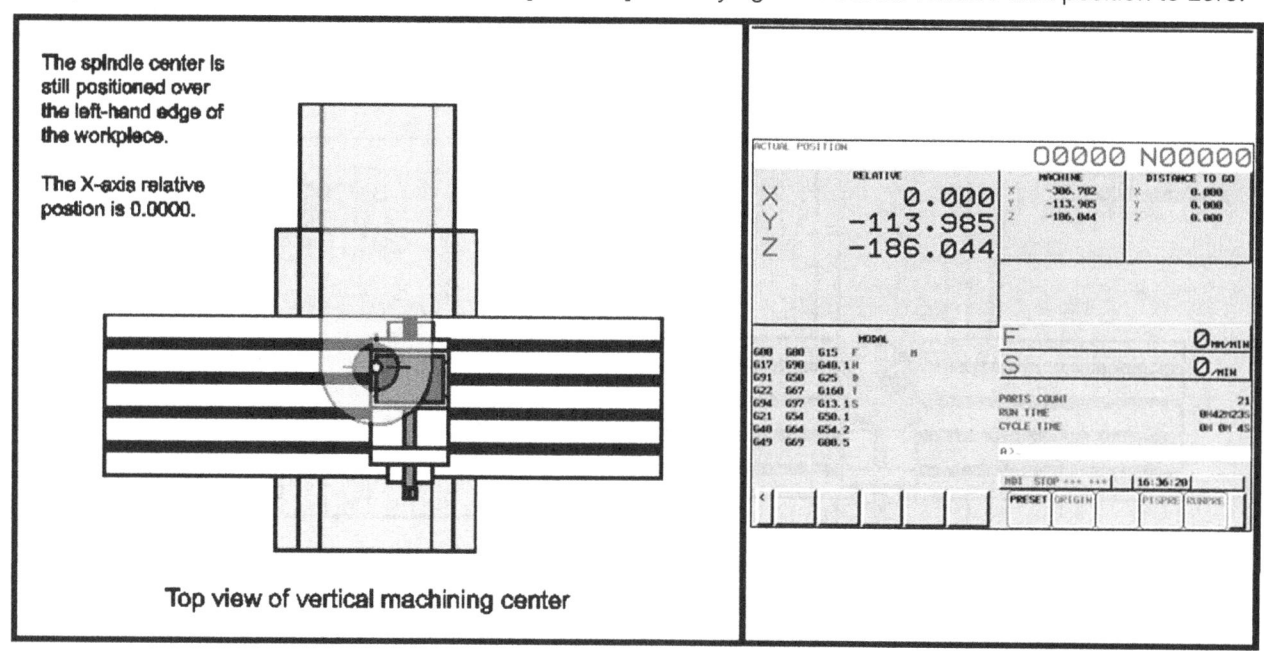

Figure 1.34: Spindle center positioned over the left-hand edge of the part and relative position zeroed

Step 6: Touch off the workpiece edge in Y

Now repeat steps 2 through 5 for the Y-axis. First, using continues jog, incremental jog or the hand-wheel, move the table so that the edge-finder comes in gentle contact with the edge of the workpiece in the Y-axis– top side of our example workpiece as shown in the drawing below.

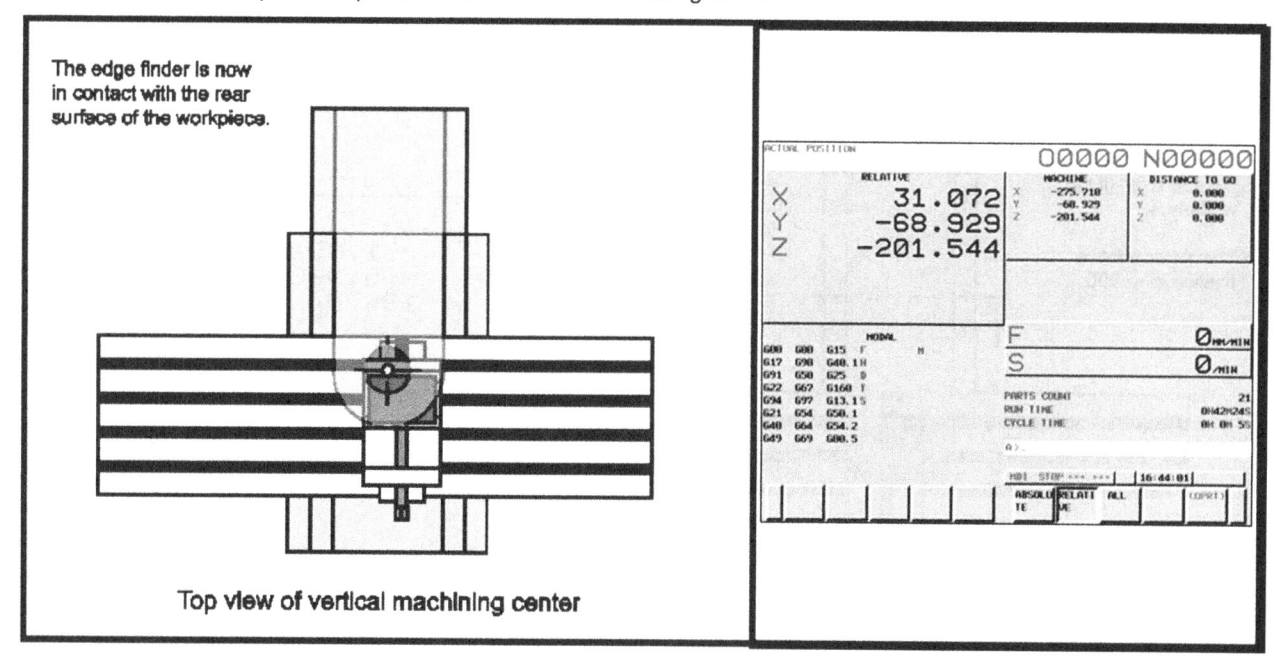

Figure 1.35: Edge-finder touching the rear edge of the part

Key Concept 1: Know Your Machine From A Programmer's Viewpoint

Step 7: Set the relative position for the Y-axis to zero

Press the Y-axis letter address and then the [ORIGIN] soft key to set the Y-axis relative position.

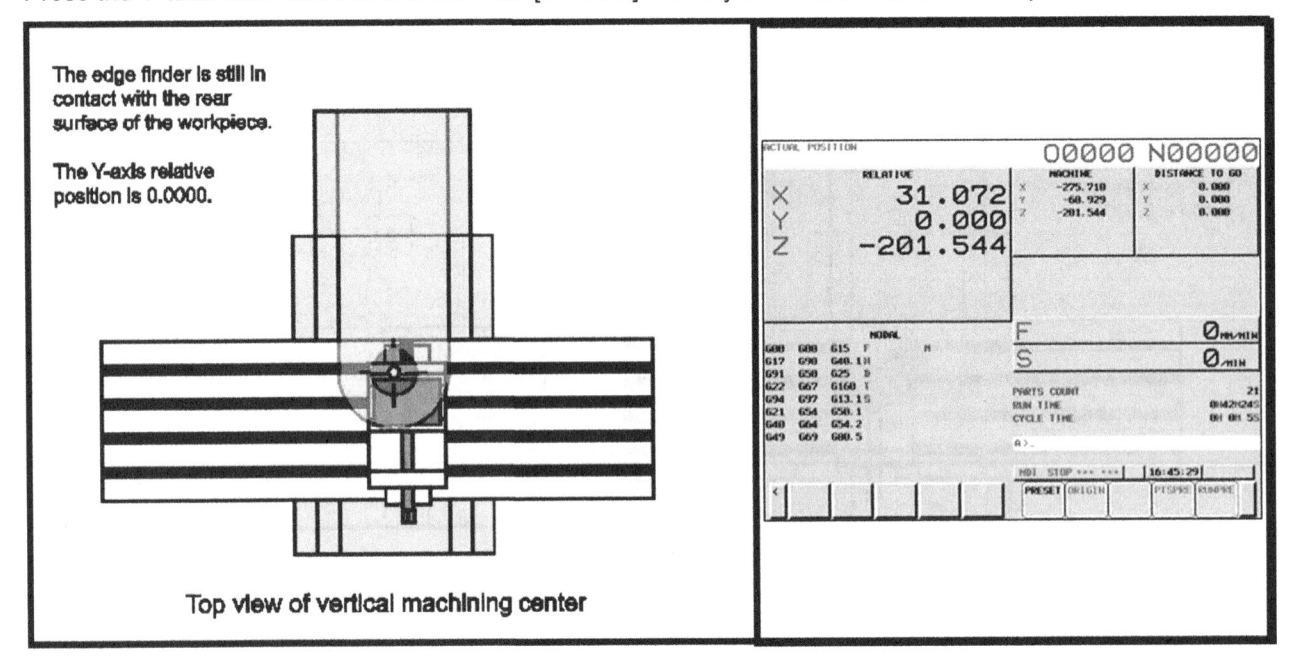

Figure 1.36: Edge-finder touching the rear edge of the part and Y-axis relative position zeroed

Step 8: Back off in Z, move over -5.0 millimeters in Y

Using incremental jog or the hand-wheel, move the edge-finder up (plus) in the Z-axis to clear the top of the workpiece. Then move the Y-axis precisely 5.0-millimeters in the minus direction using a suitable increment and watching the Y-axis relative position display. The centerline of the spindle will now be right over the top of the edge of the part in Y, which is the workpiece coordinate system zero in Y as shown in the drawing below.

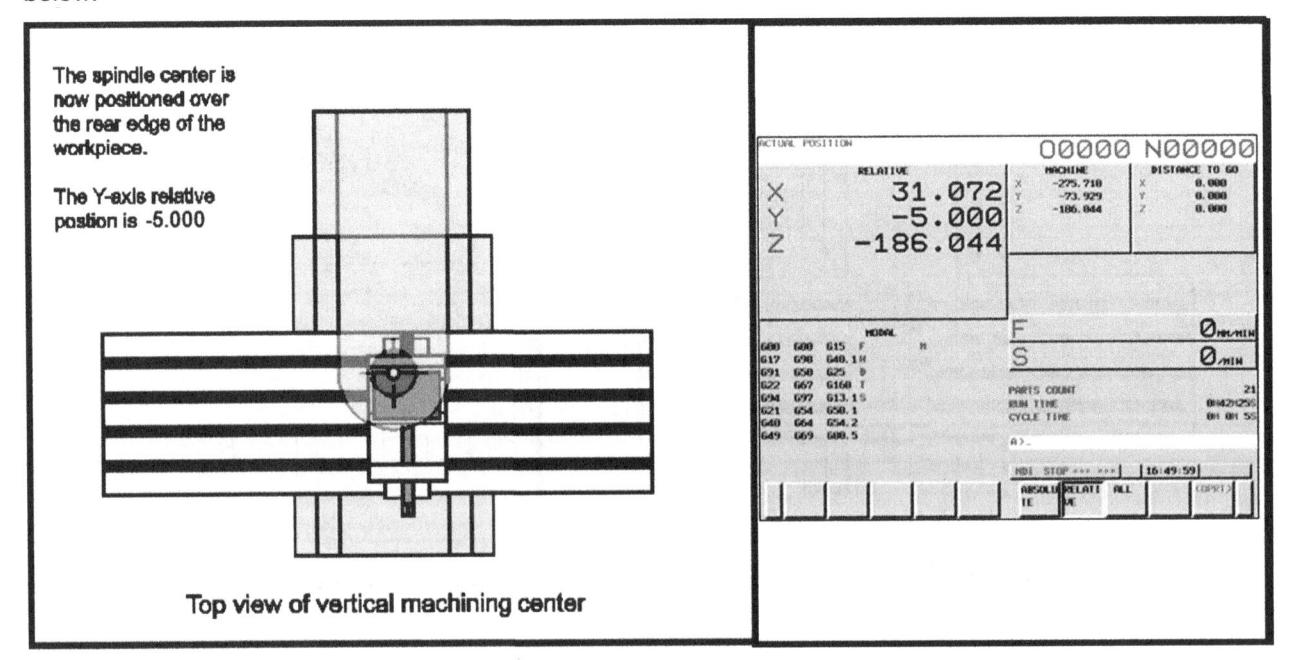

Figure 1.37: Spindle center positioned over the rear edge of the part

Step 9: Reset the Y-axis relative position again
Now press the Y-axis letter address and the [ORIGIN] soft key again to set the relative axis position to zero.

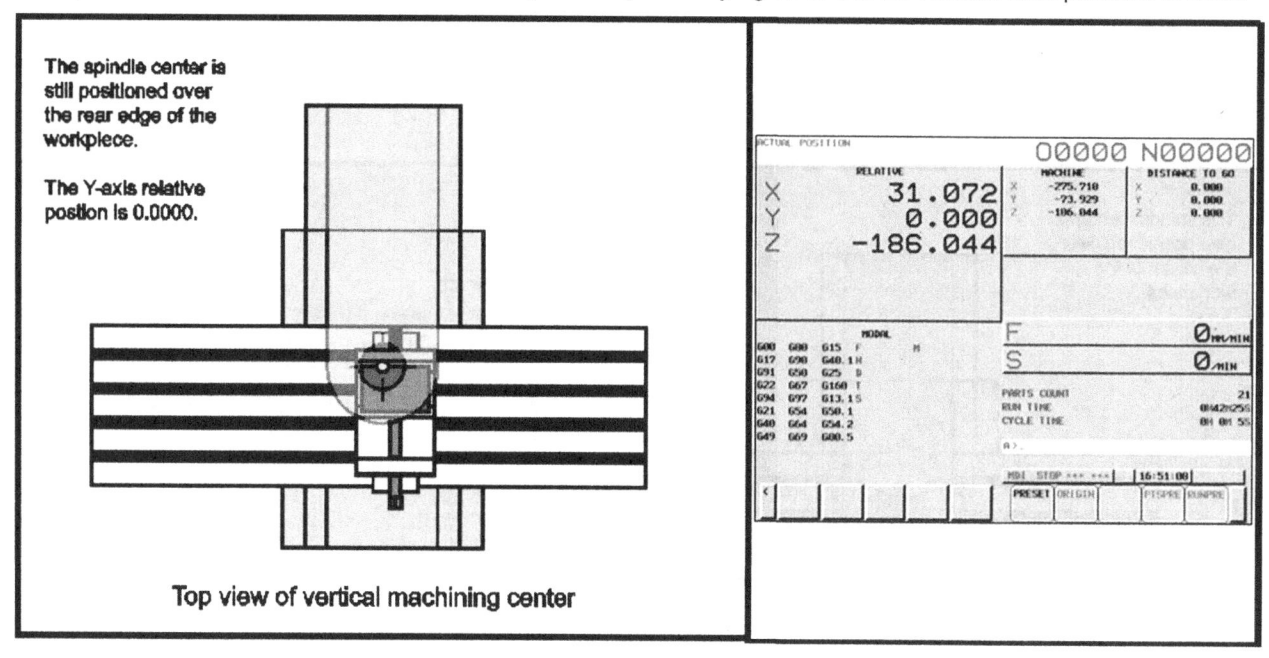

Figure 1.38: Spindle center positioned over the left-hand edge of the part and relative position zeroed

Step 10: Send the X and Y axes to the zero return (reference) position
Send the X and Y axes to the zero return (reference) position. When the X and Y axes are resting at the zero return position, the X and Y relative positions show the workpiece coordinate system offset values (distance between the workpiece coordinate system zero and spindle center in X and Y while the machine is resting at the zero return position).

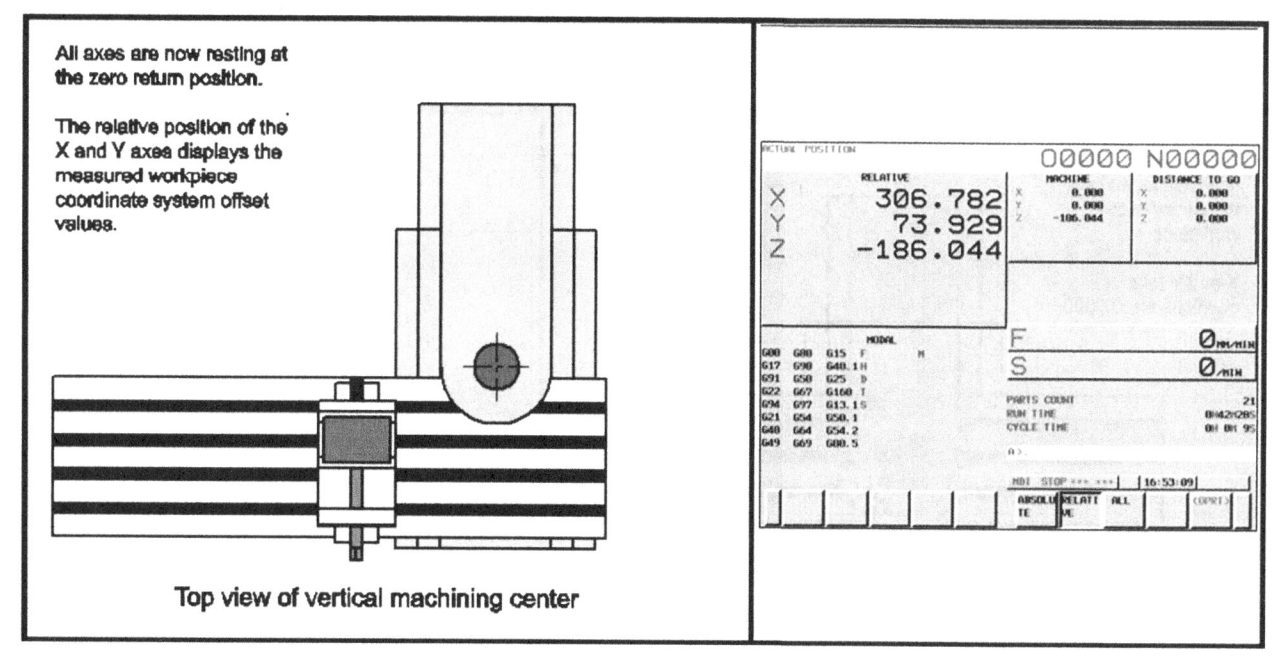

Figure 1.39: X and Y axes at the zero return position and relative position displays show measured values

Measuring the X and Y Axes for a Circular-Shaped Workpiece
This procedure involves using a dial indicator mounted in the spindle. You'll be using it to pick up the center of the round workpiece (commonly called indicating the workpiece). Again, this involves the need for basic machining practice experience.

Key Concept 1: Know Your Machine From A Programmer's Viewpoint

Step 1: Load the workpiece and dial indicator
Begin by making the work holding setup and load a workpiece. Place the dial indicator in the spindle.

Step 2: Indicate the center of the workpiece.
Using continuous jog or the hand-wheel, position the X and Y axes so that there is the same dial deflection at any position on the circular part diameter. When finished, the center of the spindle will be directly over the center of the workpiece as shown in the drawing below.

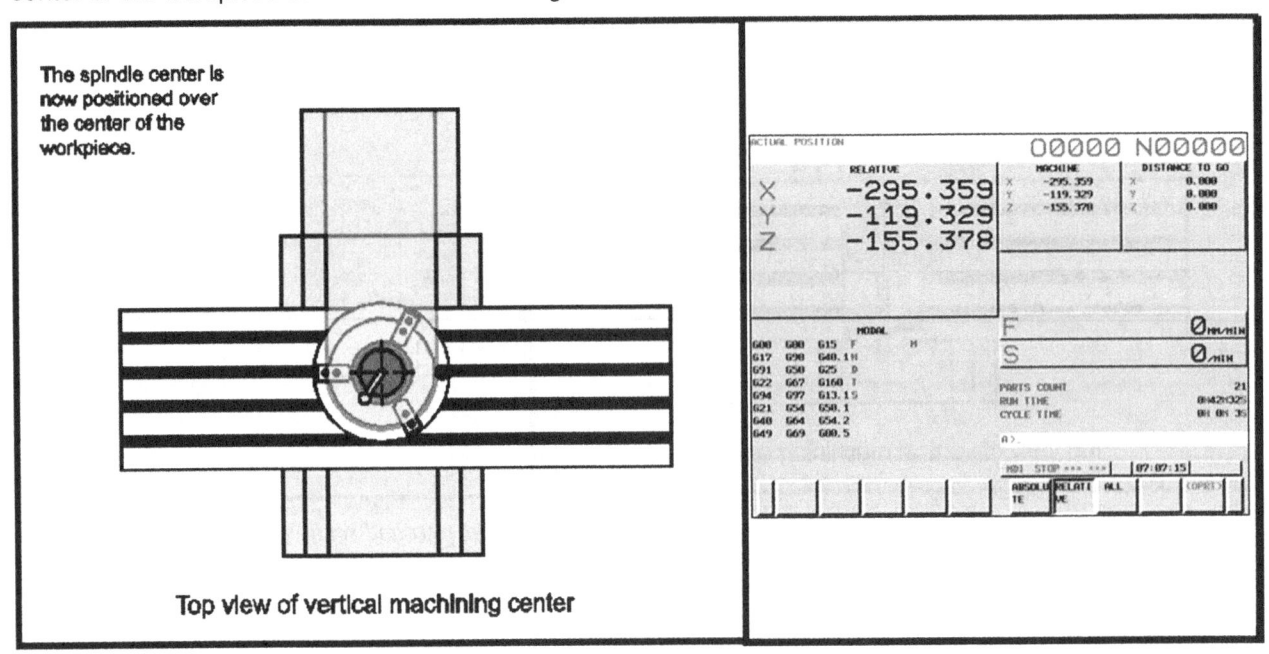

Figure 1.40: Indicate circular part

Step 3: Zero the X and Y relative positions
Press the X-axis letter address and then the [ORIGIN] soft key to set the X-axis relative position. Press the Y-axis letter address and then the [ORIGIN] soft key to set the Y-axis relative position.

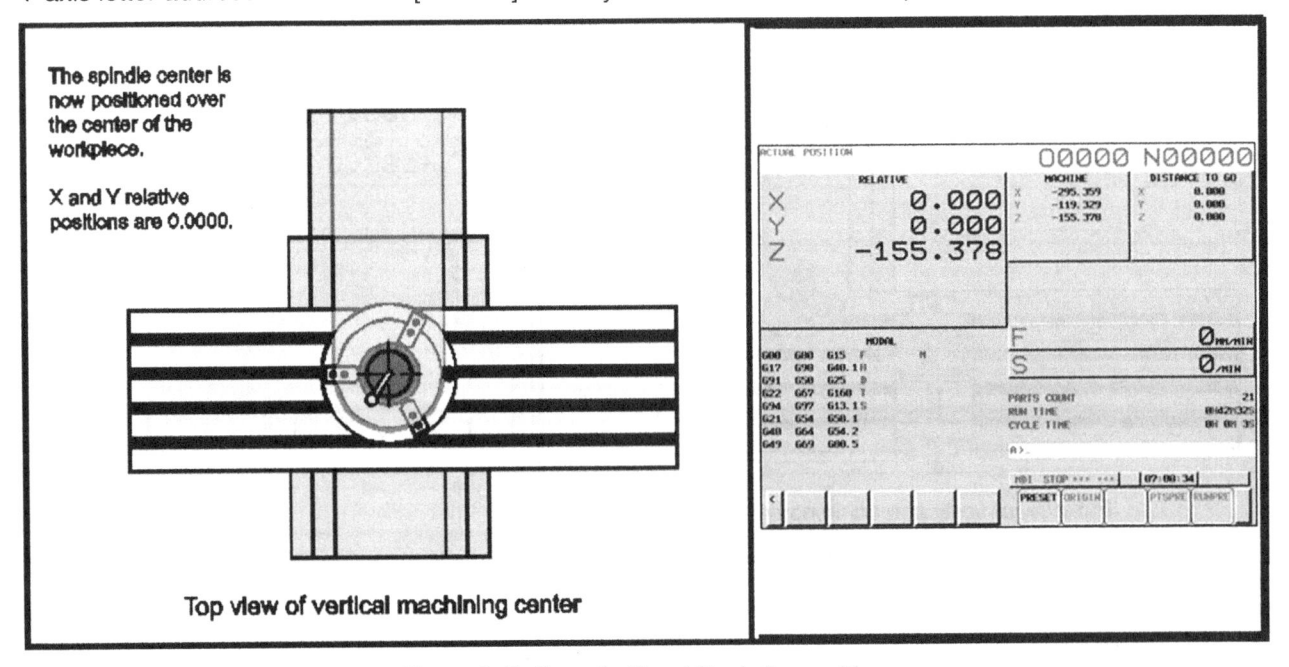

Figure 1.41: Zero the X and Y relative positions

Lesson 1.5: Determining Workpiece Coordinate System Offsets

Step 4: Send the X and Y axes to the zero return position

When you send the X and Y axes to the zero return (reference) position, the axis position display will follow along. When the X and Y axes are resting at the zero return position, the X and Y-axis displays will be showing the workpiece coordinate system offset values (distance between workpiece coordinate system zero and spindle center in X and Y while the machine is resting at the zero return position).

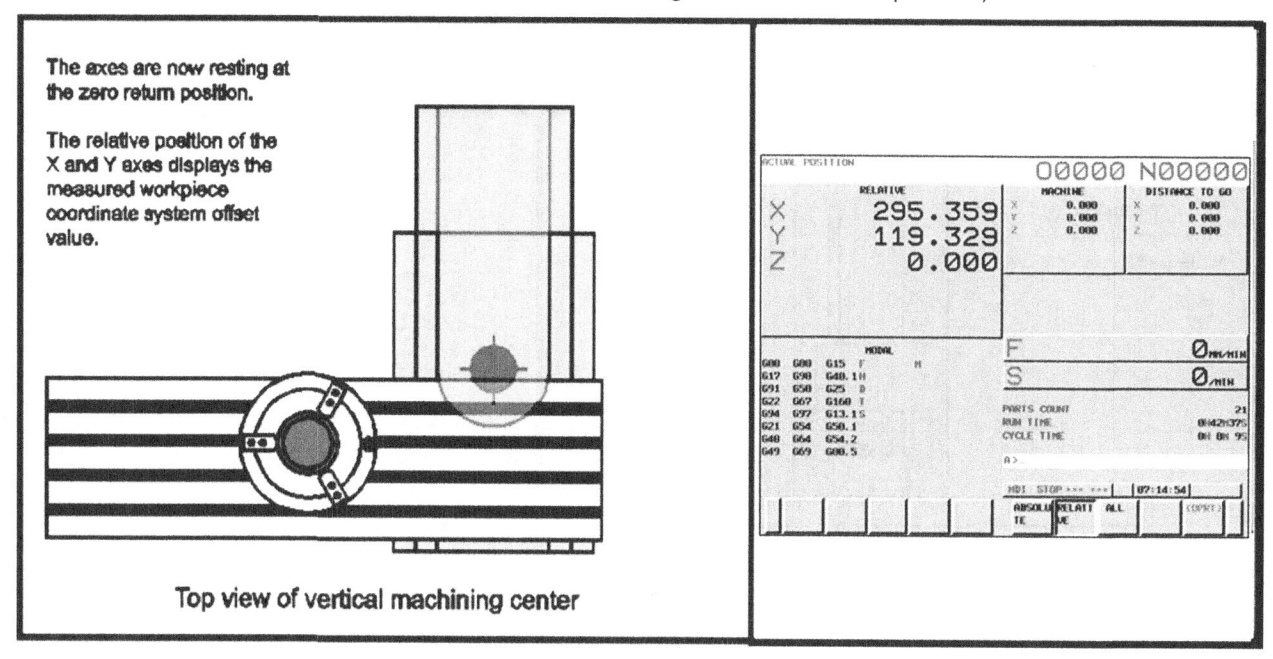

Figure 1.42: The relative positions of the X and Y axes at the zero return position

Measuring the Z-axis

For the Z-axis we must measure the distance between the workpiece coordinate system zero in Z (commonly the top of the workpiece) and the spindle nose while the machine is resting at the Z-axis zero return position. We recommend using a gauge block or a thin piece of paper between the workpiece and spindle nose to keep from scuffing the spindle nose (in the example below, we are using a 75.0-mm gauge block).

Step 1: Make the work holding setup and load a workpiece

The top of the workpiece is workpiece coordinate system zero. Place a gauge block on top of the workpiece. The gauge block is, of course, resting on the Z-axis workpiece coordinate system zero surface and the top of the gauge block is a known distance above that surface (75.0-millimeters).

Key Concept 1: Know Your Machine From A Programmer's Viewpoint

Step 2: Touch the spindle nose to the gauge block

Using continuous jog or the hand-wheel, move the Z-axis down until the spindle nose is close to the top of the gauge block. Select the smallest axis movement increment. Move the Z-axis down one increment and see if the gauge block can be inserted between the workpiece surface and the spindle nose. If it can, move down one more increment and test again. Repeat this process, moving down one increment and testing to see if the gauge block can be placed between the workpiece and the spindle nose, until you can no longer insert the gauge block between the workpiece and the spindle nose. Move the Z-axis up exactly one increment.

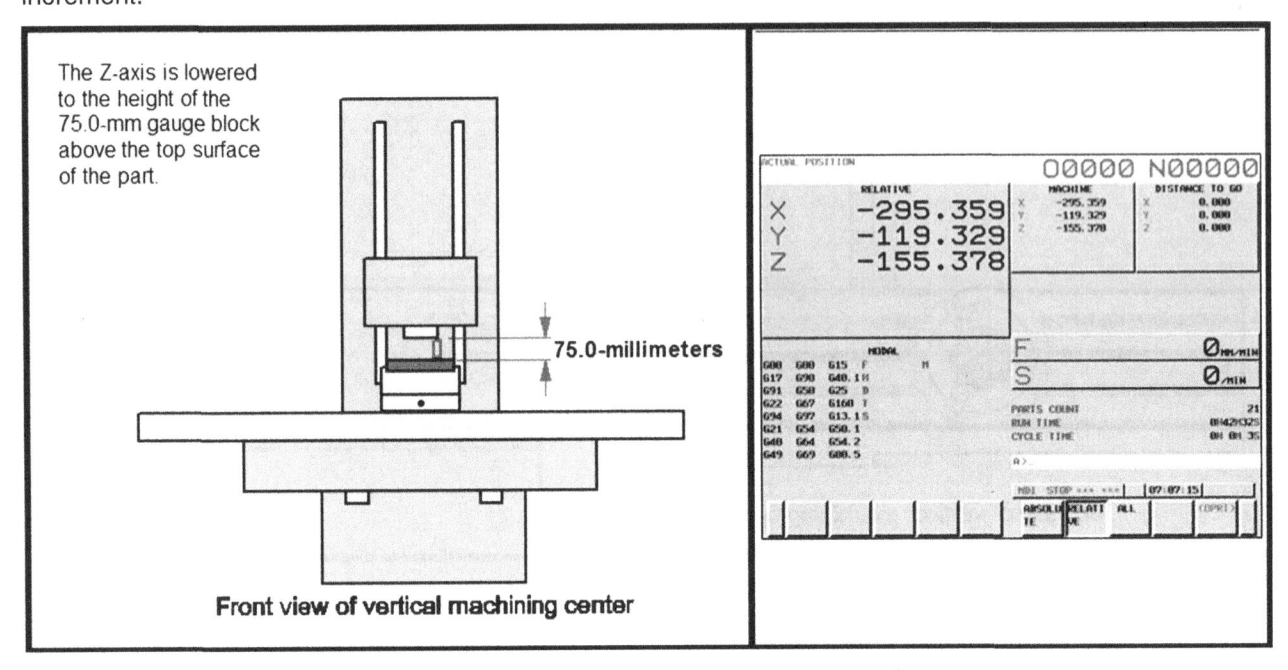

Figure 1.43: Z-axis move to the height of the gauge block

Step 3: Preset the Z-axis relative position to the height of the gauge block

Now set the Z-axis relative position to the thickness of the gauge block 75.0-millimeters in our case). The typical procedure on the 0i-MD CNC is to select the relative position page on the CNC, press the [(OPRT)] operations soft key until the [PRESET] is soft key is displayed, enter Z75.0 in the CNC key input buffer, and press the [PRESET] soft key.

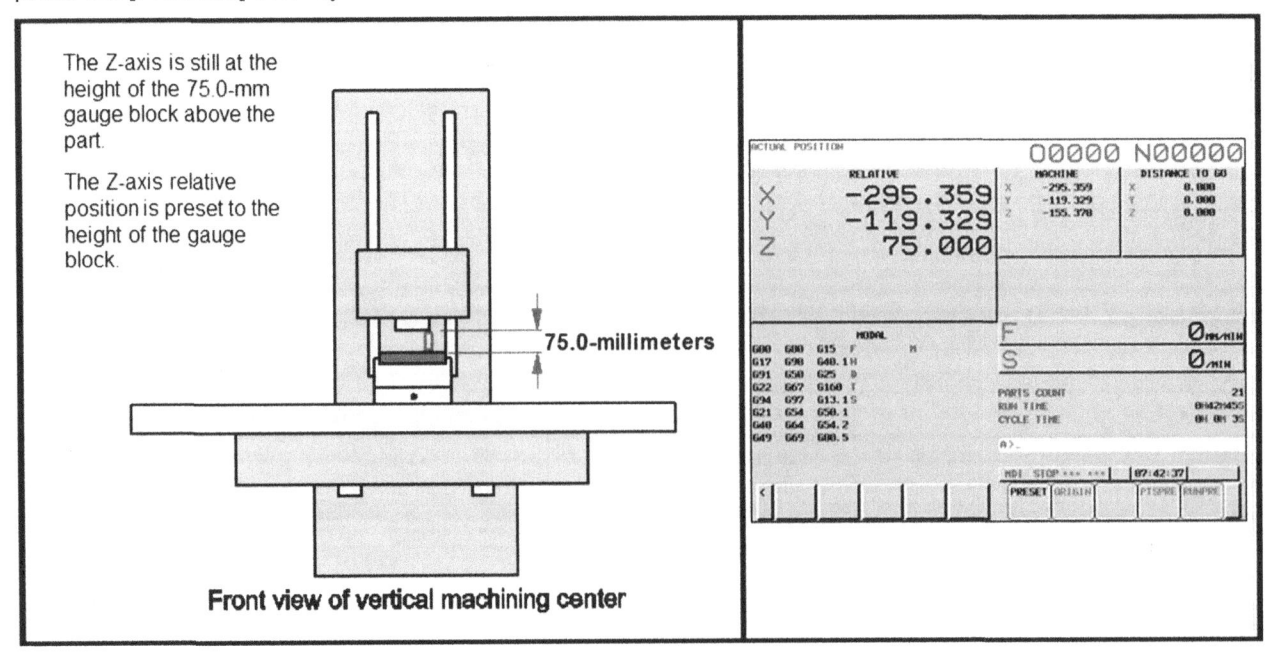

Figure 1.44: Z-axis relative position preset to the height of the gauge block

Step 4: Send the Z-axis to the zero return position

Send the Z-axis to the zero return (reference) position. When the Z-axis is resting at the zero return position, the Z-axis relative position will display the workpiece coordinate system offset value for the Z-axis (distance between program zero and the spindle nose in Z while the machine is resting at the zero return position).

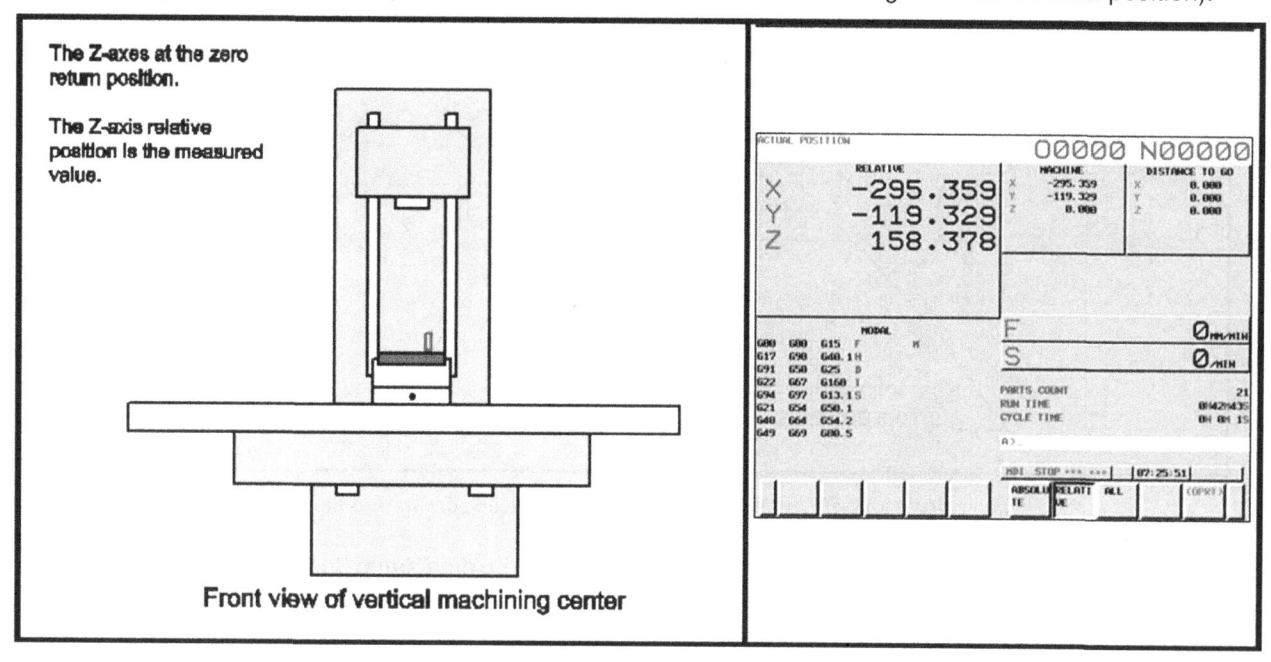

Figure 1.45: Z-axis at the zero return position displaying the measured value

What If I'm Using Inch Mode?

Manually measuring workpiece coordinate system assigning values is nearly identical regardless of which measurement system you use. Here we show the minor differences for measuring workpiece coordinate system assignment values when using inch mode. In all cases, of course, axis displays will be showing values in inches.

X and Y for rectangular-shaped workpieces
  Step 1: The edge-finder will have a 0.2 inch diameter.
  Steps 4 and 8: The amount to move over will be 0.1 inch.
  Step 10: The X and Y relative display values are in inches.

X and Y for circular-shaped workpieces
  Step 4: The X and Y relative display values are in inches.

Measuring the Z axis
  Step 1: Use the three-inch side of a 1-2-3 block instead of a 75.0-mm gauge block.
  Step 3: Preset the Z axis relative position to 3.0-inches instead of 75.0-mm.
  Step 4: The Z relative display value is in inches.

Measuring the Workpiece Coordinate System Offsets with a Probe

If your company does not make qualified work holding setups, you will have to measure the workpiece coordinate system offset values every time you setup a job. Workpiece and tooling-producing companies, for example, tend not to make qualified work holding setups. They cannot justify the additional tooling costs. Even product making companies sometime avoid complex work holding because of cost and lead time constraints.

As you have seen from the previous discussions, manually measuring program zero using an edge-finder or dial indicator can be tedious, error prone, and time consuming. A spindle probe can dramatically simplify the task. The next illustration shows one.

Key Concept 1: Know Your Machine From A Programmer's Viewpoint

Using a spindle probe to measure the program zero point

1. The probe is manually positioned to within about 12-millimeters (0.5 inch) of the program zero point in each axis.
2. A program is then run that causes the probe to automatically measure the program zero assignment values.
3. This program also assigns program zero as well (see lesson six).

Probe at zero return position

Program zero

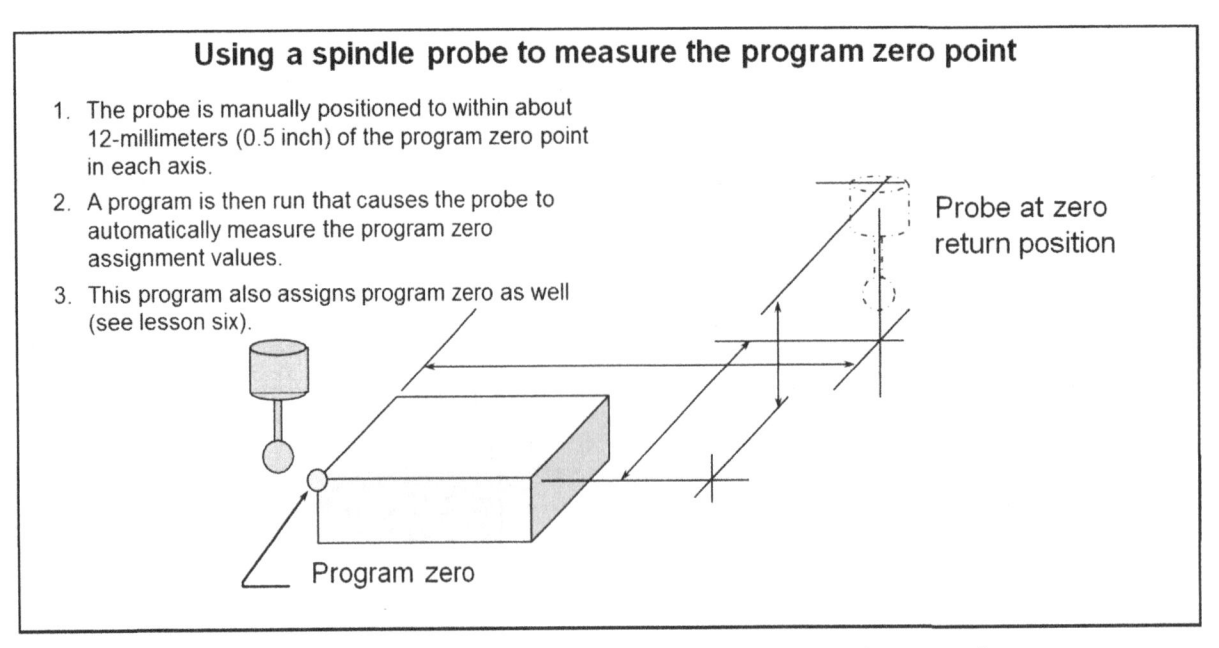

Figure 1.46: A spindle probe used to measure program zero assignment values

A spindle probe is quite simple to use. You simply make the work holding setup, load a workpiece and place the spindle probe into the spindle. Though machine tool builders vary with regard to how you use the spindle probe, most require that you first manually position it to within about 12-millimeters (0.5-inch) of the program zero point in each axis. Then you execute a special program. The program moves the probe to touch each program zero surface and measure the workpiece coordinate system offset value. The special program will also automatically set the workpiece coordinate system offsets (the topic of the next lesson). When finished, most machines will position the spindle probe back where it started.

Machine tool builders commonly supply special probing programs for a variety of program zero possibilities, including (among others) any corner of the workpiece (lower-left, lower-right, upper-left, and upper-right), the center of a hole, the center of a plug, and the center of a slot.

A spindle probe makes it much easier to measure workpiece coordinate system offset values and set the values in the CNC. Any company that does not make qualified work holding setups should be able to easily justify the cost of a spindle probe. Spindle probes are delicate and somewhat expensive instruments and must be treated with appropriate care.

How the Workpiece Coordinate System Zero Selection Affects the Setup Person

The wise selection of the workpiece coordinate system zero simplifies programming. In many cases, coordinates used in the program come directly from the workpiece engineering drawing—or at least they will come from a logical datum. We offer another (actually more important) way to determine where the workpiece coordinate system zero is placed. First determine how the setup person will be locating the workpiece in the setup. Then be sure that your workpiece coordinate system zero point is as easy to determine as possible—especially if the workpiece coordinate system zero must be measured during setup.

In most cases, the two methods will render the same results. That is, the surfaces from which dimensions begin are usually the same surfaces that will be used as location surfaces in the setup. But consider the drawing shown in the figure below.

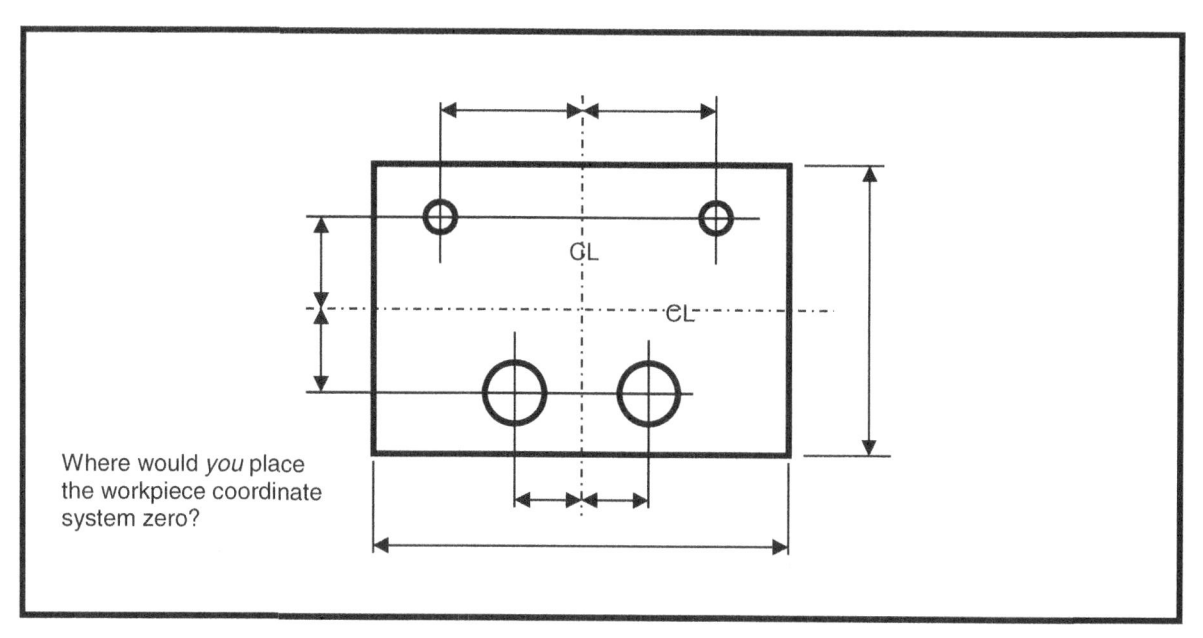

Figure 1.47: A difficult decision for program zero point placement

While we don't show actual dimensions, this should be enough to get the point across. The design engineer is specifying that the two centerlines are right in the middle of the workpiece. And all hole-locations are being specified from these two centerlines. From a programmer's perspective, programming will be quite easy if you place the workpiece coordinate system zero point at the center of the workpiece, right?

But think about the setup person for a minute. If you place the workpiece coordinate system zero in the center of the workpiece, the measurements they take for X and Y workpiece coordinate system zero assignment values must begin from the center of the workpiece. There is nothing to touch with an edge-finder – and each measurement will be more difficult to make. More time and effort will be required.

How will they locate this workpiece in the set-up? While there are special centering vises available, most CNC-using companies will use a standard table vise to hold this workpiece–meaning the setup person will be locating from two of the edges–possibly the lower-left corner. They will depend upon the overall width and length of the workpiece to be precise enough to ensure that holes are centered properly relative to the centerlines. In this case, the workpiece coordinate system zero should be placed in one of the four corners.

Again, setup ease must always take priority over programming ease when it comes to workpiece coordinate system zero placement. You only write the program once – but the job may be setup and run hundreds of times. (And hopefully you write the program while the machine is running another job–meaning programming time will not affect machine utilization.)

Key Concept 1: Know Your Machine From A Programmer's Viewpoint

Key Points for Lesson 1.5:
- ✓ Workpiece coordinate system zero assignment values must be determined before program zero can be assigned.
- ✓ Workpiece coordinate system offset values are the distances between the program zero point and the machine's zero return (reference) position.
- ✓ There are only two ways to determine workpiece coordinate system zero assignment values–calculate or measure them.
- ✓ Workpiece coordinate system zero assignment values can only be calculated if you make qualified work holding setups, if you use predictable work holding tools, and if you know some important dimensions on the machining center.
- ✓ If you make qualified work holding setups but your work holding tooling is not predictable, you will have to measure workpiece coordinate system zero assignment values the very first time a setup is made. If you document these values, you will not have to perform these measurements again.
- ✓ Without qualified work holding setups, workpiece coordinate system zero assignment values must be measured for every setup and a spindle probe may be a good option.
- ✓ You can manually measure workpiece coordinate system offsets using an edge-finder or dial indicator.
- ✓ You must consider the ease of setup as the top priority when it comes to deciding where the workpiece coordinate system zero point is placed.

Quiz

1. Using qualified work holding setups means that you never have to measure the workpiece coordinate system offset values.
    a. True
    b. False

2. Using a 10.0-mm diameter edge-finder, once the edge-finder is made to touch an edge, how far is it from the edge to the center of the edge-finder?
    a. 5.0-mm
    b. 10.0-mm
    c. 20.0-mm
    d. None of the above

3. A poorly chosen workpiece coordinate system zero point can complicate the setup person's job.
    ☐ True
    ☐ False

4. When you position the X-axis so that the edge-finder is just in contact with the workpiece, the next step is to
    a. Measure the Y-axis offset
    b. Retract the Z-axis
    c. Origin the CNCs X-axis relative position
    d. None of the above

5. A spindle probe
    a. Simplifies offset measurement
    b. Error-proofs offset measurement
    c. Simplifies and error-proofs offset measurement
    d. None of the above

Answers: 1: False (first time), 2: a 3: True, 4: c 5: c

Lesson 1.6–Setting Workpiece Coordinate System Offsets

Once the workpiece coordinate system offset values are determined as shown in the previous lesson, you must enter the values into the appropriate registers on CNC workpiece coordinate offset screen.

Objectives

After completing this lesson, students should be able to:

- ✓ Describe how to set the workpiece coordinate system offsets values measured using the relative position page in the previous lesson in the workpiece coordinate system offsets in the CNC
- ✓ Describe how G54 in the program references the workpiece coordinate system offset values in the workpiece coordinate system offset tables
- ✓ Describe the polarity of workpiece coordinate systems offset values

Introduction

In the last lesson you learned how to determine workpiece coordinate system offset values. Most setup people will always have to measure workpiece coordinate system offset values the very first time a new job is setup. If the work holding setup is qualified (it can be replaced on the table in exactly the same position), they will document these values so measurements need not be repeated every time the job is set up. If the work holding setup is not qualified, workpiece coordinate system offset values must be re-measured every time the job is setup–which is an excellent application for a spindle probe.

In this lesson, we're going to show how to set the workpiece coordinate system offsets using the previously determined workpiece coordinate system offset values. This assumes that you do not have a spindle probe to measure and set the workpiece coordinate system offsets automatically. But even if you do, you must understand the points we make in this lesson.

As stated in the previous lesson, this topic has more to do with setup than with operation. But, programmers must know enough about making setups to instruct setup people. This includes knowing how set the workpiece coordinate system offset values in the CNC.

Understanding Workpiece Coordinate System Offsets

Offsets are storage registers within the control. They store numerical values that are used when they are invoked in the program. In general, offsets are used to separate the physical attributes of a machine, tooling and setup from the CNC program–keeping you from having to know the specific attribute values when the program is written.

Workpiece coordinate system offsets are the registers used to assign workpiece coordinate system zeros. This is where the workpiece coordinate system offset values are stored. Most machining center CNCs have at least six sets of workpiece coordinate system offsets, meaning up to at least six different workpiece coordinate system zeros can be assigned and used in a program. (We haven't seen any applications for more than one workpiece coordinate system zero in a program yet. We'll show some in a later lesson.)

The figure below shows the workpiece coordinate system offset screen page of a popular CNC machining center control.

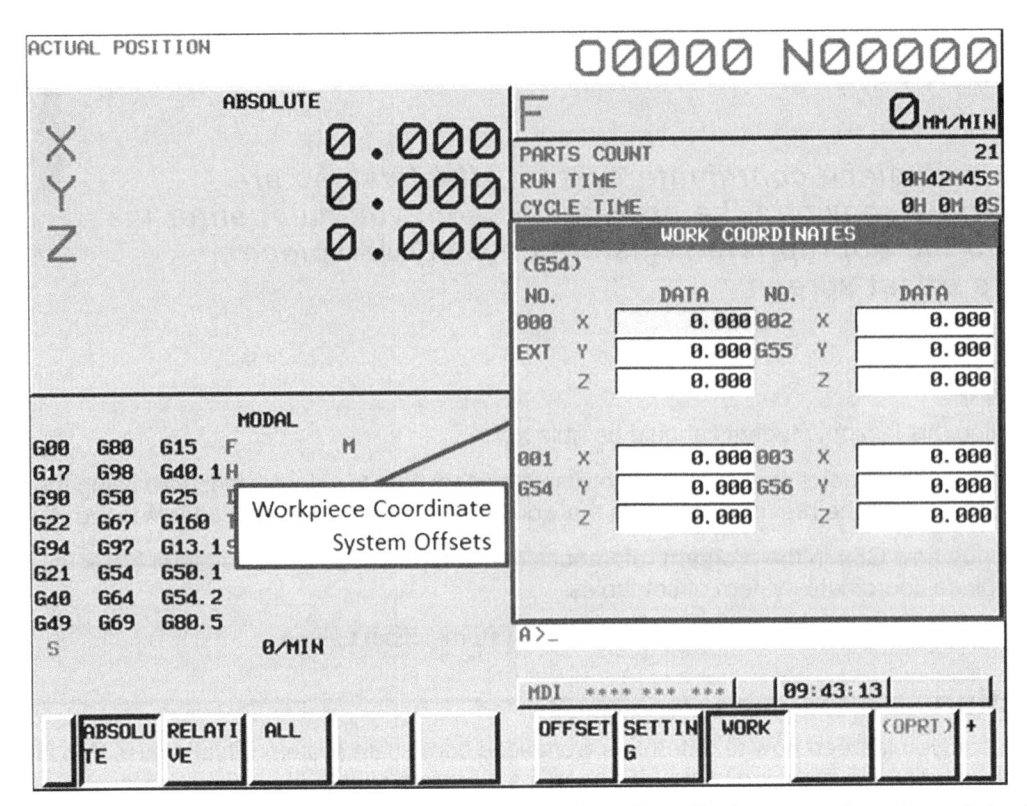

Key Concept 1: Know Your Machine From A Programmer's Viewpoint

Figure 1.48: First page of workpiece coordinate system setting offsets display screen (metric mode is shown)

Notice that work offsets are organized by number. The figure above shows the external offset (#0 on the work offset screen) as well as work offsets #1 through #3. The second page of work offsets (not shown) displays offsets #4 through #6.

For now, let's concentrate on assigning just one workpiece coordinate system zero per program. We'll use workpiece coordinate system offset #1 to do so. The work offset #1 is shown in the lower left corner of the work offset screen. Also note the G-code next to it: G54. G54 is the G-code used to invoke workpiece coordinate system setting offset number one. When G54 is active, the CNC uses the values stored in work offset #1 set of registers as the current set of workpiece coordinate system zero offset values. Note the "(G54)" in the top left of the work offset screen, it indicates that G54 is active. In the lower left corner of the screen, G54 is also shown in the list of currently instated G-codes.

Work offset values have a polarity. These values are the distances <u>from</u> the zero return (reference) position <u>to</u> the workpiece coordinate system zero in each axis. When the zero return position is close to the plus limit for each axis (as it is for most machining centers), the polarity of work offset values will always be negative.

To complete the task of assigning the workpiece coordinate system zero, the setup person must enter the previously measured workpiece coordinate system offset values into the appropriate work offset registers–with the required polarity (as negative values for most machines). As stated in the previous lesson, if a spindle probe is used to measure workpiece coordinate system offset values, the CNC will automatically enter the values into work offsets. The next illustration shows values measured during the previous lesson.

A Full Example of Using the Workpiece Coordinate System

The figure below shows the simple drawing used during our discussion of visualizing program execution. Now we'll use it to illustrate how workpiece coordinate system offsets are set.

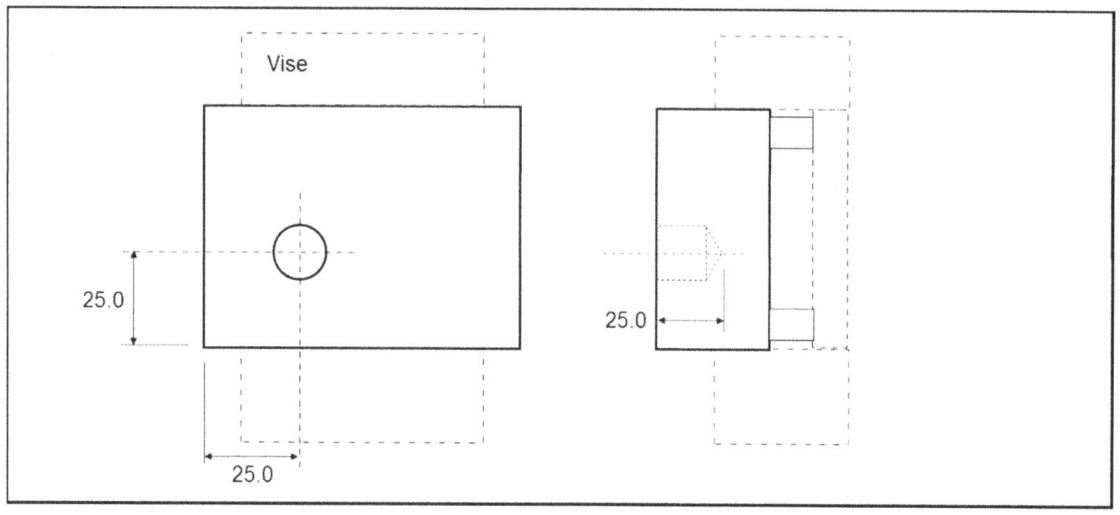

Figure 1.49: Drawing to illustrate program zero assignment (shown using metric mode)

The Program

Since the workpiece is dimensioned from the lower-left corner, we'll use this corner as the workpiece coordinate system zero point for the program. Program zero in Z is the top surface of this workpiece. Here is the metric-mode version of the program shown earlier.

```
%
O0002 (Program number)
N010 G21 G90 (Select metric & absolute programming modes)
N020 G54 (Set the workpiece coordinate system zero)
N030 T01 M06 (Load the drill into spindle)
N040 S600 M03 (turn spindle on CW at 600 RPM)
N050 G00 X25.0 Y25.0 (Move tool above the hole in X and Y)
N060 G43 H01 Z2.5 M08 (Rapid to workpiece surface, instate tool length compensation, turn coolant on)
N070 G01 Z-25.0 F85.0 (Drill hole at 85 mmpm)
N080 G00 Z2.5 M09 (Retract drill, coolant off)
N090 G91 G28 Z0. M05 (Rapid to Z-ref position, spindle off)
N100 G28 X0. Y0. (Rapid to X/Y reference positions)
N110 M30 (End of program)
%
```

In block N020, notice the G54. It is telling the CNC to use the values in work offset registers #1 to compensate for the workpiece coordinate system zero. This work holding setup (vise) is not qualified–so at the time this program is written–the CNC programmer will not know the workpiece coordinate system offset values. Only when the job is setup will the setup person be able to measure them.

While it may not seem like much, notice what the G54 in this program is doing for you. It is allowing you to write the program even though you don't know where this vise is going to be placed on the machine table. Again, it is allowing you to separate program zero assignment from the program.

Now notice the G90 in line N010. This word specifies the absolute positioning mode. Every coordinate the CNC sees from this point in the program will be specified from the workpiece coordinate system zero. In block N050, for example, the X25.0 and Y25.0 specifies a move to a position of 25-millimeters in the X and Y axes. This position, of course, is relative to the workpiece coordinate system zero. Since the machine has

Key Concept 1: Know Your Machine From A Programmer's Viewpoint

been told the distance between the workpiece coordinate system zero and the zero return position (in the work offset #1), and since it can keep constant track of its current position relative to the zero return position, it will be able to determine how far to make the X and Y axes move in order to get to the programmed position (X25.0 Y25.0).

What is the G91 Doing in Block N090?

You know that G91 specifies incremental positioning mode. And we said you should always program in the absolute positioning mode. So what's going on?

G28 is the CNC word that sends the machine to its zero return (reference) position in any axes specified within the G28 command. In our case, Z is included in this command, so when this command is finished, the Z-axis will be resting at the zero return position. The modal G91 is still active in block N100, and the block includes the X and Y axes. After block N100 is finished, the X and Y axes will also be resting at their zero return positions.

While we don't want to get into too much detail yet (we fully explain the G28 command during Key Concept 5), the G28 command requires that you specify something called an intermediate point through which the tool will pass on its way back to the zero return position. In block N090, the G91 along with Z0 are telling the CNC that the intermediate position is incrementally nothing from the current position in the Z-axis. Similarly, block N100 is telling the CNC that the X and Y intermediate positions are incrementally nothing from the current positions of the X and Y axes. In essence, this will send the tool straight to the zero return position in each axis. We don't feel that using a G91 within the G28 is really programming incrementally—we're simply using it to send the machine straight to its zero return position. However, the G90 is important at the beginning of the program (N010) to switch the CNC back to absolute mode the next time the program is executed.

The Setup

Before this program can be run, of course, several setup related tasks must be completed. At this point, we will only address those tasks that are related to workpiece coordinate system zero assignment (we'll discuss more setup-related tasks as the course progresses).

First of all, the work holding must be set up. For this setup, the vise must be mounted on the table. It must also be aligned so that it is square with the table. (If you have never seen this done, ask an experienced person in your company or school to demonstrate the squaring of a vise.) An end stop must be added to the vise to locate the left side of the workpiece. If the vise jaws do not have a step, parallels must be placed under the workpiece.

After loading a workpiece in the vise, the program zero assignment values must be determined. For us, this means actually measuring the workpiece coordinate system offset values. If we have a spindle probe, this will be as easy as positioning it about 12-millimeters (0.5 inch) away from each program zero surface and activating the appropriate program. If we don't have a spindle probe, program zero assignment values must be manually measured as shown in the previous lesson.

These workpiece coordinate system offset values must be entered in the appropriate work offset table. Again, if we have a spindle probe, this is done automatically. But if we don't, we must manually enter the values into work offset #1 (as negative values). This is done on the work offset CNC screen.

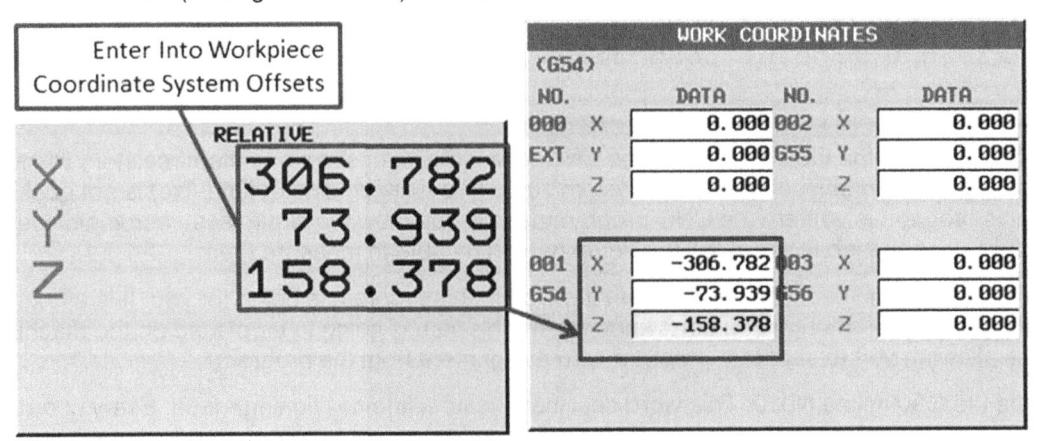

Figure 1.50: CNC relative positions after work coordinate system offset measurement and work offset screen

Lesson 1.6: Setting Workpiece Coordinate System Offsets

In the figure above, on the left you can see the CNC relative position page after we measure the workpiece coordinate system offset values in the X, Y and Z axes. These positions are entered in work offset #1 as negative values. We are using metric mode, so all values are entered in millimeters. If you are using inch mode, of course, you would enter the related values in inches. With G54 active, the CNC absolute position becomes X306.782, Y73.939 and Z158.378. Hopefully, you can visualize that if you program G90 G00 X0.0 Y0.0 Z0.0, the axes will move to the workpiece coordinate system zero on the workpiece.

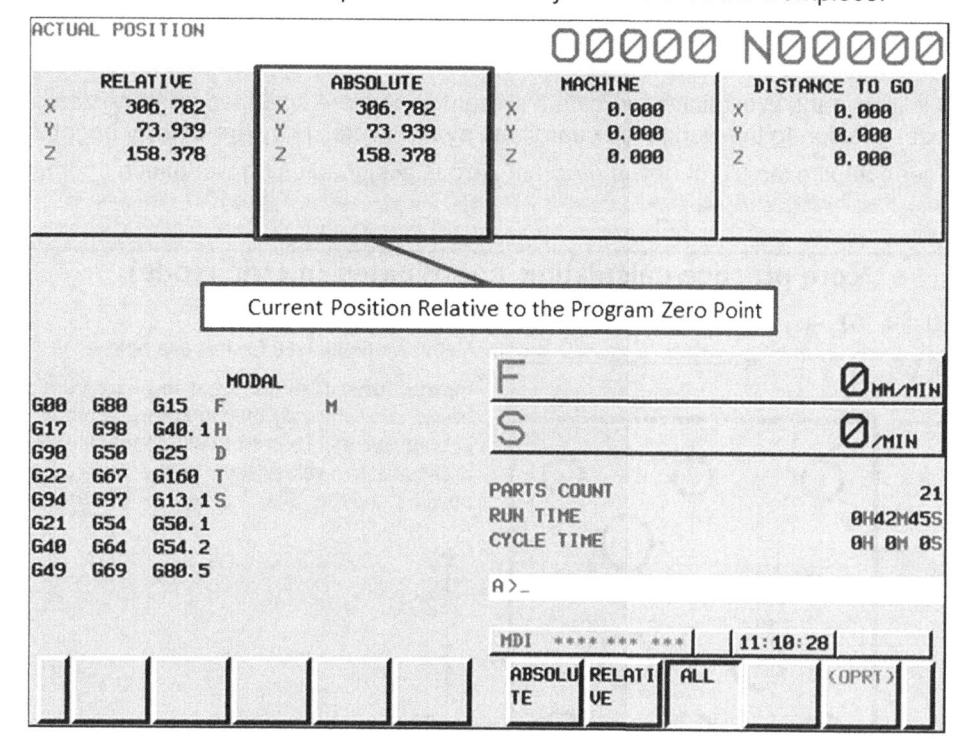

Figure 1.51: CNC all positions screen with the workpiece coordinate system offset values loaded in work offset #1

The program must also be loaded into the control. We can type it through the keyboard of the control–or better–we can use some kind of communications software system to quickly load the program–assuming we've already typed it on the text editor of a computer. Current FANUC CNCs support Ethernet and RS-232 files transfer. They also support transferring files using PCMCIA flash cards or USB memory sticks.

There are some other setup related tasks that must be completed–mostly related to cutting tools–but since we haven't addressed them as of yet in this course, we will not discuss them here. With these exceptions, we're ready to test run the program.

You should now see how setting the workpiece coordinate system offset values on the CNC work offset screen marries the work holding setup to the program. It tells the machine where program zero is located so that the machine can position the cutting tools appropriately during the program's execution.

What is the External/Common Work Coordinate System Offset?

On the CNC work coordinate offset screen shown above, the first workpiece coordinate system setting offset (upper-left on the screen) is labeled as "00 EXT". On some older FANUC CNCs this may be labeled "#0 COMMON". You're probably wondering what it is for. The goal in this lesson is simply to show you the most common way that the workpiece coordinate system zero is assigned. But there are other ways. We will address them in detail during Key Concept 4.

The external or common offset provides a way to shift the point of reference for workpiece coordinate system offset values. When set to zero (as it is above–and as it probably is on your machines), the point of reference for work offsets is the zero return (reference) position. So program zero assignment values in workpiece coordinate system offsets #1 through #6 are specified from zero return to workpiece coordinate system zero in each axis–just as we say in the previous lesson.

But as you'll see in Key Concept 4, the zero return position doesn't always make the best point of reference for workpiece coordinate system offsets. There are applications when shifting the point of reference to a

Key Concept 1: Know Your Machine From A Programmer's Viewpoint

different position can make it easier to assign the program zero. The external/common offset allows you to do so.

Key Points for Lesson 1.6:

- ✓ If your machine has workpiece coordinate system offsets, use them to assign program zero.
- ✓ To tell the CNC where the workpiece coordinate system zero is located, the workpiece coordinate system offset values measured in the previous lesson are placed into the workpiece coordinate system offset table.
- ✓ Polarity is important. Workpiece coordinate system offsets are specified *from* the zero return (reference) position *to* the workpiece coordinate system zero. They are usually negative.
- ✓ You've seen a full example of how a program zero is assigned using workpiece coordinate system offsets.

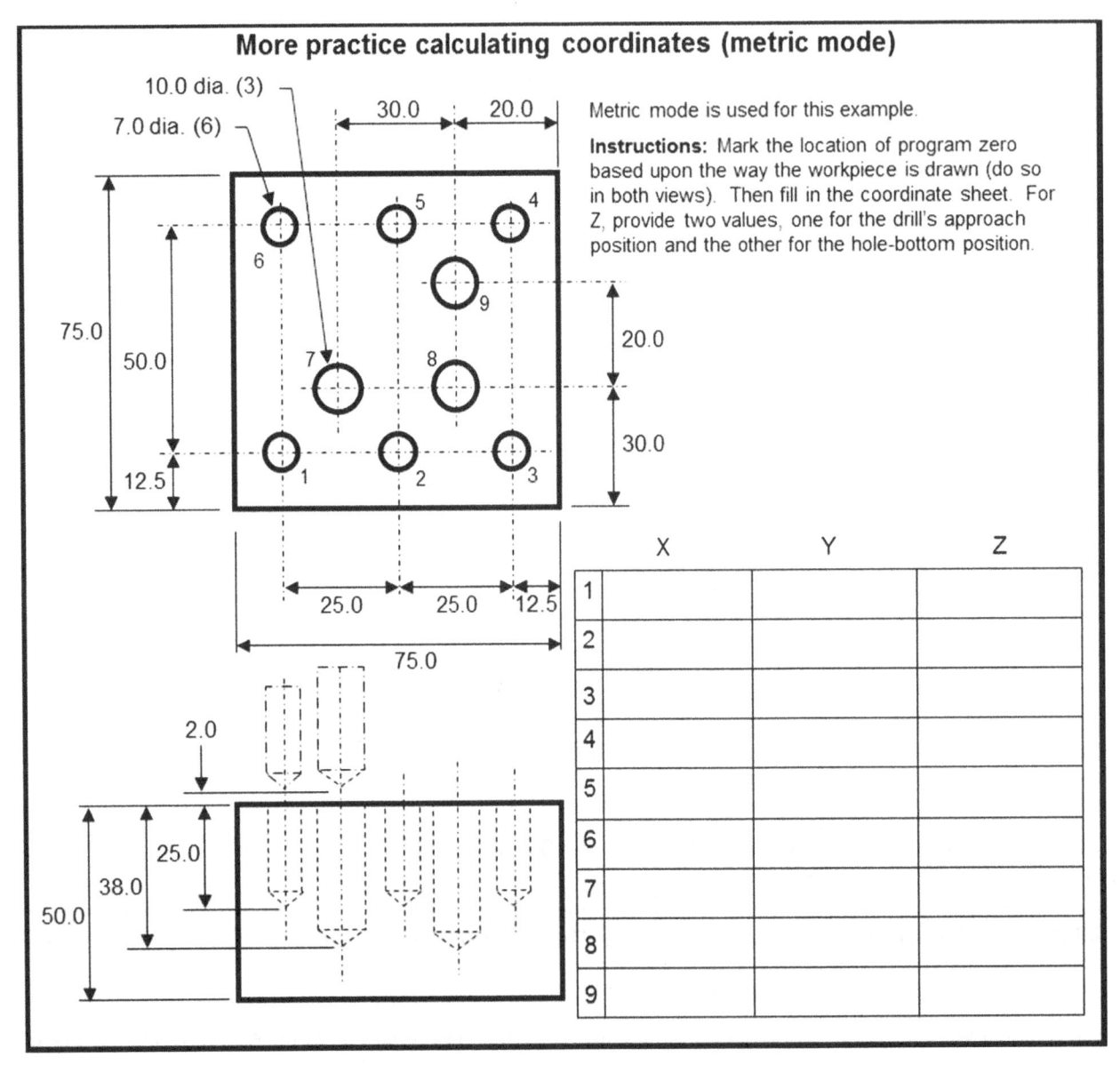

More practice calculating coordinates (metric mode)

Metric mode is used for this example.

Instructions: Mark the location of program zero based upon the way the workpiece is drawn (do so in both views). Then fill in the coordinate sheet. For Z, provide two values, one for the drill's approach position and the other for the hole-bottom position.

	X	Y	Z
1			
2			
3			
4			
5			
6			
7			
8			
9			

Answers (metric mode exercise):
In the top view, program zero is in the lower-right corner. In the front view, program zero is in the upper right corner (Z zero is the top of the workpiece).

1: X-62.5 Y12.5 Z2.0 & Z-25.0
2: X-37.5 Y12.5 Z2.0 & Z-25.0
3: X-12.5 Y12.5 Z2.0 & Z-25.0
4: X-12.5 Y62.5 Z2.0 & Z-25.0

5: X-37.5 Y62.5 Z2.0 & Z-25.0
6: X-62.5 Y62.5 Z2.0 & Z-25.0
7: X-50.0 Y30.0 Z2.0 & Z-38.0
8: X-20.0 Y30.0 Z2.0 & Z-38.0
9: X-20.0 Y50.0 Z2.0 & Z-38.0

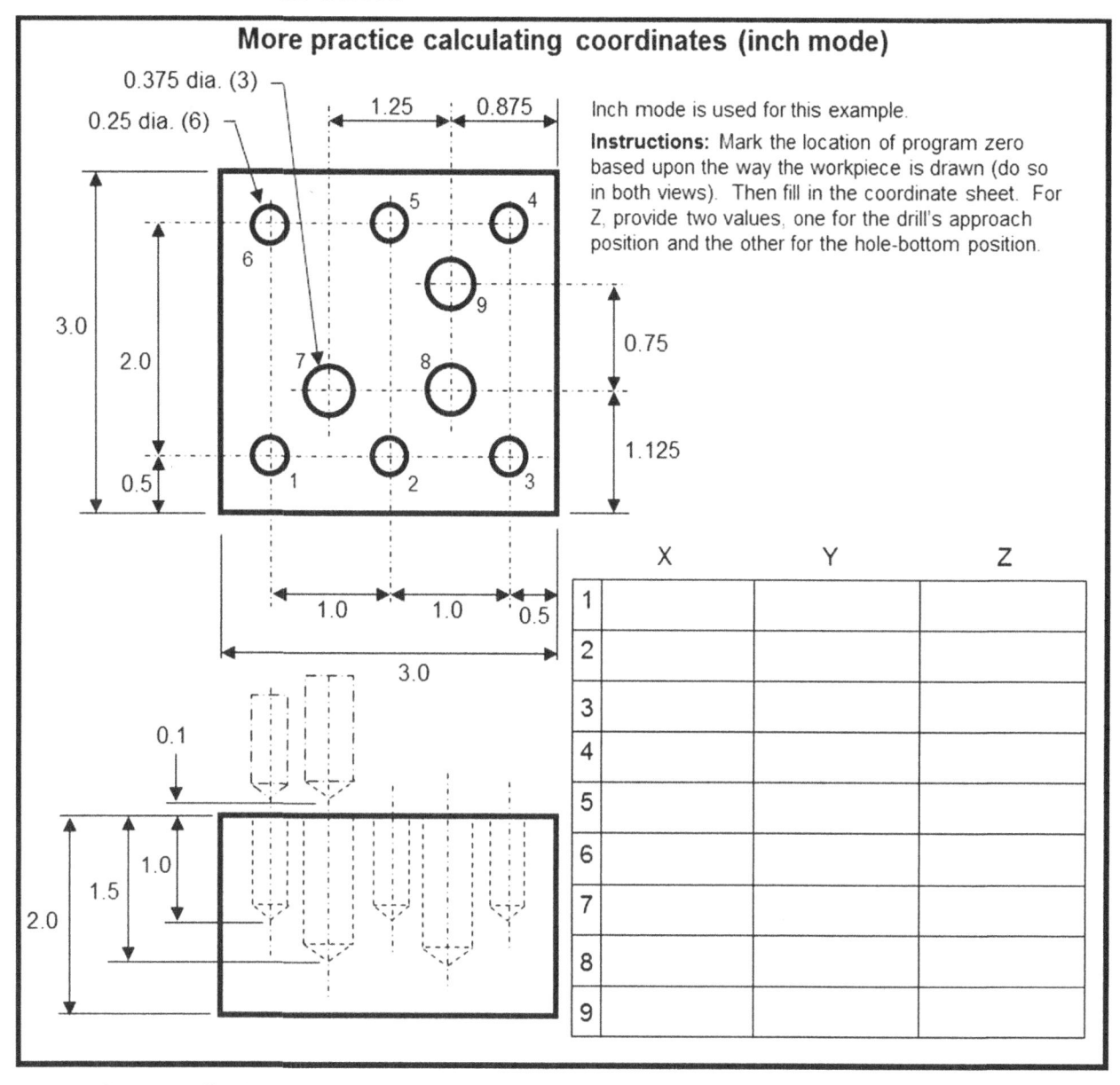

More practice calculating coordinates (inch mode)

Inch mode is used for this example.

Instructions: Mark the location of program zero based upon the way the workpiece is drawn (do so in both views). Then fill in the coordinate sheet. For Z, provide two values, one for the drill's approach position and the other for the hole-bottom position.

	X	Y	Z
1			
2			
3			
4			
5			
6			
7			
8			
9			

Answers (inch mode exercise):
In the top view, program zero is in the lower-right corner. In the front view, program zero is in the upper right corner (Z zero is the top of the workpiece).

1: X-2.5 Y0.5 Z0.1 & Z-1.0
2: X-1.5 Y0.5 Z0.1 & Z-1.0
3: X-0.5 Y0.5 Z0.1 & Z-1.0
4: X-0.5 Y2.5 Z0.1 & Z-1.0
5: X-1.5 Y2.5 Z0.1 & Z-1.0
6: X-2.5 Y2.5 Z0.1 & Z-1.0
7: X-2.125 Y1.125 Z0.1 & Z-1.5
8: X-0.875 Y1.125 Z0.1 & Z-1.5
9: X-0.875 Y1.875 Z0.1 & Z-1.5

Lesson 1.7–Introduction to Programming Words

As you know, all CNC words include a letter address and a numerical value. The letter address identifies the word type. You should be able to quickly recognize the most common ones.

Objectives

After completing this lesson, students should be able to:

- ✓ Describe the primary programming words and their functions (O, N, G, X, Y, Z, R, M, S, T, D, H and EOB)
- ✓ Understand that some words have different meanings depending how they are used
- ✓ Understand how to use this lesson as a reference for the remainder of the course
- ✓ Describe some of the limitations of G-codes
- ✓ Describe some of the limitations of M-codes

Introduction

CNC programs are made up of blocks–and blocks are made up of words.

Following instructions

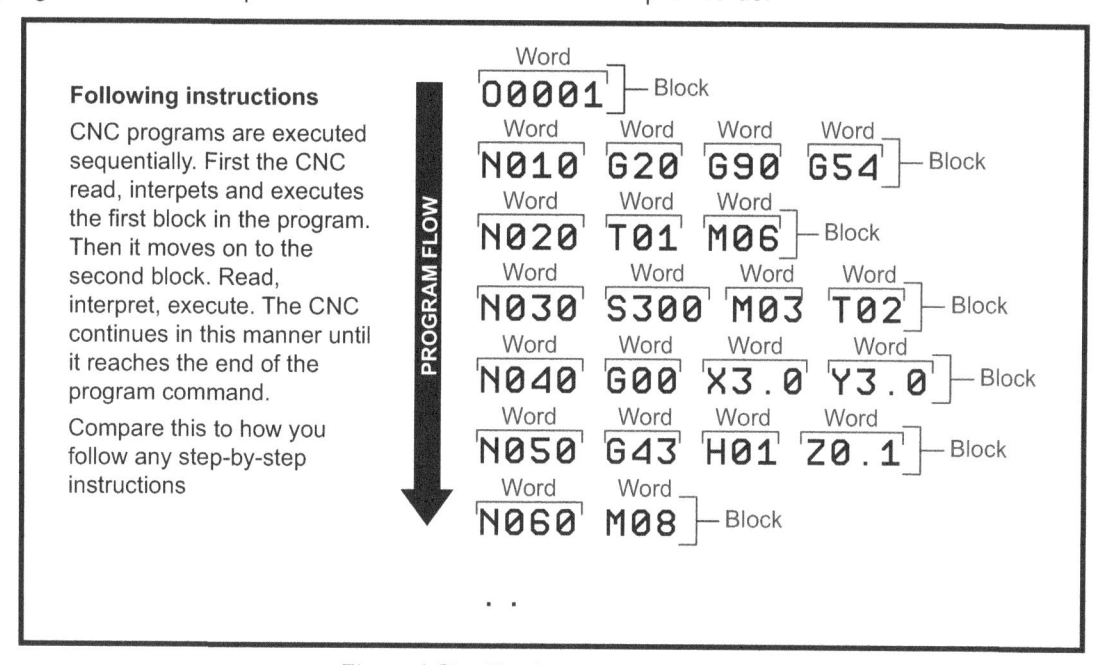

CNC programs are executed sequentially. First the CNC read, interpets and executes the first block in the program. Then it moves on to the second block. Read, interpret, execute. The CNC continues in this manner until it reaches the end of the program command.

Compare this to how you follow any step-by-step instructions

```
O0001                         — Block
N010  G20   G90   G54         — Block
N020  T01   M06               — Block
N030  S300  M03   T02         — Block
N040  G00   X3.0  Y3.0        — Block
N050  G43   H01   Z0.1        — Block
N060  M08                     — Block
```

Figure 1.52: The flow of a CNC program

Each word type has a special meaning to the CNC. Each is designated by a specific letter address. You already know a few of the letter addresses, like N for sequence number, G for preparatory function, X, Y, and Z for axis designators, S for spindle speed, F for feedrate, and M for miscellaneous (or machine) functions. In this lesson, we're going to introduce several other word types.

If you are a beginner looking at the word types for the first time, you may want to read this section a few times to get better acquainted with each word type. Note that we are not asking you to memorize the word types–just to get familiar with them. In Key Concept 5–program formatting–we will provide you with a way to remember each word's function.

Also, this lesson is only intended to introduce each word, not to give you an in-depth description. When appropriate, we'll point you to the lessons that discuss each word type in more detail.

Key Concept 1: Know Your Machine From A Programmer's Viewpoint

You will find that certain words are seldom used, meaning you will have little or no need for them. Other words are constantly used, and you will soon have them memorized after writing a few programs.

Some CNC words have more than one function, depending on other words in the block. We will show you the primary (most common) function of the word following the "A" description and the secondary use for the word following the "B" description.

Once you see a word type a few times, it should not be too difficult to remember its function since most word types are identified with a logical letter address. Additionally, only about fifty different words are used consistently in programs, so consider learning to program a CNC machining center like learning a foreign language that contains only fifty words.

As you continue with this course, use this lesson as a reference. If you come across a word or word type you don't recognize, remember to come back to this lesson. You've probably already noticed that we provide a quick reference for CNC words at the end of this text, but information in this lesson is more detailed.

Decimal Points in Words

Remember that current CNCs allow you to include a decimal point in those words that are used to specify *real numbers* (values that require a portion of a whole number). You must remember to include a decimal point with these words or the control may interpret it to have the fixed format (as discussed in lessons 1.3 and 1.4). Word types that allow a decimal point include:

A, B, C, G (newer CNCs), X, Y, Z, I, J, K, F, Q, and R

Certain CNC words are used to specify *integers* (whole numbers). These word types do *not* allow a decimal point:

O, N, G (older CNCs), P, L, S, T, M, D, and H

O-Word

The O-word specifies a program number that the CNC uses to identify each unique program in the CNCs memory. The CNCs discussed in this course allow the user to store multiple programs in the CNC memory. You will be assigning every program a number from 0001 through 9999 (O0001 through O9999). The O-word will be the very first word in the program. A decimal point is not allowed with the O-word. Program numbers are discussed in more detail in lesson 5.2.

N-Word

This word specifies a sequence number. This is the word the CNC uses to label or identify blocks in a program or subprogram. Some advanced CNC commands can use these labels, but for basic programming they are optional. They also allow you to organize each block in the program by a number sequence. Sequence numbers are not required to be in any particular order and can even be repeated in the program. Actually, they need not be in the program at all. But for the sake of organization purposes, we recommend that you include them in your programs and place them in a logical numerical order. We use them for all examples in this course. We skip ten numbers for each sequence number (going by tens). This allows new blocks to be inserted in the program later without having to renumber all the blocks that follow. A decimal point is not allowed with the N-word. Sequence numbers are introduced in lesson 1.3 and will be discussed in more detail in lesson 6.3. Sequence numbers allows the operator to uniquely recognize specific sections in a program to simplify modification of the program at the CNC.

G-Word

This word specifies a preparatory function. Preparatory functions prepare the control for what is coming–in the current command–and often in upcoming commands. They set modes (though some G-codes are non-modal). There are many G-words, but only a few that are used on a consistent basis. For a list of all G-codes, see the list at the end of this lesson. A decimal point is not allowed with most G-words. But do note that with some FANUC CNCs, some G-codes do use a decimal point (like G84.1 for rigid tapping). A decimal point used in this fashion is simply part of the G-code's designation–and is not the true application for a decimal point. Preparatory functions are introduced in lesson 1.1 and discussed often throughout this text.

X-Word

A. The primary use for the X-word is to designate a coordinate along the X-axis. That is, it is the X-axis designator. The X-word allows a decimal point. An X-axis position of 10-inches or 10-millimeters will be specified X10.0. Axis designators were introduced in lesson 1.1 and discussed often during this course.

B. The secondary use for the X-word is that it can be used to specify dwell time in seconds in a dwell command (G04). Dwell commands are used to make axis motion (for all axes) pause for a specified length of time. The dwell command is discussed in lesson 6.3.

Y-Word

The Y-word is the Y-axis designator. The Y-word allows a decimal point.

Z-Word

A. The primary use for the Z-word is as the Z-axis designator. The Z-word allows a decimal point.

B. The secondary use for the Z-word is to specify the hole-bottom position in a canned cycle command. Canned cycles are discussed in lesson 6.1.

A-Word

A. For machines that have a true rotary axis mounted parallel to the X-axis, letter address A is the A-axis designator. By parallel to the X-axis, we mean the centerline of the rotary axis is parallel to the X-axis. Most vertical machining centers that are equipped with a rotary table have the rotary table mounted in this fashion. A decimal point is allowed with the A-word when it is used as the A-axis designator. The A-word can be programmed to three places, meaning the rotary axis has 360,000 positions. A true rotary axis allows machining during rotary axis motion. Rotary axes are discussed lesson in 6.4.

B. For machines equipped with a one-degree indexer and when the indexer is mounted on the machine with its center parallel to the X-axis, the A-word is the indexer activator. You are not allowed to use a decimal point with the A-word when it is used to activate an indexer. A50 specifies a fifty degree index. With an indexer, machining can only occur after the indexer rotates (not during rotation). Indexers are discussed in lesson 6.4.

B-Word

A. For machines that have a rotary axis mounted parallel to the Y-axis, B is the B-axis designator. By parallel to the Y-axis, we mean the centerline of the rotary axis is parallel to the Y-axis. This is always the case when a rotary axis is mounted within the table of a horizontal machining center. A decimal point is allowed with the B-word when it is used as the B-axis designator. The B-word can be programmed to three places, meaning the rotary axis has 360,000 positions. A true rotary axis allows machining during rotary axis motion. Rotary axes are discussed in lesson 6.4.

B. For machines equipped with a one-degree indexer and when the indexer is mounted on the machine with its center parallel to the Y-axis, the B-word is the indexer activator. You are not allowed to use a decimal point with the B-word when it is used to activate an indexer. B50 specifies a fifty degree index. With an indexer, machining can only occur after the indexer rotates (not during rotation). Indexers are discussed in lesson 6.4.

C-Word

A. For machines that have a rotary axis mounted parallel to the Z-axis, C is the C-axis designator. By parallel to the Z-axis, we mean the centerline of the rotary axis is parallel to the Z-axis. This may be the case when a machining center has two rotary axes (it is a five-axis machining center). A decimal point is allowed with the C-word when it is used as the C-axis designator. The C-word can be programmed to three places, meaning the rotary axis has 360,000 positions. A true rotary axis allows machining during rotary axis motion. Rotary axes are discussed in lesson 6.4.

B. For machines equipped with a one-degree indexer and when the indexer is mounted on the machine with its center parallel to the Z-axis, the C-word is the indexer activator. You are not allowed to use a decimal point with the C-word when it is used to activate an indexer. C50 specifies a fifty degree index. With an

Key Concept 1: Know Your Machine From A Programmer's Viewpoint

indexer, machining can only occur after the indexer rotates (not during rotation). Indexers are discussed in lesson 6.4.

Note About Rotary Axis Designators and Indexer Activators

Not all machine tool builders adhere to the rotary axis and indexer activator naming conventions we've just introduced. What is more important than naming conventions, however, is that you know how the rotary axis or indexer is designated for the machines with which you must work. If your machining center is equipped with a rotary axis or one-degree indexer, consult your machine tool builder's programming manual to determine its designating letter address.

R-Word

A. The primary use for the R-word is to specify the radius of a circular move. The R-word allows a decimal point. Circular motion is discussed in lesson 3.1.

B. The secondary use for the R-word is to specify the rapid or retract plane for a canned cycle command. Canned cycles are discussed in lesson 6.1.

I, J, K-Words

A. I, J, and K-words can be used to specify the arc center point of a circular interpolated motion. While they are still effective, we strongly recommend that beginners concentrate on using the R-word to specify the arc size in a circular move because it is much easier. I, J and K allow a decimal point. Circular motion (including the use of directional vectors) is discussed in lesson 3.1.

B. The secondary function for I and J is with canned cycles to specify the move over amount at the bottom of a fine boring cycle (G76). I and J used in this fashion allow a decimal point. Canned cycles are discussed in lesson 6.1.

Q-Word

The Q-word is used with peck drilling canned cycles (G73 and G83) to specify peck depth. The Q-word allows a decimal point. Canned cycles are discussed in lesson 6.1.

P-Word

A. The P-word can be used to specify dwell time in milliseconds for a dwell command (G04). Dwell commands are used to make axis motion (for all axes) pause for a specified length of time. A time of three seconds is specified as P3000 (a decimal point is not allowed). Note the fixed format of the P-word. There are three places to the right of the automatically placed decimal point position. Other examples: P2500 is 2.5 seconds, P500 is 0.5 second, and P10000 is 10 seconds. Note that the X-word can also be used to specify the time for a dwell command–and since it allows a decimal point–most programmers prefer using it instead of the P-word. Again, a decimal point is not allowed with the P-word. The dwell command is discussed in lesson 6.3.

B. The secondary use for the P-word is with sub-programming to specify the subprogram program number of the program to be called. A decimal point is not allowed with the P-word. Sub-programming is discussed in lesson 6.2.

L-Word

A. The L-word is used with sub-programming to specify the number of executions for the subprogram. A decimal point is not allowed with the L-word. Subprograms are discussed in lesson 6.2.

B. The L-word can be used with canned to specify the number of holes to machine. A decimal point is not allowed with the L-word. Canned cycles are discussed in lesson 6.1.

F-Word

The F-word specifies feedrate–which is the motion rate for machining operations. It is used with straight line and circular motion commands (see G01, G02, and G03), along with any other interpolation types equipped with your machining centers. Feedrate is affected by the currently instated measurement system mode (inch or metric). Inch mode is specified with G20 and metric mode is specified with G21. With today's machining centers, feedrate can be specified in distance-per-minute–either inches-per-minute or millimeters-per-minute or in distance-per-revolution (inches-per-revolution or millimeters-per-revolution). The feedrate mode is selected with two G-codes, G94 for distance-per-minute feedrate mode and G95 for distance-per-revolution feedrate mode. The F-word allows a decimal point. A feedrate of 3-1/2 millimeters-per-minute in metric mode (or 3-1/2 inches per minute in inch mode) is specified as F3.5. Feedrate is introduced in lesson 1.1 and discussed in more detail in lessons 3.1 and 6.4.

There is another feedrate mode (G93) on some CNCs call inverse time feed. Here the feedrate is specified as 1/time (mins) or 1/time (secs). The feedrate is 1 divided by the time that all axes should arrive at the destination point. This is an old method to solve a problem of programming machines with complex axes configurations.

M-Word

An M-word specifies a miscellaneous function (also called a machine function). You can think of M-words as programmable on/off switches that control functions like coolant and spindle activation. For a list of many common M words, see the list at the end of this lesson. Note that machine tool manufacturers will select their own set of M-codes. While there are many standard M-codes, you must consult your machine tool builder's programming manual to find the exact list for a particular machine. A decimal point is not allowed with the M-word. Miscellaneous functions are introduced in lesson 1.1 and discussed numerous times throughout this text.

S-Word

The S-word specifies spindle speed. Most current CNCs allow you to specify spindle speed in one rpm increments. A spindle speed of 350 rpm is specified with S350. A decimal point is not allowed with the S-word. The spindle is activated with M-codes. M03 turns the spindle on in a forward direction. M04 turns the spindle on in a reverse direction. M05 turns the spindle off. Spindle control is introduced in lesson 1.1 and discussed in more detail during lessons 5.1 and 5.2.

T-Word

With most machining centers, the T-word specifies the ready position tool station. When a T-word is commanded, the automatic tool changer magazine will rotate, bringing the specified tool station to the ready or waiting position. This allows one tool to be cutting while the tool changer magazine is getting ready with the next. With these machines, an M06 word is used to command the tool change. The command T05 M06, for example, will first cause the machine to bring tool station 5 to the ready position–then make the tool change–placing tool 5 in the spindle. There are machining centers, however, with which the T-word by itself commands the entire tool change. A decimal point is not allowed with the T-word. Automatic tool changers are introduced in lesson 1.1 and discussed in more detail in lessons 5.1 and 5.2.

H-Word

The H-word specifies the tool length compensation offset number to be activated. In this offset, the setup person will store the cutting tool's tool length compensation geometry and/or wear values. A decimal point is not allowed with the H-word. Tool length compensation is discussed in lesson 4.2.

D-Word

The D-word specifies the cutter radius compensation offset number to activate. In this offset, the setup person stores the milling cutter's radius geometry and/or wear values. A decimal point is not allowed with the D-word. Cutter radius compensation is discussed in lesson 4.3.

EOB (end of block character)

EOB stands for end-of-block. Though it is not actually a CNC word, it is an important part of a CNC program. It is called a block terminator. On a CNC screen, it usually appears as a semicolon (;). When you type a program using a computer, it is automatically inserted at the end of each command when you press the Enter key (though it will not appear on the computer monitor). Though it is automatically inserted into your program when typing programs on a computer, it is not automatically entered if you type (or modify) programs through the keyboard and display of the CNC machining center. A special key on the control's keyboard labeled EOB must be pressed to insert the end-of-block character into the program. This must be done at the end of every command. Program structure is discussed in lessons 5.1 and 5.2.

/ (slash code)

This is called the **block delete** word (also called **optional block skip**). It is the slash character on your keyboard (under the question mark on most keyboards). It works in conjunction with an on/off switch on the control panel (labeled block delete or optional block skip). If the switch is on when the control reads the slash code, the control will ignore the words to the right of the slash code. If the switch is off, the control will execute the command in the normal manner. Block delete, including several applications for this feature, is discussed in lesson 6.3.

G-codes and M-codes

Here we list most of the G-codes and M-codes that can be used in programming, providing little more than the name for each word. Rest assured that the most often used G-codes and M-codes are discussed in detail in this course. You can find documentation for lesser used G-codes and M-codes in your CNC manufacturer's programming manual.

G-codes

As you know, G-codes specify preparatory functions. They prepare the machine for what is to come—in the current command—and possibly in up-coming commands. They set modes.

G-code Limitations:

Only one G-code from a G-code group should be included in a single program block or line. For example G20 and G21 select inch mode and metric mode respectively. They belong to the same G-code group (group 6 for the FANUC 0i-MD CNC-see table below). It is logical that you cannot select both inch mode and metric mode at the same time, so only one of them should be programmed in a single program block. If programmed, the last one specified in the block will be the active G-code.

With some older CNCs, only three G-codes are allowed per block. If you exceed this limitation, most CNCs will not generate an alarm. They will simply execute the last three G-codes in the block–ignoring any additional G-codes. For example, in the command

```
G90 G80 G40 G21 (Select absolute mode, cancel canned cycles,
   cancel cutter radius compensation, select metric mode)
```

Only the G80, G40, and G20 codes will be executed. The G90 code will be ignored. If needed in a program, these four G-codes must be broken into two commands.

Optional G-Codes

Some of the G-codes in the up-coming list are optional G-codes. It is impossible to tell whether a given option is included in your CNC or not by just looking at our list since machine tool builders include a package of options when they purchase controls from FANUC. When your company purchased the machining center from your machine tool builder, certain G-codes came with the machine. If there is any question as to whether your machining center has any a particular option G-code, you can perform a simple test at the machine to find out if it is available you (or you can call your builder to find out if the G-code was included).

To make the test for an option G-code, simply command the G-code in the MDI mode (techniques given in lesson 9.1). You need not even specify the correct format for the G-word. If you receive the alarm Unusable G-code or G-code not available, your machine does not have the G-code. If you receive no alarm or if the alarm is related to the format of the G-code, the option G-code should be available for you to use.

What Does Initialized Mean?

An initialized word is one that is automatically instated when the machine's power is turned on.

What Does Modal Mean?

Most G-codes are modal, meaning once they are invoked, they remain in effect until they are changed or cancelled. This means you do not have to keep repeating modal G-codes in every command. There are some G-codes that are non-modal (also referred to as *one-shot G-codes*). Non-modal G-codes only have an effect on the command in which they are included.

Lesson 1.7: Introduction To Programming Words

Commonly Used G-Codes

As you look at this list, it may seem a little intimidating. There are a lot of G-codes. Here are the most popular G-codes (about thirty), along with where they are discussed in this text.

G00, G01, G02, G03: motion types–discussed in lesson 3.1

G04: dwell command–discussed in lesson 6.4

G20, G21: inch and metric mode–discussed in lessons 1.1, 1.4, and 5.1

G28: zero return command–introduced in lesson 1.5 and discussed in more detail in lessons 5.1 and 5.2

G40, G41, G42–cutter radius compensation–discussed in lesson 4.3

G43: tool length compensation–discussed in lesson 4.2

G54: invoke workpiece coordinate system setting offset number one–introduced in lesson 1.6, discussed in more detail in lesson 4.4

G55 though G59: invoke other workpiece coordinate system setting offsets–discussed in lesson 4.4

G73 through G89: hole-machining canned cycles–discussed in lesson 6.1

G90, G91: absolute and incremental positioning modes–discussed in lesson 1.4

G98, G99: initial and rapid plane used with hole-machining canned cycles–discussed in lesson 6.1

0*i*-F G-codes

G-code	Group	Function
G00	01	Positioning (rapid traverse)
G01		Linear interpolation (cutting feed)
G02		Circular interpolation CW or helical interpolation CW
G03		Circular interpolation CCW or helical interpolation CCW
G04	00	Dwell
G05.1		AI contour control / Nano smoothing / Smooth interpolation
G05.4		HRV3 on/off
G07.1 (G107)		Cylindrical interpolation
G09		Exact stop
G10		Programmable data input
G11		Programmable data input mode cancel
G15	17	Polar coordinates command cancel
G16		Polar coordinates command
G17	02	XpYp plane selection
G18		ZpXp plane selection
G19		YpZp plane selection
G20	06	Input in inch
G21		Input in mm
G22	04	Stored stroke check function on
G23		Stored stroke check function off
G27	00	Reference position return check
G28		Automatic return to reference position
G29		Movement from reference position
G30		2nd, 3rd and 4th reference position return
G31		Skip function
G33	01	Threading
G37	00	Automatic tool length measurement
G38		Tool radius/tool nose radius compensation : preserve vector
G39		Tool radius/tool nose radius compensation : corner circular interpolation
G40	07	Tool radius/tool nose radius compensation : cancel
G41		Tool radius/tool nose radius compensation : left
G42		Tool radius/tool nose radius compensation : right

Key Concept 1: Know Your Machine From A Programmer's Viewpoint

G-code	Group	Function
G40.1	19	Normal direction control cancel mode
G41.1		Normal direction control on : left
G42.1		Normal direction control on : right
G43	08	Tool length compensation +
G44		Tool length compensation -
G45	00	Tool offset : increase
G46		Tool offset : decrease
G47		Tool offset : double increase
G48		Tool offset : double decrease
G49	08	Tool length compensation cancel
G50	11	Scaling cancel
G51		Scaling
G50.1	22	Programmable mirror image cancel
G51.1		Programmable mirror image
G50.9	00	Auxiliary function output in moving axis
G52	00	Local coordinate system setting
G53		Machine coordinate system setting
G53.1		Tool axis direction control
G53.6		Tool center point retention type tool axis direction control
G54	14	Workpiece coordinate system 1 selection
G54.1		Additional workpiece coordinate system selection
G55		Workpiece coordinate system 2 selection
G56		Workpiece coordinate system 3 selection
G57		Workpiece coordinate system 4 selection
G58		Workpiece coordinate system 5 selection
G59		Workpiece coordinate system 6 selection
G60	00	Single direction positioning
G61	15	Exact stop mode
G62		Automatic corner override
G63		Tapping mode
G64		Cutting mode
G65	00	Macro call
G66	12	Macro modal call
G67		Macro modal call cancel
G68	16	Coordinate system rotation start or 3-dimensional coordinate conversion mode on
G68.2		Tilted working plane command
G68.3		Tilted working plane command by tool axis direction
G68.4		Tilted working plane command (incremental multi-command)
G69		Coordinate system rotation cancel or 3-dimensional coordinate conversion mode off
G73	09	Peck drilling cycle
G74		Left-handed tapping cycle
G75	01	Plunge grinding cycle
G76	09	Fine boring cycle
G80	09	Canned cycle cancel
		Electronic gear box : synchronization cancellation
G80.4	34	Electronic gear box: synchronization cancellation
G81.4		Electronic gear box: synchronization start
G81	09	Drilling cycle or spot boring cycle
		Electronic gear box : synchronization start
G81.1	00	Chopping

G-code	Group	Function
G82	09	Drilling cycle or counter boring cycle
G83		Peck drilling cycle
G84		Tapping cycle
G84.2		Rigid tapping cycle (FS15 format)
G84.3		Left-handed rigid tapping cycle (FS15 format)
G85		Boring cycle
G86		Boring cycle
G87		Back boring cycle
G88		Boring cycle
G89		Boring cycle
G90	03	Absolute programming
G91		Incremental programming
G91.1	00	Checking the maximum incremental amount specified
G92		Setting for workpiece coordinate system or clamp at maximum spindle speed
G92.1		Workpiece coordinate system preset
G93	05	Inverse time feed
G94		Feed per minute
G95		Feed per revolution
G96	13	Constant surface speed control
G97		Constant surface speed control cancel
G98	10	Canned cycle : return to initial level
G99		Canned cycle : return to R point level

M-codes

Common M-codes Used on a CNC Machining Center

Machine tool builders vary with regard to the M-codes they provide–so this is just a partial list. You must reference your machine tool builder's programming manual for a complete list of the M-codes available for a particular CNC machining center.

Some CNC models (like many supplied by FANUC) only allow one M-code per block. In this case, if you include more than one M-code, it's hard to predict what will happen for all machines. Some machines will stop executing the program while others will generate an alarm. Yet others will continue executing the program, ignoring all but the last M-code in the command.

M-code	Description
M00	Program stop
M01	Optional stop
M02	End of program (does not rewind memory for most machines–use M30)
M03	Spindle on in the forward direction
M04	Spindle on in the reverse direction
M05	Spindle stop
M06	Tool change command
M07	Mist coolant on (option)
M08	Flood coolant on
M09	Coolant off
M30	End of program (rewinds memory)
M98	Subprogram call
M99	End of subprogram

Key Concept 1: Know Your Machine From A Programmer's Viewpoint

Other M Codes for Your Machine (found in your machine tool builder's manuals)

____ _____
____ _____
____ _____
____ _____
____ _____
____ _____
____ _____
____ _____

Key Points for Lesson 1.7:
- ✓ While there are many CNC words available to programmers, there are only about fifty words that are used on a regular basis. Look at learning CNC programming as like learning a foreign language that has only fifty words.
- ✓ The letter addresses for many CNC word types are easy to associate with their usage (F for feedrate, T for tool, S for spindle speed, etc.). But other word types are more difficult to remember.
- ✓ You've been exposed to all of the word types used with CNC machining center programming in this lesson–as well as where (in this course) you can find more information about the most often-used words.
- ✓ There are only about thirty G-words used on a regular basis.
- ✓ Only one M-code is allowed per command with some CNC controls (including many FANUC models).
- ✓ You must reference your machine tool builders programming manual to find the complete list of M-codes for your machining center/s.

Key Concept 2: You Must Prepare to Write Programs

While this Key Concept does not involve any programming words or commands, it is among the most important of the Key Concepts. The better prepared you are to write a CNC program, the easier it will be to develop a workable program.

Key Concept Number Two is made up of but one lesson:

2.1: Preparation for Programming

This Key Concept can be applied to any CNC related task, including setup and operation. Truly, the better prepared you are to perform any task, the easier it will be to correctly complete the task. When it comes to setting up a machine, for example, if you have all of the needed components to make the setup at hand (fixtures, cutting tools, cutting tool components, program, documentation, etc.), making the setup will be much easier and will go faster. So gathering needed components is always a great preparation step to perform.

The same goes for completing a production run. If the CNC operator has all needed components (raw material, inserts for dull tool replacement, a place to store completed workpieces, gauging tools, etc.), they will be able to smoothly complete the production run.

In this Key Concept, however, we're going to limit our discussion to preparation steps that you can perform to get ready to write a CNC program. Remember, the actual task of writing a CNC program is only a part of what you must do. Some things must be done before you are adequately prepared to write the program.

Preparation for programming is especially important for entry-level programmers. For the first few programs you write, you will have trouble enough remembering the various CNC words–remembering how to structure the program correctly–and in general–you'll have trouble getting familiar with the entire programming process. The task of programming is infinitely more complicated if you are not truly prepared to write the program in the first place.

Preparation and Time

Without adequate preparation, writing a CNC program can be compared to working on a jigsaw puzzle. A person doing the puzzle has no idea where each individual piece will eventually fit. The person makes a guess and attempts to fit the pieces together. Since the person has no idea as to whether pieces will fit together, it is next to impossible to predict how long it will take to finish the puzzle.

In similar fashion, if you attempt to write a CNC program without adequate preparation, you will have a tendency to piece-meal the program together in much the same way as a person doing a jigsaw puzzle. You will not be sure that anything will work until it is tried. The program may be half finished before it becomes obvious that something is seriously wrong. Worse, the program may be completed and being verified on the CNC machining center before a critical error is found.

CNC machine time is much more expensive than your time. There is no excuse for wasting precious machine time for something as avoidable as a lack of preparation.

You can also liken the preparation that is required to write a CNC program to the preparation needed for giving a speech. The better prepared the speaker, the easier it will be to make the presentation, and the more effective the speech. Truly, the speaker must think through the entire presentation (probably several times) before the speech can be presented. Similarly, the CNC programmer must think through and visualize the entire CNC process, setup, and program before the program can be written.

Key Concept 2: You Must Prepare To Write Programs

With adequate preparation, writing the program will be much easier. Most experienced programmers will agree that the actual task of writing a CNC program is the easy part of the programming process. The real work is done in the preparation stages. If preparation is done properly, writing the program will be a simple matter of translating what you want the machine to do (from English) into the language a CNC machine can understand and execute.

Though preparation is so very important, it is amazing to see how many so-called expert programmers muddle through the writing of a program without adequate preparation. While an experienced person may be able to write workable programs for simple applications with minimal preparation (and even then they only gain this ability through trial-and-error practice), even expert programmers have problems with more complex programs.

To think that you are saving time by not preparing to write a CNC program is a terrible mistake. In reality, you usually add time to programming and program-verification time if you do not prepare adequately. The period of time saved by not preparing will be quickly lost when you consider the problems caused by the lack of preparation. The waste is compounded if the added time caused by lack-of-preparation results in wasted machine time.

Preparation and Safety

Wasted time is but one of the symptoms of poor preparation. Indeed, it may be the least severe one. Poor preparation will often result in all kinds of mistakes.

A CNC machining center will follow a CNC program's instructions exactly. While the CNC may display an alarm if it cannot recognize a given command, it will give absolutely no special consideration to motion mistakes. Indeed, a CNC cannot detect most motion mistakes. The level of problem encountered because of motion mistakes ranges from minor to catastrophic.

Minor motion mistakes usually do not result in any damage to the machine or tooling, and the operator is not exposed to a dangerous situation. However, the workpiece will not be machined correctly. For example, say you intend to drill a hole at 120 millimeters-per-minute with a feedrate word of F120.0 in the drilling command. But, you place the decimal point in the wrong location. You specify a feedrate of F12.0 instead of F120.0. In this case, the control is being told to run the drill at a much slower feedrate than you intend. No damage to the tool or machine will result–but machining time will be much longer than it should be.

This mistake will be more serious, of course, if the programmer incorrectly specifies a feedrate of F1200.0 instead of F120.0. This time the tool will probably break–and the workpiece and/or the machine may be damaged. If the machine does not have a fully enclosed work area, flying debris could injure the operator or someone else.

Catastrophic mistakes can result in damage to the machine and injury to the operator. For example, say you intend to position a cutting tool at the machine's rapid rate (the machine's fastest motion rate–which for some machines is well over 25,000 millimeters per minute). The correct position for your command is Z2.0, which is 2.0-millimeters above the work surface. But you make a mistake and include a minus sign for this motion word (Z-2.0 instead of Z2.0). The tool is being told to crash into the part at the rapid rate. Depending on what kind of tooling is being used, this will, at the very least, cause the tool to break. Worse, the workpiece could be pushed out of the setup. Possibly, if the setup is very sturdy and the tool is very rigid, damage to the machine's way system and/or axis drive system could result. If the tool breaks and parts fly out of the work area, the operator or someone else could be injured.

We're not saying these things to scare you. There are program verification procedures that, if followed, almost guarantee that no crash will occur, regardless of how severe the positioning mistake. (These verification techniques are shown in lesson 10.1). However, as the programmer of a CNC machining center, you must constantly recognize the potential for dangerous situations. Due to this constant potential for personal injury and/or damage to the machine tool, everyone involved with CNC equipment must treat all machine tools with great respect. You must be constantly aware of this possible danger and do everything you can to avoid dangerous situations. Being adequately prepared to write a CNC program is the single most important thing you can do to achieve this end.

We cannot overstress the importance of preparation. Just as the well-prepared speaker is less apt to make mistakes during his or her presentation, so will the well-prepared CNC programmer be less apt to make mistakes while writing a CNC program.

Typical Mistakes

It helps to know the mistakes a beginning programmer is prone to make. Knowing these tendencies should help you avoid them. There are actually four categories of mistakes that are common.

Syntax Mistakes

Syntax mistakes are mistakes related to the basic structure of your program. CNC programming is extremely structured. If a command is encountered that the CNC does not recognize or cannot process, it will generate an alarm. Since a beginning programmer is still getting acquainted with the various CNC words and commands, they are very prone to making syntax mistakes.

For example, a beginner might intend to give the command G54 to invoke workpiece coordinate system setting offset number one. But by mistake the command G5 is given. Since most CNC controls do not have a function commanded by G5, a syntax alarm will be generated. That is, the machine will go into alarm state, halting the program's execution and placing an alarm message on the display screen to help you diagnose the alarm. For this reason, syntax mistakes are usually quite easy to find and diagnose since the message will tell you the reason for the alarm. Other examples of syntax mistakes include:

```
N055 G02 X30.0 Y20.0 F140.0 (Missing arc radius designator [R]
for circular move)

N055 G43 P01 Z2.0 (Incorrect offset designator for tool length
compensation [P01 should be H01])

N055 G98 P1000 (Incorrect call to subprogram [G98 should be
M98])
```

Motion Mistakes

Motion mistakes can be harder to find and diagnose since no alarm will be detected unless the move causes an axis over travel. They can also be much more severe since the control will perform the *incorrect* motions, regardless of how severe the consequences. Minor motion mistakes may not cause any damage, but can be very difficult to catch–even when cautious program verification procedures are followed. For example, if a small mistake of less than 0.5-millimeters is made in a cutting command, everything will appear to be just fine throughout all program verification procedures. Yet when the first workpiece is machined, it will not pass inspection. More severe motion mistakes can cause more severe damage and personal injury, but they are usually easier to find during program verification. An incorrect positioning movement of more than about 10.0-millimeters is usually pretty easy to spot. Motion mistakes are commonly caused by miscalculations (incorrect arithmetic) or transposition of numbers. For example, the programmer may have meant to specify X125.0, but instead specified X215.0. As stated, the control will interpret and execute the transposed word, regardless of how severe the consequences.

Mistakes of Omission

This is the kind of mistake a beginner is most prone to making. Beginners tend to forget things. They forget to turn the spindle on. They forget to turn the coolant on. They forget to include a feedrate word. They forget to drill a hole before tapping it. They forget to include a decimal point in CNC words that require it. Unfortunately, only experience will help you spot your own tendencies when it comes to mistakes of omission. As you write and verify your first few programs, you'll likely find things you have forgotten to do.

Process Mistakes

A CNC programmer must have a good understanding of basic machining practice. Since machining centers are multi-tool machines, you must be able to develop a sequence of machining operations by which a workpiece can be properly machined (the process). Even a poorly written CNC program can eventually be made to work if the sequence of machining operations is good. However, even a perfect program (one that does exactly what the programmer intends) will fail if the sequence of machining operations is bad. Beginning programmers who may be a little weak in their machining skills should seek the help of experienced machinists when developing processes. Here is an example of a poor process. See if you can spot the two mistakes before reading on to the next paragraph.

1) Rough face mill the top surface of the workpiece (100.0-mm rough face mill)

2) Finish face mill the top surface of the workpiece 100.0-mm finish face mill)

3) Rough mill the ends of the workpiece (25.0-mm rough end mill)

Key Concept 2: You Must Prepare To Write Programs

    4) Finish mill the ends of the workpiece (25.0-mm finish end mill)

    5) Drill (3) 6.0-mm holes (6.0-mm drill)

While processing is a <u>very</u> subjective topic, most experienced machinists would agree that this process breaks an important rule of basic machining practice: *"You should rough everything before you finish anything."* Prior to finish milling the top surface of the workpiece, the rough milling of the workpiece ends should be done. Also, most machinists would say you should spot drill or center drill all holes under about 15.0-mm in diameter.

As with most powerful machining operations, it is likely that the workpiece will shift a small amount during the roughing operations in this process. If it shifts after the top is finished, during the rough milling of the ends, the top and ends will not be perpendicular to one another. Even if the program is written with perfect syntax, using this sequence of machining operations, the workpiece may not pass inspection.

Lesson 2.1–Preparation for Programming

Any complex project can be simplified by breaking it down into small pieces. This can make seemingly insurmountable tasks much easier to handle. CNC machining center programming is no exception. Learning how to break up this complex task will be the primary focus of this lesson.

Objectives

After completing this lesson, students should be able to:

- ✓ Describe value of preparing for programming
- ✓ Describe preparing the sequence of machining operations
- ✓ Describe preparing the cutting conditions
- ✓ Describe using math to prepare the program coordinates
- ✓ Describe preparing the cutting tools that you plan to use in the program
- ✓ Describe preparing the work holding setup that you plan to use in the program
- ✓ Describe and complete the documentation required to prepare to write a program including marking-up the workpiece engineering drawing, entering information in a machining process planning form, a coordinate sheet, and a universal setup sheet

Introduction

Good preparation makes programming easier, safer, and less error-prone. We will introduce some specific steps you can perform to prepare to write CNC programs.

Plan the Sequence of Machining Operations

Process sheets, also called routing sheets, are used by most manufacturing companies to specify the sequence of machining operations that must be performed on a workpiece during the manufacturing process. The person who actually prepares the process sheet must have a good understanding of machining practice, and must be well acquainted with the various machine tools the company owns. This person determines the best way to produce the workpiece in the most efficient and inexpensive possible way, given the company's available resources.

In most manufacturing companies, this involves routing the workpiece through a series of different machine tools and processes. Each machine tool along the way will perform only those operations the process planner intends, as specified on the routing sheet. This commonly means that non-CNC machine tools are needed to complete a given workpiece. For example, the simple square workpiece shown in the practice exercise at the end of lesson 1.6 might have the following routing sheet.

Op. #	Operation	Machine
10	Procure material	Vender ID #12322
20	Cut bar stock to 80 mm long	Cut off saw
30	Clean and de-burr	Cleaning tanks
40	Mill contour, drill (9) holes	CNC machining center
50	Clean and de-burr	Cleaning tanks
60	Plate with nickel	Finishing tanks

Note that only operation 40 of this process requires a CNC machine tool. When a CNC machine is involved in the process plan, often the CNC machine will be required to perform several machining operations on the workpiece (as is the case in our example). All true CNC machining centers are multi-tool machines, meaning

Key Concept 2: You Must Prepare To Write Programs

several tools can be used during one program. In some companies, the process sheet will clearly specify the order of machining operations that must be performed by the CNC machine. However, the vast majority of companies do not get so specific with their routing sheets. Instead, the sequence of machining operations to be performed on the CNC machine is left completely to the CNC programmer. If the programmer must develop the sequence of machining operations, they must possess a good knowledge of basic machining practice.

In any event, the step-by-step sequence of machining operations required to produce a workpiece on the CNC machine must be developed before the CNC program can be written. With a simple process, an experienced programmer may elect to develop the process as the program is being written. While some experienced programmers have the ability to do this, beginning programmers will initially find it necessary to develop a sequence of machining operations for even the most basic parts.

The sequence of machine operations used to machine the workpiece will have a dramatic impact on the success of the program. If the process is correct, the workpiece will be machined efficiently and pass inspection. If the process is poor, the workpiece will not be machined correctly no matter how well the program is written. If you are new to developing a sequence of machining operations, you should seek help whenever there is a question as to whether your planned machining process will work.

Developing a sequence of machining operations before the program is written will serve several purposes:

Check the process for errors - You will be forced to think through and visualize the entire process before the first CNC command is written. You will have the opportunity to spot a problem with the process that will be difficult to spot and/or repair later.

Make you concentrate on your machining skills separate from your CNC programming skills - While developing the sequence of machining operations, you concentrate just on machining practices in order to develop a workable process. Your mind is occupied only with the task at hand. Later, when programming, you concentrate just on programming skills, translating your sequence of machining operations into a language that the CNC machine can understand.

Correct mistakes before sending the program out to the machine - If you spot an error in your thinking during the preparation stage, you will be able correct the mistake before sending the program out to the machine. Indeed, you should make every effort to spot mistakes before you write the program. There is never an excuse for wasting machine time for something as basic as inadequate preparation.

Provide you with documentation – You will be able to use your sequence of operations documentation as a checklist while programming to ensure that you don't forget an operation. And anyone that works on your program in the future will be able to quickly familiarize themselves with your process using this documentation. The figure below shows an example planning form that you can use to develop and document your sequence of machining operations.

Part no:	Date:				
Part name:	Programmer:	**Machining Process Planning Form**			
Machine:	Material:				
Seq.	Operation Description	Tool	Station	Speed	Feed	Note
1						
2						
3						
4						
5						
6						
7						

Figure 2.1: Example sequence of operations form

Notice how this form guides you to document the sequence of machining operations that your program will use. Months or years after a CNC program is developed, there may be a need to revise it. If the person doing the revision can review a completed machining process planning form for the workpiece, it will be much easier to make the necessary changes.

The last reason we will give to plan the process first is to simply help you remember the operations to perform during programming. Remember, beginners tend to make mistakes of omission. You will have enough to think about when it comes to remembering the various commands needed in the program. The process planning form is a step-by-step set of instructions to machine the workpiece. It can be used as a check-list. Without this form, you will be prone to omitting important operations from the CNC program.

Develop the Cutting Conditions

Before a program can be written, cutting conditions must be determined for all the cutting tools used in the program. Each cutting tool will need a spindle speed in revolutions-per-minute (rpm) and a feedrate in millimeters-per-minute (mmpm) or inches-per-minute (ipm). For roughing tools, like rough milling cutters and rough boring bars, you must also determine a depth-of-cut for the tool–as well as how much finishing stock you will leave for the finishing tool. You must also determine whether or not to use coolant and, if so, what kind–flood, mist, through-the-tool, or high pressure, based upon the workpiece and cutting tool materials.

It is important to come up with the cutting conditions needed in the program while developing the sequence of machining operations–before you write the program. This will keep you from having to break out of your train of thought while programming. Using a machining process planning form like the one above, you will be able to document the speeds and feeds needed for programming.

Here are a few terms used when calculating cutting conditions:

smpm (surface meters per minute) or **sfm** (surface-feet-per-minute) is the speed used for the machining operation–this is the linear amount of material that will pass by the cutting tool's cutting edges during one minute. You find the recommended speed in reference books related to cutting conditions like tooling manufacturers' technical books or the cutting tool manufacturer's catalog.

rpm (revolutions-per-minute)–this is the number of spindle revolutions that occur in one minute.

mmpt (millimeters per tooth or **ipt** (inches-per-tooth)–for milling cutters, this is the distance a cutting tool will move for each tooth/insert/flute of the cutting tool. Like smpm/sfm, this value is found in the tooling manufacturer's reference handbooks related to cutting conditions.

mmpr (millimeters per revolution feedrate) or **ipr** (inches-per-revolution feedrate)–for any cutting tool, this is the distance a cutting tool will move during one revolution of the tool. For milling cutters, this value is determined by multiplying the number of flutes/teeth/inserts times the mmpt/ipt value. For most other cutting tools, this value can be found in the tooling manufacturer's reference handbooks related to cutting conditions.

mmpm (millimeters per minute feedrate) or **ipm** (inches per minute feedrate)–this is the distance the cutting tool moves during one minute.

CNC machining centers require that you specify spindle speed in rpm. Feedrate is specified in either feed-per-minute (ipm or mmpm) or feed-per-revolution (ipr or mmpr), though older machine may only allow feed-per-minute feedrate specification.

Here are a few formulae used when calculating cutting conditions:

rpm = 318.3 * smpm / tool diameter (in millimeters)   **rpm** = 3.82 * sfm / tool diameter (in inches)

mmpr = mmpt * number of teeth   **ipr** = ipt * number of teeth

mmpm = mmpr * rpm   **ipm** = ipr * rpm

pitch (for inch mode tapping) = 1 / number of threads per inch (for metric threads, pitch is specified in the thread designation)

The data provided by cutting tool manufacturers typically includes the cutting speed (in surface-feet-per-minute) and feedrate (in either inch-per-revolution or inch-per-tooth). This information is based upon the cutting tool material (high-speed steel, carbide, ceramic, etc.) and the workpiece material (mild steel, medium carbon steel, high carbon steel, stainless steel, aluminum, etc.). When appropriate, cutting tool manufacturers will also specify whether or not you should use coolant–as well as the recommended depth of

Key Concept 2: You Must Prepare To Write Programs

cut for a roughing tool. In some cases, they will even provide recommendations about how the cutting tool should move as it machines the workpiece.

For machining center programs, you must of course, calculate the speed in rpm and per-minute feedrate. For rpm, you must know the recommended speed in surface meters per minute (metric mode) or surface feet per minute (inch mode) and the cutting tool diameter. For the per-minute feedrate, you must know the rpm and millimeters-per-revolution (metric mode) or inches-per-revolution feedrate (inch mode).

An Example

Say you must machine a mild steel workpiece with this machining process:

Part no:	Date:				
Part name: exercise part	Programmer: your name		**Machining Process Planning Form**		
Machine: machining center	Material: mild steel				
Seq.	Operation Description	Tool	Station	Speed	Feed	Note
1	Rough face mill the top surface	100-mm face mill				
2	Rough mill the two ends	25.0-mm end mill				
3	Finish face mill the top surface	100-mm face mill				
4	Finish mill the two ends	25.0-mm end mill				
5	Center drill 5 holes	#3 center drill				
6	Drill (2) 6.0 mm holes	6.0-mm drill				
7	Drill (2) 13.0 mm holes	13.0-mm drill				

Figure 2.2: Example Machining Process Planning Form

You must first determine some important information about your cutting tools, including what they are made of (cutting edge material) and how many flutes, inserts, or teeth they have. We also have to look up the manufacturers recommended feeds and speeds. Here is an example of what you may find:

Tool	inserts/flutes	Material	Mild Steel
100-mm face mill	8 inserts	Carbide	Rough: 130 smpm / 0.1 mmpt Finish: 160 smpm / 0.063 mmpt
25.0-mm end mill	4 flutes	hss	Rough: 25 mmpm / 0.13 mmpt Finish: 30 smpm / 0.1 mmpt
#3 center drill	2 flutes	hss	25 smpm / 0.1 mmpr
6.0-mm drill	2 flutes	hss	22 smpm / 0.1 mmpr
13.0-mm drill	2 flutes	hss	22 smpm / 0.18 mmpr

We can now calculate the feeds and speeds for each tool and operation:

100-mm face mill roughing:
  spindle speed (rpm) = 318.3 x 130 (smpm) / 100 (mm) = 414 rpm
  feedrate (ipm) = 414 (rpm) * 8 (teeth) * 0.1 (mmpt) = 331.2 mmpm

100-mm face mill finishing:
  spindle speed (rpm) = 318.3 x 160 (smpm) / 100 (mm) = 509 rpm
  feedrate (ipm) = 509 (rpm) * 8 (teeth) * 0.063 (mmpt) = 256.5 mmpm

25.0-mm end mill roughing:
  spindle speed (rpm) = 318.3 x 25 (smpm) / 25 (mm) = 318 rpm
  feedrate (ipm) = 318 (rpm) * 4 (teeth) * 0.13 (mmpt) = 165.6 mmpm

25.0-mm end mill finishing:
   spindle speed (rpm) = 318.3 x 30 (smpm) / 25 (mm) = 382 rpm
   feedrate (ipm) = 382 (rpm) * 4 (teeth) * 0.1 (mmpt) = 152.8 mmpm

#3 Center drill:
   spindle speed (rpm) = 318.3 x 25 (smpm) / 4.75 (mm) = 1675 rpm (4.75 is largest diameter of #3)
   feedrate (ipm) = 1675 (rpm) * 0.1 (mmpr) = 167.5 mmpm

6.0-mm drill:
   spindle speed (rpm) = 318.3 x 22 (smpm) / 6.0 (mm) = 1167 rpm
   feedrate (ipm) = 1167 (rpm) * 0.1 (mmpr) = 116.7 mmpm

13.0-mm drill:
   spindle speed (rpm) = 318.3 x 22 (smpm) / 13.0 (mm) = 537 rpm
   feedrate (ipm) = 537 (rpm) * 0.18 (mmpr) = 96.6 mmpm

These values can now be entered in the process planning form:

Part no:	Date:				
Part name: exercise part	Programmer: your name		**Machining Process Planning Form**		
Machine: machining center	Material: mild steel				
Seq.	Operation Description	Tool	Station	Speed	Feed	Note
1	Rough face mill the top surface	100-mm face mill		414	331.2	
2	Rough mill the two ends	25.0-mm end mill		509	256.5	
3	Finish face mill the top surface	100-mm face mill		318	165.6	
4	Finish mill the two ends	25.0-mm end mill		382	152.8	
5	Center drill 5 holes	#3 center drill		1675	167.5	
6	Drill (2) 6.0 mm holes	6.0-mm drill		1167	116.7	
7	Drill (2) 13.0 mm holes	13.0-mm drill		537	96.6	

Figure 2.3: Example Machining Process Planning Form with Feeds and Speeds

Cutting Conditions Can Be Subjective

Many programmers, indeed many machinists, choose feeds and speeds using a seat-of-their-pants approach. They do so by watching and listening to the cutting operation—and manually overriding the programmed spindle speed and feedrate until the machining operation looks right. Admittedly, expert machinists can do this pretty well, and it's hard to argue with success, but the majority of operators using these techniques don't even come close to efficient cutting conditions. Instead, they tend to be well under the cutting tool manufacturer's recommendations. While many factors determine how quickly a cutting tool can machine (like rigidity of the setup, length of the cutting tool, and sharpness of the cutting tool), we urge beginners to at least *start with* the recommendations made by the manufacturer of the cutting tools they use.

Key Concept 2: You Must Prepare To Write Programs

Practice calculating speeds and feeds (metric mode)

Using the formulae shown above, calculate the speed (in rpm) and feedrate (in mmpm) for these tools based upon the recommendations in smpm and mmpt/mmpr.

No.	Tool description	Speed	Feedrate	Speed in rpm	Feed in mmpm
1)	75-mm rough face mill (6 carbide inserts)	150 smpm	0.12 mmpt	_____	_____
2)	75-mm finish face mill (6 carbide inserts)	175 smpm	0.1 mmpt	_____	_____
3)	20-mm hss rough end mill (4 flutes)	25 smpm	0.15 mmpt	_____	_____
4)	20-mm hss finish end mill (4 flutes)	30 smpm	0.11 mmpt	_____	_____
5)	10.0 mm high speed steel twist drill	25 smpm	0.12 mmpr	_____	_____
6)	16.0 mm high speed steel twist drill	25 smpm	0.18 mmpr	_____	_____
7)	22.0 mm high speed steel twist drill	25 smpm	0.25 mmpr	_____	_____

Practice calculating speeds and feeds (inch mode)

Using the formulae shown above, calculate the speed (in rpm) and feedrate (in ipm) for these tools based upon the recommendations in sfm and ipt/ipr.

No.	Tool description	Speed	Feedrate	Speed in rpm	Feedrate in ipm
1)	3.0" rough face mill (6 carbide inserts)	450 sfm	0.005 ipt	_____	_____
2)	3.0" finish face mill (6 carbide inserts)	525 sfm	0.004 ipt	_____	_____
3)	0.75" hss rough end mill (4 flutes)	70 sfm	0.006 ipt	_____	_____
4)	0.75" hss finish end mill (4 flutes)	90 sfm	0.0045 ipt	_____	_____
5)	0.375" high speed steel twist drill	75 sfm	0.005 ipr	_____	_____
6)	0.625" high speed steel twist drill	75 sfm	0.007 ipr	_____	_____
7)	0.875" high speed steel twist drill	75 sfm	0.010 ipr	_____	_____

Answers (metric mode):
1: 637 rpm, 458.64 mmpm
2: 743 rpm, 445.8 mmpm
3: 398 rpm, 238.8 mmpm
4: 477 rpm, 209.88 mmpm
5: 796 rpm, 95.52 mmpm
6: 497 rpm, 89.46 mmpm
7: 362 rpm, 90.5 mmpm

Answers (inch mode):
1: 573 rpm, 17.19 ipm
2: 668 rpm, 16.032 ipm
3: 356 rpm, 8.544 ipm
4: 458 rpm, 8.244 ipm
5: 764 rpm, 3.82 ipm
6: 458 rpm, 3.206 ipm
7: 327 rpm, 3.27 ipm

Do the Math and Mark-up the Print

As stated in the Preface of this text, the word numerical in computer numerical control implies a strong emphasis on numbers and math. Most college curriculums related to CNC do require a strong math background. However, most forms of CNC equipment require less math than you might think. Believe it or not, many CNC machining center programs can be completely prepared solely with simple addition and subtraction. A basic knowledge of right angle trigonometry is also helpful, but not always mandatory.

While there are times when a manual programmer must apply trigonometry, there are many reference books that give all the formulae related to right-angle trigonometry in a very simple format. This makes it relatively easy to solve trig problems, even for a person who knows little about trigonometry. The formulae shown in the figure below, for example, offer the solutions you will need to solve almost all trig problems related to CNC programming. While a serious CNC programmer will eventually want to thoroughly learn trigonometry, it may not be mandatory at the start of your programming career.

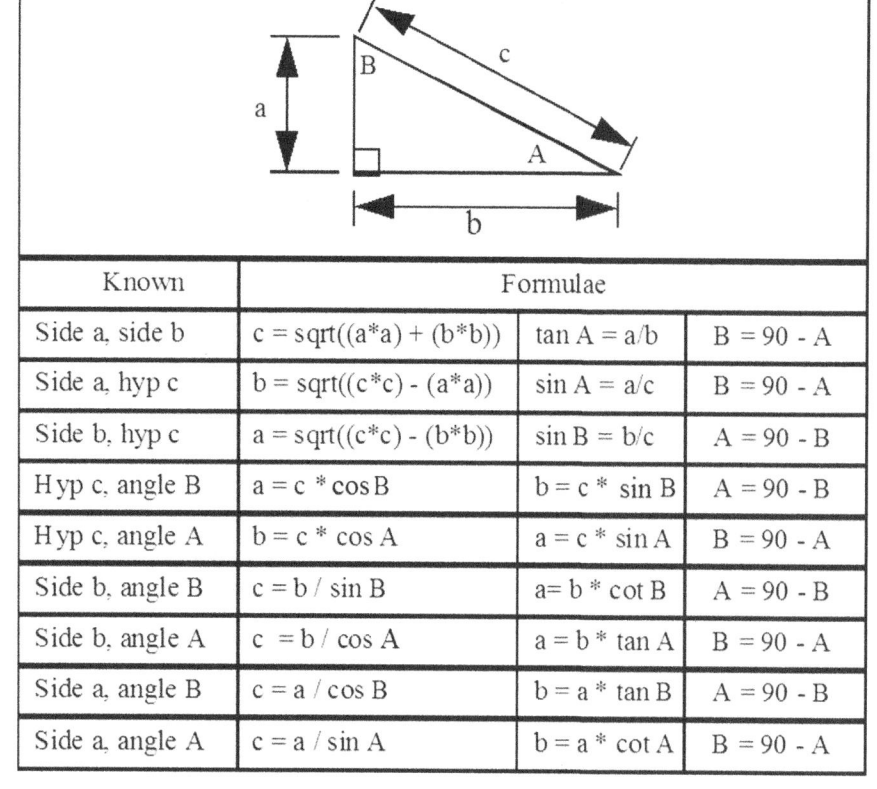

Known	Formulae		
Side a, side b	c = sqrt((a*a) + (b*b))	tan A = a/b	B = 90 - A
Side a, hyp c	b = sqrt((c*c) - (a*a))	sin A = a/c	B = 90 - A
Side b, hyp c	a = sqrt((c*c) - (b*b))	sin B = b/c	A = 90 - B
Hyp c, angle B	a = c * cos B	b = c * sin B	A = 90 - B
Hyp c, angle A	b = c * cos A	a = c * sin A	B = 90 - A
Side b, angle B	c = b / sin B	a = b * cot B	A = 90 - B
Side b, angle A	c = b / cos A	a = b * tan A	B = 90 - A
Side a, angle B	c = a / cos B	b = a * tan B	A = 90 - B
Side a, angle A	c = a / sin A	b = a * cot A	B = 90 - A

Figure 2.4: Chart to help you solve right-angle trigonometry problems

The trick to using the chart above is finding the appropriate formula to solve your problem. It is based upon knowing two things about the triangle in question. If you know two things about any right-angle triangle, this chart will help you determine anything else you need to know about it.

Look at the drawing below for an example.

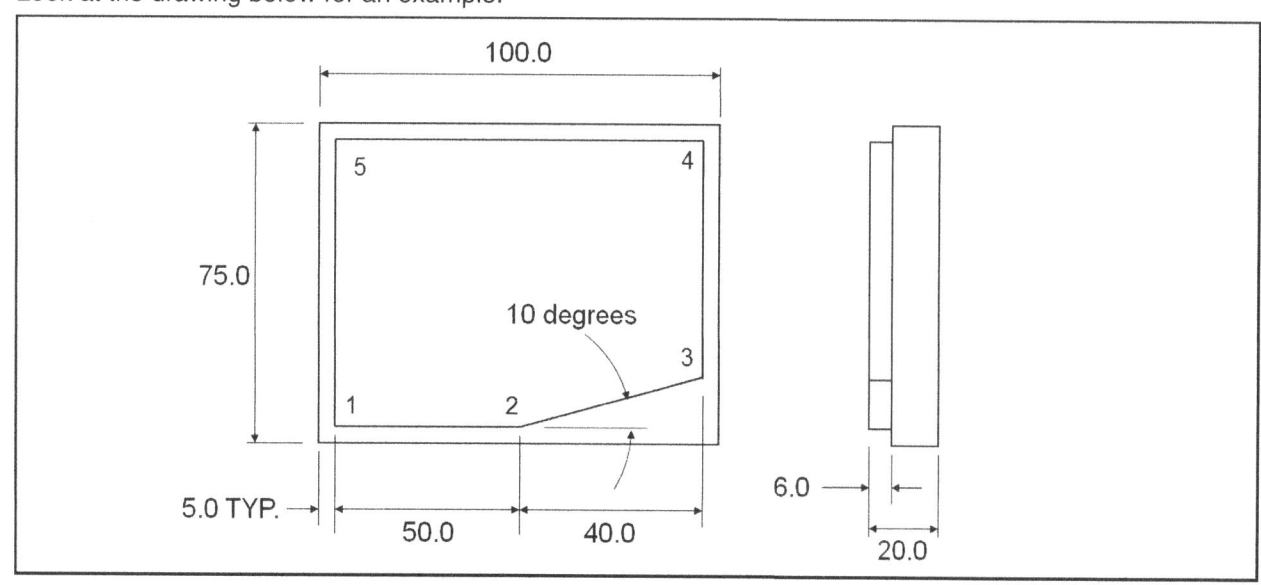

Figure 2.5: Drawing to illustrate the use of the trigonometry chart (shown using metric mode)

Simple addition and subtraction can be used to calculate the X and Y coordinates for all points on this drawing except the Y-coordinate of point 3. Notice that it is specified with an angular dimension of 10°. For this right triangle, we know the long side (40.0-millimeters) and the adjacent angle (10°). We must now determine which formula to use from the chart provided above. While the chart shows the mirror image of our triangle, hopefully you agree that we know the values of side b and angle A. We are looking for side a. So in the Known column, first find side b and angle A. It happens to be the seventh one from the top.

Key Concept 2: You Must Prepare To Write Programs

Notice that there are three formulae to choose from, and we must choose the correct one. Since we're looking for side a, we'll need to use the middle formula, a=b*tan A, which restated is side a is equal to side b (40.0 millimeters) times the tangent of angle A (ten degrees). The tangent of ten degrees is 0.17633 (found using an electronic calculator that has trig functions). When this value is multiplied times 40.0 (the length of side b), the result is 7.053 mm. The Y-coordinate of point 3 is 12.053 (our answer must be added to the 5.0 millimeter step).

The reason why you should do the math needed to calculate coordinates before attempting to write the program is the same as the reason why you should first come up with a sequence of machining operations. With the coordinates needed in your CNC program calculated, you won't have to break out of your train of thought while writing the program. You will be able to concentrate on programming separately from calculating coordinates.

Documenting your math also helps when mistakes are made. When you go back to check a mistake, if you have your math documented, you may be able to easily determine how the mistake was made to keep you from repeating the mistake in the future.

Marking up the Workpiece Engineering Drawing

You must have a good copy of the workpiece engineering drawing. You should have your own working copy, and be allowed to do whatever you need to with the print to help you with the programming task. Your marked-up copy of the print should be kept with the program as part of the documentation for the program.

Depending on the complexity of the workpiece to be produced, interpreting a workpiece drawing can range from quite simple to very difficult. Once you study the workpiece engineering drawing and understand the machining operations that must be performed, you should mark-up the print in any way that makes programming easier. The first thing we recommend is to take a high-lighting pen of a bright color and mark those surfaces on the print that require machining operations by your CNC program. Especially helpful for complicated workpieces, this helps you narrow down just what the program must do.

Additionally, you should indicate any information required for programming on the print. The location of workpiece coordinate system zero, the placement of work holding devices (fixtures, vises, etc.), and clamps should be included on the marked-up print. If there is room on the print, you can also include coordinates needed for programming.

The figure below shows an example of a marked-up print

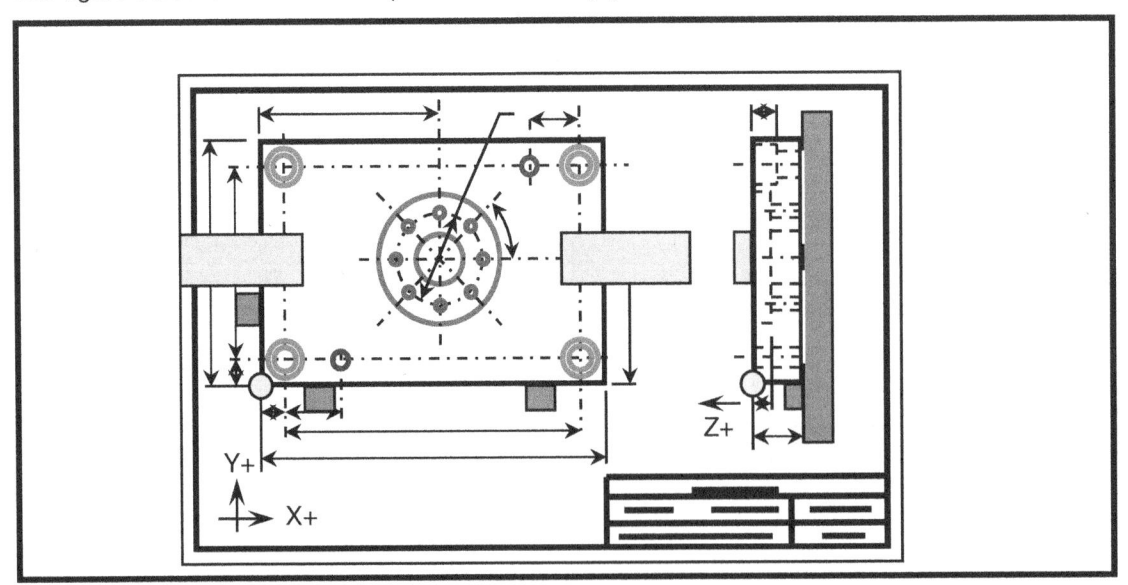

Figure 2.6: A marked-up print showing the location of workpiece coordinate system zero, the position of clamps, the machining operations to be performed, and how the workpiece is oriented in the fixture

Doing the Math

How dimensions are described on the workpiece engineering drawing will determine with how much math is required to write the program. In progressive companies, design engineers use *datum surface dimensioning* techniques. When datum surface dimensioning is used, each dimension on the print will be specified from one surface in each axis (the datum surface). This dramatically reduces the amount of math required to write a program.

Unfortunately, not all design engineers use datum surface dimensioning techniques. You may have to do a great deal of math in order to calculate the coordinates required in a CNC program.

When doing the math required, you will be calculating the axis coordinates used in the program. As discussed in Key Concept 1, each coordinate is the distance from the workpiece coordinate system zero point to a position through which a cutting tool must move. With a 3-axis machining center, each cutting tool position will have three coordinate values, one for each axis (X, Y, and Z).

How the programmer documents the coordinates needed in a program depends on the workpiece engineering drawing and how many coordinates must be calculated. If the print is large and roomy, and if there are only a few coordinates to be calculated, often you can easily write all of the coordinates right on the print close to the cutting tool position required in the program.

On the other hand, if the print is small and crowded, and/or if many coordinates are required, you should use a *coordinate sheet* to document the needed coordinates. On the print itself, draw a dot and place a number near each position to be included in the program. Each number represents a point number to be filled in on the coordinate sheet. The columns in the coordinate sheet will include point number and each axis letter address (X, Y, & Z). We have been using this technique throughout this course so it should be familiar to you.

For the Z-axis, most points will require multiple positions. Consider, for example, a drilling operation. The drill first moves above the hole in X and Y. It then moves down to an approach position above the hole in Z. Then the drill feeds down to another position in Z to machine the hole. Finally, it retracts from the hole back to the approach position in Z. So, at least two Z positions are needed to drill (the approach position and the bottom position).

Also, certain points will be used by several cutting tools. Again consider hole-machining operations. A single hole may be center-drilled, drilled, and counter-bored. The X and Y coordinates will be the same for each tool, but each tool will require its own Z positions. Though the approach position in Z will usually be the same for each hole-machining tool (usually 2.0-millimeters [or 0.1 inch] above the work surface), each tool will require its own hole-bottom position.

While it may be enough for experienced programmers to calculate just the X and Y positions prior to programming (they'll determine Z positions as they write the program), we recommend that you calculate <u>all</u> coordinates needed in the program before you start writing the program. This includes all of the Z positions needed by hole-machining tools.

To document these Z values in your coordinate sheet, we recommend writing them in the same order you will need them in a program. Separate them with a comma. If the approach position is the same for all tools, just write it once (first). For example, here are the Z coordinates needed for a hole that must be center-drilled, drilled, and counter-bored.

    2.0, 3.5, -31.75, -6.0

Again, these values are written in the Z column for any point (hole) requiring these three machining operations. 2.0 (mm) is the approach position for all three tools. -3.5 is the hole-bottom position for the center-drill. -31.75 is the hole-bottom position for the drill. -0.6.0 is the hole-bottom position for the counter-bore (end mill). The figure below shows a more elaborate example.

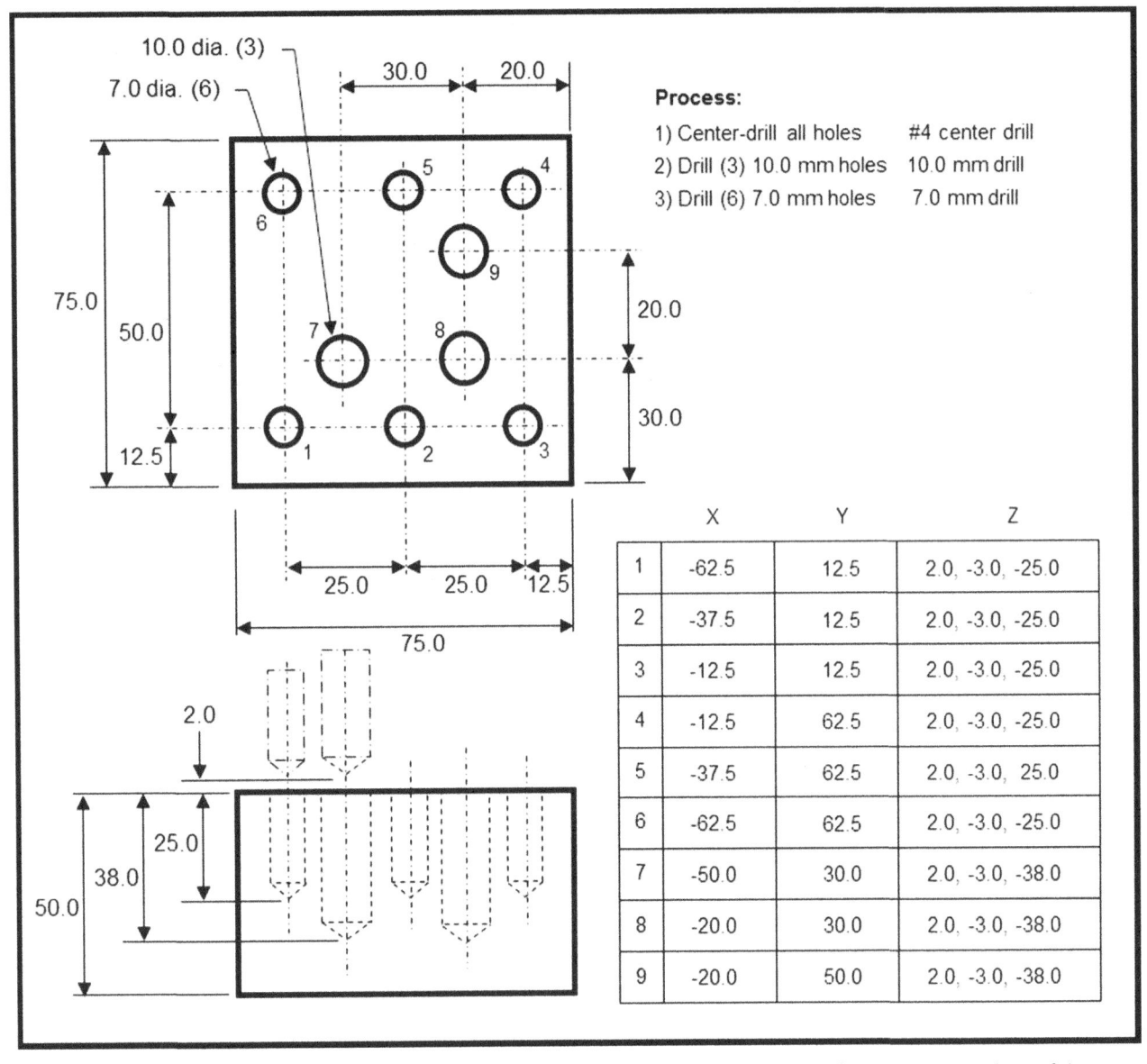

Key Concept 2: You Must Prepare To Write Programs

Process:
1) Center-drill all holes   #4 center drill
2) Drill (3) 10.0 mm holes   10.0 mm drill
3) Drill (6) 7.0 mm holes   7.0 mm drill

	X	Y	Z
1	-62.5	12.5	2.0, -3.0, -25.0
2	-37.5	12.5	2.0, -3.0, -25.0
3	-12.5	12.5	2.0, -3.0, -25.0
4	-12.5	62.5	2.0, -3.0, -25.0
5	-37.5	62.5	2.0, -3.0, 25.0
6	-62.5	62.5	2.0, -3.0, -25.0
7	-50.0	30.0	2.0, -3.0, -38.0
8	-20.0	30.0	2.0, -3.0, -38.0
9	-20.0	50.0	2.0, -3.0, -38.0

Figure 2.7: Example showing how to document the math needed in a program (shown using metric mode)

In the Z column of the coordinate sheet in the figure above, notice there are three values. The first (2.0) is the approach position. The same approach distance is used for all tools. The second (-3.0) is the hole bottom position for the center drill, and again, is the same for all holes. The third (-25.0 or -38.0) is the hole bottom position for the drill.

Our suggestions for documenting the math you do prior to programming are only recommendations. You can use any documentation methods that make sense to you as long as you get the desired result–calculating all coordinates needed for you program. You may be able to improve the method described here. You will quickly know how well you have prepared as you write your program. If you are constantly calculating (more) coordinates while you write the program, you haven't done very well.

What About Contour Milling Operations?

To this point, we have only discussed how to calculate coordinates for hole-machining operations. But you must also have the ability to do so for milling operations. This means, of course, that you must be able to plan the tool path for the milling cutter. It also means you must know *which* path the tool will use, the centerline path or the work-surface path. At this early point in the text, we can only show how to plan the milling cutter's centerline tool path. Later, in lesson 4.3, we'll discuss a feature called cutter radius compensation, and we'll show how to plan the (simpler) work-surface tool path.

Planning a Milling Cutter's Centerline Tool Path

This involves determining every position the cutter must move through as it mills the workpiece. And the contour being milled, of course, determines where these positions will be. Again, this requires basic machining practice experience. Consider the drawing shown in the figure below.

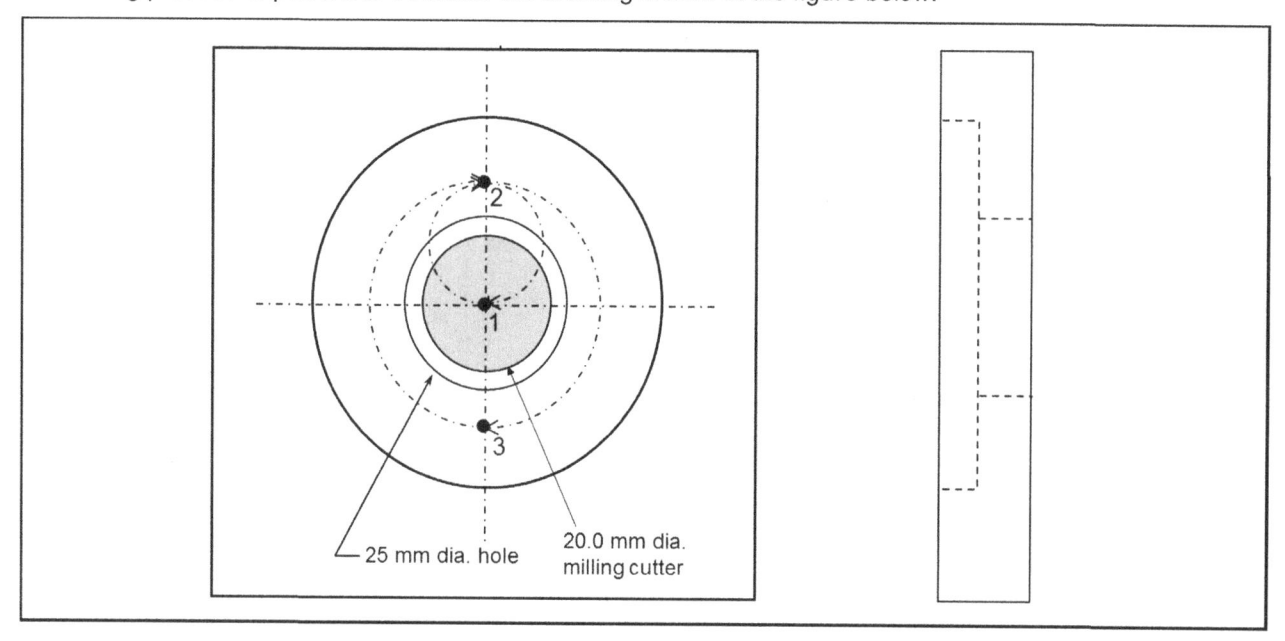

Figure 2.8: Calculating coordinates for a milling cutter's centerline tool path

The 20.0-millimeter diameter milling cutter is milling a round pocket. The tool path being used first sends the tool to point number one. The tool will then approach in Z–into the previously drilled (25.0-mm diameter) hole and to the work surface. It will then form a half circle to point two–arcing into the pocket edge. Next it will form another half circle to point three milling the right half of the pocket. Next, the cutter will make a half circle back to point two, milling the left half of the pocket. Finally the cutter will make a half circle motion back to point one, arcing off the surface being milled and getting back to a clearance position in XY.

Admittedly, planning a milling cutter's tool path requires an understanding of the kinds of motions that the machining center can make–and we haven't talked about motion types yet (motion types are discussed in lesson 3.1). For now, just keep in mind that you must be able to determine the tool path and come up with coordinates for *all* cutting tools you use in your program.

Check the Required Tooling

The next preparation step is related to the cutting and work holding tools used by your program. Tooling problems can cause even a perfectly written CNC program to fail. You can avoid production delays by considering potential tooling problems while you prepare for programming. Again, there is no excuse for machine down-time for as avoidable a reason as poor preparation.

First, you will need to confirm that the various cutting tools you plan to use in the program are available. You may incorrectly assume, for instance, that common tools are always in your company's inventory. It is always wise to double-check, even with relatively common tools. And always be sure your company has lesser-used tools in stock.

If a given tool is not in stock, of course, it must be ordered. Or you may have to make do with what your company does have available. If this is the case, your entire machining process may have to be changed (another reason to check *before* you begin programming).

Second, you must confirm that each cutting tool is capable of *reaching* the surfaces to be machined. Castings can present special problems in this regard. Say for example, a hole must be machined very close to a tall wall of a casting, as the figure below shows.

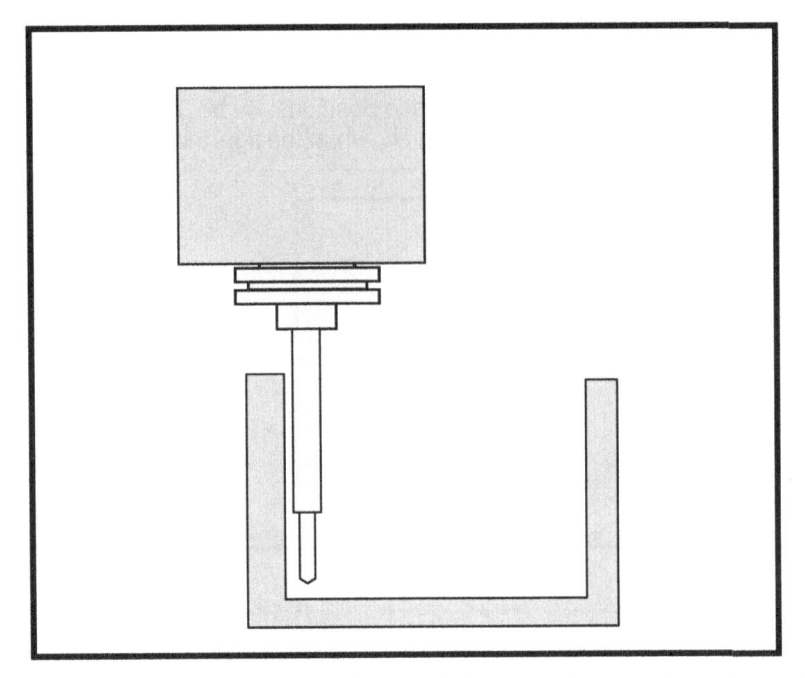

Key Concept 2: You Must Prepare To Write Programs

Figure 2.9: Tooling related problem requiring a special extension to be used

In this case, the tool must be long enough to reach the surface being machined. Additionally, the tool extension must be of a diameter small enough to clear the wall of the workpiece. Without being told how to assemble this cutting tool, most setup people will keep it as short and rigid as possible. For this reason, you must ensure that each cutting tool is made correctly by documenting special cutting tool considerations in the setup documentation. There is a column in the machining process planning form described previously named Note in which you can document special tooling considerations.

Another example of a tooling problem that can cause hold-ups during program verification has to do with the tool's minimum length. All machining centers have a Z-axis travel limitation. On a large vertical machining center, for example, the spindle nose of the machine may only come down in Z to within about nine inches of the table top. With this machine, consider a very short setup–possibly a thin plate resting on short parallels. The top of the workpiece may only be about two inches above the table top.

In this situation, a cutting tool must be at least seven inches long just to reach the top of the workpiece. If the cutting tool is not long enough, the Z-axis will over-travel when the tool attempts to approach the surface. If this problem isn't found until the program is being run on the machine for the first time, machine time will be wasted while the setup person remakes all of the cutting tools. This is yet another example of a down-time-causing problem that can be avoided with adequate preparation.

Plan the Work Holding Set-up

The programmer is usually responsible for developing the work holding setup required to hold the workpiece during machining. Even for simple work holding setups, you should make a drawing or sketch indicating how the setup is to be made. For example, a sketch showing where a vise is placed on the machine's table may adequately instruct the setup person.

For more complicated setups, you may not be intimately involved in special fixture design. In most cases, a tool designer will actually design the work holding fixture and supervise its construction. Once the fixture drawing is made, you will need it to determine how the workpiece will be held, and will write the program accordingly. Though you may not actually design the fixture, you will be responsible for instructing the setup person with regard to how the fixture will be mounted on the machine's table.

Most companies use a setup sheet to help the setup person understand everything they need to know about how a given setup must be made. Most setup sheets will include a sketch of the setup (possibly even a photograph of the setup once it has been made), the location of program zero, a list of cutting tools (including a list of components needed for each tool), and in general, any other instructions necessary for getting the job up and running. The figure below shows an example of a universal setup sheet. We call it a universal setup sheet because this form is used for all setups made on a given CNC machine tool.

Part no.:	Date:		**Setup Sheet**
Part name:	Programmer:		
Machine:	Program no:		

Stat.	Tool description	Offset	Insert	Instructions:
				Sketch of setup:

Fixture/vise:
Clamps:
Other notes:

Figure 2.10: An example universal setup sheet

This universal setup sheet does not include a detailed list of components for the cutting tools used in the program. Many companies do include a more complete tool list, possibly on a separate page. This setup sheet also does not include a complete list of the work holding tools. Again, many companies do include this kind of information so someone (probably *other* than the setup person) can be gathering all needed components even before the setup is made.

You should plan the setup and create the setup sheet before you write the CNC program. How the setup is made has a dramatic impact on how the program must be written. While you write the program, you'll need to know the position of work holding clamps and other obstructions in the setup so you can make sure that cutting tools don't hit them during their motions. You'll also need to know the orientation of the workpiece in the work holding devices. A workpiece is not always oriented in the same position as you see it in the top view of a workpiece drawing. When it comes to cutting tools, you'll need to know where they are placed in the machine's automatic tool changer magazine (station numbers). Without having planed the setup, you really can't write the CNC program.

Other Documentation Needed for the Job

The programmer is often responsible for providing all documentation needed for the jobs they program. The documentation should be aimed at the lowest skill level of people performing the related tasks. As with setup documentation, we recommend that all documentation related to a job be done prior to writing the program. You never know when you'll come across something that affects the way the program must be written.

Production Run Documentation

Production run documentation should be aimed at the CNC operator, providing instructions for how to load and unload workpieces, how often to take sampling measurements, approximately how long cutting tools will last before they must be replaced, and a list of all perishable tools (inserts, twist drills, taps, etc.) used in the job so they can be quickly found and replaced when they get dull. With longer production runs, many companies turn the machine over to another person, the CNC operator, once the setup is made and the first workpiece passes inspection. In some cases, more than one operator will complete the job (first and second shift operators, for example).

Key Concept 2: You Must Prepare To Write Programs

Program Listing

Most companies supply a printed copy of the program with the documentation that goes out to the setup person and operator who will be running the job. This will help them make modifications to the program if they are required.

Is it all Worth It?

You have seen that there is a great amount of work involved with *preparing* to write a CNC program. Again, experienced programmers will agree that this is where the real work lies. Once preparation is done, you will have a clear understanding of what the program must do. There will be no questions left to ponder while you write the program–and writing the program will be relatively easy–especially after you have written a few programs. Also, the work you do in preparation to write programs will pay special dividends when it comes time to run your program on the machine–there will be fewer mistakes to find and correct.

Key Points for Lesson 2.1:

- ✓ The better prepared you are, the easier it will be to write the program and the fewer mistakes you will make.
- ✓ There is no excuse for wasting machine time for something that could be avoided with better preparation.
- ✓ Beginners are prone to making mistakes in four categories: syntax mistakes, motion mistakes, mistakes of omission, and process mistakes.
- ✓ First, study and mark up the print to become familiar with what must be done.
- ✓ Second develop a machining process, including cutting tools and cutting conditions to be used.
- ✓ Third, do the math calculations, coming up with all coordinates needed in the program.
- ✓ Fourth, check that cutting tools are available and capable of performing their machining operations.
- ✓ Fifth, plan the work holding setup.
- ✓ Sixth, create any other documentation that is required for the job (like production run documentation).

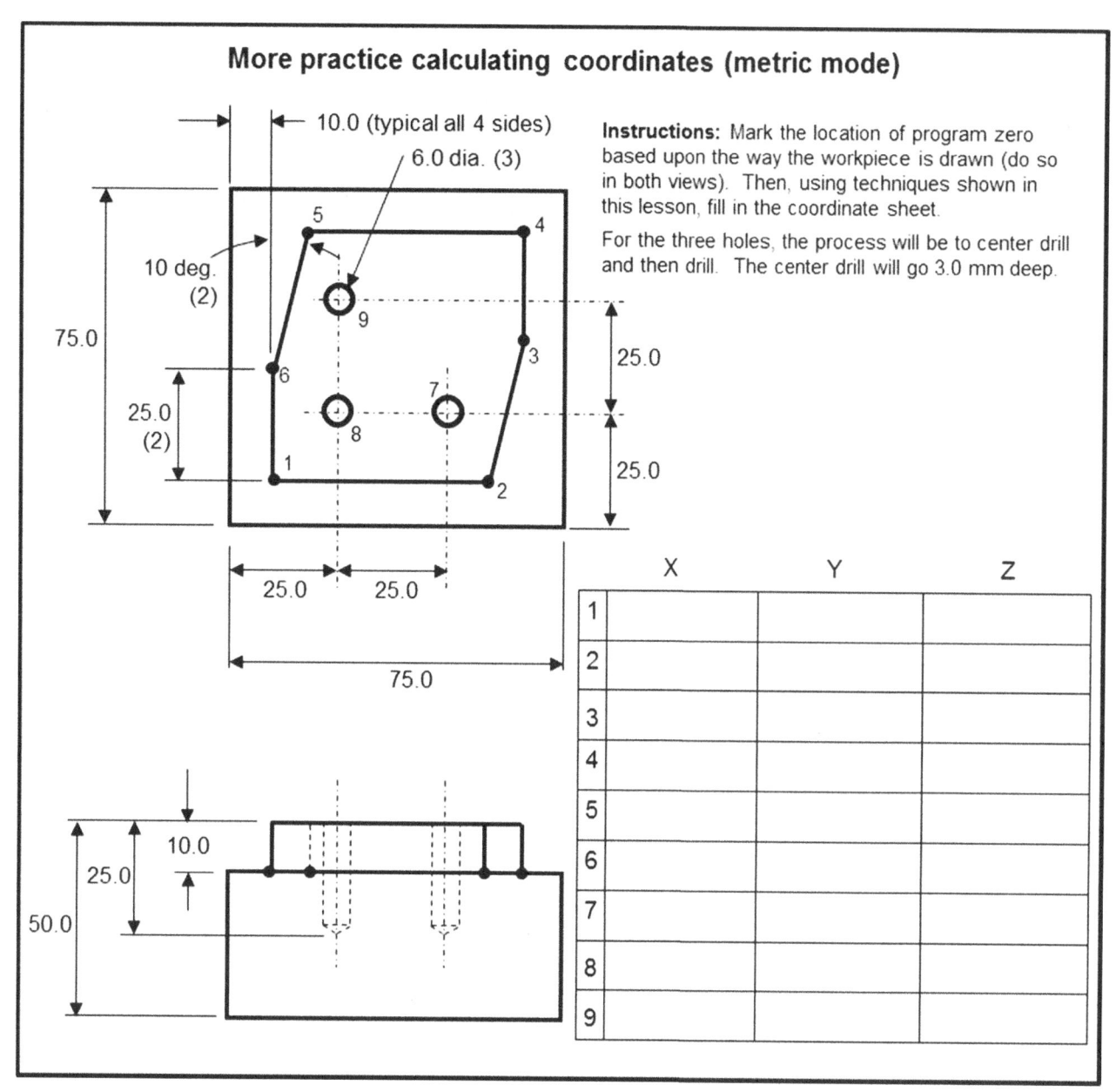

More practice calculating coordinates (metric mode)

Instructions: Mark the location of program zero based upon the way the workpiece is drawn (do so in both views). Then, using techniques shown in this lesson, fill in the coordinate sheet.

For the three holes, the process will be to center drill and then drill. The center drill will go 3.0 mm deep.

	X	Y	Z
1			
2			
3			
4			
5			
6			
7			
8			
9			

Answers:

In X and Y, program zero is the lower-left corner of the top view. In Z, program zero is the top of the front view (top of the workpiece).

1: 10.0, 10.0, -10.0
2: 59.71, 10.0, -10.0
3: 65.0, 40.0, -10.0
4: 65.0, 65.0, -10.0
5: 15.29, 65.0, -10.0
6: 10.0, 35.0, -10.0
7: 50.0, 25.0, Z: 2.0, -3.0, -25.0
8: 25.0, 25.0, Z: 2.0, -3.0, -25.0
9: 25.0, 50.0, Z: 2.0, -3.0, -25.0

More practice calculating coordinates (inch mode)

Instructions: Mark the location of program zero based upon the way the workpiece is drawn (do so in both views). Then, using techniques shown in this lesson, fill in the coordinate sheet.

For the three holes, the process will be to center drill and then drill. The center drill will go 0.12 deep.

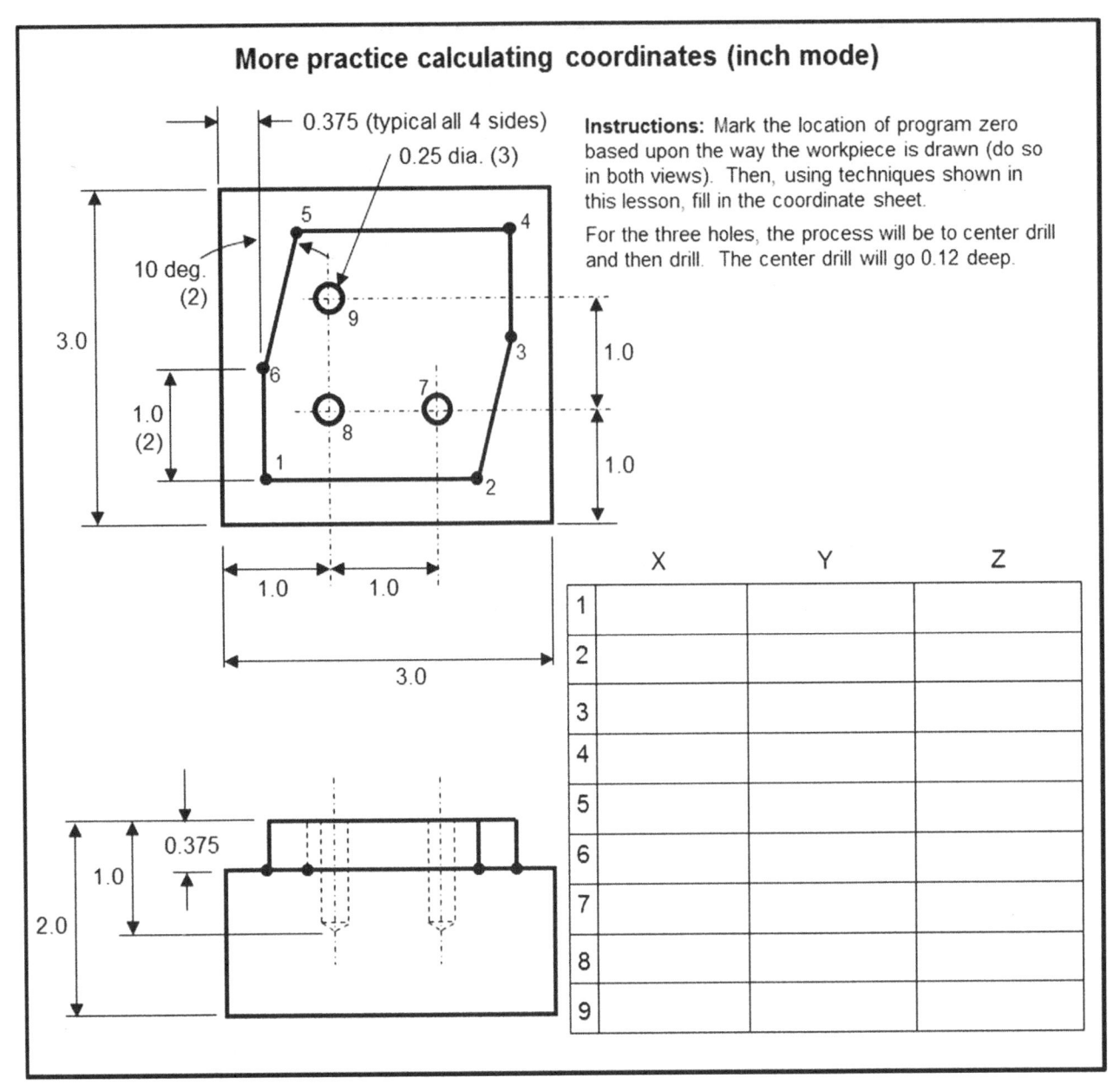

	X	Y	Z
1			
2			
3			
4			
5			
6			
7			
8			
9			

Answers:
In X and Y, program zero is the lower-left corner of the top view. In Z, program zero is the top of the front view (top of the workpiece).

1: 0.375, 0.375, -0.375
2: 2.4046, 0.375, -0.375
3: 2.625, 1.625, -0.375
4: 2.625, 2.625, -0.375
5: 0.5954, 2.625, -0.375
6: 0.375, 1.375, -0.375
7: 2.0, 1.0, Z: 0.1, -0.12, -1.0
8: 1.0, 1.0, Z: 0.1, -0.12, -1.0
9: 1.0, 2.0, Z: 0.1, -0.12, -1.0

Key Concept 3: Understanding Motion Types

Motion control is at the heart of any CNC machine tool. CNC machining centers have at least three ways that motion can be commanded. Understanding the motion types you can use in a program will be the focus of Key Concept 3.

Key Concept Number 3 is made up of but one lesson:

   3.1: Motion Types

All CNC machining centers have at least three axes (X, Y, and Z) along which you can specify positions or coordinates relative to the workpiece coordinate system zero. These coordinates are positions through which your cutting tools move. In Key Concept 3, we are going to discuss the ways to specify how cutting tools move from one position to another.

Understanding Interpolation

When a single linear axis is moving (X, Y, or Z), the motion will be along a perfectly straight line. For example, look at the figure below.

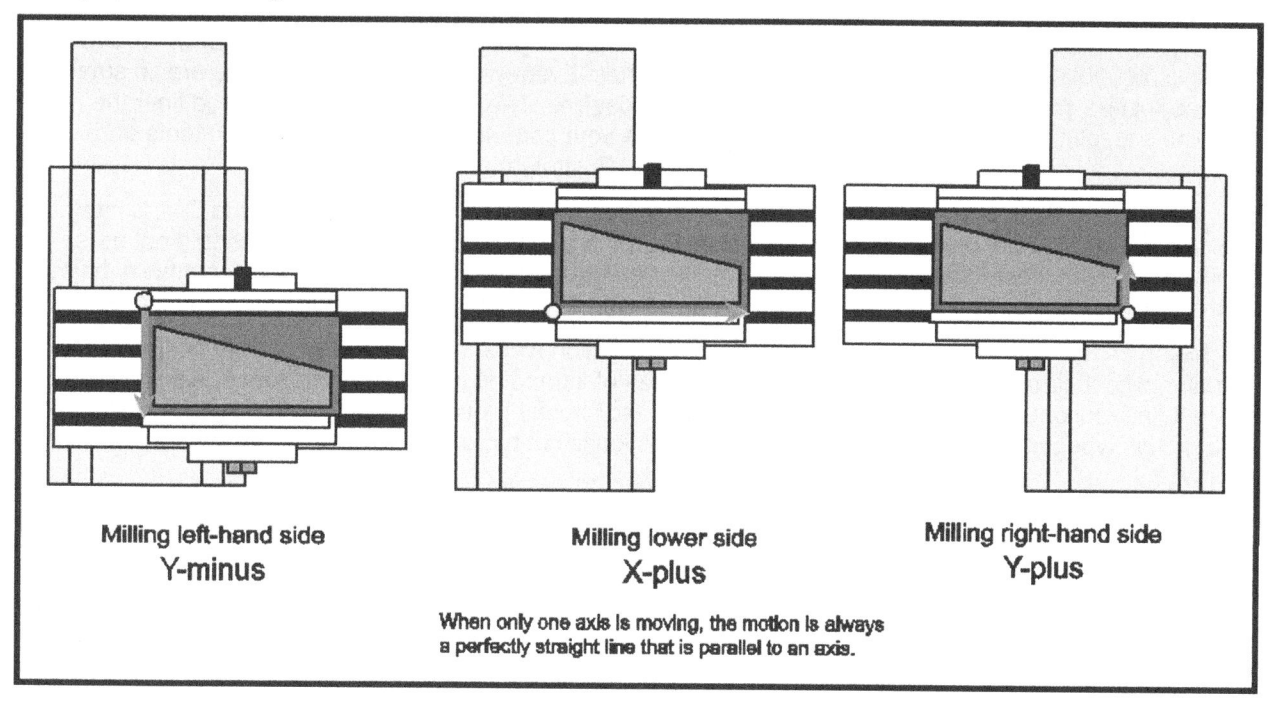

Figure 3.53: A perfectly straight motion will occur if only one axis is moving

When milling the left side of the workpiece (left view above), only the Y-axis is moving. And since the Y-axis is a linear axis, the motion is perfectly straight and parallel to the Y-axis. When milling the lower surface (middle view above), only the X-axis is moving and the motion is perfectly straight and parallel to the X-axis. The same goes for milling the right side (right view above)–only the Y-axis is moving and the motion has to be perfectly straight.

Key Concept 3: Understanding Motion Types

But notice that the upper side of this workpiece is tapered. It will require that both the X and Y axes to move in a coordinated manner, as shown in the figure below.

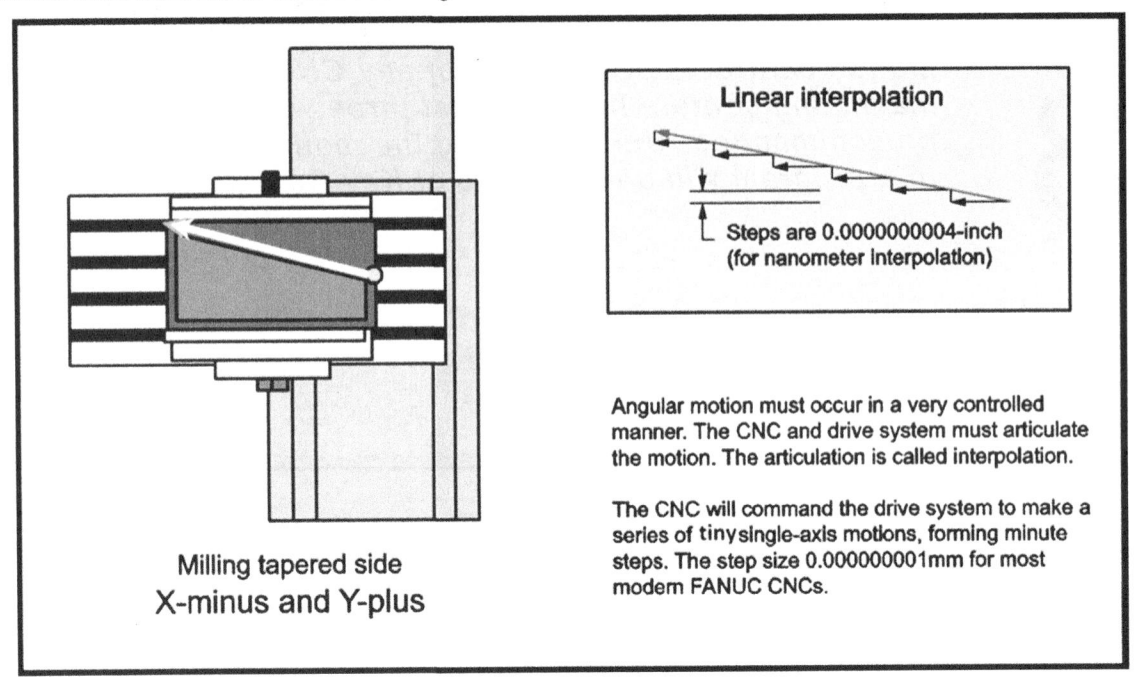

Figure 3.54: Milling the upper side of this workpiece requires both the X and Y axes move in a controlled manner

Two axes must move if the tool motion is at an angle as shown in the figure above. In CNC terms, this kind of motion is called interpolation.

The CNC breaks the two-axis motion up in to a series of very tiny steps. The step size for current FANUC CNCs is 0.000000001 millimeters or 0.0000000004-inches. Even with older CNCs the steps are so small that you cannot see or measure them with most measuring devices. The smaller the step size, the finer the machine's resolution and the more precisely it will follow your commanded motions. For all intents-and-purposes, all machined surfaces will appear to be perfectly straight and without steps.

The step size during interpolation is referred to as the CNC resolution. It is a measure of the CNC's internal calculation and memory resolution and the resolution of the axis feedback devices. This should not be confused with the least-input-increment (the smallest programmable increment), which is commonly be set by the machine tool builder to 0.0001 inches in the inch mode or 0.001 millimeters in the metric mode.

Machining center manufacturers offer a variety of interpolation types, based upon what their customers will be doing with their machines. All offer at least two types of interpolation–linear interpolation (also called straight-line motion) and circular interpolation (also called circular motion). All machining centers come with a third motion type, called rapid motion–though rapid motion is not typically interpolated, as we shall see shortly.

Depending upon your company's specific needs, there may be other interpolation types available with your machining centers. However, these additional interpolation types will only be used for special applications. Other interpolation modes may include:

Helical interpolation–used for ramp milling to machine or enter a pocket and for thread milling operations (not simple tapping).

Cylindrical interpolation–used with a rotary axis to machining contours around the outside of a round cylinder or tube.

Exponential interpolation–used with a rotary axis to produce tapered grooves with a constant helix angle.

Involute interpolation–used machine involute curves.

Spiral and conical interpolation–used machine spirals or conical curves.

NURBS interpolation–used machine non-uniform rational B-spline curves.

Actually, it should be comforting to know that there are only three primary ways to cause axis motion–rapid, linear interpolation, and circular interpolation. Just about every motion a CNC machining center makes can be divided into one of these categories. Once you master these three motion commands, you will be able to generate the motions required to machine a workpiece. The figure below illustrates these 3 primary motion types.

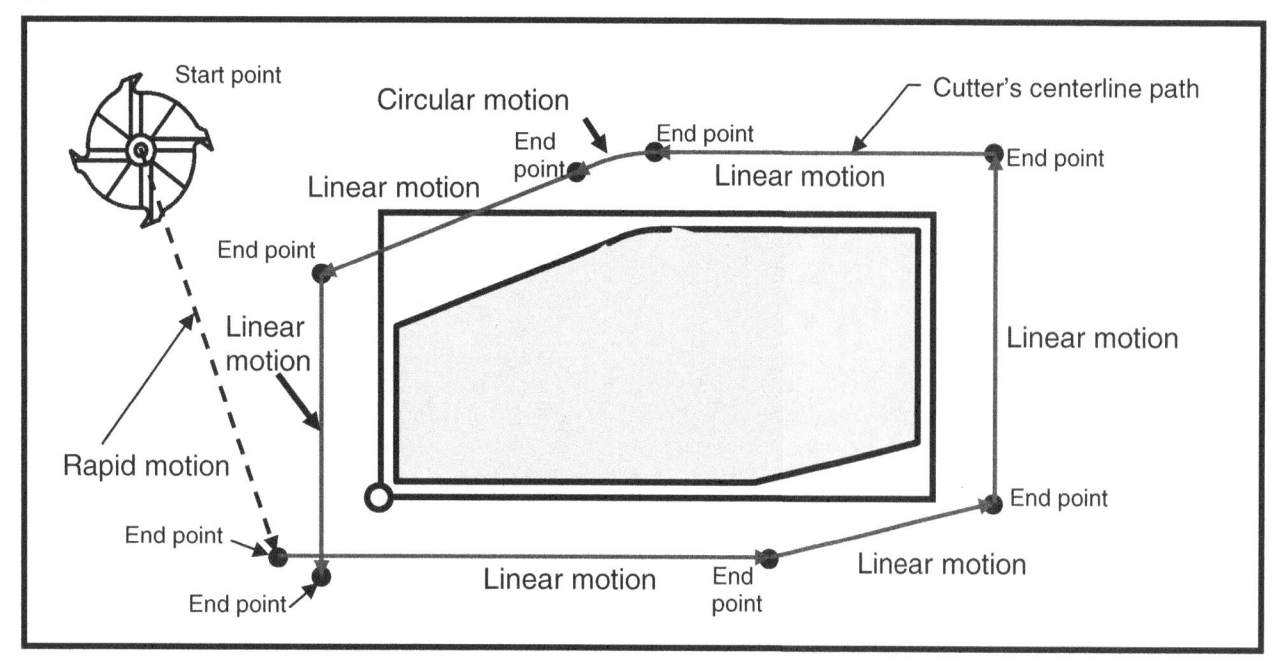

Figure 3.55: The three kinds of motion available with all CNC machining centers

As you will see in the lesson that follows, it is actually quite easy to specify motion commands within a CNC program. In general, each motion will require you to specify the kind of motion (rapid, linear interpolation, or circular interpolation) along with the motion's end-point (the coordinates at the end of the motion). Linear and circular interpolated motion additionally require that you specify the rate at which the axes will move (feedrate). Circular interpolation also requires that you specify either the radius of the arc or the coordinates of the center of the arc.

(this page intentionally left blank)

Lesson 3.1–Programming the Three Basic Motion Types

There are only three motion types used in CNC machining center programs on a regular basis–rapid, straight-line, and circular motion. You must understand how they are commanded.

Objectives

After completing this lesson, students should be able to:

- ✓ Describe the G00 rapid motion type
- ✓ Describe the G01 linear interpolation motion type
- ✓ Describe the G02/G03 circular interpolation motion type
- ✓ Calculate the end-point coordinates for commonly used tools
- ✓ Determine appropriate safe approach distances when using G00
- ✓ Calculate the end-points, radius, and I, J and K values when contour milling using the tool centerline method

Introduction

While it helps to understand how CNCs interpolate motion, it is not as important as knowing how to specify the appropriate motion commands in a program. This is the focus of this lesson. Let's begin our discussion by showing those things that all motion types share in common.

Motion Type Commonalities

All motion types share five things in common:

Modal - this means that once programmed a motion type will remain in effect until it is changed. When more than one consecutive movement of the same motion type is programmed, you need only include the motion type G-code in the first block of the series of movements.

End-point specification - each motion command requires the coordinates of the end-point of the motion to be specified. The CNC assumes the start-point for the motion is the current tool position. Think of the motion commands that form a tool-path as being a series of connect-the-dots.

Absolute or incremental modes - all motion commands are affected by whether or not you specify coordinates in the absolute or incremental positioning mode. In the absolute positioning mode (G90), the end-points are specified relative to the workpiece coordinate system zero. In the incremental positioning mode (G91), the end-points are specified relative to the tool's current position. As stated in lesson 1.4, beginning programmers should concentrate on specifying coordinates in the absolute positioning mode.

Specify only moving axes – only specify the axes that will move in a motion command block. If specifying a motion in only one axis, only one axis specification (X, Y, or Z) needs be included in the motion command. Axes that are not moving can and should be left out of the command.

Leading zero suppression - the leading zeros can be left out of the G-codes related to motion types commands. This means the actual G-codes used to instate the motion types can be programmed in one of two ways. G00 and G0 (stated G-zero-zero and G-zero) mean exactly the same thing to the control, as do G01 and G1, G02 and G2, G03 and G3. However, all examples in this course include the leading zero (G00, G01, G02, G03).

Key Concept 3: Understanding Motion Types

Understanding the Point Programmed for Each Cutting Tool

Hole-Making Tools

In order to generate correct motions with your cutting tools, you must understand the location on the tool that you are programming. In some cases, you will need to calculate the coordinates programmed based on the tool's geometry–not just the blueprint dimensions. The figure below shows a series of center-cutting tools used to perform hole-making operations.

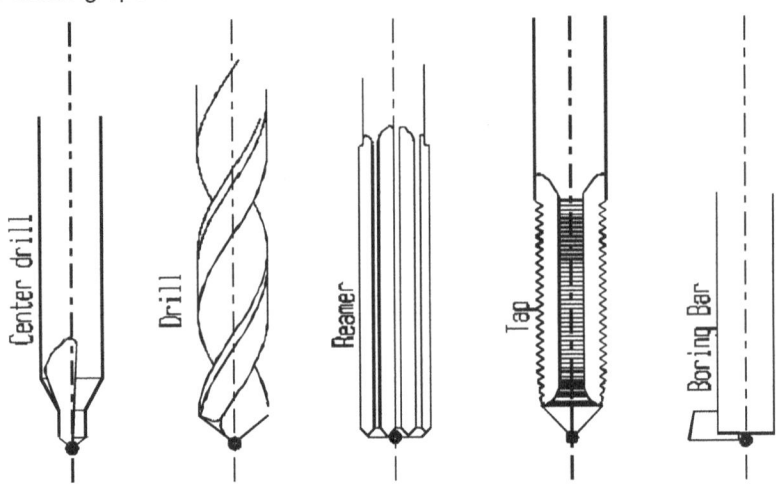

Figure 3.1: The point you must program on each hole-machining tool

For all hole-making tools shown in the figure above, you program the center of the tool in X and Y. This makes programming hole-locations quite easy, since the center of holes are dimensioned on the workpiece engineering drawing. For the Z-axis, the extreme end (tip) of the tool is programmed. However, you must often consider the lead of the tool when calculating hole-depth positions. This means the Z-axis programmed coordinates may not exactly match print dimensions. Here are some suggestions for determining the amount of lead for the most common hole-making tools.

Center Drill

Most programmers only center drill deep enough to make clearance for the web of the up-coming drill. The web is the center part of the drill body that joins the lands. That is, they do not try to center drill deep enough to form a chamfer on the drilled hole. If a chamfer is required on the hole, most programmers will use a spot drill (not shown above) instead of a center drill to start the hole.

Spot Drill (not shown above)

Spot drills are used to form a chamfer on a hole that will eventually be drilled. Since spot drills have a ninety degree point angle, the depth of the spot drill will be half the diameter of the chamfer to be machined. If machining a 13.0-millimeter (0.512-inch) diameter chamfer for a 12.0-millimeter (0.472-inch) diameter hole (actually a 0.5-mm or 0.02-in chamfer), the depth for the spot drill must be 6.5-mm (0.256-in), which is half of chamfer diameter, 13.0-mm (0.512-in).

Drill

Most high speed steel drills (including twist drills) have a 118-degree point angle. The lead for these drills is calculated by multiplying the drill diameter by 0.3. A 12.0-mm diameter twist drill, for example, has a 3.6-mm lead (12.0 x 0.3 = 3.6). A 0.5-in diameter twist drill has a 0.15-in lead (0.5 x 0.3 = 0.15).

Another consideration for drilling is related to through-holes. These are holes that pass all the way through a workpiece surface. (By the way, holes that do not pass through a workpiece are commonly called blind-holes). In addition to compensating for the drill's lead, you must add a small amount to the drill depth (usually about 0.75-mm or 0.03-inches) to ensure that the drill machines the hole all the way through the surface. This is especially important when machining ductile materials like certain steels or aluminums.

Reamer

The lead for a reamer is a small forty-five degree chamfer on the reamer's end. For metric reamers under about 13.0-mm, the lead is usually about 0.5-mm. For reamers over 13.0-mm, the lead is larger–usually about 1.0-mm. If working in inch mode, reamers under about 0.5-in diameter have a lead that is usually about 0.03-inches. For reamers over 0.5-in diameter, the lead is larger–usually about 0.06-inches. You must consult the reamer's technical specifications for confirmation.

Tap

The lead for a tap is usually specified by the number of imperfect threads on the end of the tap. The plug tap shown above, for example, may have as many as four imperfect threads. To calculate a tap's lead, you must multiply the number of imperfect threads time the pitch of the tap. For metric taps, pitch is specified in the thread's designation. An M8-1.25 tap, for example, is 8.0-mm in diameter and has a pitch of 1.25-mm. If the tap has a lead of six imperfect thread, the tap's lead is 5.0-mm (4 x 1.25).

For inch taps, pitch is equal to one divided by the number of threads per inch. For a 1/2-13 tap having a lead of four imperfect threads, for instance, the lead will be 0.3077-in (1 / 13 tpi x 4 imperfect threads = 0.3077).

Boring Bar

Most boring bars do not have a lead, meaning you will be specifying the hole-depth directly from the print dimension.

Milling Cutting Tools

When performing milling operations, there are two locations on the tool from which you may work.

When performing hole-machining operations with an end mill (using an end mill for counter-boring, for example), you will still work from the tool's center line position for the X and Y coordinates.

When performing face milling operations, you also work from the cutter's centerline for the X and Y coordinates.

When performing side milling operations (milling on the periphery of the milling cutter), you can calculate and specify coordinates based upon the cutter's centerline (called the centerline tool-path) or use the coordinates right off the workpiece engineering drawing (called the work surface tool-path).

It is not very convenient to work from the cutter's centerline, especially with complicated contours because it requires calculating the location of the center of the tool at each coordinate position. As you will see in Key Concept 4, a feature called cutter radius compensation makes it possible to use coordinates that are on the work surface (right from workpiece engineering drawing). However, since we have not introduced cutter radius compensation yet, the side milling examples in this lesson will be based upon the milling cutter's centerline tool-path.

Key Concept 3: Understanding Motion Types

G00 Rapid Motion (also called positioning)

Rapid motion is used to position a cutting tool at high-speed. Under normal conditions, G00 (stated G-zero-zero) will cause the machine to move at its fastest possible rate–which is called the machine's rapid rate. The rapid rate will vary from one machine to another–but it is always very fast. Several current CNC machining centers boast rapid rates well over 40 meters per minute (1,574 inches-per-minute), which means they move 635-mm (25-in) in just one second.

Due to this very fast–and somewhat scary–motion rate, most machine tool builders will allow you to override the machine's rapid rate during a program's verification using a multi-position switch called rapid traverse override. Though this feature varies from machine to machine, most machining centers allow you to slow the rapid rate significantly. This relieves some of the stress of running a program for the first time, and minimizes the potential for problems if a mistake has been made in a rapid motion command.

The figure below shows an example of rapid motion.

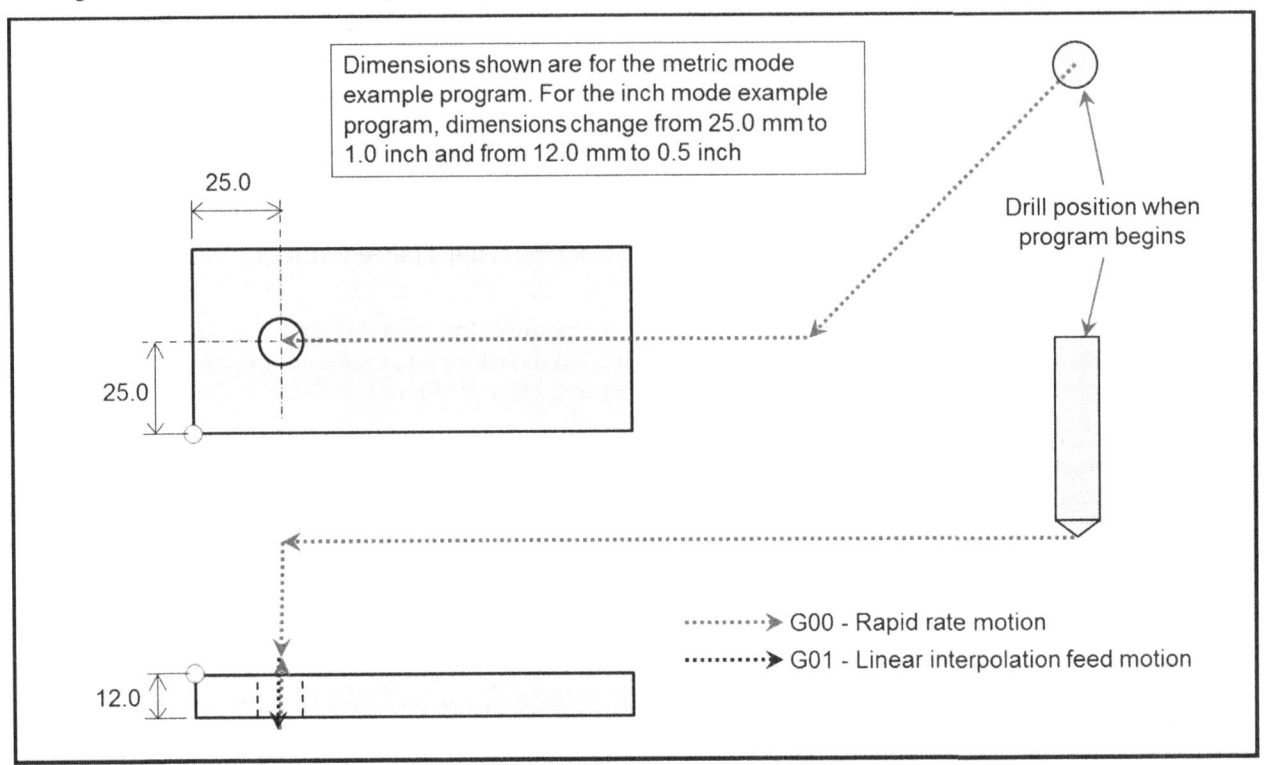

Figure 3.2: Rapid motion example

Program with comments (metric-mode example program):

```
%
O0003 (Program number)
N010 G21 G90 G54 (Select metric & absolute modes, workpiece coordinate system setting offset #1)
N020 T01 M06 (select drill in tool location 1)
N030 S400 M03 (start spindle fwd at 400 rpm)
N040 G00 X25.0 Y25.0 (Rapid to hole-location)
N050 G43 H01 Z2.0 (Instate tool length compensation, rapid to just above work surface)
N060 G01 Z-16.35 F140.0 (Feed linear to hole-bottom at 140 mmpm)
N070 G00 Z2.0 (Rapid retract from hole)
N080 G91 G28 Z0 (Rapid to Z-axis zero return position)
N090 G28 X0.0 Y0.0 (Rapid to X/Y axes zero return position)
```

Lesson 3.1: Programming The Three Basic Motion Types

```
N100 M30 (End of program)
%
```

The tool is well away from the workpiece when the program in the figure above starts (possibly at the zero return position). In block N010, G21 tells the CNC that the coordinates are specified in millimeters. The G90 tells the CNC that all up-coming coordinates will be specified from the workpiece coordinate system zero. The G54 word tells the CNC to look in workpiece coordinate system offset 1 to find the workpiece coordinate zero offset values. Block N020 selects the tool. In block N030, the S400 M03 starts the spindle at 400 rpm in the forward direction.

The G00 word in block N040 specifies the rapid motion mode, so all motions from this point will be at rapid until the motion type is changed. In this block, the drill will move at the rapid rate to above the hole in XY (X25.0 Y25.0).

With most machining centers, this rapid motion path is not along a straight line. Instead, most machines will allow each axes (X and Y in our case) to move as fast as they can. With most machining centers, the rapid rate is the same for all axes. If the distance moved by each axis is different, one of the axes will arrive at the destination coordinate first (the Y-axis in our example). The other axes (just X in our example) will continue to move at rapid until they also arrive at the destination coordinate. Some programmers refer to this motion as a dog-leg motion (taken from the game of golf).

Now look at block N050. Though it will not be discussed in depth until Key Concept 4, the G43 H01 are invoking something called tool length compensation. However, the Z-position coordinate specified is Z2.0. Notice that this block does not include a motion type designation. Since G00 is modal, the motion to Z2.0 will also move at the machine's rapid rate. While it would not hurt to include a G00 in block N050, it would not help either. We recommend leaving out redundant words to make your programs shorter and to minimize the potential for data entry (writing/typing) mistakes.

As we will discuss shortly, the G01 in block N060 specifies a linear interpolation motion. It changes the motion type from rapid to straight-line motion at a specified feedrate. The tool will now move at a feedrate of 140 millimeters-per-minute in during this motion, drilling the hole to a Z-coordinate of Z-16.35.

In block N070, the tool will once again move at the machine's rapid rate (note the G00 in the block), retracting out of the hole and back to a safe initial position of Z2.0.

Program with comments (<u>inch-mode</u> example program):

```
%
O0003 (Program number)
N010 G20 G90 G54 (Select inch & absolute modes, workpiece coordinate system setting offset #1)
N020 T01 M06 (select drill in tool location 1)
N030 S400 M03 (start spindle fwd at 400 rpm)
N040 G00 X1.0 Y1.0 (Rapid to hole-location)
N050 G43 H01 Z0.1 (Instate tool length compensation, rapid to just above work surface)
N060 G01 Z-0.7 F5.5 (Feed linear to hole-bottom at 5.5 ipm)
N070 G00 Z0.1 (Rapid retract from hole)
N080 G91 G28 Z0 (Rapid to Z-axis zero return position)
N090 G28 X0.0 Y0.0 (Rapid to X/Y axes zero return position)
N100 M30 (End of program)
%
```

Please read the descriptions under the metric mode program for detailed information about what is happening in this program. Differences to the inch mode program include: In block N010, inch mode (G20) is being selected, all X, Y, and Z coordinates are specified in inches, and the feedrate for the linear interpolation motion (block N060) is specified in inches per minute.

Key Concept 3: Understanding Motion Types

How Many Axes Can Be Included in a Rapid Motion Command?

One, two, or three (even four if the machine has a rotary axis) axes can be included in the G00 command. Block N040 above, for example, causes a two-axis rapid motion (in X and Y). The Z-axis will remain stationary during this command. (Remember, you need only include the moving axes in each motion command—X and Y in this example).

While we recommend approaching in XY first, then Z, you can approach in all three axes simultaneously. To do so in our example, replace blocks N040 and N050 with:

```
N040 G00 X25.0 Y25.0 G43 H01 Z50.0
```

While this minimizes cycle time, it will be much scarier for the operator. If you do elect to use this method for approaching, be sure that the Z-coordinate you include in this block is above all obstructions in the setup. Most programmers that use this technique will keep Z at least 50.0-millimters (or 2-inches) above all obstructions (Z50.0 in our example) to keep from concerning the operator during the approach move. Once the cutting tool is in position, they then rapid down to the final approach position in Z, as shown in this new version of the program (only the metric version is shown):

```
%
O0004 (Program number)
N010 G21 G90 G54 (Select metric & absolute modes, workpiece coordinate system setting offset #1)
N020 T01 M06 (select drill in tool location 1)
N030 S400 M03 (start spindle fwd at 400 rpm)
N040 G00 X25.0 Y25.0 G43 H01 Z50.0 (Instate tool length compensation, rapid to 50-millimeters above work surface)
N050 Z2.0 (rapid to just above work surface)
N060 G01 Z-16.35 F140.0 (Feed linear to hole-bottom at 140.0 mmpm)
N070 G00 Z2.0 (Rapid retract from hole)
N080 G91 G28 Z0 (Rapid to Z-axis zero return position)
N090 G28 X0.0 Y0.0 (Rapid to X/Y axes zero return position)
N100 M30 (End of program)
%
```

About Dog-Leg Motion...

It is very important that you understand the dog-leg style of motion most machines will use when two or more axes are included in a rapid motion command (as shown in block N040 in the example). You must always be concerned with obstructions between the starting and ending point for the rapid motion. The dog-leg motion will sometimes cause a tool to contact an obstruction that it would not if the motion occurs along a straight line. In these cases, you must break the rapid motion into two or more commands to go around the obstruction.

When Do You Use Rapid Motion?

Though it may be obvious, you should use the rapid motion command whenever the cutting tool is not machining the workpiece during the motion. In this way, you can minimize a program's air-cutting time (reducing cycle time). This includes approaching surfaces to be machined, retracting tools to the machine's tool changing position, and any other non-cutting operation that occurs (getting from one cutting position to another). A good rule-of-thumb is "If the tool is not cutting, it should be moving at rapid".

Certain other commands automatically cause the machine to move at its rapid rate. A G28 command, which sends the axes to the zero return position, will also be done at the rapid rate. The command,

```
N100 G91 G28 Z0.0
```

for example, will send the machine to its Z-axis zero return position. This is the tool change position for most vertical machining centers. Though a G00 is not included in this command, the machine will move at its rapid rate during this move.

What Is a Safe Approach Distance?

As stated, the rapid motion command minimizes motion time during your program. However, as the programmer, you must determine how closely you position each tool relative to a surface to be machined. Note that some machining operations occur at a very slow feedrate (at least when compared to the rapid rate), meaning a great deal of cycle time will be taken if you rapid a tool too far away from the work surface before the tool starts its cutting motion.

Though beginning programmers will need to be extra careful with rapid approach positioning movements, you must also be concerned with creating efficient programs. In the beginning of your programming career, the priority must be creating safe and easy-to-use programs. But as you gain experience, you'll need to create more efficient programs. Safety and efficiency seldom go together when it comes to CNC machine tool usage.

Consider the effect that rapid approach distance has on cycle time. Most programmers use a 2.0-millimeter (or about 0.1-in if using inch mode) approach distance when tools approach a qualified surface. By qualified surface, we mean a surface that will not vary by more than about 0.25-mm (or 0.010-in) from one workpiece to the next. Examples of qualified surfaces include surfaces machined during or prior to the CNC machining center operation, cold finished surfaces, precision die cast surfaces, and other accurate surfaces formed prior to the CNC operation.

On the other hand, if the surface is not qualified (it varies widely from one workpiece to another), the rapid approach distance must be increased. For sand castings, forgings, and other raw materials that are prone to variance, an approach distance of at least 7.0-millimeters (or 0.25-inches) is recommended.

A good rule-of-thumb is to make the approach distance about ten times the amount that a surface is varying from one workpiece to the next (if the surface varies about 0.2-millimeters (or 0.01-inch), use an approach distance of 2.0-millimeters (or 0.100 inch). Even with this rule-of-thumb, remember that as you continue your career in programming, you may need to consider reducing rapid approach distance (at least for qualified surfaces) as production quantities–and your skills–grow. Metric-mode example programs shown in this course use a rapid approach distance of 2.0-millimeters. Inch-mode examples use a rapid approach distance of 0.1-inches).

Consider this example: You have fifty holes to drill into a qualified surface. You are using a high speed steel drill, requiring a feedrate of just 127.0 mmpm (5.0 ipm). If using a rapid approach distance of 2.5-millimeters (about 0.100-inch), the cycle will experience 125-millimeters (about 5.0 inches) of air cutting motion (50 holes times 2.5-mm [0.1-in] of approach distance). At 127 mmpm (5 ipm), that equates to <u>one minute</u> of cycle time. By reducing the rapid approach distance to 1.25-mm (0.050-in), thirty seconds of cycle time could be saved. We do not encourage beginners to reduce rapid approach distance. Only consider doing so after you have written and verified several programs, and if cycle time is of major concern.

Key Concept 3: Understanding Motion Types

G01 Linear Interpolation (straight-line motion)

This motion type causes the machine to move along a perfectly straight path in one, two, or three axes. The control will calculate and interpolate the path between the start-point and the end-point of the motion automatically, no matter what angle of motion is required. So you simply specify the end-point for the motion. The G-code used to command this motion type is G01 (stated G-*zero*-one).

The motion rate for a linear interpolation move is programmable. It is specified with an F-word (specifying a feedrate for the motion). With most machining centers, the feedrate is typically specified in per-minute fashion (inches-per-minute in the inch mode or millimeters-per-minute in the metric mode). Today's machining centers also allow feedrate to be specified in distance-per-revolution (inches-per-revolution or millimeters-per-revolution). If your machining center allows both feedrate types, two G-codes will be used to specify which feedrate mode you desire (G94 for distance-per-minute, G95 for distance-per-revolution).

Like motion types, the feedrate word is modal. If a series of cutting motions will be machined at the same feedrate, the F-word need only be included in the first cutting motion block.

The linear interpolation motion command is used primarily to machine straight surfaces. Examples of when a G01 command can be used include drilling a hole to depth in the Z-axis and milling a straight or angular surface.

As with rapid motion, you will be able to override the motion rate for all cutting motions, including straight-line motions. A special multi-position switch, called *the feedrate override switch*, allows you to do so. With most machines, you'll be able to stop the motion rate (0%) and increase it to double the programmed feedrate (200%) in ten percent increments on the switch. This switch will help you verify the cutting feedrate/s used for each cutting tool.

The figure below shows an example of linear interpolation cutting commands.

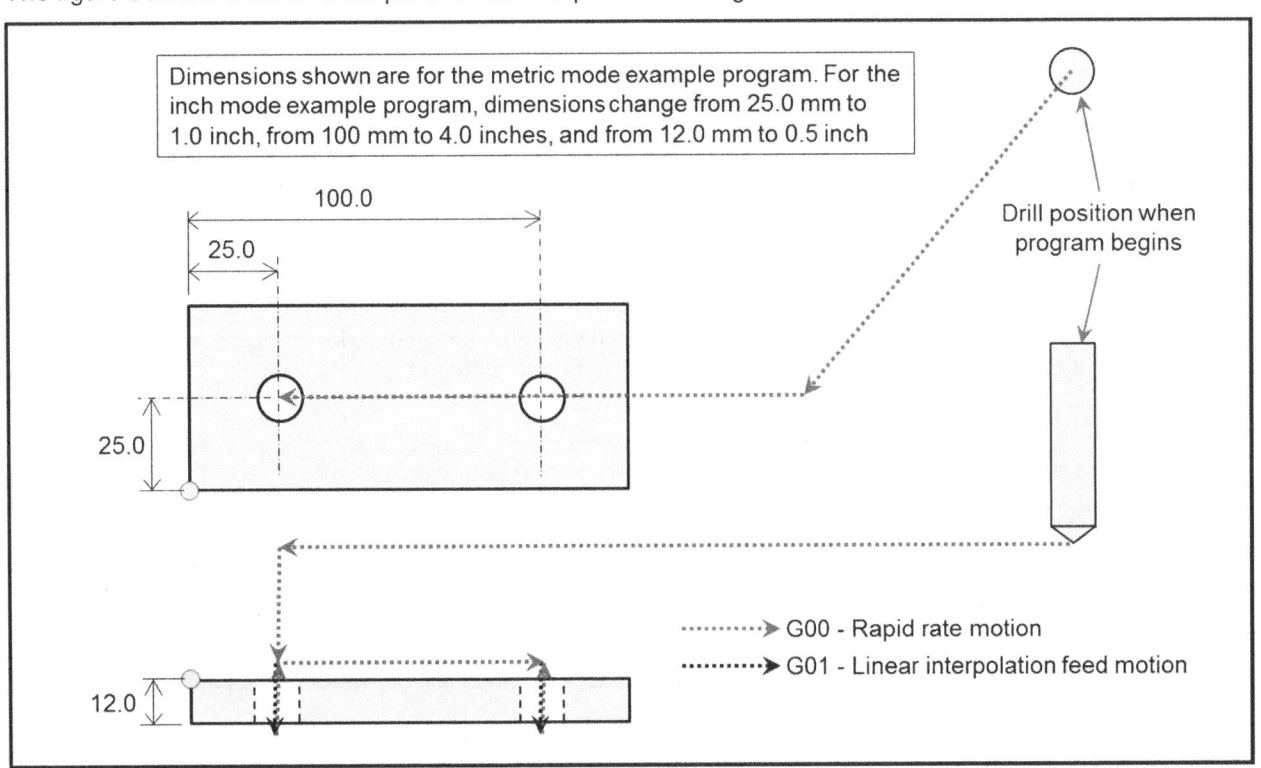

Figure 3.3: Linear interpolation cutting command example (shown using metric mode)

Program with comments (metric-mode example program):

```
%
O0005 (Program number)
N010 G21 G90 G94 G54 (Metric, absolute, mmpm modes, select work offset #1)
```

```
N020 S600 M03 (Start spindle fwd at 600 rpm)
N030 G00 X25.0 Y25.0 (Rapid to first hole-location)
N040 G43 H01 Z2.0 M08 (Instate tool length compensation, rapid
to just above work surface, start coolant)
N050 G01 Z-16.35 F140.0 (Drill hole to bottom at 140.0 mmpm)
N060 G00 Z2.0 (Rapid retract from hole)
N070 X100.0 (Rapid to second hole)
N080 G01 Z-16.35 (Drill second hole, still at 140.0 mmpm)
N090 G00 Z2.0 (Rapid retract from hole)
N100 G91 G28 Z0 (Rapid to Z-axis zero return position)
N110 G28 X0.0 Y0.0 (Rapid to X/Y axes zero return position)
N120 M30 (End of program)
%
```

In block N010, the G20 tells the CNC that the coordinates are in inch. G90 tells the CNC that all up-coming coordinates will be specified from the workpiece coordinate system zero. The G94 tells the CNC that the feedrate for interpolated moves will be provided in per-minute fashion. Since metric mode is selected, feedrate will be specified in millimeters-per-minute. The G54 word tells the control to look in workpiece coordinate system offset #1 to find the workpiece coordinate system offset values. (Note that older controls allow only three G codes per command, meaning this command would have to be divided into two commands.)

In block N020, S600 and M03 start the spindle at 600 rpm in the forward direction.

The G00 in block N030 specifies the rapid mode, so motions will be at rapid until the motion mode is changed. In this command, the drill will rapid to the hole-position in XY (X25.0 Y25.0). Again, notice the dog-leg style of motion caused by the G00 command.

In block N040, the drill will approach in Z (still at rapid), instating tool length compensation and also turning on the coolant (M08).

Block N050 is the first linear interpolation cutting move, causing the drill to machine the hole at a feedrate of 140.0 mmpm.

Block N060 retracts the drill from the hole at rapid back to a Z position of Z2.0.

Block N070 retains the rapid mode, moving the drill to its new hole position in X and Y.

Block N080 machines the second hole. The G01 in this command is very important. If it is left out, the drill will rapid into the workpiece and break the drill. Notice that there is no feedrate word in the block. Feedrate is modal. It will even be retained (at 140.0 mmpm in our case) even after motion mode changes have been made. So the second hole will be machined at the same feedrate as the first, 140.0 mmpm.

Block N090 retracts the drill from the second hole (at rapid) back to a position of Z2.0.

Though the motion is not shown in the diagram above for clarity, block N100 sends the machine to its Z-axis zero return position. In this particular example, the machine happens to be in the rapid mode at this point from block N090. But even if it is not, remember that G28 will automatically invokes the rapid mode. Similarly, block N110 sends the X and Y axes to their zero return positions at rapid.

The M30 in block N120 is an end-of-program word. It will cause the machine to turn off anything that is still running (spindle and coolant in our example)–then rewind to the beginning of the program and stop the executing the program.

Program with comments (inch-mode example program):

```
%
O0006 (Program number)
N010 G20 G90 G94 G54 (Inch, absolute, ipm modes, select work
offset #1)
```

```
N020 S600 M03 (Start spindle fwd at 600 rpm)
N030 G00 X1.0 Y1.0 (Rapid to first hole-location)
N040 G43 H01 Z0.1 M08 (Instate tool length compensation, rapid
to just above work surface, start coolant)
N050 G01 Z-0.72 F4.0 (Drill hole to bottom at 4.0 ipm)
N060 G00 Z0.1 (Rapid retract from hole)
N070 X4.0 (Rapid to second hole)
N080 G01 Z-0.72 (Drill second hole at 4.0 ipm)
N090 G00 Z0.1 (Rapid retract from hole)
N100 G91 G28 Z0 (Rapid to Z-axis zero return position)
N110 G28 X0.0 Y0.0 (Rapid to X/Y axes zero return position)
N120 M30 (End of program)
%
```

Please read the descriptions under the metric mode program for detailed information about what is happening in this program. Differences to the inch mode program include: In block N010, inch mode (G20) is being selected, all X, Y, and Z coordinates are specified in inches, and the feedrate for the linear interpolation motions (block N050 and N080) is specified in inches per minute.

Using G01 for a Fast-Feed Approach

For the most part, you will use G01 linear interpolation whenever you are machining a straight surface. However, there may be times when you elect to use G01 even when the tool is not actually machining. For example, many programmers are reluctant to rapid a tool past the top surface of the workpiece. They may be afraid that doing so will scare the person running the machine. Instead, they program approach movements below the work surface in Z with a fast feedrate. This relieves fear and still allows relatively efficient movements. Here is the beginning a program that uses the fast feedrate technique. In this program, the fast feedrate done in block N050 and at 1250 mmpm (about 50 ipm).

```
%
O0007 (Program number)
N010 G21 G90 G94 G54 (Metric, absolute, mmpm modes, select work
offset #1)
N020 S600 M03 (Start spindle fwd at 600 rpm)
N030 G00 X-7.0 Y-15.0 (Rapid to XY position)
N040 G43 H01 Z2.0 (Instate tool length compensation, rapid to
just above workpiece)
N050 G01 Z-7.0 F1250.0 (Fast feed to work surface-tool is clear
of workpiece at this position)
N060 G01 Y82.0 F90.0 (Mill to point 2)
N070 X115.0 (Mill to point 3)
```

Just as you when you transition from rapid motion to straight-line cutting motion (and you must specify the G01 for the cutting motion), so must you remember to specify the cutting feedrate when transitioning from the fast feedrate to the cutting feedrate—as this program shows in block N060 with F90.0.

A Milling Example

The figure below shows another example using G01—this time for a milling operation.

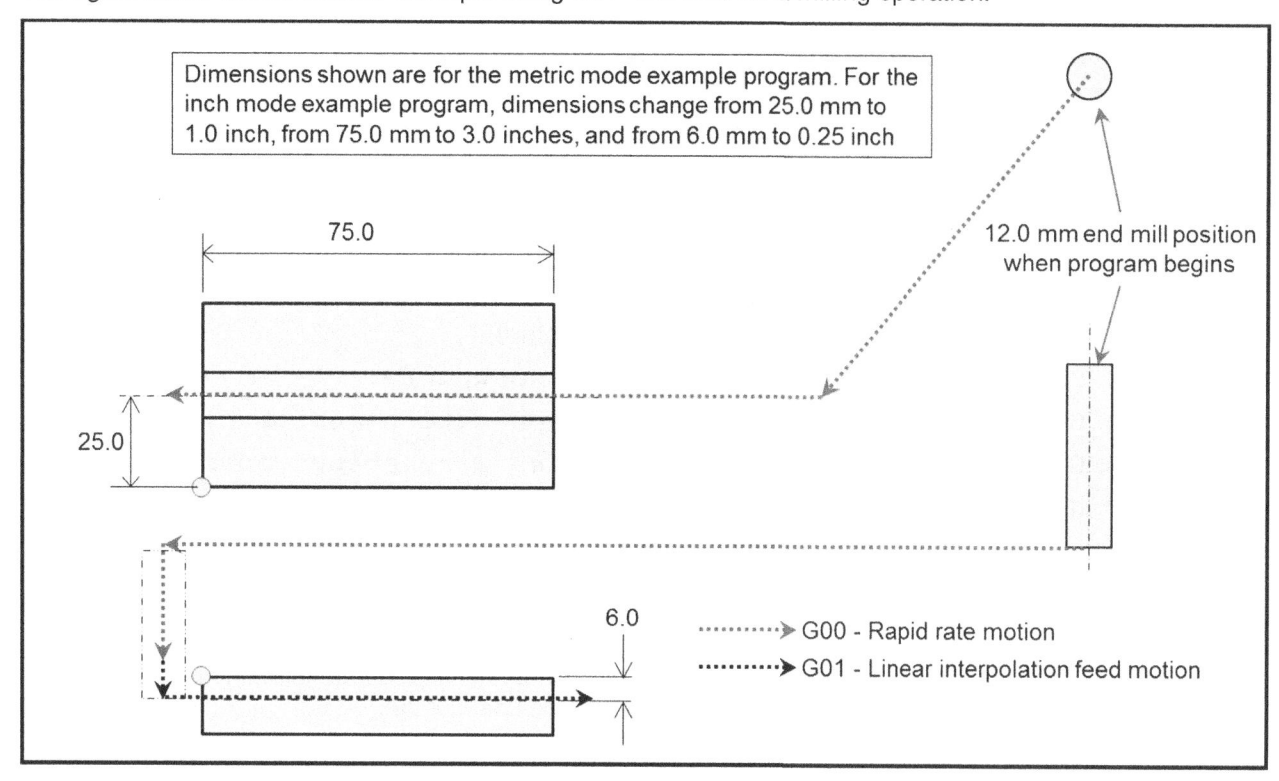

Figure 3.4: Another G01 example

Program with comments (metric-mode example program):

```
%
O0008 (Program number)
N010 G21 G90 G94 G54 (Metric, absolute, mmpm modes, select work offset #1)
N020 S800 M03 (Start spindle fwd at 800 rpm)
N030 G00 X-8.0 Y25.0 (Rapid to XY approach position)
N040 G43 H01 Z2.0 (Instate tool length compensation, rapid to just above workpiece)
N050 G01 Z-6.0 F1250.0 (Fast feed to work surface at 1250 mmpm)
N060 X83.0 F75.0 (Mill slot at 75.0 mmpm)
N070 G00 Z2.0 (Rapid to just above workpiece)
N080 G91 G28 Z0 (Rapid to Z-axis zero return position)
N090 G28 X0.0 Y0.0 (Rapid to X/Y axes zero return position)
N100 M30 (End of program)
%
```

Again, notice the fast feed approach in block N050. Also notice that in block N060, the milling cutter is milling the slot in the center of the workpiece. The motion type for block N060 will still be linear interpolation, carrying over from block N050. Notice the new feedrate in block N060 (F3.0), which is very important. Without this feedrate, the motion will occur at 50.0 ipm, which would probably break the cutter.

Key Concept 3: Understanding Motion Types

Program with comments (<u>inch-mode</u> example program):

```
%
O0008 (Program number)
N010 G20 G90 G94 G54 (Inch, absolute, ipm modes, select work
offset #1)
N020 S800 M03 (Start spindle fwd at 800 rpm)
N030 G00 X-0.35 Y1.0 (Rapid to XY approach position)
N040 G43 H01 Z0.1 (Instate tool length compensation, rapid to
just above workpiece)
N050 G01 Z-0.25 F50.0 (Fast feed to work surface at 50 ipm)
N060 X3.35 F3.0 (Mill slot at 3.0 ipm)
N070 G00 Z0.1 (Rapid to just above workpiece)
N080 G91 G28 Z0 (Rapid to Z-axis zero return position)
N090 G28 X0.0 Y0.0 (Rapid to X/Y axes zero return position)
N100 M30 (End of program)
%
```

Please read the descriptions under the metric mode program for detailed information about what is happening in this program. Differences to the inch mode program include: In block N010, inch mode (G20) is being selected, all X, Y, and Z coordinates are specified in inches, and the feedrate for the linear interpolation motions (block N050 and N060) is specified in inches per minute.

Drill Holes with G01?

We have already mentioned that linear interpolation can be used for drilling holes. While this will be demonstrated several times during the course (including in your exercises and programming activities), there are special programming features, called Canned Cycles for Drilling, that dramatically simplify the programming of hole-machining operations, including drilling, tapping, reaming and boring. Canned cycles are discussed in Key Concept 6.

G02 and G03–Circular Interpolation (circular motion)

Milling operations commonly require the machining of circular workpiece attributes. Consider, for example, the milling of a circular pocket. Frankly speaking, just about the only time you'll need to command a circular motion is when side milling. When circular motion is commanded, two axes (usually X and Y) will be moving together to form the motion.

Circular motion can be either clockwise or counter-clockwise. Two G-codes are involved–G02 specifies clockwise motion while G03 specifies counter-clockwise motion.

To determine whether a given XY circular motion is clockwise or counter-clockwise (G02 or G03), view the motion from the perspective of the cutter. In most cases, this means viewing the motion from above the workpiece engineering drawing. See the figure below for an example of clockwise and counter-clockwise motion. Notice that we're simply viewing the motion from above the workpiece engineering drawing.

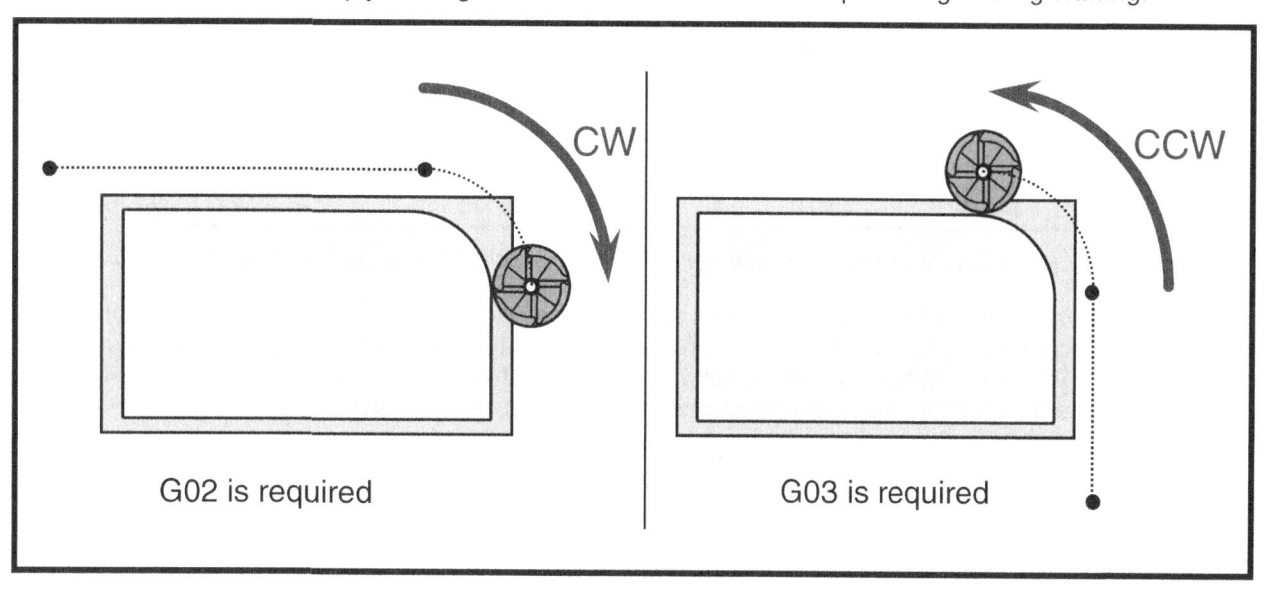

Figure 3.5: Clockwise versus counter-clockwise circular motion

Like linear interpolation, circular interpolation requires that a feedrate be specified (with an F-word). Feedrate is specified in distance-per-minute (inches- or millimeters-per-minute). And, feedrate is modal. Even if a feedrate is originally specified in a linear interpolations command, it will remain effective during subsequent circular interpolation commands.

Also as with linear interpolation motion, circular interpolation commands require that the end point of the circular motion be specified. The tool's position prior to the circular interpolation motion is the starting point for the circular motion.

Which Positions to Program

As stated at the beginning of this lesson, you can either program the shape of the work surface being machined (work surface coordinates) or the centerline path of the milling cutter (centerline coordinates). And also as stated, work surface coordinates are easier for a manual programmer to calculate, since they mostly are available on the workpiece engineering drawing. The feature cutter radius compensation–not discussed until Key Concept 4–makes it possible for programmers to specify work surface coordinates. But until Key Concept 4, we will show all examples using the milling cutter's centerline tool path.

Specifying Arc Size with the R-word

Circular motion commands also require that you specify the arc size of the circular path you are commanding. Today's CNCs allow you to do so with a simple R-word. With the R-word, you specify the radius of the arc being machined.

Note that the value of the R-word must correspond to the path you are programming-tool centerline or work surface. When programming the cutter's centerline path (which we will demonstrate until we present cutter radius compensation), the R-word must reflect the radius of the cutter's centerline path. When programming

Key Concept 3: Understanding Motion Types

the work surface path (using cutter radius compensation), the R-word must reflect the workpiece radius being machined.

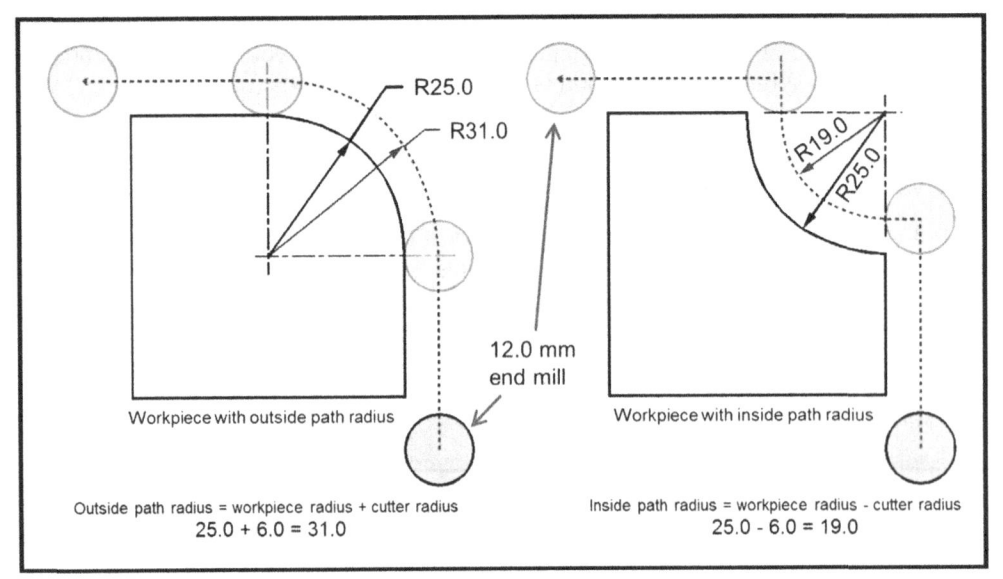

Figure 3.6: Outside radius tool centerline path versus inside radius tool centerline path (shown using metric mode)

With an outside radius as shown above, you must add the cutter radius to the workpiece radius to come up with the cutter's centerline path radius. If milling an inside radius (as would be the case when milling a circular pocket), you must subtract the milling cutter's radius from the workpiece radius to come up with the cutter's centerline path radius. The figure below shows a full example of circular motion.

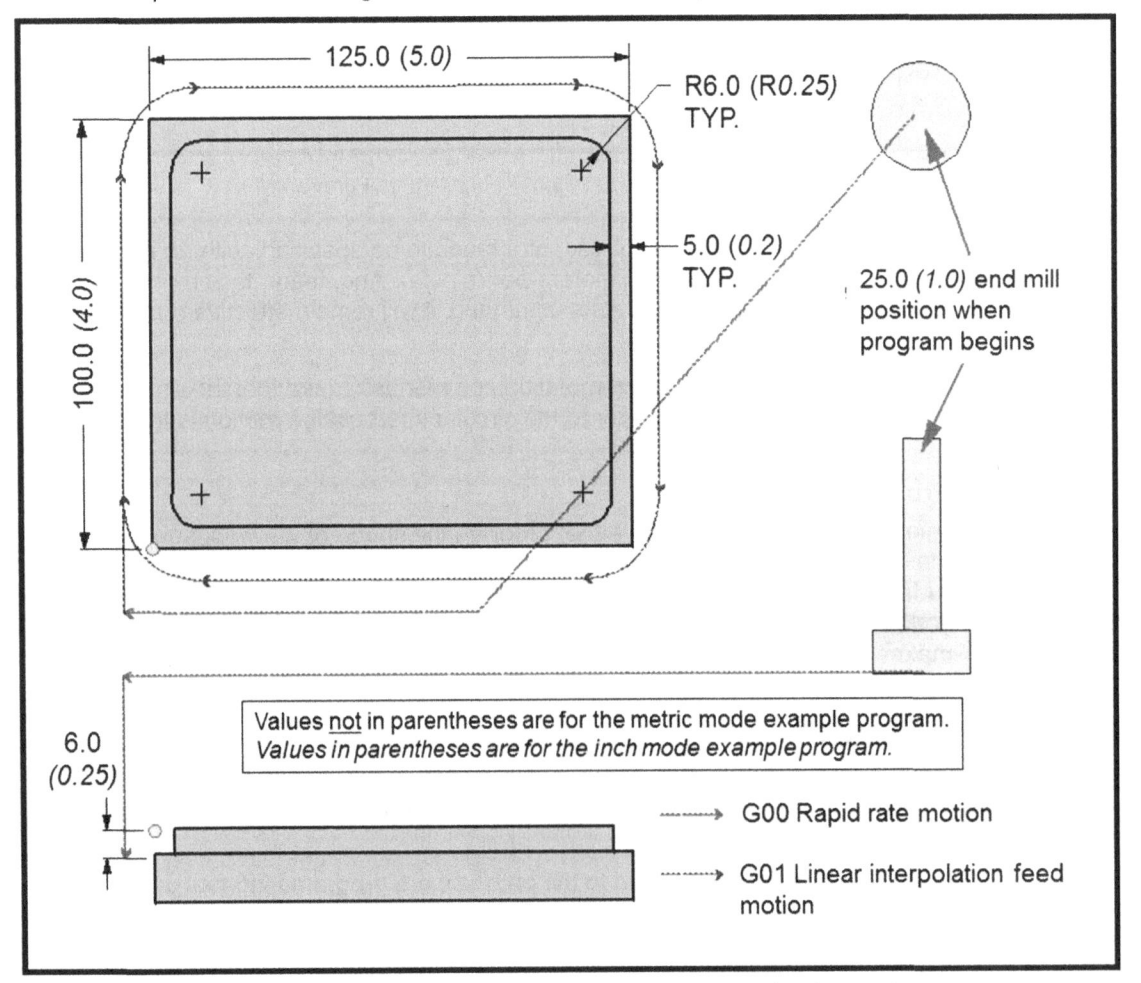

Figure 3.7: Drawing for example program showing circular motion

Program with comments (metric-mode example program):

```
%
O0009 (Program number)
N010 G21 G90 G94 G54 (Metric, absolute mmpm modes, work offset #1)
N020 T05 M06 (select tool #5)
N030 S400 M03 (Start spindle fwd at 400 rpm)
N040 G00 X-14.5 Y-7.5 (Rapid to approach position in XY)
N050 G43 H01 Z2.0 (Instate tool length compensation, move to just above workpiece)
N060 G01 Z-6.0 F1250.0 (Fast feed to work surface)
N070 X114.0 F127.0 (Mill lower surface at 127.0 mmpm)
N080 G03 X132.5 Y11.0 R18.5 (Mill lower-right radius)
N090 G01 Y89.0 (Mill right surface)
N100 G03 X114.0 Y107.5 R18.5 (Mill upper-right radius)
N110 G01 X11.0 (Mill upper surface)
N120 G03 X-7.5 Y89.0 R18.5 (Mill upper-left radius)
N130 G01 Y11.0 (Mill left surface)
N140 G03 X11.0 Y-7.5 R18.5 (Mill lower-left radius)
N150 G00 Z2.0 (Rapid to just above workpiece)
N160 G91 G28 Z0.0 (Rapid to the Z-axis zero return position)
N170 G28 X0.0 Y0.0
N180 M30 (End of program)
%
```

We're using a 25-mm diameter end mill. Notice that each of the four radii being machined are R6.0 mm. This means the cutter's centerline path radius will be R18.5 (6.0 workpiece radius + 12.5 cutter radius). Also notice that we're milling in a counter-clockwise direction around this workpiece, so all of the circular motions are in this program are commanded with G03.

In block N040, the cutter is being positioned (at rapid) clear of the left side of the workpiece in X (by 2.0-mm) and flush with the lower surface to mill in Y (5.0 - 12.5 cutter radius = Y-7.5).

In block N050, the cutter is brought to within 2.0-mm of the top of the workpiece (again, at rapid) and tool length compensation is being instated.

In block N060, the cutter fast feeds to the work surface (Z-6.0) at the fast feedrate of 1250.0 mmpm.

Machining begins in line N070, as the lower surface is milled. The tool ends this motion in a position ready to machine the first radius (the lower-right radius). That is, the end point for this straight-line motion is the start point for the lower-right radius.

In block N080, the first radius is milled. G03 specifies counter-clockwise motion, and the end point in XY brings the cutter to a position to begin the milling of the right side of the workpiece. And the R18.5 specifies the size of the radius to be machined (again, this is the cutter's centerline path radius: R18.5).

The motions switch between G01 for the straight sections and G03 for the circular sections from this point as the cutter moves around the balance of the contour. In block N150 the cutter retracts from the work surface back to a clearance position above the workpiece in Z (Z2.0).

Key Concept 3: Understanding Motion Types

Program with comments (<u>inch-mode</u> example program):

```
%
O0009 (Program number)
N010 G20 G90 G94 G54 (Inch, absolute ipm modes, work offset #1)
N020 T05 M06 (select tool #5)
N030 S400 M03 (Start spindle fwd at 400 rpm)
N040 G00 X-0.6 Y-0.3 (Rapid to approach position in XY)
N050 G43 H01 Z0.1 (Instate tool length compensation, move to just above workpiece)
N060 G01 Z-0.25 F50.0 (Fast feed to work surface)
N070 X4.55 F5.0 (Mill lower surface at 5.0 ipm)
N080 G03 X5.3 Y0.45 R0.75 (Mill lower-right radius)
N090 G01 Y3.55 (Mill right surface)
N100 G03 X4.55 Y4.3 R0.75 (Mill upper-right radius)
N110 G01 X0.45 (Mill upper surface)
N120 G03 X-0.3 Y3.55 R0.75 (Mill upper-left radius)
N130 G01 Y0.45 (Mill left surface)
N140 G03 X0.45 Y-0.3 R0.75 (Mill lower-left radius)
N150 G00 Z0.1 (Rapid to just above workpiece)
N160 G91 G28 Z0.0 (Rapid to the Z-axis zero return position)
N170 G28 X0.0 Y0.0
N180 M30 (End of program)
%
```

Please read the descriptions under the metric mode program for detailed information about what is happening in this program. Differences to the inch mode program include: We're using a 1.0-in end mill, in block N010, inch mode (G20) is being selected, all X, Y, and Z coordinates are specified in inches (as are the related calculations, the R word in blocks N080, N100, N120, and N140 is calculated by adding the workpiece radius (0.25) to the cutter radius (0.5), totaling R0.75, and the feedrate for the interpolation motions (block N060 through N140) is specified in inches per minute.

The R-Word is Not Modal

Note that the R-word is not modal. Even though all of the radii in the program are the same (R18.5-mm or 0.75-in), every circular interpolation command requires the radius word to be specified. Each G03 circular interpolation command follows a G01 linear interpolation command. The G01 for each straight line motion is required. Since circular commands are modal, if the G01 commands are left out, the CNC retains the circular interpolation mode (an alarm will likely be sounded).

More About Clockwise Versus Counterclockwise Circular Interpolation

Almost all of the circular motions you command will occur in the XY-plane (as the example program shows). However, circular motions can also be performed in the two other planes (XZ and YZ). In these somewhat rare occurrences, the method by which you determine clockwise versus counter clockwise will change. Additionally, three plane selection G-codes specify which plane is being used for the circular motion. G17 selects the XY-plane. G17 is initialized at power-up, meaning if you are working exclusively in the XY plane (as you almost always will), there will be no need to specify G17 in your program. Notice that we're making this assumption in the example program. There is no G17 in this program.

G18 selects the XZ plane and G19 selects the YZ plane.

A better way of determining whether a motion is clockwise or counterclockwise is to view the motion from the plus side of the uninvolved perpendicular axis. When making an XY circular motion, as we have been, you view the motion from the plus side of the Z-axis (again from above the print). But when making an XZ circular motion, you must view the motion from the plus side of the Y-axis (this is the column side of a vertical machining center). When making a YZ circular motion, you must view the motion from the plus side of the Y-axis (the right side of a vertical machining center).

Specifying Circular Interpolation with Directional Vectors (I, J, and K)

CNCs also allow you to specify circular motion with another, somewhat older and more cumbersome method–using something called directional vectors. With this method, I, J, and K are used to designate the center-point position of the arc to be machined.

I, J, and K are incremental distances from the starting-point of the arc to the center of the arc, irrespective of the G90/G91 mode. Using directional vectors requires that you know the distance and direction from the start-point of the arc to the center of the arc along the X and Y axes. Additional calculations are commonly required as compared to using the simpler R-word. I is the distance along the X-axis, J is the distance along the Y-axis, and K is the distance along the Z-axis.

- **I:** the incremental distance and direction from the starting-point of the arc to the center of the arc along the X-axis.
- **J:** the incremental distance and direction from the starting-point of the arc to the center of the arc along the Y-axis.
- **K:** the incremental distance and direction from the starting-point of the arc to the center of the arc along the Z-axis.

The word direction implies a polarity–the polarity of I, J, and/or K could be plus or minus (and as with other CNC words, plus is assumed). To determine whether a directional vector should be plus or minus, draw an arrow on the workpiece engineering drawing from the start-point of the arc to the arc center along each axis. Note that there might be more than one arrow, depending upon the arc to be programmed. Now ask yourself if each arrow is pointing in the plus or minus axis direction. This determines whether the related I, J, and/or K will be plus or minus.

If any of the arrows you draw points along the X-axis, you will use an I-word in the circular interpolation command. If any of the arrows points along the Y-axis, a J-word will be used.

The distance is the actual calculated value from the start-point of the arc to the center of the arc. If the distance from the start-point to the center of the arc is zero (start point and arc center are in the same location), it can be left out of the circular command, or included with a zero value. The next illustration shows how to determine which directional vector/s must be used and the polarity of each.

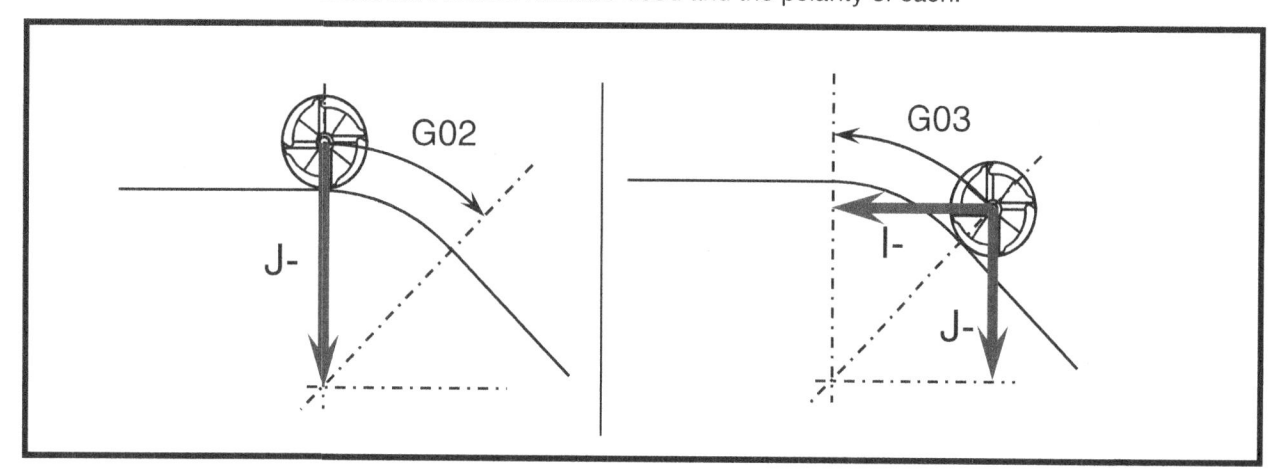

Figure 3.8: Use of directional vectors

Key Concept 3: Understanding Motion Types

Here is a revised program for the drawing shown in Figure 3.7 using directional vectors within circular commands instead of the R-word (only the metric-mode example program is being shown).

Program with comments (only the <u>metric-mode</u> example is shown):

```
%
O0010 (Program number)
N010 G21 G90 G94 G54 (Metric, absolute mmpm modes, work offset #1)
N020 T05 M06 (select tool #5)
N030 S400 M03 (Start spindle fwd at 400 rpm)
N040 G00 X-14.5 Y-7.5 (Rapid to approach position in XY)
N050 G43 H01 Z2.0 (Instate tool length compensation, move to just above workpiece)
N060 G01 Z-6.0 F1250.0 (Fast feed to work surface)
N070 X114.0 F127.0 (Mill lower surface at 127.0 mmpm)
N080 G03 X132.5 Y11.0 J18.5 (Mill lower-right radius)
N090 G01 Y89.0 (Mill right surface)
N100 G03 X114.0 Y107.5 I-18.5 (Mill upper-right radius)
N110 G01 X11.0 (Mill upper surface)
N120 G03 X-7.5 Y89.0 J-18.5 (Mill upper-left radius)
N130 G01 Y11.0 (Mill left surface)
N140 G03 X11.0 Y-7.5 I18.5 (Mill lower-left radius)
N150 G00 Z2.0 (Rapid to just above workpiece)
N160 G91 G28 Z0.0 (Rapid to the Z-axis zero return position)
N170 G28 X0.0 Y0.0
N180 M30 (End of program)
%
```

Notice that the only difference between this program and the previous one is R-words have been replaced with I and J-words. Also, since all of the arcs in this program are full 90 degree arcs, only one of I or J are required in each block. Do not get confused and think that I/J will always be the same value as the radius.

Arc Limitations

Simultaneously Specifying R with I, J or K
If I, J, K, and R addresses are specified in the same block, the arc specified by the R-word takes precedence and the I, J and K-words are ignored.

Specifying a Semicircle
When an arc has a center angle that approaches 180° or 360°, the calculated center coordinates when using an R-word may contain an error. In such a case, specify the center of the arc with I, J, and K.

Specifying Arcs Greater than 180° Using the R-word
When specifying an arc greater than 180° using the R-word, a negative value must be specified.

Programming a Full Circle
If you are making a full circle, the start point of the arc is the same as the end point.

To program a full circle in one block you must use directional vectors (I, J, or K) to specify arc center for the full circle being generated and leave out the end-point (X, Y, Z).

Since directional vectors are more difficult to use, it may be just as easy to break the full circle motion into two commands and use the simpler R-word for the radius designation.

Lesson 3.1: Programming The Three Basic Motion Types

Example Using I, J and K

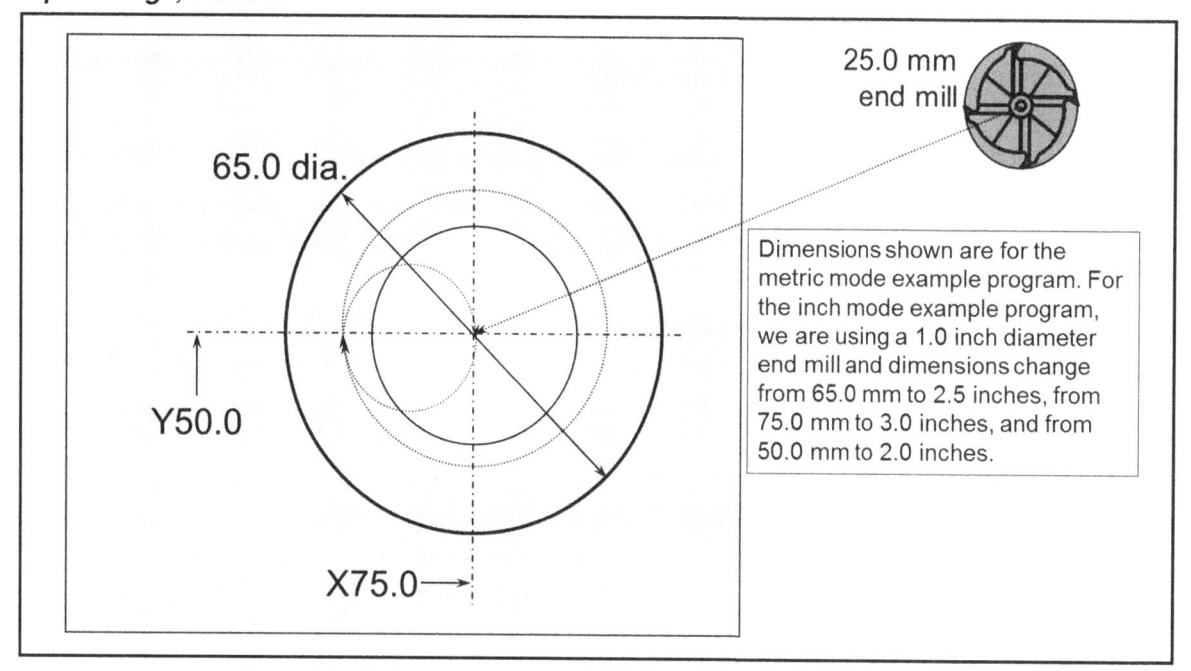

Figure 3.9: Full circular motion in one block using directional vectors

Program with comments (metric-mode example program):

```
%
O0011 (Program number)
N010 G21 G90 G94 G54 (Metric/absolute/mmpm modes, select work offset #1)
N020 T01 M06 (select tool)
N030 S400 M03 (Start spindle fwd at 400 rpm)
N040 G00 X75.0 Y50.0 (Rapid to center of circular pocket)
N050 G43 H01 Z2.0 (Instate tool length compensation, rapid to just above workpiece)
N060 G01 Z-6.0 F1750.0 (Fast feed to work surface-a hole has previously been drilled in this location)
N070 G02 X55.0 I-10.0 F125.0 (Make a half-circle arc-in to the pocket surface-Y start and end point are the same [Y50.0] so Y is omitted)
N080 I20.0 (Make full circle in one command-X and Y start point and end point are the same, so they are omitted-directional vector points in X plus direction)
N090 X75.0 I10.0 (Make half-circle arc-out back to center of pocket)
N100 G00 Z2.0 (Rapid to just above workpiece)
N110 G91 G28 Z0.0 (Rapid to zero return position in Z)
N120 G28 X0.0 Y0.0 (Rapid to zero return position in XY)
N130 M30 (End of program)
%
```

Key Concept 3: Understanding Motion Types

Blocks N070 and N090 are half circles to arc-in and arc-out of the circle pocket to be milled. Note they are programmed using the I-word directional vector. Block N080 mills a full circle. Again, notice the use of the I-word. X and Y are not programmed in this block because the starting and ending point are the same.

Program with comments (<u>inch-mode</u> example program):

```
%
O0011 (Program number)
N010 G20 G90 G94 G54 (Inch/absolute/ipm modes, select work offset #1)
N020 T01 M06 (select tool)
N030 S400 M03 (Start spindle fwd at 400 rpm)
N040 G00 X3.0 Y2.0 (Rapid to center of circular pocket)
N050 G43 H01 Z0.1 (Instate tool length compensation, rapid to just above workpiece)
N060 G01 Z-0.25 F50.0 (Fast feed to work surface-a hole has previously been drilled in this location)
N070 G02 X2.25 I-0.375 F5.0 (Make a half-circle arc-in to the pocket surface-Y start and end point are the same [Y2.0] so Y is omitted)
N080 I0.75 (Make full circle in one command-X and Y start point and end point are the same, so they are omitted-directional vector points in X plus direction)
N090 X3.0 I0.375 (Make half-circle arc-out back to center of pocket)
N100 G00 Z0.1 (Rapid to just above workpiece)
N110 G91 G28 Z0.0 (Rapid to zero return position in Z)
N120 G28 X0.0 Y0.0 (Rapid to zero return position in XY)
N130 M30 (End of program)
%
```

Example Using R-word

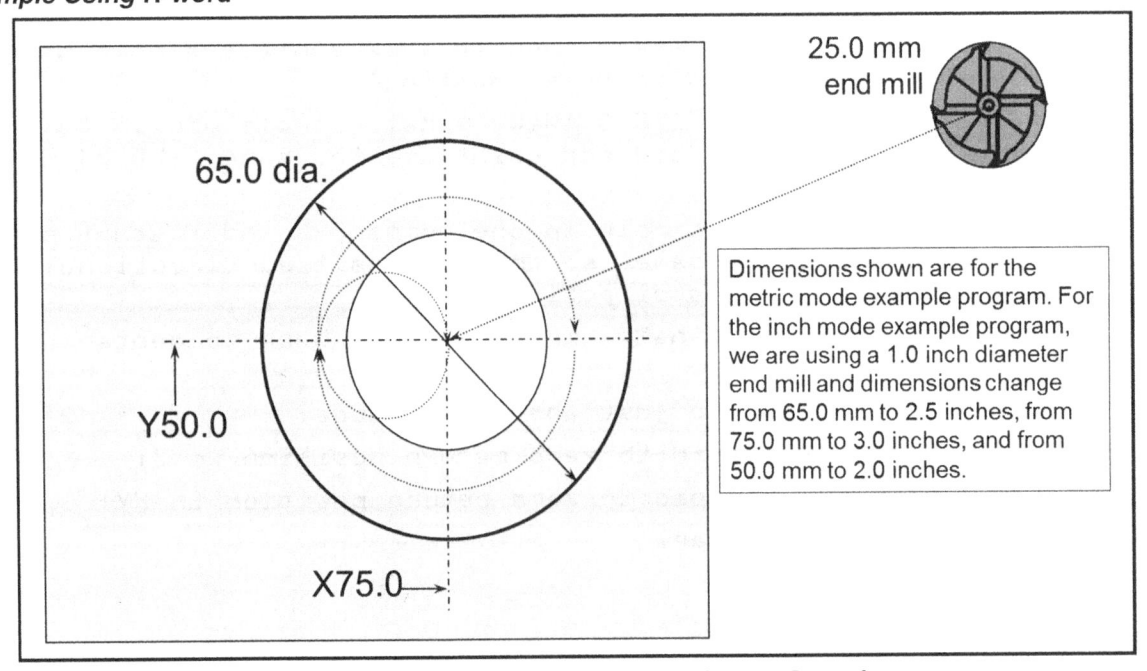

Figure 3.10: Full circle in two commands using R-word

Lesson 3.1: Programming The Three Basic Motion Types

Program with comments (<u>metric-mode</u> example program):

```
%
O0012 (Program number)
N005 G54 G21 G90 S400 M03 (Select workpiece coordinate system setting offset #1, metric & absolute modes, start spindle fwd at 400 rpm)
N010 G00 X75.0 Y50.0 (Rapid to center of circular pocket)
N015 G43 H01 Z2.0 (Instate tool length compensation, rapid to just above workpiece)
N020 G01 Z-6.0 F1750.0 (Fast feed to work surface-a hole has been drilled in this location)
N025 G02 X55.0 R10.0 F5.0 (Make a half-circle arc-in to the pocket surface-Y start and end point are the same [Y2.0] so Y is omitted)
N030 X95.0 R20.0 (Mill top side of circle-one half the circle)
N035 X55.0 R20.0 (Mill bottom side of circle)
N040 X75.0 R10.0 (Make half-circle arc-out back to center of pocket)
N045 G00 Z2.0 (Rapid to just above workpiece)
N050 G91 G28 Z0 (Rapid to zero return position in Z)
N055 M30 (End of program)
%
```

Program with comments (<u>inch-mode</u> example program):

```
%
O0012 (Program number)
N005 G54 G20 G90 S400 M03 (Select workpiece coordinate system setting offset #1, inch & absolute modes, start spindle fwd at 400 rpm)
N010 G00 X3.0 Y2.0 (Rapid to center of circular pocket)
N015 G43 H01 Z0.1 (Instate tool length compensation, rapid to just above workpiece)
N020 G01 Z-0.25 F50.0 (Fast feed to work surface-a hole has been drilled in this location)
N025 G02 X2.25 R.375 F5.0 (Make a half-circle arc-in to the pocket surface-Y start and end point are the same [Y2.0] so Y is omitted)
N030 X3.75 R0.75 (Mill top side of circle-one half the circle)
N035 X2.25 R0.75 (Mill bottom side of circle)
N040 X3.0 R0.375 (Make half-circle arc-out back to center of pocket)
N045 G00 Z0.1 (Rapid to just above workpiece)
N050 G91 G28 Z0 (Rapid to zero return position in Z)
N055 M30 (End of program)
%
```

Key Concept 3: Understanding Motion Types

Planning Your Own Tool Paths

Planning tool paths for hole-machining tools, like drills, is pretty simple. To drill a hole, you will first rapid the drill to the hole location in X and Y. Next you'll rapid the drill down to just above the work surface in Z. You'll machine the hole with G01 and then rapid the drill out of the hole in Z. If there are more holes to drill, you'll rapid the drill to the next hole-location (clearing any obstructions along the way) then machine the hole with G01, and retract with G00. You'll then repeat this for all other holes to be drilled.

While there are other types of holes that must be machined (peck-drilling, tapping, reaming, for example), their tool paths are still pretty easy to plan. And as you'll see in a later lesson, there are several hole-machining canned cycles that dramatically simplify the programming of holes.

But the tool paths for milling operations can be more difficult to plan–even for experienced machinists. You've already seen one complication based upon whether or not you'll be using cutter radius compensation (work surface tool path or cutter centerline tool path).

Milling tool paths can be difficult to plan. In Figure 3.7, the contour must be milled. You must plan an appropriate approach position. We chose a position left of the workpiece by a clearance distance in X and flush with the lower surface to mill in Y. This approach position, must of course, be based upon the kind of milling you will be doing (climb or conventional) and must be clear of any clamps or obstructions in the setup. From this point, the tool path for the workpiece may be pretty easy to understand–but do you see anything wrong with it?

Notice that in our example program, after milling the final radius in block N140, we simply retracted the milling cutter in Z. While this will work, it will probably leave a nasty mark called a *witness mark* on the milled surface. Unless there is nothing critical about this milled surface, it will be better to arc-off the last surface with a circular command. This will eliminate the witness mark. This means an additional circular command is required after block N140:

In the metric-mode example program:

```
N145 G02 X17.0 Y-13.5 R6.0 (Arc-off the work surface)
```

In the inch-mode example program:

```
N145 G02 X0.7 Y-0.55 R0.25 (Arc-off the work surface)
```

Again, planning tool paths for milling cutters can be challenging. In many cases, it helps to actually draw the tool paths you intend to use. Use a computer aided drawing (CAD) system or circle template and straight edge, and be sure to draw everything to scale.

Key Points for Lesson 3.1:

- ✓ CNC control manufacturers provide interpolation types for the motions that their machines must make.
- ✓ You must know the point on the cutting tool that is being programmed.
- ✓ The three most basic motion types are rapid motion–G00, linear motion–G01 (also called straight-line motion), and circular motion–G02 (clockwise) and G03 (counter-clockwise).
- ✓ All motion types are modal.
- ✓ All motion types require that you specify the end point for the motion in the motion command.
- ✓ Use rapid motion whenever the tool is not cutting to minimize cycle time.
- ✓ Linear and circular motions additionally require that you specify a feedrate for the motion–and feedrate is also modal.
- ✓ Circular motion additionally requires that you specify the arc size for the motion–and the R-word is the simplest way to do so.

Exercise: Work On Your First CNC Program

Practice calculating tool center coordinates and selecting motion types. The workpiece engineering drawing has been marked up with the planned machining process and a sketch has been made of the tool path for the ¾ milling cutter. Use the appropriate blank spaces on the workpiece engineering drawing to calculate and enter the axis coordinates for the holes and the 20-mm (3/4-in) milling cutter moves.

Then complete the G-code program that follows considering the feeds and speeds required for each tool, modal G-codes and other modal words and only programming moving axes.

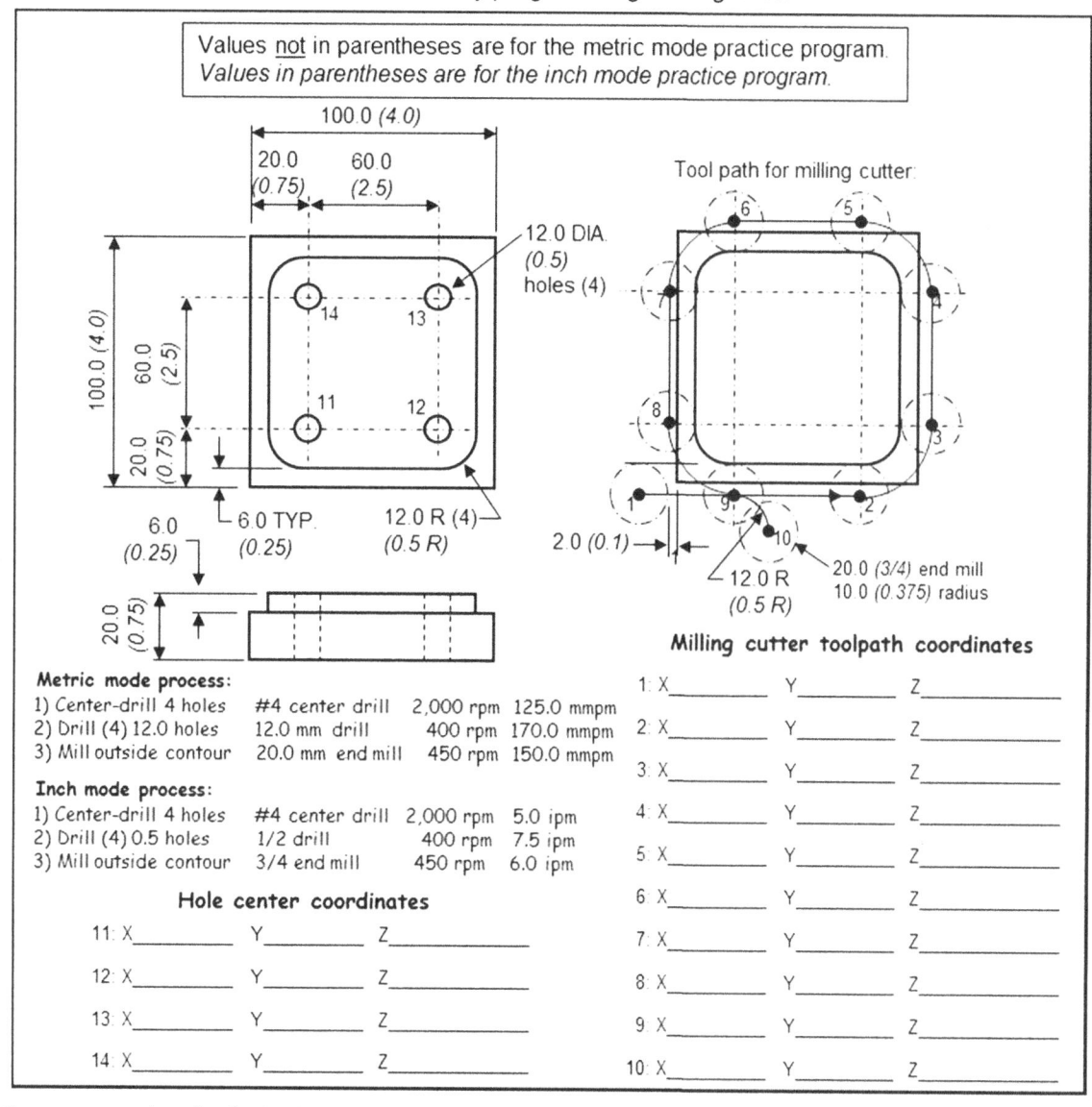

Values **not** in parentheses are for the metric mode practice program.
Values in parentheses are for the inch mode practice program.

Metric mode process:
1) Center-drill 4 holes   #4 center drill   2,000 rpm   125.0 mmpm
2) Drill (4) 12.0 holes   12.0 mm drill   400 rpm   170.0 mmpm
3) Mill outside contour   20.0 mm end mill   450 rpm   150.0 mmpm

Inch mode process:
1) Center-drill 4 holes   #4 center drill   2,000 rpm   5.0 ipm
2) Drill (4) 0.5 holes   1/2 drill   400 rpm   7.5 ipm
3) Mill outside contour   3/4 end mill   450 rpm   6.0 ipm

Hole center coordinates
11: X_____ Y_____ Z_____
12: X_____ Y_____ Z_____
13: X_____ Y_____ Z_____
14: X_____ Y_____ Z_____

Milling cutter toolpath coordinates
1: X_____ Y_____ Z_____
2: X_____ Y_____ Z_____
3: X_____ Y_____ Z_____
4: X_____ Y_____ Z_____
5: X_____ Y_____ Z_____
6: X_____ Y_____ Z_____
7: X_____ Y_____ Z_____
8: X_____ Y_____ Z_____
9: X_____ Y_____ Z_____
10: X_____ Y_____ Z_____

Practice program (works for both metric mode and inch mode):

```
%
O0013 (EXERCISE MOTION TYPES)
N010 G___ G90 G94 G54 (Specify measurement system mode)
N020 T01 M06 (select #4 center drill)
N030 S____ M03 T02 (start spindle)
N040 G00 X____ Y____ (rapid above hole 11)
N050 G43 H01 Z____ M08 (rapid to workpiece)
N060 G__ Z____ F___ (center drill hole)
N070 G__ Z____ (rapid retract)
N080 X____ (rapid above hole 12)
```

FANUC CERT – Machining Center   ©2017 CNC Concepts, Inc.

Key Concept 3: Understanding Motion Types

```
N090 G__ Z_____ (center drill hole)
N100 ___ ___ (rapid retract)
N110 _____ (rapid above hole 13)
N120 ___ _____ (center drill hole)
N130 ___ ___ (rapid retract)
N140 _____ (rapid above hole 14)
N150 ___ _____ (center drill hole)
N160 ___ ___ M09 (rapid retract)
N163 G91 G28 Z0 (retract to tool change position)
N165 M01 (optional stop)

N170 T02 M06 (select 12.0 mm [or 1/2] inch drill)
N180 G90 G54 S_____ M03 T03 (start spindle)
N190 G__ X_____ Y_____ (rapid to hole 11)
N200 G43 H02 Z_____ M08 (rapid to workpiece)
N210 G__ Z_____ F_____ (drill hole 11)
N220 ___ Z___ (rapid retract)
N230 _____ (rapid to hole 12)
N240 ___ ___ (drill hole)
N250 ___ ___ (rapid retract)
N260 _____ (rapid to hole 13)
N270 ___ ___ (drill hole)
N280 ___ ___ (rapid retract)
N290 _____ (rapid to hole 14)
N300 ___ _____ (drill hole)
N310 ___ _____ M09 (rapid retract)
N313 G91 G28 Z0 (retract to tool change position)
N315 M01 (optional stop)

N320 T03 M06 (select 20.0 mm [or 3/4 inch] end mill)
N330 G90 G54 S_____ M03 T01 (start spindle)
N340 ___ _____ _____ (rapid to milling start pt. 1)
N350 G43 H03 Z_____ M08 (rapid to workpiece)
N360 G01 Z_____ F50.0 (fast-feed move to cutting depth)
N370 _____ F_____ (machine to pt. 2)
N380 G__ _____ _____ R_____ (machine to pt. 3)
N390 ___ _____ (machine to pt. 4)
N400 ___ _____ _____ _____ (machine to pt. 5)
N410 ___ _____ (machine to pt. 6)
N420 ___ _____ _____ _____ (machine to pt. 7)
N430 ___ _____ (machine to pt. 8)
N440 ___ _____ _____ _____ (machine to pt. 9)
N450 ___ _____ _____ _____ (exit cut to pt. 10)
N460 G00 Z_____ M09 (rapid retract)
N470 G91 G28 Z0.0
N480 G28 X0.0 Y0.0
N490 M30
%
```

```
%
O0013 (METRIC MODE EXERCISE MOTION TYPES AMSWERS)
N010 G21 G90 G94 G54
N020 T01 M06 (select #4 center drill)
N030 S1200 M03 T02 (start spindle)
N040 G00 X20.0 Y20.0 (rapid above hole 11)
N050 G43 H01 Z2.0 M08 (rapid to workpiece)
N060 G01 Z-3.0 F125.0 (center drill hole)
N070 G00 Z2.0 (rapid retract)
N080 X80.0 (rapid above hole 12)
N090 G01 Z-3.0 (center drill hole)
N100 G00 Z2.0 (rapid retract)
N110 Y80.0 (rapid above hole 13)
N120 G01 Z-3.0 (center drill hole)
N130 G00 Z2.0 (rapid retract)
N140 X20.0 (rapid above hole 14)
N150 G01 Z-3.0 (center drill hole)
N160 G00 Z2.0 M09 (rapid retract)
N163 G91 G28 Z0 (retract to tool change position)
N165 M01 (optional stop)

N170 T02 M06 (select 12.0 mm [or ½ inch] drill)
N180 G90 G54 S400 M03 T03 (start spindle)
N190 G00 X20.0 Y20.0 (rapid to hole 11)
N200 G43 H02 Z2.0 M08 (rapid to workpiece)
N210 G01 Z-24.35 F170.0 (drill hole 11)
N220 G00 Z2.0 (rapid retract)
N230 X80.0 (rapid to hole 12)
N240 G01 Z-24.35 (drill hole)
N250 G00 Z2.0 (rapid retract)
N260 Y80.0 (rapid to hole 13)
N270 G01 Z-24.35 (drill hole)
N280 G00 Z2.0 (rapid retract)
N290 X20.0 (rapid to hole 14)
N300 G01 Z-24.35 (drill hole)
N310 G00 Z2.0 M09 (rapid retract)
N313 G91 G28 Z0 (retract to tool change position)
N315 M01 (optional stop)

N320 T03 M06 (select 20.0 mm [or 3/4 inch] end mill)
N330 G90 G54 S450 M03 T01 (start spindle)
N340 G00 X-12.0 Y-4.0 (rapid to milling start pt. 1)
N350 G43 H03 Z2.0 M08 (rapid to workpiece)
N360 G01 Z-6.0 F1750.0 (fast-feed move to cutting depth)
N370 X82.0 F150.0 (machine to pt. 2)
N380 G03 X106.0 Y18.0 R22.0 (machine to pt. 3)
N390 G01 Y82.0 (machine to pt. 4)
N400 G03 X82.0 Y106.0 R22.0 (machine to pt. 5)
N410 G01 X18.0 (machine to pt. 6)
N420 G03 X-4.0 Y82.0 R22.0 (machine to pt. 7)
N430 G01 Y18.0 (machine to pt. 8)
N440 G03 X18.0 Y-4.0 R22.0 (machine to pt. 9)
N450 G02 X30.0 Y-16.0 R12.0 (exit cut to pt. 10)
N460 G00 Z0.1 M09 (rapid retract)
N470 G91 G28 Z0.0
N480 G28 X0 Y0
N490 M30
%
```

Key Concept 3: Understanding Motion Types

```
%
O0013 (INCH MODE EXERCISE MOTION TYPES AMSWERS)
N010 G20 G90 G94 G54
N020 T01 M06 (Center drill)
N030 G90 G54 S1200 M03 T02 (start spindle)
N040 G00 X0.75 Y0.75 (rapid above hole 11)
N050 G43 H01 Z0.1 M08 (rapid to workpiece)
N060 G01 Z-0.12 F5.0 (center drill hole)
N070 G00 Z0.1 (rapid retract)
N080 X3.25 (rapid above hole 12)
N090 G01 Z-0.12 (center drill hole)
N100 G00 Z0.1 (rapid retract)
N110 Y3.25 (rapid above hole 13)
N120 G01 Z-0.12 (center drill hole)
N130 G00 Z0.1 (rapid retract)
N140 X0.75 (rapid above hole 14)
N150 G01 Z-0.12 (center drill hole)
N160 G00 Z0.1 M09 (rapid retract)
N163 G91 G28 Z0 (retract to tool change position)
N165 M01 (optional stop)

N170 T02 M06 (select 1/2 drill)
N180 G90 G54 S400 M03 T03 (start spindle)
N190 G00 X0.75 Y0.75 (rapid to hole 11)
N200 G43 H02 Z0.1 M08 (rapid to workpiece)
N210 G01 Z-1.05 F7.5 (drill hole 11)
N220 G00 Z0.1 (rapid retract)
N230 X3.25 (rapid to hole 12)
N240 G01 Z-1.05 (drill hole)
N250 G00 Z0.1 (rapid retract)
N260 Y3.25 (rapid to hole 13)
N270 G01 Z-1.05 (drill hole)
N280 G00 Z0.1 (rapid retract)
N290 X0.75 (rapid to hole 14)
N300 G01 Z-1.05 (drill hole)
N310 G00 Z0.1 M09 (rapid retract)
N313 G91 G28 Z0 (retract to tool change position)
N315 M01 (optional stop)

N320 T03 M06 (select 3/4 end mill)
N330 G90 G54 S450 M03 T01 (start spindle)
N340 G00 X-0.475 Y-0.125 (rapid to milling start pt. 1)
N350 G43 H03 Z0.1 M08 (rapid to workpiece)
N360 G01 Z-0.25 F50.0 (move to cutting depth)
N370 X3.25 F6.0 (machine to pt. 2)
N380 G03 X4.125 Y0.75 R0.875 (machine to pt. 3)
N390 G01 Y3.25 (machine to pt. 4)
N400 G03 X3.25 Y4.125 R0.875 (machine to pt. 5)
N410 G01 X0.75 (machine to pt. 6)
N420 G03 X-0.125 Y3.25 R0.875 (machine to pt. 7)
N430 G01 Y0.75 (machine to pt. 8)
N440 G03 X0.75 Y-0.125 R0.875 (machine to pt. 9)
N450 G02 X1.25 Y-0.625 R0.5 (exit cut to pt. 10)
N460 G00 Z0.1 M09 (rapid retract)
N470 G91 G28 Z0.0
N480 G28 X0.0 Y0.0
N490 M30
%
```

Key Concept 4: Know the Compensation Types

CNC machining centers provide three kinds of compensation to help you deal with tooling related problems. In essence, each compensation type allows you to create your CNC program without having to know every detail about your tooling. The setup person will be entering certain tooling information into the machine separately from the program.

Key Concept 4 contains four lessons:

- 4.1: Introduction to compensation
- 4.2: Tool length compensation
- 4.3: Cutter radius compensation
- 4.4: Workpiece coordinate system offsets

The fourth Key Concept is that you must know the three kinds of compensation designed to let you ignore certain tooling problems as you develop CNC programs. In lesson 4.1, we'll introduce you to compensation, showing the reasons why compensation is needed on CNC machining centers.

Two of the compensation types are related to cutting tools: tool length compensation and cutter radius compensation. We'll discuss them in lessons 4.2 and 4.3. The third compensation type, workpiece coordinate system offsets, is related to work holding devices–and we'll discuss workpiece coordinate system offsets in lesson 4.4. While we discuss the use of workpiece coordinate system offsets for setting the workpiece coordinate system zero in Key Concept 1, we will show more applications for them in lesson 4.4.

Not all of the material presented in Key Concept 4 may be of immediate importance to you. If you do not perform any contour milling on your machines (maybe your machines only perform hole-machining operations), then you will not have much need for cutter radius compensation. Similarly, you may not need to know any more about workpiece coordinate system offsets than what was discussed in Key Concept 1. Depending upon your needs, you may want to simply skim these lessons to gain an understanding of when and why the related features are needed–then move on. You can always come back for further study should your needs change.

On the other hand, tool length compensation is a very important feature that is used in every tool of every program you write. You will need to completely master its use.

(Page intentionally left blank)

Lesson 4.1–Introduction to Compensation

An airplane pilot must compensate for wind direction and velocity when setting a heading. A race-car driver must compensate for track conditions as they negotiate a turn. A marksman must compensate for the distance to the target when firing a rifle. And a CNC programmer must compensate for certain tooling-related issues as programs are written.

Objectives

After completing this lesson, students should be able to:

- ✓ Describe the need for CNC compensations
- ✓ Describe the three primary compensation types
- ✓ Describe the need for trial machining
- ✓ Describe the target value for tolerances that must be held
- ✓ Describe the use of offsets

What are Compensations and Why are They Needed?

When you compensate for something, you are allowing for some unpredictable (or nearly unpredictable) variation. A race car driver must compensate for the condition of the race track before a curve can be negotiated. In this case, the unpredictable variation is the condition of the track. An airplane pilot must compensate for the wind direction and velocity before a heading can be set. For them, wind direction and velocity are the unpredictable variations. A marksman must compensate for the distance to the target before a shot can be fired–and the distance to the target is the unpredictable variation. The marksman analogy is remarkably similar to what happens with most forms of CNC compensation. Let's take it further...

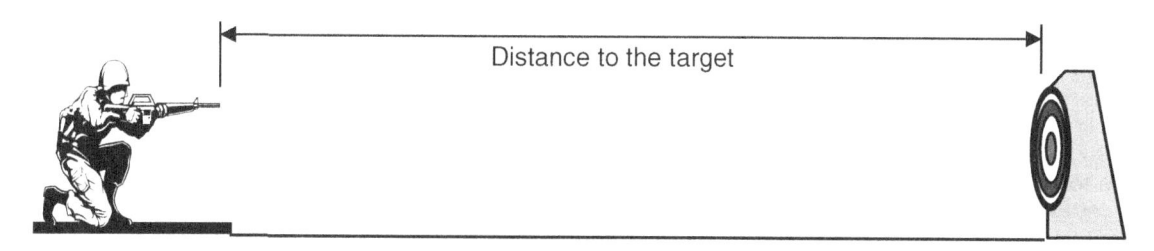

Before a marksman can fire a rifle, they must judge the distance to the target. If the target is judged to be fifty yards away, the sight on the rifle will be adjusted accordingly. When the marksman adjusts the sight, they are compensating for the distance to the target. But even after this preliminary adjustment and before the first shot is fired, the marksman cannot be absolutely sure that the sight is adjusted perfectly. If they've incorrectly judged the distance–or if some other variation (like wind) affects the sight adjustment–the first shot will not be perfectly in the center of the target.

After the first shot is fired, the marksman will know more. If the shot is not perfectly centered, another adjustment will be needed. And the second shot will be closer to the center of the target than the first. Depending upon the skill of the marksman, it might be necessary to repeat this process until the sight is perfectly adjusted.

With all forms of CNC compensation, the setup person will do their best to determine the compensation values needed to perfectly machine the workpiece. But until machining actually occurs, the setup person cannot be sure that their initial compensation values are correct. After machining, they may find that another variation (like tool pressure) is causing the initial adjustment to be incorrect. Depending upon the tolerances for the surfaces being machined, a second adjustment may be required. After this second adjustment, machining will be more precise.

Key Concept 4: Know The Compensation Types

There is even a way to make an initial adjustment (prior to machining) that ensures excess material will remain on the machined surface after the first machining attempt (this technique is called trial machining). This guarantees that the workpiece will not be scrapped when the cutting tool machines for the first time–and is especially important for very tight (small) tolerances. With tight tolerances, even a small machining imperfection will cause a scrap workpiece.

Once the cutting tool has machined for the first time, the setup person will stop the cycle and measure the surface. If they have used the trial machining technique, there will be more material yet to remove. They will then make the appropriate adjustment and re-run the cutting tool. The second time the cutting tool machines, the surface will be within its tolerance band, probably right at the target dimension.

More On Tolerances

All dimensions have tolerances. And from lesson 1.4, you know that you must program the mean value of the tolerance band for every coordinate you include in your programs. The mean value, of course, is right in the middle of the tolerance band.

Companies vary when it comes to how tight (small) the tolerances are that they machine on their CNC machining centers. Generally speaking, overall tolerances over about 0.25 mm (about 0.010 inch) are considered pretty open (easy to hold with today's CNC machining centers). Tolerances between 0.050-0.25 mm (0.002 and 0.010 inch) are common, and still not considered to be very tight. But tolerances under 0.050 mm (0.002 inches) can be more difficult to hold. And under 0.0013 mm (0.0005 inch)–which many companies do regularly hold on their CNC machining centers–can be quite challenging–especially when many workpieces must be produced.

The Initial Setting for Compensation

The setup person will do their best to assemble and measure certain cutting tool attributes (like length and diameter). They will then enter their measured values into the CNC (into something called tool offsets). But even if they perfectly measure and enter tooling values, and even if the programmer specifies the mean value for every tolerance in the program, there is no guarantee that every cutting tool will perfectly machine each dimension to the mean value of its tolerance band.

Tool pressure, which is the tendency for a cutting tool to deflect from the workpiece during machining, will always affect the way a cutting tool machines. It usually has the tendency of pushing the cutting tool away from the surface being machined. While the surface being machined should be close to its mean value (again, assuming the setup person perfectly measures and enters tooling information), it may not be perfectly at the mean value.

How tight is the tolerance? The tighter the tolerance, the more likely it will be that the deflection caused by tool pressure will cause the machined surface to be outside the tolerance band when the cutting tool machines for the first time. This could cause a scrap workpiece. This is the reason why *trial machining* is required–to ensure that the first workpiece gets machined correctly.

And by the way, we have assumed that the setup person has perfectly measured and entered tooling information. Any mistakes will, of course, increase the potential for problems holding size on the first workpiece to be machined.

When is Trial Machining Required?

First let's talk about when trial machining is not required. Trial machining is not required for most roughing operations–like rough milling and rough boring. While a measurement should still be taken right after the roughing operation to confirm that the appropriate amount of finishing stock is being left for the finishing tool (and an adjustment must be made if not), trial machining is not necessary.

There are certain cutting tools and attributes that do not allow adjustments, and trial machining cannot be done for these tools. With drills, for example, you cannot control the diameter that the drill machines. Hole-diameter is based upon the diameter of the drill. The same goes for taps, reamers and counter-boring tools. So for machining operations that cannot be adjusted, you will not be trial machining.

Also, whenever tolerances are greater than about 0.075 mm (0.003 inch) or so, it is not unreasonable to expect the setup person (or tool-setter) to assemble and measure cutting tools accurately enough so that the cutting tool will machine within the tolerance band (even considering tool pressure) on the first try. While an adjustment may be necessary after initial machining to bring the dimension precisely to the target value, trial machining should not be necessary. Examples include the depth of many (blind) holes and a certain milling operations.

As stated, trial machining is only required when machining tight tolerances. Though the actual cut-off point for when trial machining is required varies based upon the skill of the setup person and the operation being performed, it is not uncommon to trial machine for dimensions that have tolerances under about 0.050 mm (0.002 inch). This includes finish boring operations and many finish milling operations.

What Happens as Tools Begin to Wear?

So say the setup person uses trial machining techniques for critical dimensions and the first workpiece passes inspection. Now the production run begins. As cutting tools continue to machine workpieces, of course, they will begin to show signs of wear. Certain tools, like drills, will continue to machine properly for their entire lives without having an impact on the dimensions they machine (hole size and depth).

But as certain tools begin to wear, like milling cutters and boring bars, a small amount of material will wear away from the tool's cutting edge (a boring bar or an end mill will actually get a little smaller in diameter). Depending upon the tolerance band for the dimension being machined by the cutting tool, it may be necessary to make additional sizing adjustments during the cutting tool's life to ensure that the cutting tool continues to machine acceptable workpieces.

What do You Shoot For?

The value for each dimension that you are aiming for as you make adjustments is called the *target value*. In some companies, CNC setup people and operators are told to always target the mean value of the tolerance band (the same value the programmer includes in the program). But to prolong the time between needed adjustments, some companies ask their CNC people to target a value that is closer to the high or low limit of the tolerance band, whichever will prolong the time between adjustments. In any event, CNC people must know the target value for each dimension they must machine (this information should be in the production run documentation).

Why do Programmers Have to Know About Compensations?

Admittedly, much of this discussion is more related to setup and operation than it is to programming. But a programmer must understand enough about machine usage to be able to direct people that run the machine. And these discussions go to the heart of one of the most important reasons why compensation is required on CNC machining centers: to allow sizing adjustments without needing to modify the program.

Understanding Offsets

All three compensation types use offsets. Offsets are storage locations for values. They are very much like memories in an electronic calculator. With a calculator, if a value is needed several times during your calculations, you can store the value in one of the memories. When the value is needed, you simply type one or two keys and the value returns. In similar fashion, the setup person or operator can enter important tooling-related values into offsets. When they are needed by the program, a command within the program will invoke the value of the offset. And by the way, just as a calculator's memory value has no meaning to the calculator until it is invoked during a calculation, neither does a CNC offset have any meaning to the CNC control until it is invoked by a CNC program.

Like the memories of most calculators, offsets are designated with offset numbers. Offset number one may have a value of 166.215 in metric mode or 6.5439 in inch mode. Offset number two may have a value of 159.911 (metric mode) or 6.2957 (inch mode). For cutting tool related offsets, offset numbers are made to correspond in some way to tool station numbers. For example, the tool length compensation value for the cutting tool placed in station number one is commonly entered in one of the registers for offset number one.

Unlike the memories of an electronic calculator that will be lost when the calculator's power is turned off, CNC offsets are more permanent. They will be retained even after the machine's power is turned off–and until an operator or setup person changes them.

Offsets are used with each compensation type to tell the control important information about tooling. From the marksman analogy, you can think of offset values as being like the amount of sight adjustment a marksman must make prior to firing a shot. Tooling related information entered into offsets includes each cutting tool's length, each milling cutter's radius, and program zero assignment values for work holding devices.

Key Concept 4: Know The Compensation Types

Offset Organization

Machining center controls vary with regard to how many offsets are available and how they are organized. First of all, rest assured that you will always have enough offsets to handle your applications. Most machine tool builders supply many more offsets than are required even in the most elaborate of applications (an exception to this statement may sometimes be with regard to workpiece coordinate system setting offsets). Machining centers have two distinct sets of offsets, one for cutting tools (tool offsets) and another for the workpiece coordinate system zero (work offsets).

Offsets Related to Cutting Tools

All cutting tools will require a tool length compensation offset. Milling cutters that perform side milling operations will additionally require a cutter radius compensation offset. Today's CNCs also separate the tool offsets into geometry (the value measured at initial setup) and wear (the value determined by trial machining or tool wear and used when making sizing adjustments). CNC manufacturers vary when it comes to how offsets are organized and displayed.

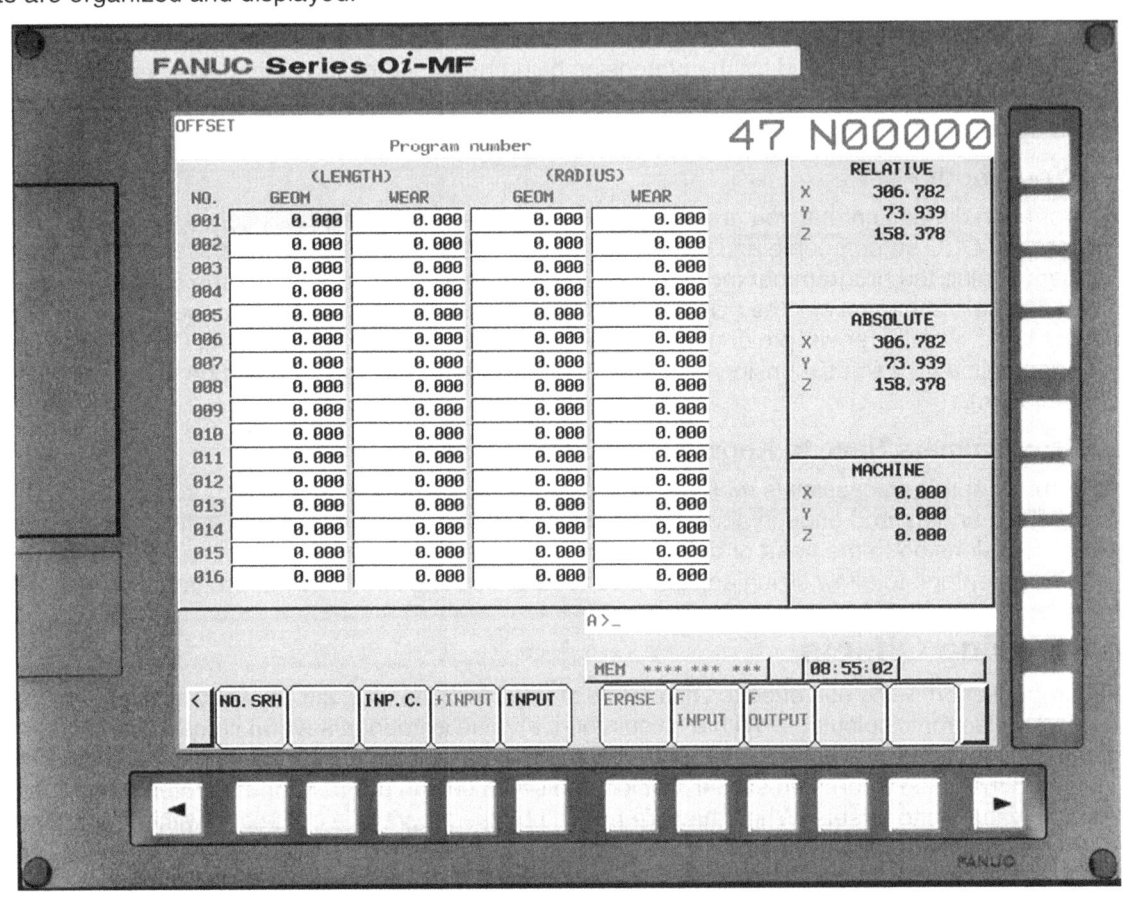

Figure 4.1: Display screen having four values per offset number (shown in metric mode)

Today's CNCs typically have four fields of data available for each tool pocket in the tool changer magazine. That means the tool length offset invoking word (H-word – example, H01) has the same number as the cutter radius offset (D-word, example, D01) and the tool number (T-word – example, T01). The tool length offset has two components, geometry and wear. The tool length geometry offset is the nominal length of the tool measured by the setup person. The tool length wear offset is used for fine tuning due to trial machining or tool wear. The cutter radius offset also has two components, geometry and wear. The tool wear geometry offset is the nominal radius of the cutter measured by the setup person. The cutter radius wear offset is used for fine tuning due to trial machining or tool wear. The CNC adds the respective geometry and wear offsets together to compensate for the tool length or the cutter radius.

Older CNC may only have one offset register per offset. The value stored in the register may be a tool length offset number or a cutter radius offset. This complicates tool offset assignments and you cannot have a tool number designator, tool length offset designator and cutter radius offset designator, all with the same value (for example, T01 H01 D01). Figure 4.1 shows the display screen for this kind of offset organization.

Figure 4.2: Display screen having only one value per offset number (shown in inch mode)

There is yet another offset organization method that provides that provides two values per offset number– one for tool length compensation and the other for cutter radius compensation. The same number can be used for tool number, tool length offset and cutter radius, but the setup person and operator have to combine the geometry and wear values each time an offset us updated (there are no separate columns for geometry and wear values). This can lead to calculation and data entry errors.

Offsets Related to Work Coordinate System Zero

Workpiece coordinate system offset values are placed in work offsets, as is introduced in lesson 1.6. Each offset has at least three registers, one for X, one for Y, and one for Z. If the machining center has any rotary axes, there will also be a register for each rotary axis. The figure below shows the first page of workpiece coordinate system setting offset display screen for the FANUC 0iF with the 10.4-inch display.

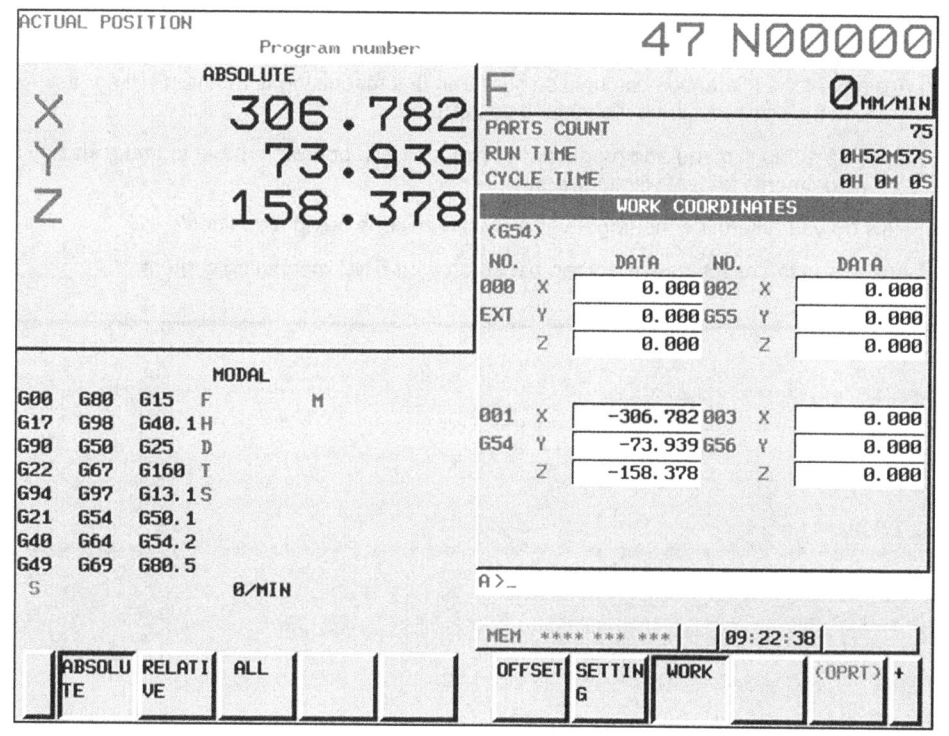

Figure 4.3: Workpiece coordinate system offset display screen (shown in metric mode)

Key Concept 4: Know The Compensation Types

How Are Offsets Instated?

Offsets have no meaning to the control until they are instated by a program command. For offsets used with tool length compensation, an H-word is used to instate the offset (we'll discuss tool length compensation in lesson 4.2). H01, for example, instates the tool length value (totaling geometry and wear registers) in offset number 1.

For cutter radius compensation, most controls use a D-word to instate the related offset (though some controls do use the H-word for this as well). D01 instates the value in offset number one. We'll discuss cutter radius compensation in lesson 4.3.

For controls that have but one register per offset, there is a bit of a problem. This requires the same table of offsets to be shared for both tool length and cutter radius compensation. You cannot use the same offset number for both tool length compensation and cutter radius compensation (as you can if there are two or four offset registers per tool offset). Say, for example, tool number 1 is an end mill that will be using cutter radius compensation. You pick offset number 1 in which to store the tool length compensation value for this tool. This means, of course, that you must use another offset number in which to store the cutter radius compensation value. We'll provide some suggestions for handling this problem in lessons 4.2 and 4.3.

Key Points for Lesson 4.1:

- ✓ Compensation types allow you to ignore certain tooling-related problems as you write programs.
- ✓ Compensation types allow sizing and trial machining for critical workpiece attributes.
- ✓ Setup people and operators must know the target value for each dimension they must machine.
- ✓ Critical dimensions require trial machining for the first workpiece being machined.
- ✓ Tool wear can affect the dimensions machined by a tool. In some cases, adjustments must be made during the tool's life.
- ✓ Offsets are used in which to store compensation values.
- ✓ The offset tables vary from one machining center to another.

Talk to experienced people in your company...
...to learn more about how tight tolerances are held.

1) What are the tightest tolerances held on your CNC machining centers?

2) Are any of the tolerances you hold so tight that trial machining is required? If so, ask to see the trial machining techniques being used.

3) Once as setup is made and production is run, do CNC operators have to make sizing adjustments to deal with tool wear?

4) How do you determine the target value for dimensions being machined?

5) Ask to see the offset display screen pages on your CNC machining centers.

Lesson 4.2–Tool Length Compensation

Tool length compensation allows a programmer to ignore the precise length of each tool as a program is written. It is used for every tool in every program you write–so you must understand this important CNC feature.

Objectives

After completing this lesson, students should be able to:

- ✓ Describe the reasons tool length compensation is needed
- ✓ Describe how to program tool length compensation
- ✓ Describe how to select which tool compensation offset to use for each tool
- ✓ Describe two methods for measuring tool length offset values
- ✓ Describe the advantages to using the recommended method of measuring tool length offset values
- ✓ Describe the typical mistakes programmers, setup people and operators make with tool length compensation
- ✓ Describe how tool length compensation values can be measured on a CNC machining center
- ✓ Describe trial machining, how to do it and when it is needed

Introduction

You know that workpiece coordinate system offset values are entered into work offsets. For the X and Y axes, the values are the distances between the workpiece coordinate system zero and the spindle center (while the machine is at its zero return position). So when you specify a position of X25.0 Y25.0 in a program, the machine will be able to send the spindle center (and tool center) to this position–relative to program zero. For the Z-axis, the value is the distance between the Z-axis workpiece coordinate system zero surface and the spindle nose (again, while the machine is at its zero return position).

But for cutting tool positioning, you don't want to specify Z-axis positions from the spindle nose. This would be very cumbersome–and it would require that you know the precise length of each tool before you could even write the program. Instead, you want to specify Z-axis positions from the tip of each cutting tool so that when you specify a position of Z2.0, the tool tip will move to this position. In order to program the tool tip in the Z-axis, a feature called tool length compensation must be used. Mastering tool length compensation is the focus of this lesson.

Reasons Why Tool Length Compensation is Needed

Cutting tools used on machining centers differ from one another. For one thing, there are a variety of cutting tool types that are used on machining centers, including center drills, spot drills, drills, taps, reamers, boring bars, end mills, and face mills (among others). Each type of tool requires a different way of gripping the cutting tool in its holder. Some tools (like some straight shank tools) use a collet system. Others (like end mills) use a set-screw to hold the cutting tool in place. Yet others (like face mills and taps) require a special style of tool holder–designed especially for the cutting tool.

No Two Tools Will Have Exactly the Same Length

Given the vast assortment of cutting tools and cutting tool components available for use on CNC machining centers, it is unlikely that any two tools used in a program will be exactly the same length. And you will not know precisely how long each tool will be when you write the program. The figure below shows five different types of cutting tools to illustrate this point.

Key Concept 4: Know The Compensation Types

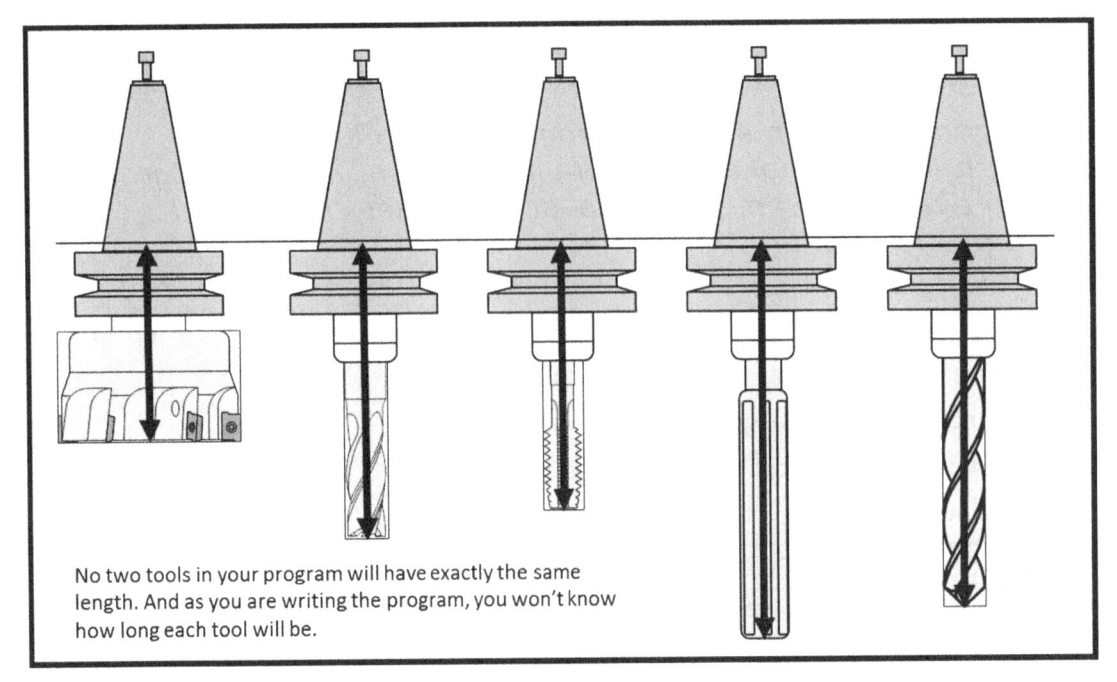

Figure 4.5: Five cutting tools that might be used in by a CNC machining center program

Tool length compensation will allow you to write programs even though you don't know how long the cutting tools will be at production time.

Tool's Length Will Vary Each Time it is Assembled

When a cutting tool is assembled more than once (even with the same components), its length will usually vary. Consider, for examples, straight shank tools that are placed in collet holders. Each time you assemble the tool, it will be of a different length. Tool length compensation will allow you to use the same program over and over again, even though each tool's length changes from one time the job is run to the next.

Tool Data is Entered Separately From the Program

The same program will work regardless of how long each cutting tool is. The program tells the control where to look for the length of each tool. During setup, the setup person (or someone) assembles and measures each cutting tool. The length of each tool is then placed in the appropriate location (a tool offset register).

Sizing and Trial Machining Must Often be Done

In lesson 4.1, we discuss the importance of being able to trial machine in order to machine the first workpiece correctly. And during a given tool's life a tool will wear and cause the surface being machined to change. Tool length compensation allows the setup person and operator to easily hold size for Z-axis related dimensions (pocket depths, hole-depths, etc.). The program need not be changed when workpiece dimensions must be adjusted.

What About Interference and Reach?

As is discussed in lesson 2.1, you must be concerned with whether or not your cutting tools will reach the surfaces to be machined–without over-traveling–and without interfering with the fixture of other obstructions on the work holding device. Tool length compensation will <u>not</u> help with these concerns.

Programming Tool Length Compensation

There are two popular methods for measuring and setting tool length compensation. As long as you use workpiece coordinate system offsets to assign program zero, the programming remains exactly the same regardless of which method you choose.

Tool length compensation is instated with a G43 word. Included within the G43 command is an H-word that specifies the offset number in which the tool length compensation value is stored. You must also include a Z-word in the G43 command, telling the machine where you want the tool tip to be positioned. This initial move allows the tool offset value to be instated without an unexpected Z-axis motion.

The G43 command will always be the cutting tool's first Z-axis motion. Said another way, you instate tool length compensation during each tool's first Z-axis motion as the tool approaches the workpiece in the Z-axis.

Once tool length compensation is instated, it remains in effect until the next tool is selected (and another G43 command is specified). All Z-axis motions you need the tool to make will be relative to the tool tip.

Since you will instate tool length compensation during the next tool's first Z-axis motion–using the appropriate offset of course–and since offsets are not accumulative–you need not cancel tool length compensation. G49 cancels tool length compensation, but if you the techniques shown in this course, you need not use G49 in your programs. Be warned, if you do program a G49 in a block by itself, it may cause the Z-axis to move by the tool offset amount to uninstall the tool offset.

Choosing the Offset Number to Be Used with Each Tool

As has been stated, offsets are storage registers for values. Each offset to be used with tool length compensation will contain a tool length compensation value for one tool. To keep from entering the tool length compensation value in the wrong offset, the programmer must use a logical approach for selecting offset numbers.

Keep it simple. Use the offset number that corresponds to the cutting tool's magazine station number. For example, use offset number one for the tool in tool station number one. Use offset number two for the tool in tool station number two. And so on.

The tool length compensation offset will be instated in the program during each tool's first Z-axis motion. An H-word is used to specify the offset number. And again, we recommend that you make the H-word number match the tool station number (the T-word number for each tool).

An Example Program

The figure below shows the drawing to be used for this example.

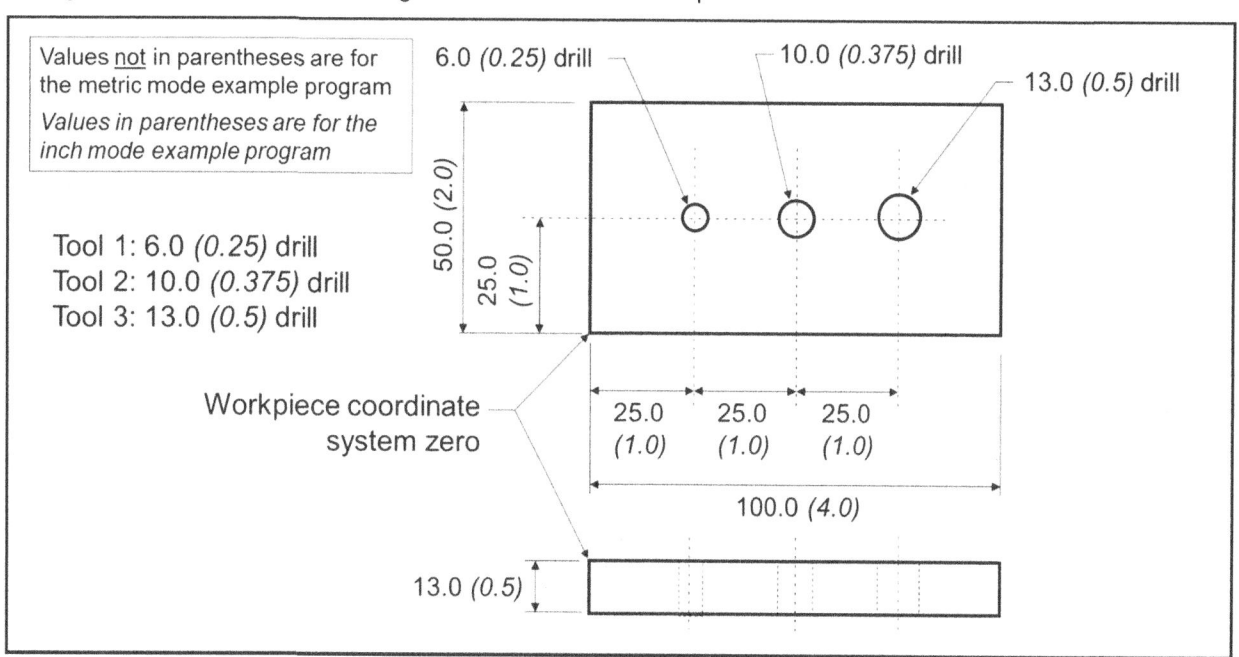

Figure 4.6: Drawing for example program

Program with comments (metric-mode example program):

```
%
O0014 (Program number)
N010 G21 G90 G54 (Select metric & absolute modes, workpiece
coordinate system setting offset #1)
N020 T01 M06 (Load tool #1 in spindle - 6.0 mm drill)
```

Key Concept 4: Know The Compensation Types

```
N030 S1200 M03 T02 (Start spindle fwd at 1200 rpm, get tool #2 ready)
N040 G00 X25.0 Y25.0 (Rapid to hole location in X and Y)
N050 G43 H01 Z2.0 (Instate tool length compensation for tool one, approach in Z to just above work surface)
N060 M08 (Turn on the coolant)
N070 G01 Z-15.55 F100.0 (Drill hole)
N080 G00 Z2.0 M09 (Rapid out of hole, turn off coolant)
N090 G91 G28 Z0 M19 (Rapid to tool change position, orient spindle)
N100 M01 (Optional stop)

N110 T02 M06 (Load tool #2 in spindle - 10.0 mm drill)
N120 G54 G90 S1000 M03 T03 (Select workpiece coordinate system setting offset #1, absolute mode, start spindle fwd at 1000 RPM, get tool number three ready)
N130 G00 X50.0 Y25.0 (Rapid to hole position in X and Y)
N140 G43 H02 Z2.0 (Instate tool length compensation for tool 2, approach in Z to just above work surface)
N150 M08 (Turn on coolant)
N160 G01 Z-16.75 F125.0 (Drill hole)
N170 G00 Z2.0 M09 (Rapid out of hole, turn off coolant)
N180 G91 G28 Z0 M19 (Rapid to tool change position, orient spindle)
N200 M01 (Optional stop)

N210 T03 M06 (Place tool number three in spindle)
(130. mm Drill)
N220 G54 G90 S800 M03 T01 (Select workpiece coordinate system setting offset #1, absolute mode, start spindle fwd at 800 RPM, get tool number one ready)
N230 G00 X75.0 Y25.0 (Rapid to hole in X and Y)
N240 G43 H03 Z2.0 (Instate tool length compensation for tool three, approach in Z to just above work surface)
N250 M08 (Turn on coolant)
N260 G01 Z-17.65 F152.0 (Drill hole)
N270 G00 Z2.0 M09 (Rapid out of hole, turn off coolant)
N280 G91 G28 Z0 M19 (Rapid to tool change position, orient spindle)
N285 G28 Y0 (Move to convenient workpiece loading position)
N290 M30 (End of program)
%
```

Blocks N050, N140, and N240 instate tool length compensation for each of the three tools. Notice that each of these commands is the first Z-axis movement for the tool (its approach movement in Z). Each instating command includes the G43 word, the appropriate H-word (that matches the tool station number that is currently in the spindle), and a Z-word. Once tool length compensation is instated, it will remain in effect until the next tool. Again, it never has to be canceled using this style of programming as the G28 automatically suspends the tool length compensation during the move to the zero return position.

Program with comments (inch-mode example program):

```
%
O0014 (Program number)
N010 G20 G90 G54 (Select inch & absolute modes, workpiece coordinate system setting offset #1)
N020 T01 M06 (Load tool #1 in spindle - 1/4 drill)
N030 S1200 M03 T02 (Start spindle fwd at 1200 rpm, get tool #2 ready)
N040 G00 X1.0 Y1.0 (Rapid to hole location in X and Y)
N050 G43 H01 Z0.1 (Instate tool length compensation for tool one, approach in Z to just above work surface)
N060 M08 (Turn on the coolant)
N070 G01 Z-0.65 F4.0 (Drill hole)
N080 G00 Z0.1 M09 (Rapid out of hole, turn off coolant)
N090 G91 G28 Z0 M19 (Rapid to tool change position, orient spindle)
N100 M01 (Optional stop)

N110 T02 M06 (Load tool #2 in spindle - 3/8 drill)
N120 G54 G90 S1000 M03 T03 (Select workpiece coordinate system setting offset #1, absolute mode, start spindle fwd at 1000 RPM, get tool number three ready)
N130 G00 X2.0 Y1.0 (Rapid to hole position in X and Y)
N140 G43 H02 Z0.1 (Instate tool length compensation for tool 2, approach in Z to just above work surface)
N150 M08 (Turn on coolant)
N160 G01 Z-0.7 F5.0 (Drill hole)
N170 G00 Z0.1 M09 (Rapid out of hole, turn off coolant)
N180 G91 G28 Z0 M19 (Rapid to tool change position, orient spindle)
N200 M01 (Optional stop)

N210 T03 M06 (Place tool number three in spindle)
(1/2 Drill)
N220 G54 G90 S800 M03 T01 (Select workpiece coordinate system setting offset #1, absolute mode, start spindle fwd at 800 RPM, get tool number one ready)
N230 G00 X3.0 Y1.0 (Rapid to hole in X and Y)
N240 G43 H03 Z0.1 (Instate tool length compensation for tool three, approach in Z to just above work surface)
N250 M08 (Turn on coolant)
N260 G01 Z-0.75 F6.0 (Drill hole)
N270 G00 Z0.1 M09 (Rapid out of hole, turn off coolant)
N280 G91 G28 Z0 M19 (Rapid to tool change position, orient spindle)
N285 G28 Y0 (Move to convenient workpiece loading position)
N290 M30 (End of program)
%
```

Key Concept 4: Know The Compensation Types

Please read the descriptions under the metric mode program for detailed information about what is happening in this program. Differences to the inch mode program include: In block N010, inch mode (G20) is being selected, all X, Y, and Z coordinates are specified in inches, and the feedrate for the linear interpolation motions (block N070, N160, and N260) is specified in inches per minute.

The Setup Person's Responsibilities with Tool Length Compensation

Before this program can be run, the setup person must perform several tasks. For example, they must mount the work holding device (probably a vise in this example), measure the workpiece coordinate system offset values (techniques shown in lesson 1.5) and enter them into work offsets (lesson 1.6). They must also load the program (this will be shown in lesson 9.1).

Cutting tools must be assembled and loaded into the machine's automatic tool changer magazine–tool station 1 for the 6-mm (or 1/4-in) drill, tool station 2 for the 10.0-mm (or 3/8-in) drill, and tool station 3 for the 13-mm (1/2-in) drill. And, tool length compensation values must also be determined and entered into the appropriate offsets. Let us first discuss how tool length compensation values are determined.

There are two popular ways to use tool length compensation. We will first show our recommended method.

Using the Tool's Length as the Tool Length Compensation (offset) Value

The reasons why this is our recommended method will be shown a bit later. With this method, the Z-axis workpiece coordinate system offset value is the distance between the Z-axis workpiece coordinate system zero and the spindle nose when it is at the Z-axis zero return (reference) position–just as is shown in lessons 1.5 and 1.6. With our recommended method, the length of each cutting tool will be its tool length compensation value.

The length of the cutting tool is the distance from the tool tip to the spindle nose and will always be a positive value. This will be the value that is entered into the tool's tool length compensation offset register. Figure 4.6 shows a typical tool's length.

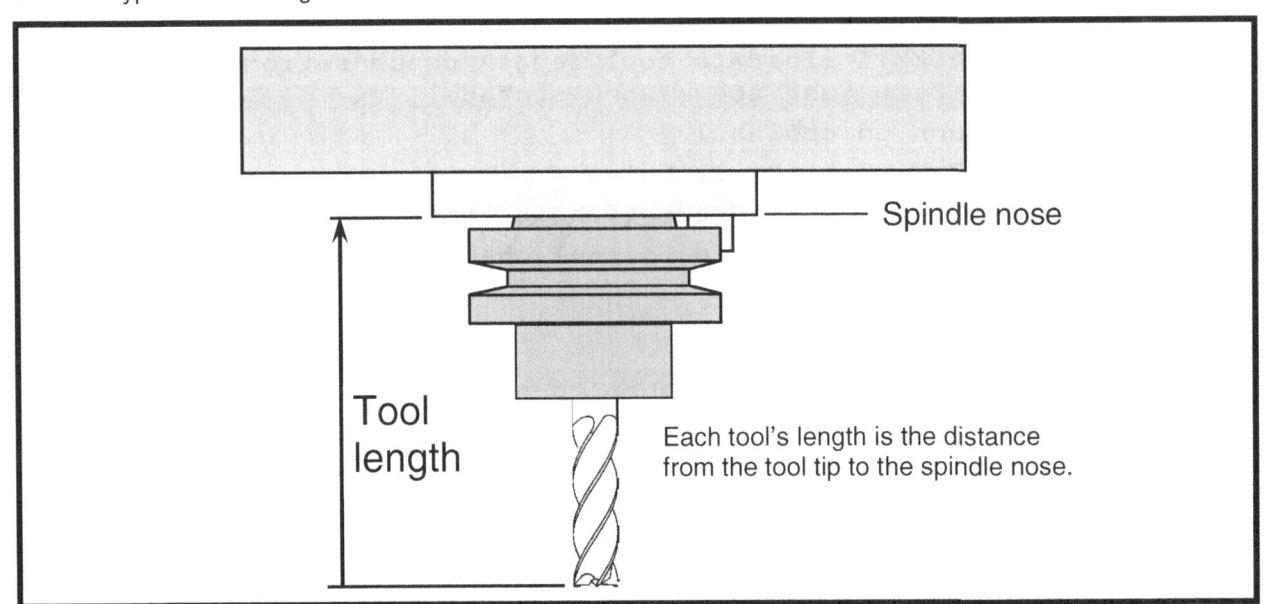

Figure 4.7: Tool length is the distance from the tool tip to the spindle nose–always a positive value

Most machining centers use tool holders that have a tapered shank (CAT-40, CAT-50, BT-40, and BT-50 tool holders are very common examples of tapered shank tool holders). Figure 4.5 shows one type. The shank taper matches the taper in the machine's spindle–and the tool locates in the spindle against this taper. From the illustration in Figure 4.7, notice that there is a small gap between the spindle nose and the flange of the tool holder. This gap is about (but not precisely) 3.17-mm (0.125-in).

We're pointing this out because you might incorrectly assume that tool length is the distance from the tool tip to the end of the flange. When holding a tool holder in your hands, this might seem logical. But again, the tool's length is the distance from the tool tip to the spindle nose, not to the end of the flange.

Determining Tool Length Compensation Values

The tool length for each tool can be measured right on the machine during setup, or it can be measured off-line using a tool length measuring device. Measuring on the machine during setup takes time–and companies that are highly concerned with reducing setup time prefer to measure cutting tools off line. Let's look at both methods.

Even if your company does measure tool lengths off-line, there will be times when you must still measure tool lengths on the machine (maybe after a dull tool is replaced)–so it is quite important that all setup people and operators know how to measure tool lengths on the machine. Techniques to do so are quite similar to those used to measure workpiece coordinate system zero assignment values (shown in lesson 1.5).

Step 1: Make the work holding setup–or place some kind of flat block on the machine table. It is important to have a nice, flat surface on which to work. The top of a vise works nicely.

Step 2: Place a gauge block on the flat surface. A 50.0-mm mm gauge block (metric mode) or the 3.0-in side of a 1-2-3 block (inch mode) works nicely. Or use a low cost dial and electronic tool touch-off gauge (a light comes on when these gauges are touched).

Step 3: Without a tool in the spindle, make the spindle nose touch the block or gauge. Use incremental jog and/or a hand-wheel to carefully touch the spindle nose to the gauge. At this point, reset (set to zero) the Z-axis relative position display–select the relative display screen, press the Z letter address and then the [ORIGIN] soft key.

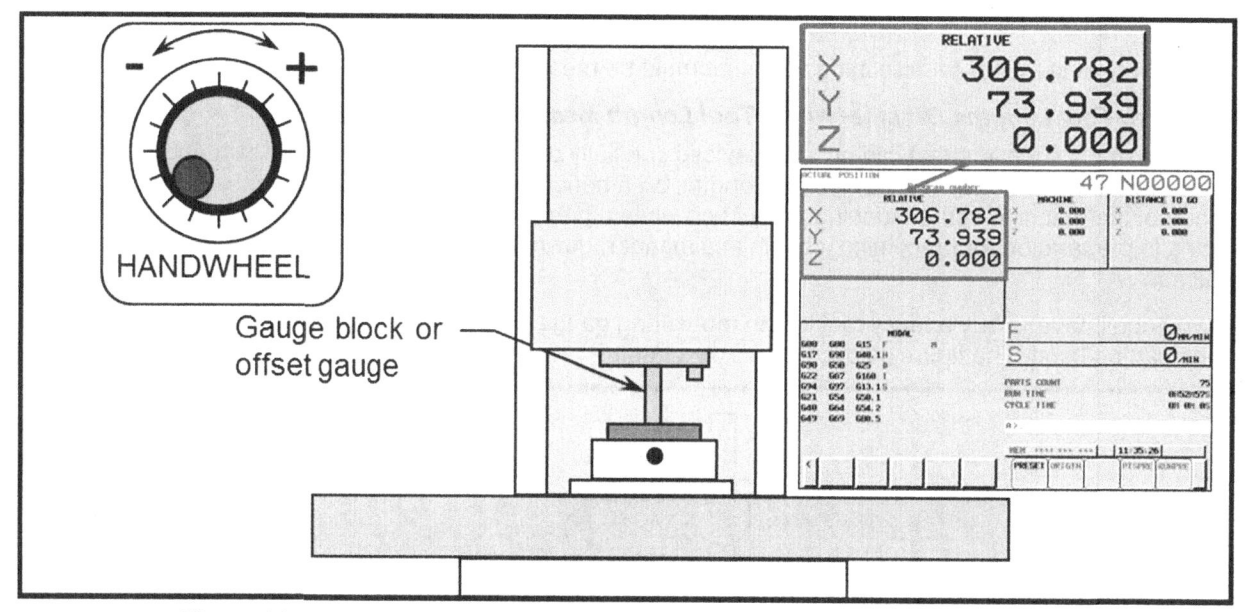

Figure 4.8: Spindle nose touching gauge block or offset gauge (metric mode is shown)

Step 4: Retract the Z-axis and load a cutting tool to be measured. Now, carefully move the tool tip to the same block that was just touching the spindle nose. The Z-axis relative display will follow along. When the tool tip is touching the block, the Z-axis display will be showing you the tool's length. This is the value that must be entered into the tool length compensation offset register for this tool. Enter the value into the appropriate tool length offset register.

> There is a way transfer the Relative Z axis position value into the appropriate tool length compensation offset register. This keeps you from having to type it.
>
> **For a 0iF CNC** (the procedure varies among FANUC CNC models):
> 1) On the offset display screen page, position the cursor to the offset register to be set (the GEOM(H) register for the offset).
>
> 2) Type Z and press the soft key under INP. C. (for input coordinate), the offset register will be set to the Z-axis RELATIVE display screen value.

Key Concept 4: Know The Compensation Types

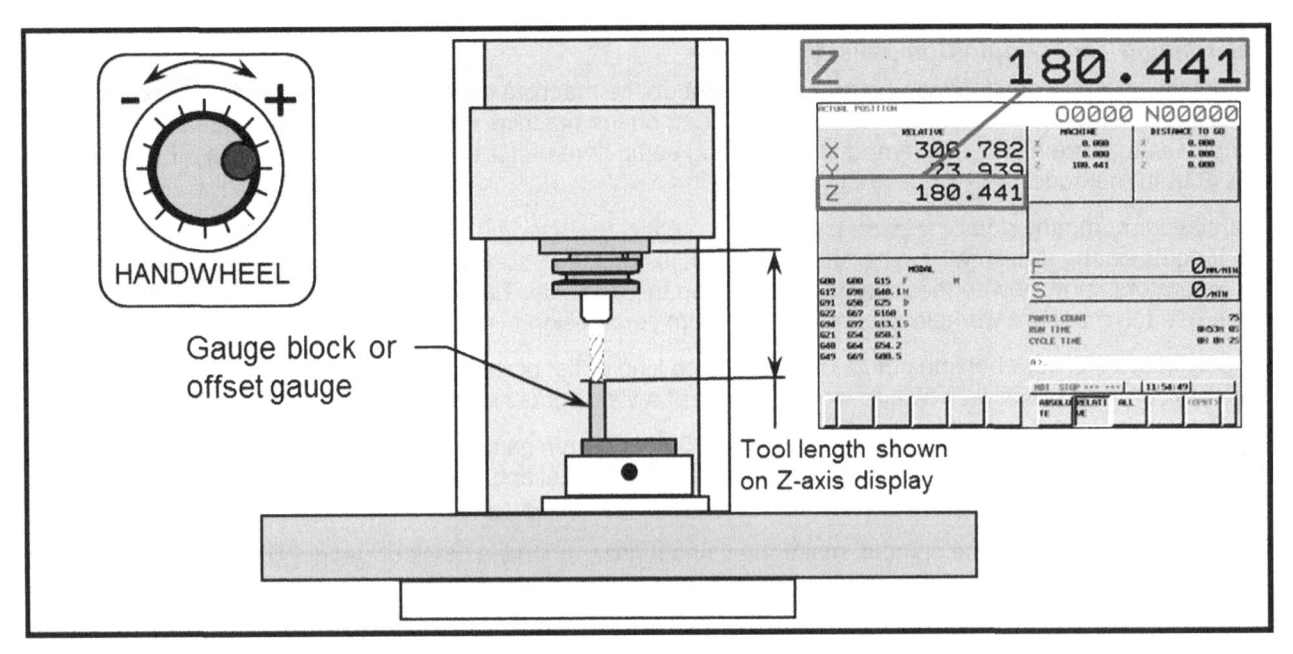

Figure 4.9: Measured tool touching gauge block or offset gauge (shown in metric mode)

Step 5: Repeat step four for all cutting tools that must be measured.

Measuring Tool Lengths Off Line–with a Tool Length Measuring Gauge

There are many suppliers that can provide devices specially designed for tool length measuring. These devices make it quite easy to measure tool lengths on a bench–away from the CNC machining center. And of course, the machine can be running production while a person assembles cutting tools and measures tool lengths in preparation for up-coming jobs. In this manner, downtime between production runs can be reduced.

But you don't have to buy a fancy tool length measuring gauge. With a little effort, just about any height gauge can be used. The figure below shows an example.

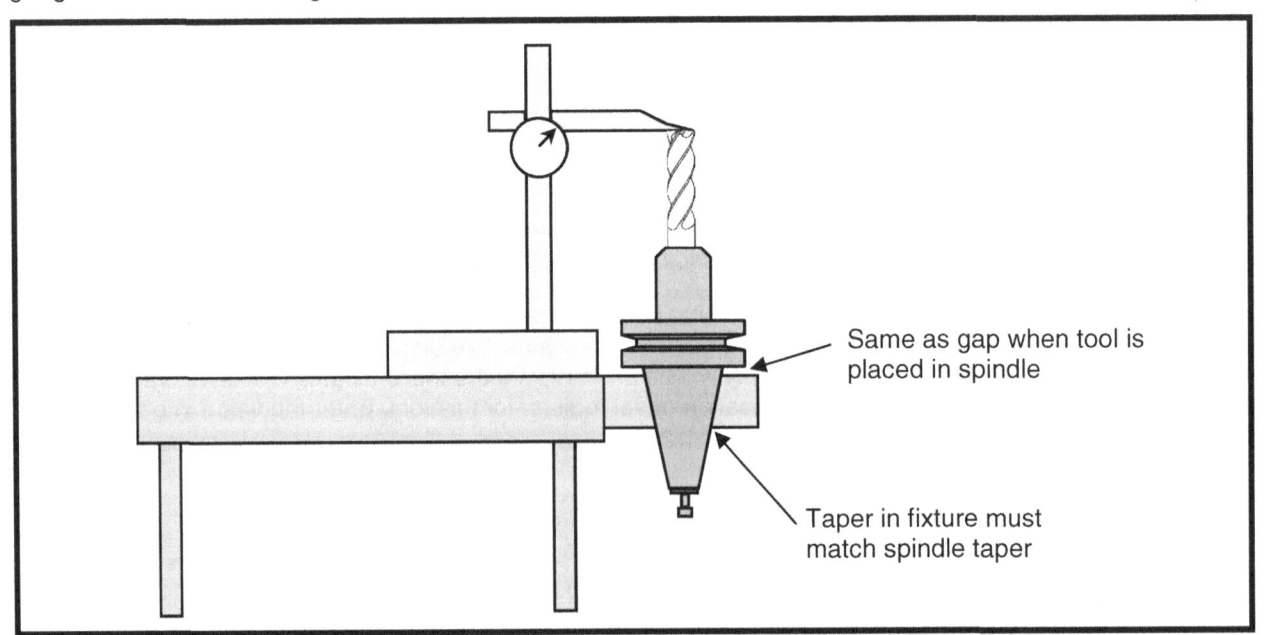

Figure 4.10: A simple and inexpensive height gauge can be used to measure tool lengths.

Entering Tool Length Compensation Offsets

Regardless of which method is used to determine tool length values (measuring on the machine or off-line), the tool length compensation values must be entered into tool offsets before the program can be run. One benefit of measuring on the machine is that most machining centers provide a way to transfer the Z-axis position display value right to the tool length compensation offset. This eliminates the possibility for data entry mistakes.

How accurate are your tool length measurements?

The accuracy of your tool length measurements directly affects the accuracy of the machining operation. While some depth dimensions have wide open (large) tolerances, others have very tight (small) tolerances.

For very tight tolerances (under about 0.05 mm [0.002 inch]), it is likely that you will not be able measure tool lengths accurately enough. In this case, if no special consideration is given to how the program is run, the dimension machined will probably *not* be machined within its tolerance band the first time the tool cuts.

Once the tool has machined, the workpiece can be measured. You can then determine how much to adjust the tool length offset in order to machine the *next* workpiece correctly. But you will have probably scrapped a workpiece. The technique **trial machining** (shown later in this lesson) will ensure that the cutting tool will machine accurately on the first workpiece.

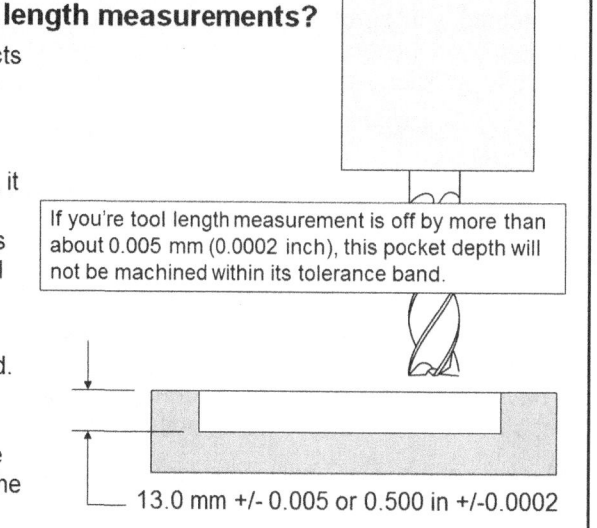

If you're tool length measurement is off by more than about 0.005 mm (0.0002 inch), this pocket depth will not be machined within its tolerance band.

13.0 mm +/- 0.005 or 0.500 in +/-0.0002

Key Concept 4: Know The Compensation Types

Using the Distance from the Tool Tip to Workpiece Coordinate System Zero as the Tool Length Compensation (offset) Value

While we don't recommend using this method, it is quite popular with vertical machining centers (it is difficult to apply this method to horizontal machining centers when rotary devices are used). If your company uses this method, you'll probably have to conform. This technique does not (feasibly) allow tool length compensation values to be determined off line, which is one of the reasons we don't recommend it. But there are many situations in which tools cannot be measured off line:

- One person is responsible for all CNC tasks–this person doesn't have time to measure tool lengths off line.
- Lot sizes are very small and cycle times are very short–and no one can keep up with the number of tools that must be measured off line.
- Short lead times–no one knows what cutting tools will be needed in upcoming jobs.
- There are not enough cutting tool components–tools needed in upcoming jobs cannot be assembled.

In these situations, tool length compensation values must be measured on the machine during setup. But do keep in mind that our recommended method can still be used. If you are new to CNC machining centers and will have control of how things are done, we urge you to use our recommended method. But again, many companies are using this second method–and in this case–you'll probably have to adapt.

With this method, the Z-axis workpiece coordinate system offset value is zero. That is, the work offset Z register will be set to a value of zero. The tool length compensation value (that is entered into the tool length compensation offset register) is the distance from the tool tip (at the Z-axis zero return position) to workpiece coordinate system zero. The polarity is negative–all of your tool length compensation offsets will be very large negative values. The next figure shows the offset value.

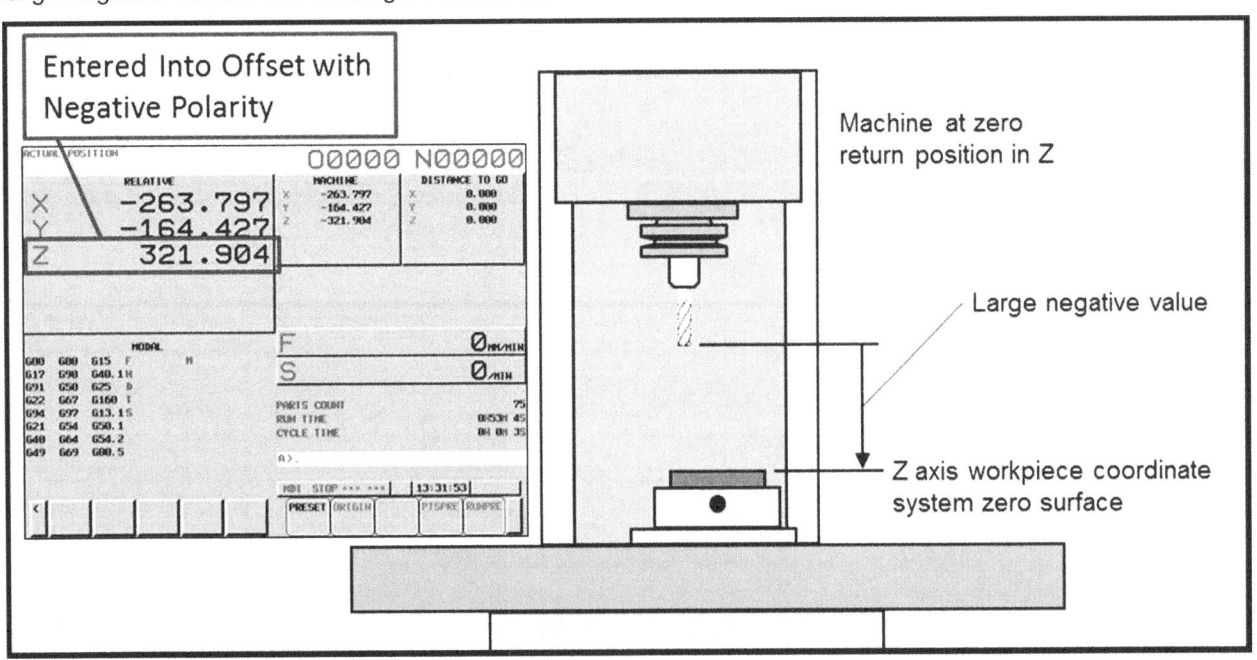

Figure 4.11: Tool length compensation value when tool tip to program zero is used as the offset (shown in metric mode)

Measuring tool length compensation values with this second method

Again, when you use this method, tool length compensation values must be measured on the machine during setup. Here's how.

Step 1: Make the work holding setup and load a workpiece. Workpiece coordinate system zero will be the top surface of the workpiece.

Step 2: Send the machine to the Z-axis zero return position and reset (set to zero) the Z-axis relative display. This sets the point of reference for your measurement.

Step 3: Load a tool to be measured and manually bring the tip of the tool to the Z-axis workpiece coordinate system zero surface. Using incremental jog and/or hand-wheel, cautiously bring the tool tip to the workpiece coordinate system zero surface. The Z-axis display follows along. With the tool tip touching the workpiece coordinate system zero surface in Z, the Z-axis relative position display will be showing you the tool length compensation value for this tool. This is the value that must be entered into the tool length compensation offset register.

> As stated earlier, there is a way transfer the Relative Z axis position value into the appropriate tool length compensation offset register. This keeps you from having to type it.
>
> **For a 0iF CNC** (the procedure varies among FANUC CNC models):
> 1) On the offset display screen page, position the cursor to the offset register to be set (the GEOM(H) register for the offset).
>
> 2) Type Z and press the soft key under INP. C. (for input coordinate), the offset register will be set to the Z-axis RELATIVE display screen value.

Reasons for Using the Recommended Method

Here we list a few reasons to use the recommended method. Note that if your company uses the second method shown, these reasons might help you convince people in your company to change.

Tool length compensation values can be measured off line - We've already mentioned this. Minimizing down time between production runs is a very important goal in many companies.

Cutting tools can be used from job to job without being re-measured–This is a very important reason. The length of a cutting tool will not change (unless it is disassembled) between production runs–but the Z-axis workpiece coordinate system zero surface probably will. If using our recommended method, offset values will remain the same for cutting tools used from job to job. If you use the second method, the tool length compensation value for *all* tools must be re-measured in every setup. Consider how many tools get used in consecutive jobs (center drills, spot drills, common drill-and-tap combinations, and common milling cutters). Indeed, many companies assign standard tool stations to most often-used cutting tools. Using the second method can result in a great deal of duplicated effort.

Cutting tools can be used from machine to machine–In similar fashion, a cutting tool's length does not change when it is placed in a different machine. If a tool is not disassembled after a production run (and if it is not dull), write down the tool's length and keep it with the tool. The next time the tool is needed, you won't have to re-measure it–regardless of which machining center uses it.

The work holding setup doesn't have to be made before tool lengths can be measured–You can measure tool lengths at any time–even on the machine.

Multiple identical tools can be setup and kept ready for action–For those tools that are most prone to wearing out–or for those tools that you use the most, you can keep several identical tools ready to go. When a tool dulls, simply replace it in the machine and enter the new tool length compensation value (that has been measured off line).

Working with multiple workpiece coordinate system zero points is much easier–Though we haven't shown any applications yet, if your company uses a rotary device to expose several surfaces to the spindle for machining during the CNC cycle, you'll find it much easier to do so with our recommended method. This is especially the case with most horizontal machining centers.

Typical Mistakes with Tool Length Compensation

Mistakes with tool length compensation can have some pretty severe consequences–so we want to prepare you for what can happen when mistakes are made.

Forgetting to Instate Tool Length Compensation

As you know, you must instate tool length compensation during every tool's first Z-axis movement, including a G43 and H-word along with the Z-axis departure. If you forget to do so, here's what can happen. Remember that tool length compensation is modal. If you forget to instate tool length compensation for the fourth tool in your program, the control will use the tool length compensation value from the third tool. If the fourth tool is shorter than the third, at least the machine will not try to crash the tool into the workpiece—but it still won't send the tool to the correct Z-axis position.

If you forget to instate for the first tool in the program (and if tool length compensation has not been instated since power-up), tool length compensation is in its cancelled state. If using our recommended method (tool length is offset value), the machine will think that the nose of the spindle is the tool tip. The machine thinks you have a tool with a zero length. The machine will bring the spindle nose to the programmed Z surface, crashing the tool into the workpiece along the way.

Forgetting to Enter the Tool Length Compensation Value

When going from job to job, the setup person must remember to enter all tool length compensation values for the up-coming job. If they forget to do so, and if there is a value in the offset from the previous job, the machine will (incorrectly) use this offset value for the new job. If the current value of the offset is zero, the machine will think your tool has zero length—and again—it will try to bring the spindle nose to the programmed Z surface.

Mismatching Offsets

Remember, the H-word must correspond to the tool station number. As you write the H-word in your program, get in the habit of looking back up into the program for the T word in the most recent tool change (M06) command. The H-word must match the tool station number.

Trial Machining with Tool Length Compensation

Tool length compensation will allow trial machining and sizing for depth (Z) dimensions that cutting tools machine. Again, trial machining is required on the first workpiece when a dimension machined by a cutting tool has a very tight tolerance. This technique ensures that the tool will not machine too much material on its very first try. If done for each tool that has a tight Z-axis tolerance, the first workpiece will pass inspection.

Trial machining for depth dimensions involves five steps:

1: Recognition of a tight tolerance that worries you—If you're worried that a dimension's tolerance is so small that your initial tool length offset measurement is not accurate enough to make the tool machine the dimension within the tolerance band, then trial machining must be done. For example, say you notice a 13-mm (metric mode) or 0.500-in (inch mode) deep pocket with a very tight depth tolerance. Say the tolerance is plus or minus 0.005-mm (metric mode) or 0.0002-in (inch mode). You're worried that your initial offset setting is off by more than 0.005-mm (metric mode) or 0.0002-in (inch mode)—or that tool pressure will cause the tool to machine improperly.

2: Increase the value of the tool length compensation offset by about 0.25-mm (0.010-in)—Again, this is done after you have measured and entered the tool length compensation offset value. In a pocket milling application, say you have measured the milling cutter's length and found it to be 133.035-millimeters (5.2376-inches) long—and you have initially placed this value 133.035 (metric mode) or 5.2376 (inch mode) in the tool length compensation offset register. You'll increase this value to 133.285 (metric mode) or 5.2476 (inch mode) to perform the trial machining operation. If the CNC has separate tool length geometry and wear offsets, you can enter the 0.25-mm (metric mode) or 0.010-in (inch mode) trial machining amount in the wear field.

3: Let the tool machine under the influence of the trial machining offset and stop the cycle after the tool is finished—In our example, you will allow the milling cutter to machine the pocket. With the increased offset, the machine will keep the milling cutter 0.25-mm (metric mode) or 0.010-in (inch mode) further away from the workpiece during machining (forcing our pocket to come out too shallow).

4: Measure the current dimension and reduce the tool length compensation offset accordingly—In our example, say you measure the pocket depth and find it to be 12.22-mm deep (metric mode) or 0.492-in deep (inch mode). It is currently 0.22-mm or 0.008-in undersize. You must reduce the offset by 0.22-mm or 0.008-in, making it 13.281 (133.285 minus 0.22 in metric mode) or 5.2396 (5.2476 minus 0.008 in inch mode). (By the way, if you had not used trial machining techniques and just let the tool cut, it would have machined the pocket about 0.03-mm (metric mode) or 0.002-in (inch mode) too deep. And since the tolerance is only plus or minus 0.005-mm or 0.0002-in, this workpiece would have been scrap.) If the CNC has separate tool length geometry and wear offsets, you can enter the variation in the wear field.

5: Re-run the tool under the influence of the adjusted offset—This time, the cutting tool will machine the pocket depth properly—very close to the target dimension (13.0-mm or 0.500-in deep in our case). Any tiny deviation you notice will be caused by the difference in tool pressure from the first time the cutter machines (the normal amount of stock is being removed) to the smaller depth-of-cut after trial machining.

When Trial Machining is Not Required

Only tight depth tolerances require trial machining with tool length compensation offsets. You typically do not need trial machining for drilling, tapping, reaming, and most other hole-machining operations.

But just because a cutting tool does not require trial machining techniques doesn't mean you don't have to measure and adjust after the tool has machined for the first time. If the depth dimension has a large tolerance, the tool will machine somewhere within the tolerance band on its first try (the workpiece will be acceptable). But it is still important to adjust the offset in such a way that the next workpiece machined will have the dimension come out to its target value.

In our pocket example, say the pocket depth dimension (13.0mm or 0.500-in) has a tolerance of plus or minus 0.125-mm–0.25-mm overall (0.005-in–0.010-in overall in inch mode). In this case trial, you feel machining is not required—so you let the milling cutter cut with the initial offset setting. Once the tool is finished you measure the pocket depth and find it to be 12.995-mm (or 0.498-in) deep. This depth is well within the tolerance band, but not right at its target value. In this case, you should still make an adjustment, and decrease the offset by 0.005-mm (or 0.002-in in inch mode) to make the milling cutter machine 0.005-mm (or 0.002-in) deeper on the next workpiece. This will make the pocket depth come out right to its target dimension. If you use this technique for all tools, all workpiece attributes will be machined to their target values when the production run begins.

Sizing with Tool Length Compensation

Sizing is required when the wear a cutting tool experiences during its life affects the surfaces it machines. Frankly speaking, this rarely occurs with depth dimensions (controlled with tool length compensation). If you do notice changes in depth dimensions during a cutting tool's life, remember that you can make the tool machine deeper by reducing the tool length compensation value by the amount of the deviation caused by tool wear.

A Tip for Remembering Which Way to Adjust the Offset

A common mistake made by beginning setup people and operators is adjusting the tool length compensation offset in the wrong direction. Here's a way to remember which way is which.

With our recommended method, think about what will happen if a tool length compensation offset is set to zero. In this case, the machine will think the cutting edge is at the spindle nose, and bring the *spindle nose* to the programmed surface. So, reducing the offset value will make the tool go deeper.

What If I Use the Second Method Shown for Tool Length Compensation?

If you use the distance from the tool tip to workpiece coordinate system zero as the offset (large negative values in the offset registers), believe it or not, techniques used for adjusting offsets are exactly the same—as long as you understand polarity.

Again, if you want a tool to go deeper you must reduce the tool length compensation offset value. If you already have a negative offset value in the offset, as is the case with the second method shown, this means you must make the already negative value more negative by the amount of your desired adjustment. If for example, the tool length compensation offset is currently -311.724 (or -12.2726 in inch mode) when you measure the pocket depth and find it to be too shallow by 0.005-mm (0.002-inch), you must reduce -311.724 by 0.005-mm (or -12.2726 by 0.002-in in inch mode), making it -311.729 (or -12.2746 in inch mode).

Do I Have to Make All These Calculations When Adjusting Offsets?

All current CNC controls allow you to modify offset values incrementally. FANUC calls this feature input plus (actually the [INPUT+] soft key on the display screen). With this feature, you need only know the amount of needed offset adjustment. If you need an offset value to be reduced by 0.005-mm (0.002-in), you simply type -0.005 (or -0.002 in inch mode) and press the [INPUT+] soft key. The control will automatically calculate the new value for the offset and enter it.

Why Can't I Just Change the Z Coordinate/s In the Program to Make Sizing Adjustments?

There are three reasons to use offsets to make all sizing adjustments. Never change programmed coordinates to make sizing adjustments.

Key Concept 4: Know The Compensation Types

All Programmed Coordinates Must Specify Mean Values
First, and maybe most importantly, it is important that all programmed coordinates specify the mean value of the tolerance band they machine. If milling a 13.0-mm (or 0.500-in) deep pocket with a tolerance of plus or minus 0.005-mm (or 0.002-in), the mean value is 13.0mm (or 0.500-in), and a position of Z-13.0 (or Z-0.5 in inch mode) must be specified in the program (assuming workpiece coordinate system zero in Z is at the top of the pocket). Programming mean values is important because it provides consistency throughout the program, it provides consistency from one time the job is run to the next, and it even provides consistency among programs when the same tool is used in different programs.

Say you machine this pocket and find it to be 0.005-mm (or 0.002-in) too shallow. It is 12.995-mm (or 0.498-in) deep. If you change the programmed Z coordinate at the pocket bottom from Z-13.0 (or Z-0.500) to Z-13.005 (or Z-0.502 in inch mode), admittedly, this pocket will be machined 0.005-mm (or 0.002-in) deeper.

But let's go a little further. What caused the 0.005-mm (0.002-in) deviation? In this case, you must not have correctly measured the cutting tool's length (to be off by this much). Additionally, tool pressure may be affecting the way the tool machines. But the program is correct. Changing the program to deal with tooling problems just doesn't make sense.

Also, consider what will happen when this tool eventually dulls and gets replaced. It's likely that you will correctly measure its new length. But when it machines, the incorrect program coordinate (Z-13.005 or Z-0.502) will cause this tool to machine the pocket too deep.

The same goes for the next time the job is run. If the program still has the Z-13.005 (or Z-0.502) coordinate from the last time it was run, the setup person will have no idea how the cutting tool will machine on its first try.

And consider times when tools are used in consecutive jobs. If Z coordinates in all programs are not specified to mean values, setup people will have to make redundant tool length compensation offset adjustments. On the other hand, if all Z coordinates are specified to mean tolerance values, cutting tools used (and properly adjusted) in one job will remain properly adjusted for the next job, and will continue machining workpiece attributes to their target dimensions.

Programs Cannot Be Changed While the Machine Is Running.
Many sizing adjustments, especially during a lengthy production run, can be done while the machine is running. Tool offsets can be modified while the machine is in cycle. Most machining centers do not allow you to modify the (active) program while the machine is running it, meaning the production run will take longer to complete.

Lots of Program Modifications May be Required
In our example, there is only one pocket to machine. But say you must machine fifty pockets. This means there will be fifty Z-13.0 words (or Z-0.500 words in inch mode) in the program to change. With offset adjustments, only one value must be changed to affect how all fifty pockets will be machined.

Again, <u>never modify programs to make sizing adjustments!</u>

Key points for lesson 4.2:
- ✓ Tool length compensation lets the programmer ignore the precise length of each tool prior to writing the program.
- ✓ Tool length compensation is used for every tool in every program and is instated on each tool's first Z-axis approach movement to the work surface.
- ✓ Tool length compensation is instated with a G43 word that includes an H-word and a Z-word. The H-word specifies the offset number to use.
- ✓ Make the H-word number (offset number) the same number as the tool station number.
- ✓ With our recommended method, the tool offset will contain a value equal to the tool's length (from tool tip to spindle nose).
- ✓ With our recommended method, the workpiece coordinate system setting offset Z value must be the distance from the spindle nose to the Z-axis workpiece coordinate system zero surface (a large negative value).

Exercise: Practice Programming Tool Length Compensation

In this exercise you will get more practice calculating tool center coordinates and selecting motion types. In addition, you will practice specifying tool length compensation commands. The workpiece engineering drawing has been marked up with the planned machining process. For the 13.0-mm (3/4-in) end mill, plunge the workpiece at point 1, feed to point 2, and then retract.

First fill in the coordinate sheet. Then complete the G-code program that follows, specifying the feeds and speeds required for each tool, filling in modal G-codes and other modal words, and only specifying moving axes.

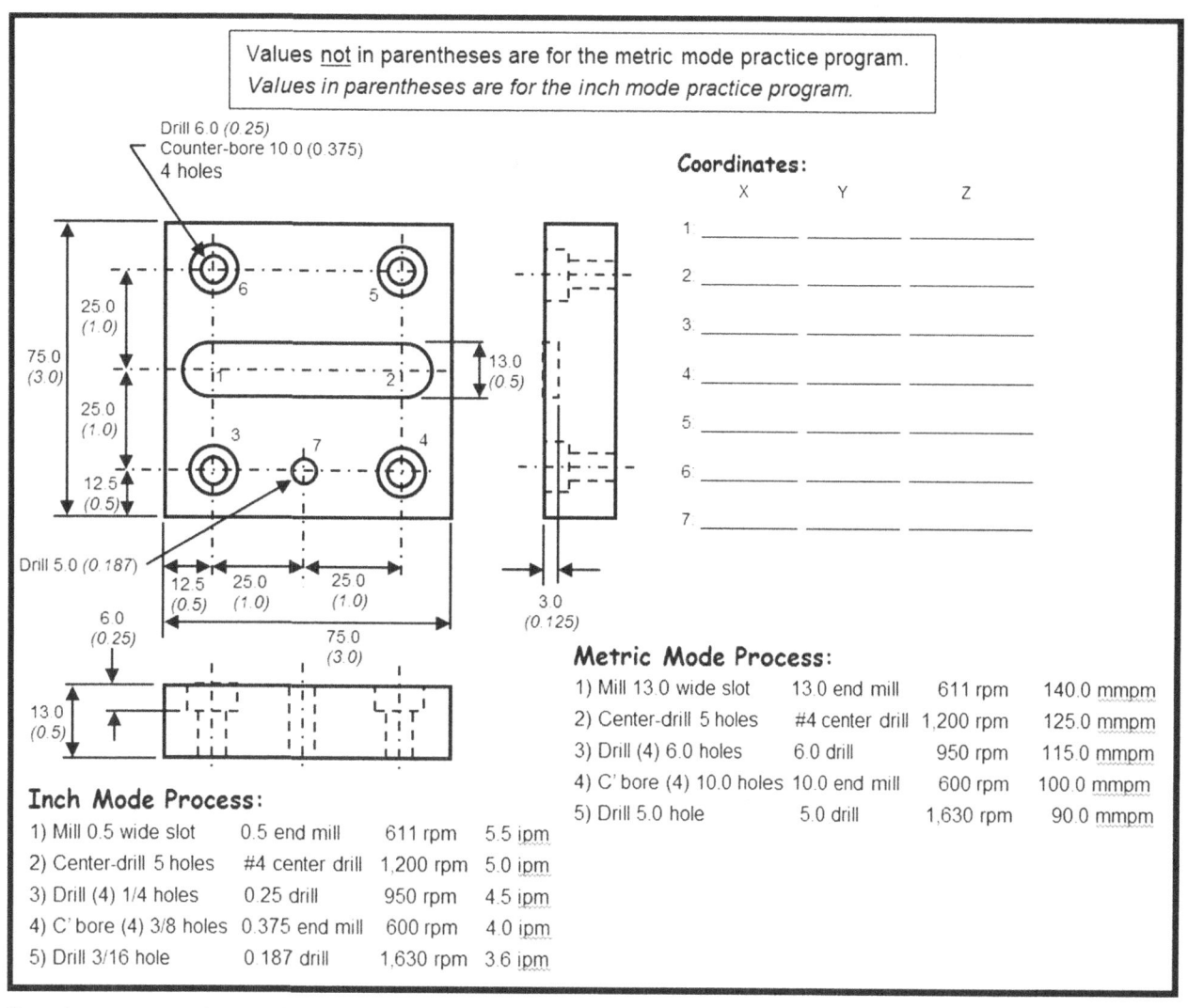

Values **not** in parentheses are for the metric mode practice program.
Values in parentheses are for the inch mode practice program.

Drill 6.0 *(0.25)*
Counter-bore 10.0 *(0.375)*
4 holes

Drill 5.0 *(0.187)*

Coordinates:

	X	Y	Z
1			
2			
3			
4			
5			
6			
7			

Metric Mode Process:
1) Mill 13.0 wide slot   13.0 end mill   611 rpm   140.0 mmpm
2) Center-drill 5 holes   #4 center drill   1,200 rpm   125.0 mmpm
3) Drill (4) 6.0 holes   6.0 drill   950 rpm   115.0 mmpm
4) C' bore (4) 10.0 holes 10.0 end mill   600 rpm   100.0 mmpm
5) Drill 5.0 hole   5.0 drill   1,630 rpm   90.0 mmpm

Inch Mode Process:
1) Mill 0.5 wide slot   0.5 end mill   611 rpm   5.5 ipm
2) Center-drill 5 holes   #4 center drill   1,200 rpm   5.0 ipm
3) Drill (4) 1/4 holes   0.25 drill   950 rpm   4.5 ipm
4) C' bore (4) 3/8 holes   0.375 end mill   600 rpm   4.0 ipm
5) Drill 3/16 hole   0.187 drill   1,630 rpm   3.6 ipm

Practice program (works for both metric mode and inch mode):

```
%
O0015 (EXERCISE TOOL LENGTH COMPENSATION)
N005 T01 M06 (13.0 mm [or 1/2 inch] end mill)
N010 G____ G54 G90 S____ M03 T02 (Select metric or inch mode)
N015 ____ ____ ____ (pt 1)
N020 ____ ____ ____ M08
N023 ____ ____ ____
N025 ____ (pt 2)
N030 ____ ____ M09
N035 G91 G28 Z0 M19
N040 M01
```

Key Concept 4: Know The Compensation Types

```
N045 T02 M06 (#4 center drill)
N050 G54 G90 _____ M03 T03
N055 G00 _____ _____ (pt 3)
N060 ____ ____ _____ M08
N065 ____ _____ _____
N070 ____ _____
N075 _____ (pt7)
N080 ____ _____
N085 ____ _____
N090 _____ (pt 4)
N095 ____ _____
N100 ____ _____
N105 _____ (pt 5)
N110 ____ _____
N115 ____ _____
N120 _____ (pt 6)
N125 ____ _____
N130 ____ _____ M09
N135 G91 G28 Z0 M19
N140 M01

N145 T03 M06 (6.0 mm [or 1/4 inch] drill)
N150 G54 G90 _____ M03 T04
N155 ____ _____ _____ (pt 3)
N160 ____ _____ _____ M08
N165 ____ _____
N168 ____ _____
N170 _____ (pt 4)
N175 ____ _____
N180 ____ _____
N185 _____ (pt 5)
N190 ____ _____
N195 ____ _____
N200 _____ (pt 6)
N205 ____ _____
N210 ____ _____ M09
N215 G91 G28 Z0 M19
N220 M01

N225 T04 M06 (10.0 mm [or 3/8 inch] end mill)
N230 G54 G90 _____ M03 T05
N235 ____ _____ _____ (pt 3)
N240 ____ _____ _____ M08
N245 ____ _____
N248 ____ _____
N250 _____ (pt 4)
N255 ____ _____
```

```
N260 ____ _____
N265 _____ (pt 5)
N270 ____ _____
N275 ____ _____
N280 _____ (pt 6)
N285 ____ _____
N290 ____ _____ M09
N295 G91 G28 Z0 M19
N300 M01

N305 T05 M06 (5.0 mm [or 3/16 inch] drill)
N310 G54 G90 _____ M03 T01
N315 ____ _____ _____ (pt 7)
N320 ____ ____ _____ M08
N325 ____ _____
N330 ____ _____ M09
N335 G91 G28 Z0 M19
N340 G28 Y0
N345 M30
%
```

Metric mode answer program for tool length compensation exercise

```
%
O0015
N005 T01 M06 (13.0 mm end mill)
N010 G21 G54 G90 S611 M03 T02
N015 G00 X12.5 Y37.5 (pt 1)
N020 G43 H01 Z2.0 M08
N023 G01 Z-3.0 F140.0
N025 X62.5 (pt 2)
N030 G00 Z2.0 M09
N035 G91 G28 Z0 M19
N040 M01

N045 T02 M06 (#4 center drill)
N050 G54 G90 S1200 M03 T03
N055 G00 X12.5 Y12.5 (pt 3)
N060 G43 H02 Z2.0 M08
N065 G01 Z-2.0 F125.0
N070 G00 Z2.0
N075 X37.5 (pt7)
N080 G01 Z-2.0
N085 G00 Z2.0
N090 X62.5 (pt 4)
N095 G01 Z-2.0
N100 G00 Z2.0
N105 Y62.5 (pt 5)
N110 G01 Z-2.0

N115 G00 Z2.0
N120 X12.5 (pt 6)
N125 G01 Z-2.0
N130 G00 Z2.0 M09
N135 G91 G28 Z0 M19
N140 M01

N145 T03 M06 (6.0 mm drill)
N150 G54 G90 S950 M03 T04
N155 G00 X12.5 Y12.5 (pt 3)
N160 G43 H03 Z2.0 M08
N165 G01 Z-15.55 F115.0
N168 G00 Z2.0
N170 X62.5 (pt 4)
N175 G01 Z-15.55
N180 G00 Z2.5
N185 Y62.5 (pt 5)
N190 G01 Z-15.55
N195 G00 Z2.0
N200 X12.5 (pt 6)
N205 G01 Z-15.55
N210 G00 Z2.0 M09
N215 G91 G28 Z0 M19
N220 M01

N225 T04 M06 (10.0 mm end mill)
N230 G54 G90 S600 M03 T05
N235 G00 X12.5 Y12.5 (pt 3)
N240 G43 H04 Z2.0 M08
N245 G01 Z-6.0 F4.0
N248 G00 Z2.0
N250 X62.5 (pt 4)
N255 G01 Z-6.0
N260 G00 Z2.0
N265 Y62.5 (pt 5)
N270 G01 Z-6.0
N275 G00 Z2.0
N280 X12.5 (pt 6)
N285 G01 Z-6.0
N290 G00 Z2.0 M09
N295 G91 G28 Z0 M19
N300 M01

N305 T05 M06 (5.0 mm drill)
N310 G54 G90 S1630 M03 T01
N315 G00 X37.5 Y12.5 (pt 7)
N320 G43 H05 Z2.0 M08
N325 G01 Z-15.25 F3.6
N330 G00 Z2.0 M09
N335 G91 G28 Z0 M19
N340 M30
%
```

Key Concept 4: Know The Compensation Types

Inch mode answer program for tool length compensation exercise

```
%
O0015
N005 T01 M06 (1/2 end mill)
N010 G21 G54 G90 S611 M03 T02
N015 G00 X0.5 Y1.5 (pt 1)
N020 G43 H01 Z0.1 M08
N023 G01 Z-0.125 F5.5
N025 X2.5 (pt 2)
N030 G00 Z0.1 M09
N035 G91 G28 Z0 M19
N040 M01

N045 T02 M06 (#4 center drill)
N050 G54 G90 S1200 M03 T03
N055 G00 X0.5 Y0.5 (pt 3)
N060 G43 H02 Z0.1 M08
N065 G01 Z-0.12 F5.0
N070 G00 Z0.1
N075 X1.5 (pt7)
N080 G01 Z-0.12
N085 G00 Z0.1
N090 X2.5 (pt 4)
N095 G01 Z-0.12
N100 G00 Z0.1
N105 Y2.5 (pt 5)
N110 G01 Z-0.12

N115 G00 Z0.1
N120 X0.5 (pt 6)
N125 G01 Z-0.12
N130 G00 Z0.1 M09
N135 G91 G28 Z0 M19
N140 M01

N145 T03 M06 (1/4 drill)
N150 G54 G90 S950 M03 T04
N155 G00 X0.5 Y0.5 (pt 3)
N160 G43 H03 Z0.1 M08
N165 G01 Z-0.605 F4.5
N168 G00 Z0.1
N170 X2.5 (pt 4)
N175 G01 Z-0.605
N180 G00 Z0.1
N185 Y2.5 (pt 5)
N190 G01 Z-0.605
N195 G00 Z0.1
N200 X0.5 (pt 6)
N205 G01 Z-0.605
N210 G00 Z0.1 M09
N215 G91 G28 Z0 M19
N220 M01

N225 T04 M06 (3/8 end mill)
N230 G54 G90 S600 M03 T05
N235 G00 X0.5 Y0.5 (pt 3)
N240 G43 H04 Z0.1 M08
N245 G01 Z-0.25 F4.0
N248 G00 Z0.1
N250 X2.5 (pt 4)
N255 G01 Z-0.25
N260 G00 Z0.1
N265 Y2.5 (pt 5)
N270 G01 Z-0.25
N275 G00 Z0.1
N280 X0.5 (pt 6)
N285 G01 Z-0.25
N290 G00 Z0.1 M09
N295 G91 G28 Z0 M19
N300 M01

N305 T05 M06 (3/16 drill)
N310 G54 G90 S1630 M03 T01
N315 G00 X1.5 Y0.5 (pt 7)
N320 G43 H05 Z0.1 M08
N325 G01 Z-0.586 F3.6
N330 G00 Z0.1 M09
N335 G91 G28 Z0 M19
N340 M30
%
```

Lesson 4.3–Cutter Radius Compensation

Cutter radius compensation is only used for milling cutters. As tool length compensation lets you ignore the precise length of cutting tools as you write programs, cutter radius compensation allows you to ignore the precise diameter of cutters used for side milling.

Objectives

After completing this lesson, students should be able to:

- ✓ Describe the reasons cutter radius compensation is needed
- ✓ Describe how to program cutter radius compensation
- ✓ Describe how to select which cutter radius offset to use for each tool
- ✓ Describe trial machining and sizing with cutter radius compensation
- ✓ Describe how to use cutter radius compensation with a CAM system

Introduction

Milling cutters (like end mills) can be used for side milling operations. Motions can be linear (straight-line) or circular. To this point, we have shown the side milling tool path based upon a milling cutter's centerline path. As you know, calculating coordinates for a milling cutter's centerline path requires that you consider the milling cutter's radius for every coordinate—which can make calculating coordinates quite difficult.

In this lesson, we are going to show how to minimize the calculations that must be done when determining coordinates needed for side milling. Instead of using the milling cutter's centerline coordinates, you will be using coordinates that are right on the work surface to be milled. And again, these coordinates are much easier to calculate. Mastering the use of cutter radius compensation will be the focus of this lesson.

Will You Need to Learn this Feature?

Unlike tool length compensation–which is used for every cutting tool in every program–cutter radius compensation is only used for milling cutters, and only when side milling (milling on the periphery of the cutter). If a company does not perform any side milling operations, they won't need cutter radius compensation.

Even if this is the case, you will still want to know the reasons why cutter radius compensation is required for side milling. So at the very least, read on until we start discussing how cutter radius compensation is programmed. You may then want to skip to lesson 4.4. If the need ever arises, you can always come back to this lesson and study further.

Reasons Why Cutter Radius Compensation is Required

Let's begin by discussing why you must master cutter radius compensation. Some of the reasons are quite similar to the ones given for tool length compensation.

Calculations Are Simplified for Manual Programmers

When performing side milling operations without cutter radius compensation (as shown in lesson 3.1), you must specify coordinates for the milling cutter's centerline path. This requires that you consider the size of the milling cutter in every calculation–complicating each calculation. With cutter radius compensation, and as a manual programmer (not using a computer aided manufacturing [CAM] system), you will specify coordinates that are right on the work surface, ignoring the size of the milling cutter. The figure below shows the two different tool paths.

Key Concept 4: Know The Compensation Types

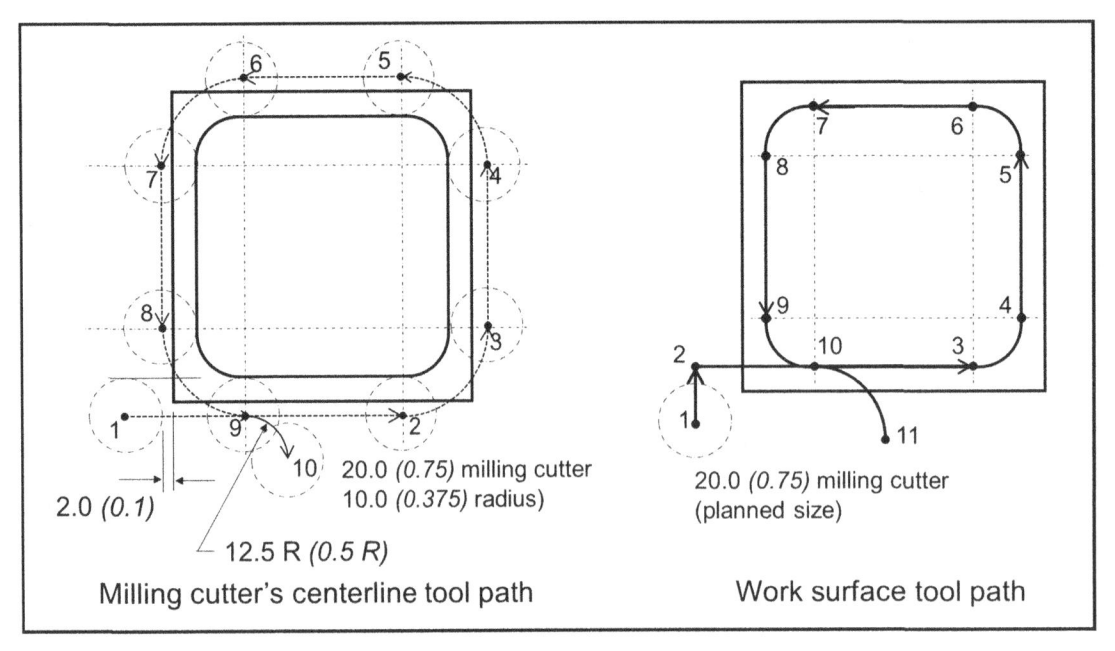

Figure 4.12: Difference between the milling cutter's centerline path and the work surface tool path

When programming the work surface path, the only time you must consider the size of the milling cutter is during its approach movement. The end point of this approach movement is called the prior position. It is point number 1 in the drawing to the right side of the figure above. The prior position must be at least the milling cutter's radius away from the first surface to mill (starting at point number 2 in the work surface tool path in the figure above).

Even with the drawing to the left in above, calculating centerline coordinates is not terribly difficult. It involves simply adding or subtracting the cutter radius to or from each work surface position. This is because every milled surface is parallel to an axis–and all circular movements are full ninety degree arcs. Consider how much more difficult it will be to calculate the cutter's centerline path for the workpiece shown in the figure below.

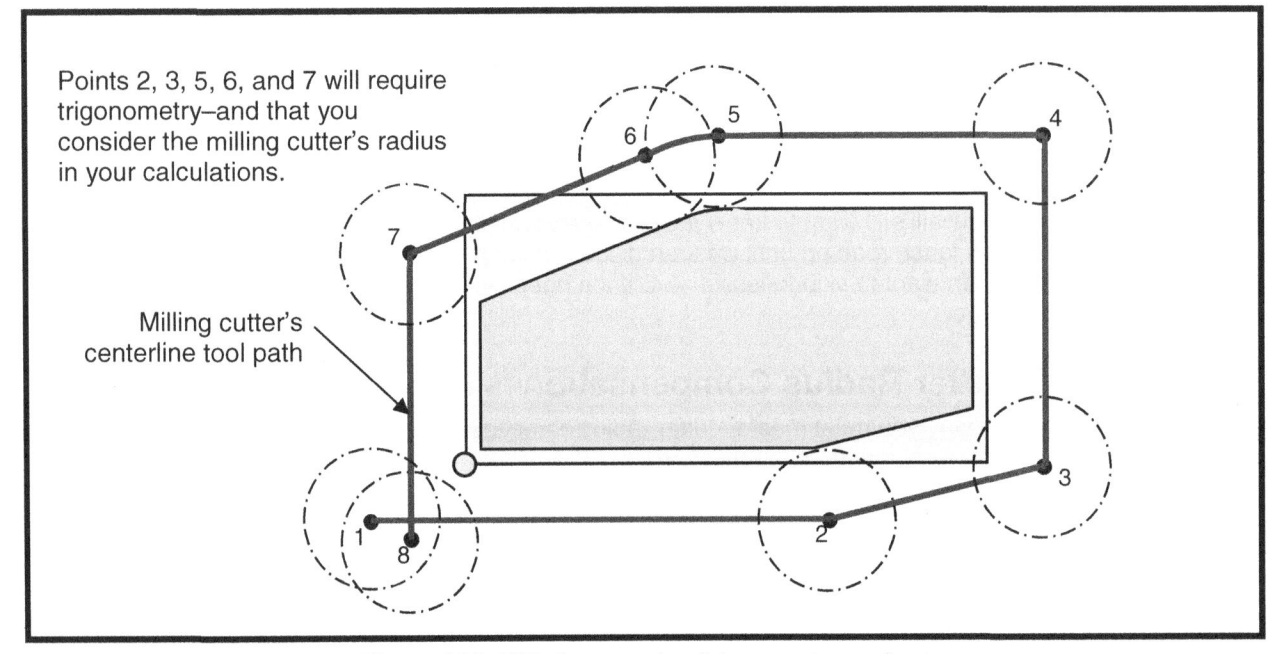

Figure 4.13: Milled contour involving angular surfaces

While cutter radius compensation will not eliminate the need for trigonometry, it will simplify the related calculations. Look at the figure below.

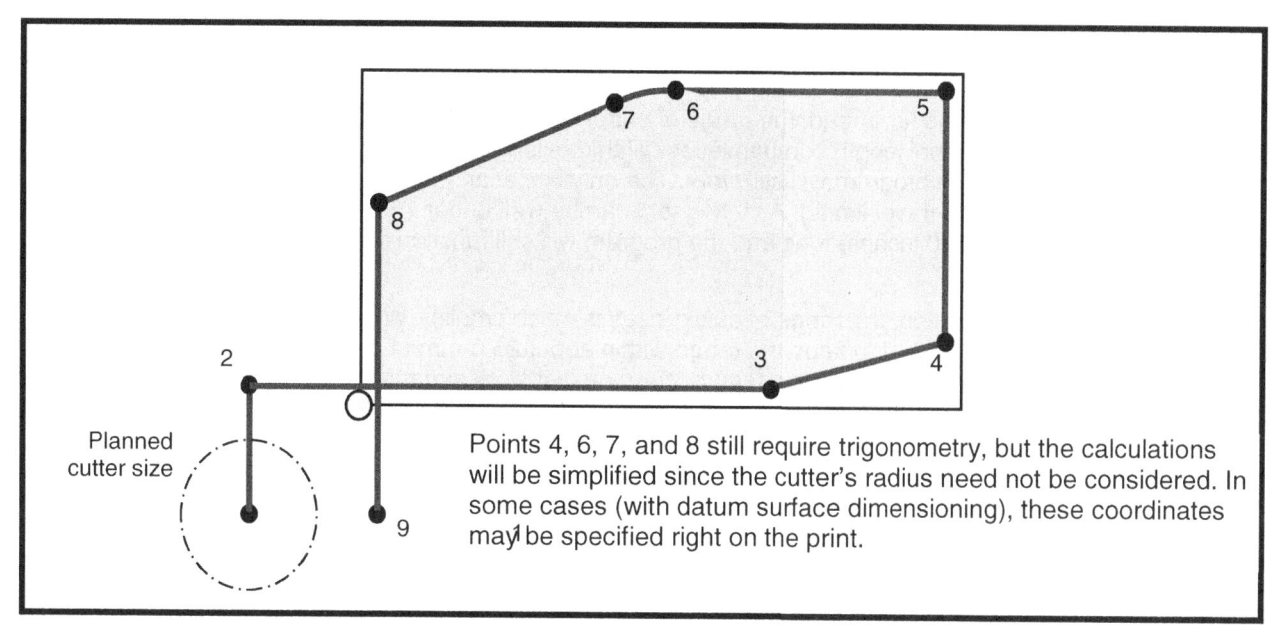

Figure 4.14: Work surface path coordinates are easier to calculate than centerline path coordinates

Simplifying calculations is a benefit only for manual programmers. If you will be using a computer aided manufacturing (CAM) system to prepare CNC programs, simplifying calculations will not be of concern to you. Your CAM system will be able calculate the milling cutter's center line path just as easily as it can calculate the work surface path. We will discuss how cutter radius compensation is used with CAM systems later in this lesson.

Range of Cutter Sizes

Just as tool length compensation allows the length of the cutting tool to vary without requiring a program change, so does cutter radius compensation allow the milling cutter's diameter (and of course, its radius) to vary without requiring a program change. This is valuable to manual and CAM system programmers, setup people, and operators (maybe more so to setup people and operators).

Consider the program shown during lesson 3.1 in figure 3.7. It uses a 25.0-mm (or 1.0-in) diameter end mill. In order for this program to machine the contour properly, the milling cutter must be precisely 25.000 millimeters (or 1.0000 inch) in diameter (a new end mill). If it is not, the contour will not be machined to its intended size. The figure below shows what will happen if you program a milling cutter's center line path (without cutter radius compensation), but do not use the intended milling cutter size.

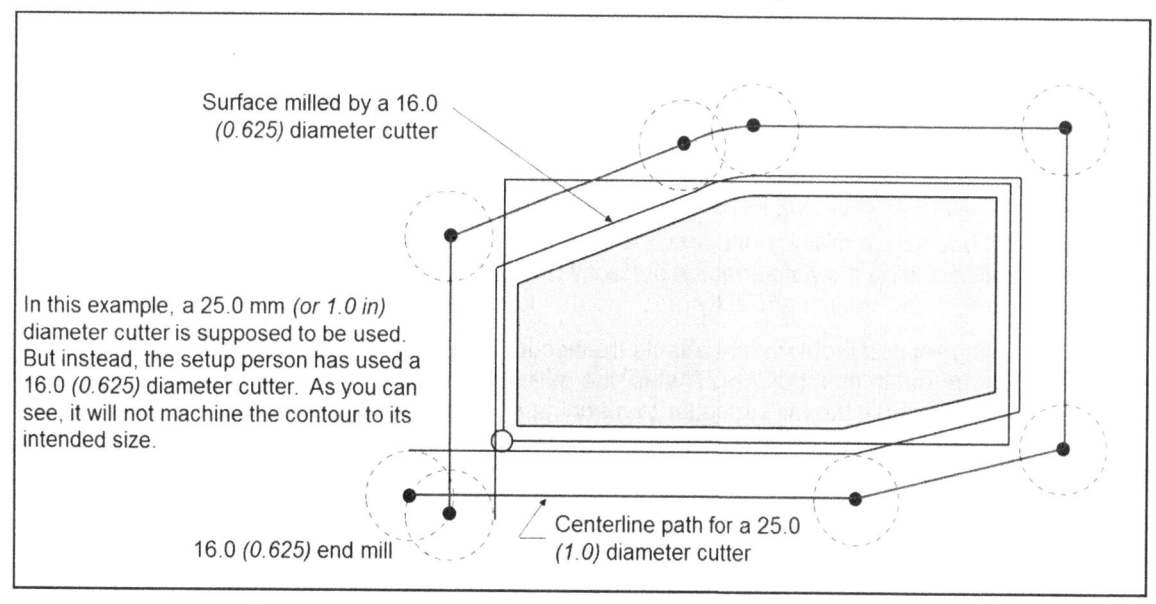

Figure 4.15: Cutter size must be exact with centerline tool paths

Key Concept 4: Know The Compensation Types

When cutter radius compensation is used, the machine will automatically adjust all motions based upon the size of the milling cutter currently being used. The setup person enters the size of the current milling cutter (its radius) in the cutter radius compensation offset.

While we don't want to get too far ahead, the range of cutter sizes allowed by cutter radius compensation is much smaller than it is with tool length compensation. With tool length compensation, a cutting tool can be just about any length and the program will still work. The only limitation (from the program's standpoint) is the Z-axis range of motion (over-travel limits). A cutting tool can be well under 75.0-mm (about 3.0-in) long or well over 400-mm (about 15.0 inches) long and the program will still function properly (as long as the Z-axis does not over-travel).

With cutter radius compensation, the range of cutter sizes is much smaller. When planning the diameter of a milling cutter to use, you will need to keep the range within about 25.0-mm (1.0-in) or so. The milling cutter's approach position in the X and Y axes (how far it is from the first work surface to mill)–its prior position–is one factor that determines how large the milling cutter can be.

Do You Use Re-Sharpened (re-ground) Cutters?

Many companies re-sharpen their dull milling cutters and use them again. When a milling cutter is re-sharpened, its diameter will become smaller by about 0.5-mm (0.020-in). The figure below shows some re-sharpened milling cutters. The milling cutter's tool path must, of course, be modified whenever a re-sharpened cutter is used–due to its smaller diameter. Again, cutter radius compensation will automatically modify the milling cutter's tool path based upon the current size of the re-sharpened cutter. If you do not use cutter radius compensation, you cannot use sharpened cutters without changing all of the cutter's motions in the program.

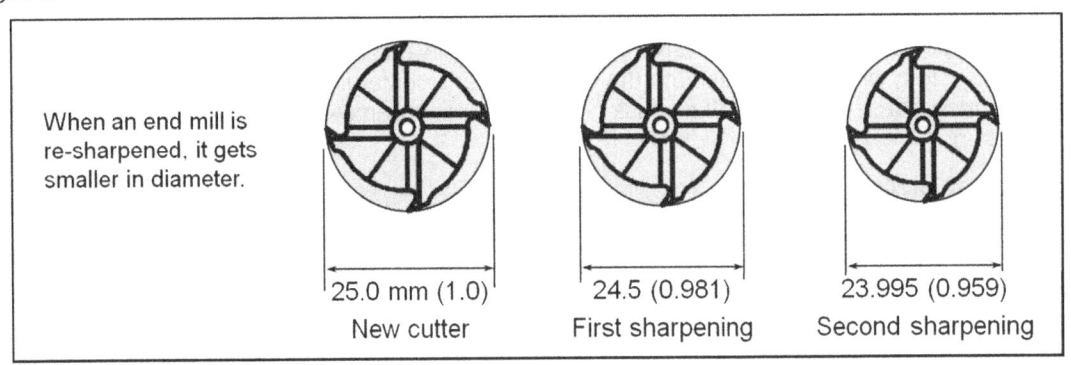

Figure 4.16: Many companies use re-sharpened cutters

Trial Machining and Sizing

Just as tool length compensation allows trial machining and sizing for depth surfaces, so does cutter radius compensation allow trial machining and sizing for XY surfaces (surfaces milled with the periphery of the milling cutter). Like depth surfaces, many XY surfaces have close tolerances. If you do not use cutter radius compensation, there will be no way to make sizing adjustments without actually changing the tool path coordinates in the program (which is usually difficult to the point of being infeasible–and we never recommend changing programmed coordinates to make sizing adjustments).

Also as with machining depth dimensions, tool pressure will affect how XY surfaces are milled. Again, just because a milling cutter's coordinates have been programmed perfectly (to the mean value for each tolerance)–and just because a milling cutter of exactly the intended size is being used–it is no guarantee that the milling cutter will machine the XY surface/s perfectly to size. The tighter the tolerance/s to be held, the more likely the milling cutter will not correctly machine the surfaces.

There is also a tooling-related problem that affects the accuracy of side milling operations. Milling cutters have a tendency to run-out in their holders. That is, the milling cutter will not be perfectly concentric with its tool holder. Run-out will cause the milling cutter to machine more material than it should–having the same effect as using a slightly larger milling cutter.

Milling cutters also have a tendency to wear during production. As the cutter wears, a small amount of material will be removed from its outside diameter. In effect, the milling cutter becomes smaller in diameter. With tight tolerances, this small amount of variance will cause problems. Sizing must be done during the tool's life to keep the milling cutter machining surfaces to size.

For these reasons, you must have the ability to perform trial machining and sizing with XY milling operations. Whether you are simply milling one straight surface (like the right end of the workpiece), or machining an elaborate contour, cutter radius compensation will allow you to do so.

Rough and Finish Milling with the Same Coordinates

This reason applies only to manual programmers. Rough milling an XY contour involves machining the contour while leaving a small amount of finishing stock. Though the amount of finishing stock will vary based upon workpiece material and the style of milling cutter used, it is usually between 0.25-mm (0.010-in) and 1.27-mm (0.050-in). This, of course, requires two milling passes–and two sets of coordinates. One pass is for the roughing operation and another is for the finishing operation. Consider the drawing shown in the figure below.

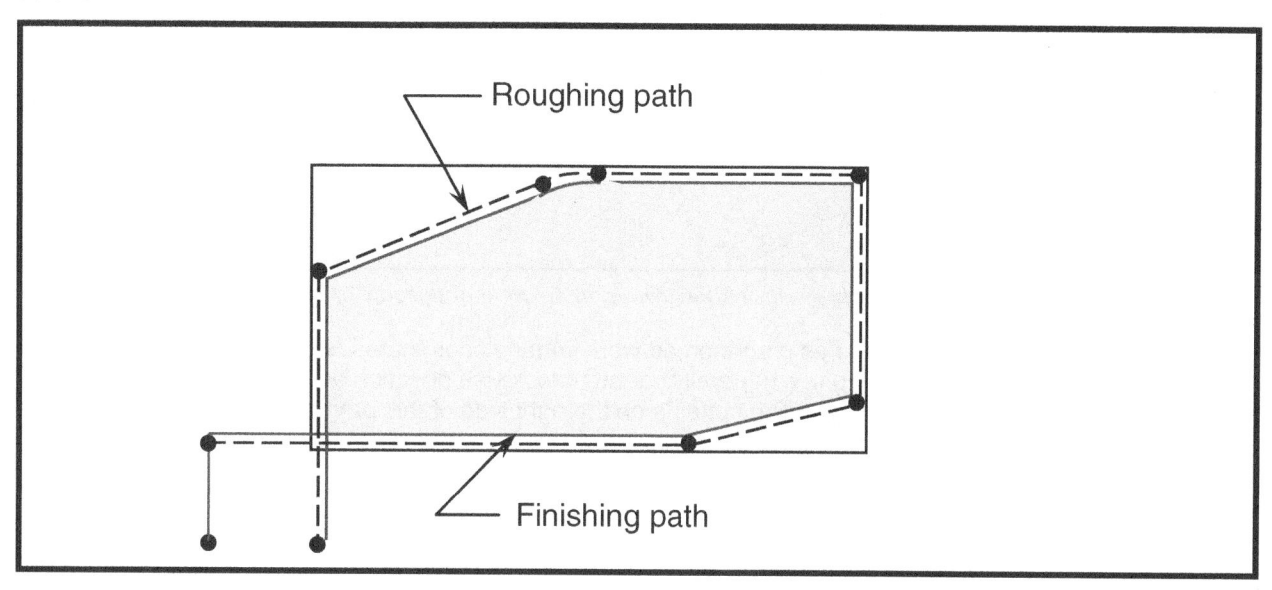

Figure 4.17: The rough and finish passes required for a contour milling operation.

It can be difficult enough to calculate the coordinates required for the finishing path–even when cutter radius compensation is used and these coordinates are on the work surface. It will be more difficult to calculate another set of coordinates that is a small distance away from the finished work surface path. With cutter radius compensation, the same set of coordinates used to perform the finish milling operation can be used to perform the rough milling operation. If you do not use cutter radius compensation (you program the milling cutter's centerline path), you must calculate the two sets of coordinates–and both must consider the size of the milling cutter you will be using.

Do You Have a CAM System?

Again, this is only important to manual programmers. If you use a computer aided manufacturing (CAM) system to create programs, roughing passes will not be of concern. The CAM system can generate the set of roughing coordinates just as easily as it can generate the set of finishing coordinates–on the work surface or based upon the planned milling cutter's centerline path.

How Cutter Radius Compensation Works

In the program, you will be telling the machine the relationship between the milling cutter and the path of the milling cutter. To do so, you specify whether the milling cutter is on the right or left side of the path. The setup person will tell the machine (in the cutter radius compensation offset) the size of the milling cutter being used (its radius). With this information, the machine will keep the milling cutter away from all programmed surfaces by cutter's radius. The next illustration shows an example.

Key Concept 4: Know The Compensation Types

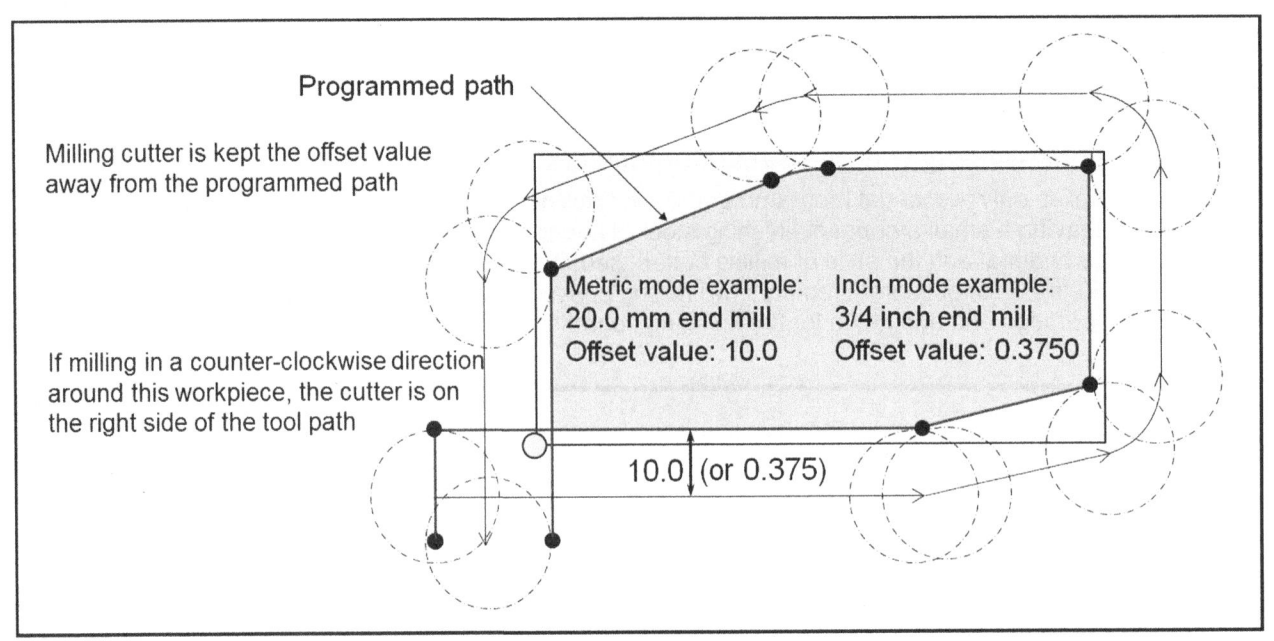

Figure 4.18: How the machine keeps the milling cutter on the right or left side of the programmed path

In the figure above, the programmer has programmed work surface coordinates (work surface tool path). Since the milling cutter will be moving in a generally counter-clockwise direction around this path, they will have specified in the program that the milling cutter is on the right side of this programmed path. (If this is confusing, look in the direction the milling cutter is moving and ask which side of the work surface the milling cutter is on? In this example, the milling cutter is on the right side of the work surface when viewed in the direction of the cut.) At the present time, a 20.0-mm (or 3/4-in) end mill is being used. So the setup person has placed a value of 10.0 (or 0.375) (the milling cutter's radius) in the appropriate cutter radius geometry offset. The machine will automatically keep this milling cutter precisely 10.0-mm (or 0.375-in) from all surfaces as it moves around this path.

Now here's the beauty of cutter radius compensation. Say the 10.0-mm (or 3/4-in) milling cutter wears out and the setup person can't find another one. All that is available is a 22.0-mm (or 0.7/8-in) end mill. They will simply change the cutter radius geometry offset to 11.0 (or 0.4375) (the radius of the 22.0-mm (or 0.875-in) end mill). When the program is run, the machine will keep this end mill 11.0-mm (or 0.4375-in) away from the programmed path–and mill the contour properly.

Steps to Programming Cutter Radius Compensation

As with tool length compensation, there are two ways to use cutter radius compensation. But unlike tool length compensation, how programs are written is directly related to the method you choose. With the method we recommend, which is especially intended for manual programmers, you program the work surface path and the cutter radius compensation offset value is the milling cutter's radius. With the other method, preferred by many computer aided manufacturing (CAM) system programmers, you program the milling cutter's centerline path and the cutter radius compensation offset value is only the difference (in radius) between the planned milling cutter size and the actual size of the milling cutter being used. We'll spend most of this lesson discussing our recommended method.

Here are the three basic steps you must master in order to program cutter radius compensation:

1. Instate cutter radius compensation
2. Program the tool path to be machined
3. Cancel cutter radius compensation

Admittedly, if you are studying cutter radius compensation for the first time, this is going to get a little complicated. Cutter radius compensation tends to be the most difficult CNC feature to fully understand and master, but stick with it. Based upon the reasons just shown, it will be well worth the time it takes to master.

Step 1: Instate Cutter Radius Compensation

Machining centers allow a great deal of flexibility when it comes to instating and using cutter radius compensation. We're going to be showing a method that will work in almost all cases.

Lesson 4.3: Cutter Radius Compensation

Instating cutter radius compensation involves at least three positioning commands that engage the milling cutter to the first surface to be milled. These motions include:

1. An XY motion to the approach position, called the prior position
2. One or more Z motions to the Z-axis work surface
3. A command instating cutter radius compensation that brings the milling cutter (in X and/or Y) to the first surface to mill

Let us discuss these commands one at a time.

The XY Motion To the Prior Position

This positioning movement brings the center of the milling cutter to a position in X and Y that makes it possible to instate the cutter radius compensation. It is usually done at rapid–with the milling cutter is well clear of the workpiece. Indeed, this motion is usually done while the milling cutter is above the workpiece in the Z-axis.

We call the end-point for this motion the prior position because it is the XY position just prior to the instating command. The next XY motion will instate cutter radius compensation.

The prior position must be at least the milling cutter's radius (the value stored in the cutter radius compensation offset) away from the surface to mill. And it is important to choose a prior position that allows the milling cutter to form a ninety degree angle (right angle) with the first surface to be milled.

That is a lot to remember, so let's look at an example. The figure below shows the prior position.

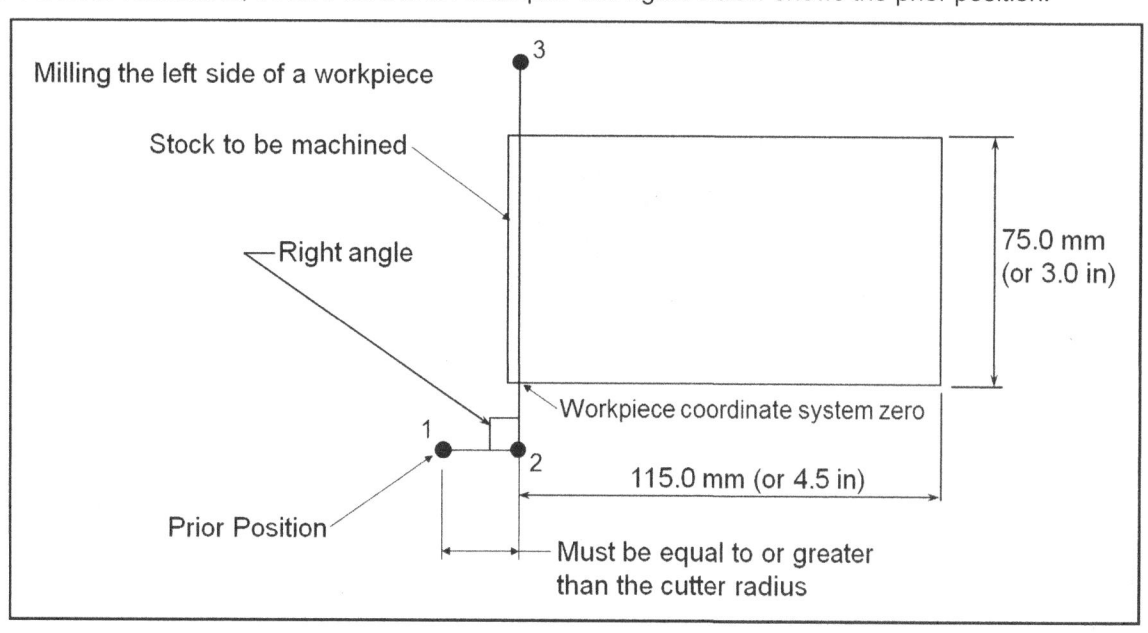

Figure 4.19: The prior position for cutter radius compensation

In the figure above, the left side of a workpiece will be milled. The prior position is point number 1. This position must be the at least the milling cutter's radius away from the first surface to mill (beginning at point number two). And notice that we've chosen this prior position in such a way that as we instate cutter radius compensation during the move to point number two, and begin machining to point 3, a ninety degree angle is formed.

While cutter radius compensation allows you to use a range of milling cutter sizes, it does not allow you to completely ignore the milling cutter size. You must, for example, specify an intended cutter size in the setup documentation. And again, your chosen prior position must be far enough away from the first milled surface that the largest allowable cutter will clear. The prior position sets the limit for maximum cutter size.

For the example in the figure above, say you intend to use a 25.0-mm (or 1.0-in) diameter end mill. The prior position (again, point one) must be at least 12.5-mm (or 0.5-in) (the radius of the intended cutter size) away from the first surface to mill (starting at point number two). Notice that the workpiece coordinate system zero point is the lower-left corner of the finished workpiece. The milling cutter is currently well above the workpiece in Z.

Key Concept 4: Know The Compensation Types

In <u>metric mode</u>, program a rapid motion to the prior position with:

```
%
O0016 (Program number)
N010 G21 G90 G94 G54
N020 T01 M06 (1.0 end mill)
N030 S350 M03 T02 (( start spindle fwd at 350 rpm, and get tool station one ready)
N040 G00 X-12.5 Y-14.5 (Rapid to prior position)
```

In <u>inch mode</u>, program a rapid motion to the prior position with:

```
%
O0016 (Program number)
N010 G20 G90 G94 G54
N020 T01 M06 (1.0 end mill)
N030 S350 M03 T02 (( start spindle fwd at 350 rpm, and get tool station one ready)
N040 G00 X-0.5 Y-0.6 (Rapid to prior position)
```

When a 25.0-mm (or 1.0-in) cutter moves to the position specified in block N040, it will be perfectly flush with the first surface to mill. While this will work as a prior position for a 25.0-mm (or 1.0-in) end mill, it does not allow for any variation (like trial machining and sizing) during the production run. We have set the maximum cutter size to a 25.0-mm (or 1.0-in) diameter end mill. If the setup person enters anything larger than 12.5-mm (or 0.500-in) in the cutter radius compensation offset, the machine will generate an alarm. (The machine will think the cutter is already violating the first surface to mill.)

Let's revise the prior positioning movement (for <u>metric mode</u>) to:

```
N040 G00 X-14.5 Y-14.5
```

Or for <u>inch mode</u>, to:

```
N040 G00 X-0.6 Y-0.6
```

Now there is plenty of clearance between our planned 25.0-mm (or 1.0-in) end mill and the first surface to machine. In fact, an end mill up to 29.0-mm (or 1.2-in) in diameter can be used without generating an alarm. So, one way to determine maximum cutter size is to double the distance from the prior position to the first surface to mill. (Note that there are some other limiters for maximum cutter size. You will see them as we continue.)

Don't forget to document the maximum cutter size—Since your prior position sets the maximum cutter size (again, anything larger and an alarm will be generated), be sure to specify the maximum cutter size in the setup documentation. While you're at it, also specify the smallest cutter that can be used (based upon the strength of the cutter versus the amount of material being machined). In our example, if there is about 2.5-mm (0.100-in) material to machine and the workpiece is 25.0-mm (or 1.0-in) thick, we may not want the setup person using an end mill smaller than about 20.0-mm (or 0.75-in) in diameter.

The Z Motions To the Z-Axis Work Surface

With the milling cutter at the prior position in the X and Y axes, it is time to move it to the Z-axis work surface. In most cases, when the milling cutter is at the prior position, it will be well clear of the workpiece in the X and Y axes, meaning you can rapid the tool to an approach distance above the Z-axis work surface. For our example, we will say the workpiece is 25.0-mm (or 1.0-in) thick. We'll want to move the tip of the end mill a little further below the bottom surface of the workpiece (say by 2.0-mm or 0.1-in). We will do so by first positioning the milling cutter just above the workpiece, then sending it to the Z-axis machining level.

In metric mode:

```
%
O0016 (Program number)
N010 G21 G90 G94 G54
N020 T01 M06 (25.0 mm mill)
N030 S350 M03 T02 (Start spindle fwd at 350 rpm, and get tool station two ready)
N040 G00 X-14.5 Y-14.5 (Rapid to prior position)
N050 G43 H01 Z2.0 (Instate tool length compensation)
N060 Z-27.0 M08 (End mill is now at Z-axis work surface and coolant is on)
```

In inch mode:

```
%
O0016 (Program number)
N010 G20 G90 G94 G54
N020 T01 M06 (1.0 inch end mill)
N030 S350 M03 T02 (Start spindle fwd at 350 rpm, and get tool station two ready)
N040 G00 X-0.6 Y-0.6 (Rapid to prior position)
N050 G43 H01 Z0.1 (Instate tool length compensation)
N060 Z-1.1 M08 (End mill is now at Z-axis work surface and coolant is on)
```

Block N040 is the movement to the prior position in the X and Y axes. Block N050 instates tool length compensation and moves the end mill to just above the work surface. Block N060 moves it to the Z-axis work surface (still at rapid). Note that, for the metric-mode example, you could include Z-27.0 in block N050 instead of Z2.0 to bring the end mill directly to the Z-axis work surface. For the inch-mode example, this would be Z-1.1 in block N050. This is a matter of programming style.

Instate Cutter Radius Compensation and Position the Cutting Tool To the First Surface To Mill

At this point, the tool is positioned and ready to instate cutter radius compensation. The instating command will include three things: 1: a G41 or G42 specifying the relationship between the milling cutter and the tool path, 2: a D-word specifying the offset number containing the cutter radius compensation value, 3: a motion to the first work surface to be machined.

G41 or G42?

One of two G-codes will be used in the instating command.

- G41–milling cutter is to the left of the its path
- G42–milling cutter is to the right of its path

One way to determine which of G41 or G42 you must use is to look in the direction the milling cutter will be moving as it machines the work surface and ask "Which side of the work surface is the milling cutter on?" If the milling cutter is on the left side of the work surface, you will use G41 in the instating command. If the cutter is on the right side of the work surface, you will use G42. Selecting the G41 or G42 tells the CNC on which side to instate the value stored in the cutter radius offset. The figure below shows some examples.

Key Concept 4: Know The Compensation Types

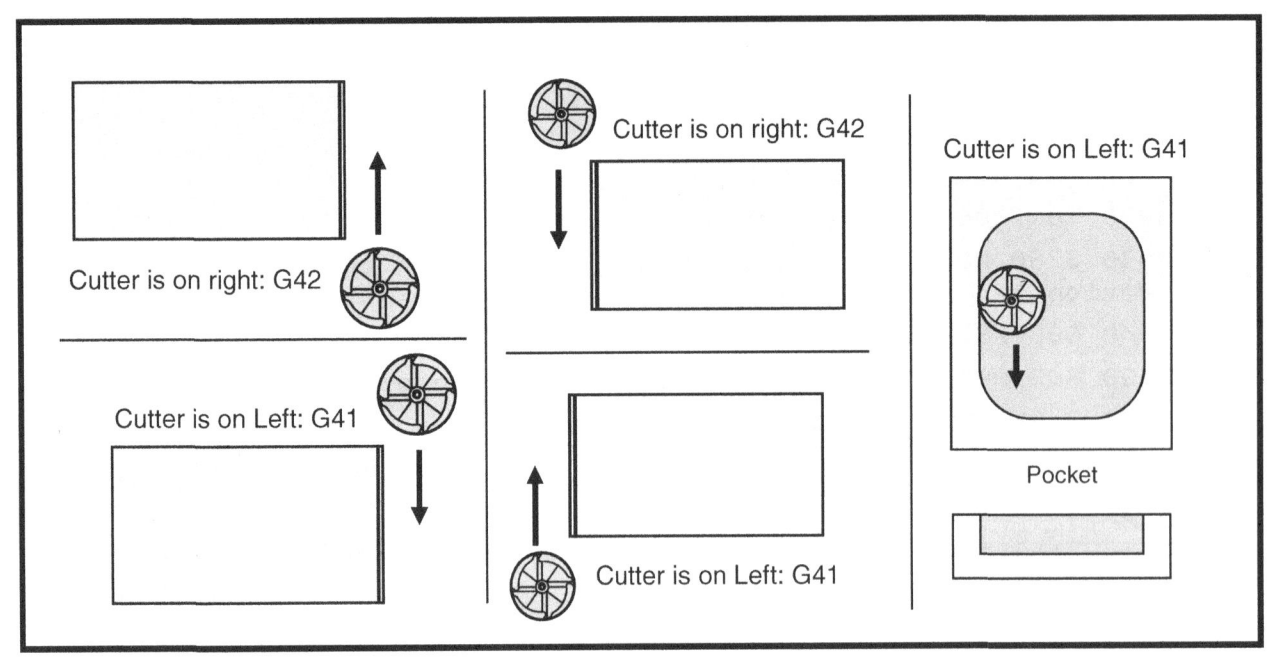

Figure 4.20: Deciding whether to use G41 or G42

If you have machining experience, there is another way to determine whether to use G41 or G42 in the instating command. As long as you're using a right-hand milling cutter (almost all milling cutters are of the right-hand type), climb milling requires G41 to instate cutter radius compensation and conventional milling requires G42 to instate. The figure below shows the same examples–but this time relative to climb versus conventional milling.

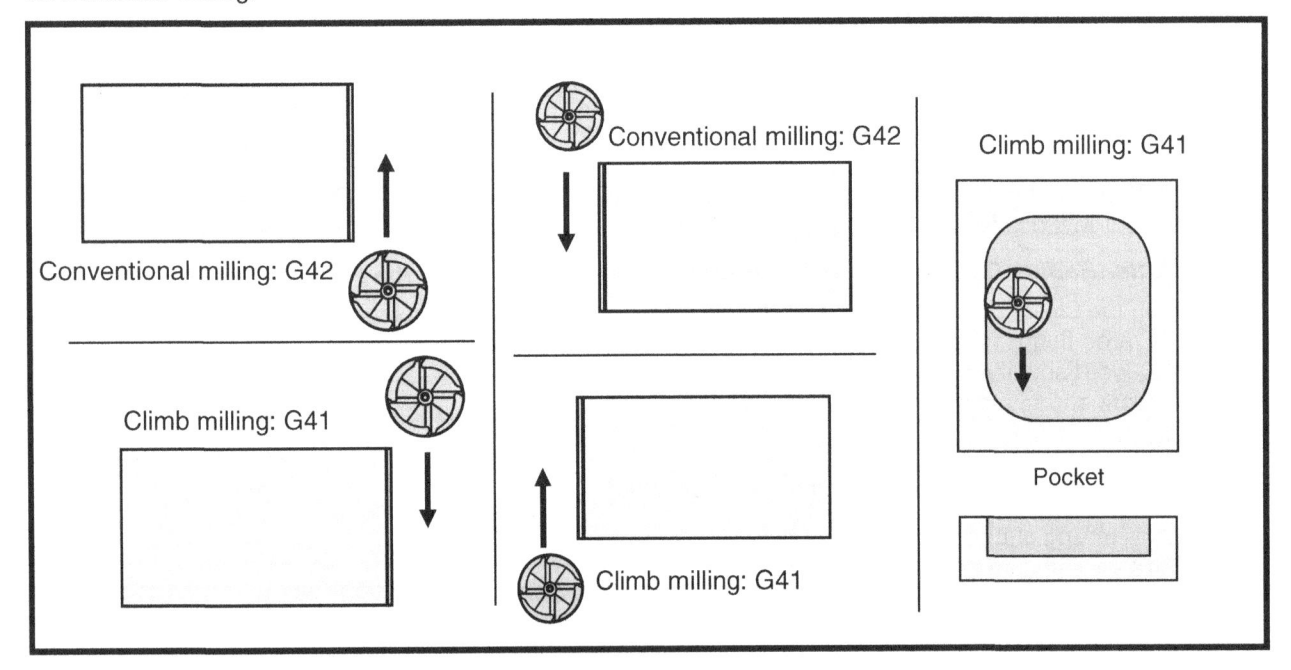

Figure 4.21: Deciding whether to use G41 or G42 based upon milling style

The Offset Used with Cutter Radius Compensation

The command that instates cutter radius compensation also requires that you specify the offset number in which the setup person will place the cutter radius compensation value. While instating, the machine will use this value to bring the cutter flush with the first surface to machine. It will also use this value to keep the cutter flush with all surfaces as the contour is being machined.

In the program, a D-word is used to specify the offset number used with cutter radius compensation. The D-word must be included in the instating command.

From lesson 4.1, remember that machining centers vary when it comes to offset organization. Most of today's CNCs have two or four register fields per offset allowing tool length compensation offset (H-word) and the cutter radius compensation offset (D-word) to have the same number. However, older CNCs may only have one register per offset, meaning they must be shared between tool length and cutter radius compensation. If you use the offset number that corresponds to the tool station number in which to store the tool length compensation value with this kind of offset table, it cannot be used again for cutter radius compensation–you must choose another.

For CNCs that have only one register per offset, we recommend adding a constant number to the tool station number to come up with the offset number to use with cutter radius compensation. This constant number must be equal to or larger than the number of cutting tools the machine's automatic tool changer can hold.

For example, if your machining center's automatic tool changer magazine can hold thirty tools, add thirty to the tool station number to come up with the cutter radius compensation offset number. For this machine, if tool number 5 is a milling cutter (T5 or T05), you will use offset number 5 in which to store its tool length compensation value (H5) and offset number 35 in which to store its cutter radius compensation value (D35).

If you have a machining center that has two or four registers per offset, choosing the cutter radius compensation offset number is much easier. Simply make it correspond to the milling cutter's tool station number. For this kind of machine, if station 5 is a milling cutter (T5 or T05), you will use offset number 5 in which to store its tool length compensation value (H5 or H05) and its cutter radius compensation value (D5 or D05).

Since CNCs with multiple values in the offset table are becoming popular, we use the same numbers for tool number, tool length compensation offsets and cutter radius offsets in this course.

> SPECIAL NOTE! There are some very old CNC models that use the H-word to specify the offset number to be used with both tool length compensation and cutter radius compensation. The Fanuc 3M, and some 0M models, for example, use the H-word in this fashion. These models do not have the letter address D on their control panels.

The Motion to the First Work Surface

Also included in the instating command is a motion that brings the milling cutter to the first work surface to be machined. With this motion, you are no longer programming the milling cutter's centerline position. The end point for this position is right on the work surface.

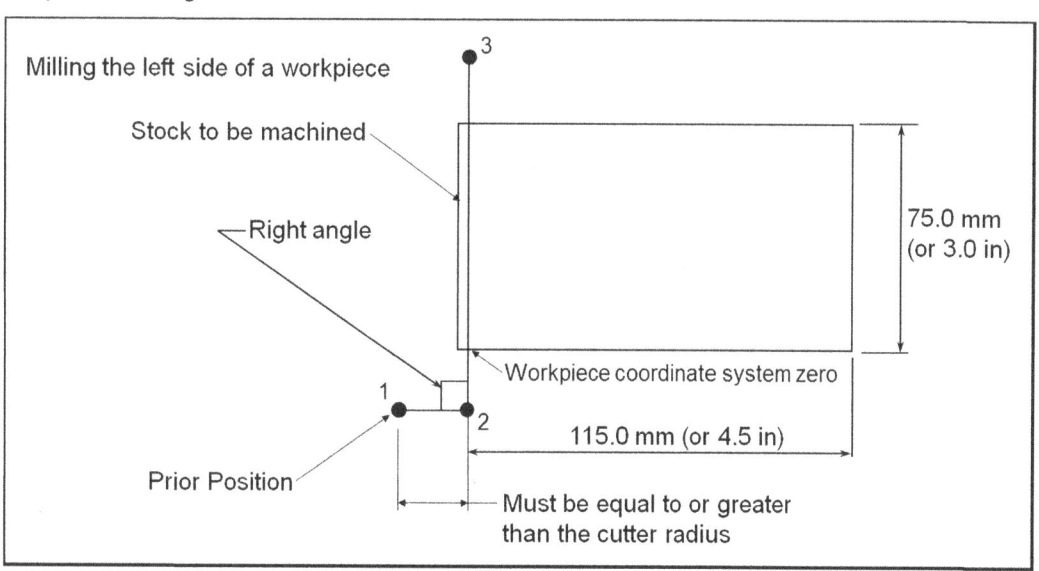

The instating command moves the milling cutter from the prior position (point number 1) to the first work surface to machine (point number 2). Again, notice that point number 2 is right on the work surface. Also, the instating command can be done at rapid (G00) if the milling cutter clears the workpiece during the motion–or in a straight line motion (G01) with a feedrate if it does not. You are not allowed to instate cutter radius compensation during a circular motion command (G02 or G03).

Key Concept 4: Know The Compensation Types

Since points one and two in our example are well clear of the workpiece in the Y-axis (with our 1.00 planned cutter size), we'll make the instating command at rapid. Here are the commands so far:

Metric-mode example:

```
%
O0016 (Program number)
N010 G21 G90 G94 G54
N020 T01 M06 (25.0 millimeter end mill)
N030 S350 M03 T02 (Start spindle fwd at 350 rpm, and get tool station one ready)
N040 G00 X-14.5 Y-14.5 (Rapid to prior position)
N050 G43 H01 Z2.0 (Instate tool length compensation)
N060 Z-27.0 M08 (End mill is now at Z-axis work surface and coolant is on)
N070 G41 D01 X0.0 (Instate cutter radius compensation)
```

Inch-mode example:

```
%
O0016 (Program number)
N010 G20 G90 G94 G54
N020 T01 M06 (1.0 inch end mill)
N030 S350 M03 T02 (Start spindle fwd at 350 rpm, and get tool station one ready)
N040 G00 X-0.6 Y-0.6 (Rapid to prior position)
N050 G43 H01 Z0.1 (Instate tool length compensation)
N060 Z-1.1 M08 (End mill is now at Z-axis work surface and coolant is on)
N070 G41 D01 X0.0 (Instate cutter radius compensation)
```

In block N070 (either example), G41 specifies that the milling cutter will be on the left side of the surface during machining. Also notice that a right-hand end mill will be climb milling–again, requiring G41 in the instating command. D01 tells the CNC to look in offset number 1 to find the cutter radius compensation value (12.5 if a 25.0-mm cutter is currently being used or 0.500 if a 1.000-in diameter milling cutter is being used). X0.0 is on the first (and in our example, the only) surface to mill. At the completion of block N070, the milling cutter's periphery will be flush with the X0.0 surface (the left side of the workpiece). A 25.0-mm end mill's center position will be at X-12.5 Y-14.5 at the end of block N070. A 1.0-in end mill's center position will be at X-0.5 Y-0.6.

Cutter radius compensation is now instated. As you might agree, we feel instating compensation is the most difficult part of programming it.

Step 2: Program the Tool Path to Be Machined

Once cutter radius compensation is instated, you must program the work surface path for the contour to be machined. It is usually important that you end the contour with the milling cutter well clear of the last machined surface. If you leave the milling cutter in contact with the last milled surface, it's likely that the milling cutter will leave a witness mark on the surface when it is retracted in the Z-axis.

The tool path in Figure 1.19 is quite simple. There is only one surface to machine:

Metric-mode example:

```
%
O0016 (Program number)
N010 G21 G90 G94 G54
```

```
N020 T01 M06 (25.0 millimeter end mill)
N030 S350 M03 T02 (Start spindle fwd at 350 rpm, and get tool station one ready)
N040 G00 X-14.5 Y-14.5 (Rapid to prior position)
N050 G43 H01 Z2.0 (Instate tool length compensation)
N060 Z-27.0 M08 (End mill is now at Z-axis work surface and coolant is on)
N070 G41 D01 X0.0 (Instate cutter radius compensation)
N080 G01 Y89.5 F5.0 (Mill left side)
```

Inch-mode example:

```
%
O0016 (Program number)
N010 G20 G90 G94 G54
N020 T01 M06 (1.0 end mill)
N030 S350 M03 T02 (Start spindle fwd at 350 rpm, and get tool station one ready)
N040 G00 X-0.6 Y-0.6 (Rapid to prior position)
N050 G43 H01 Z0.1 (Instate tool length compensation)
N060 Z-1.1 M08 (End mill is now at Z-axis work surface and coolant is on)
N070 G41 D01 X0.0 (Instate cutter radius compensation)
N080 G01 Y3.6 F5.0 (Mill left side)
```

Block N080 (either example) contains a G01 (very important here) to switch to linear interpolation cutting motion. Y89.5 (or Y3.6 in the inch-mode example) will cause the end mill to machine the left side of the workpiece, leaving it well clear of the rear surface of the workpiece.

This work surface in this example is so simple that it doesn't allow us to stress an important point. Let's look at another example, shown in the figure below.

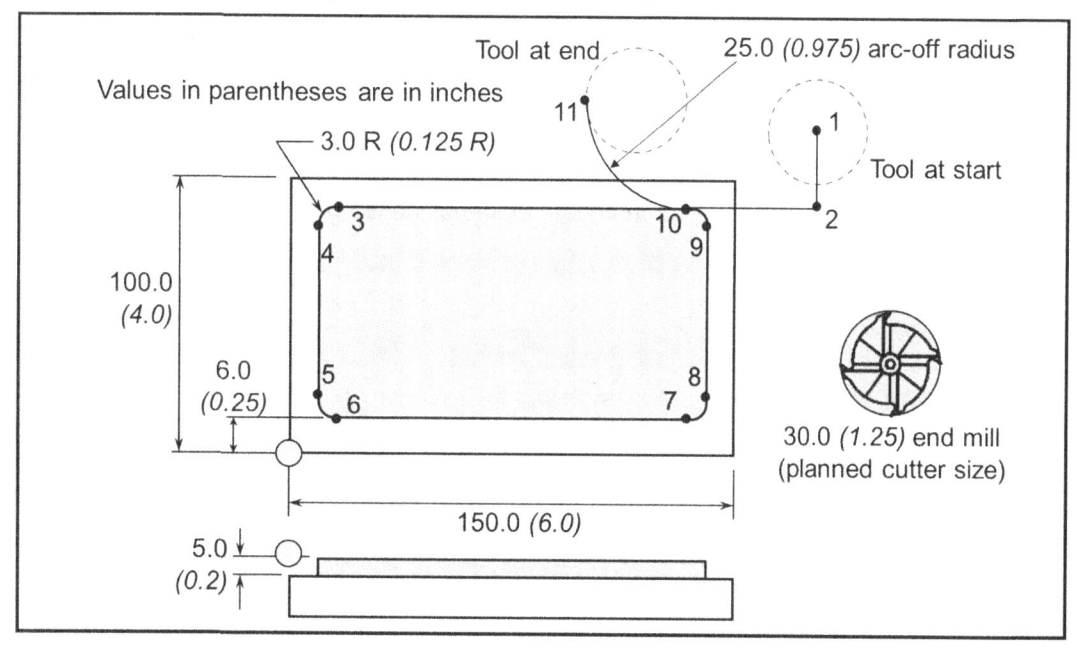

Figure 4.22: More elaborate example of a work surface tool path using cutter radius compensation

Key Concept 4: Know The Compensation Types

While we won't show the program yet (it's coming up as one of the examples later in this lesson), we want to make a point about the work surface path. To instate cutter radius compensation for this example, you send the milling cutter to the prior position in XY (point 1), move to the work surface in Z, and then, using G42 and a D-word, instate cutter radius compensation during a motion to the first milled surface (point number 2).

At this point, the milling cutter is flush with the first surface to mill (its center is above point number 2 in the Y-axis by the value placed in the cutter radius compensation offset). You're ready to program the work surface tool path.

The work surface tool path takes the milling cutter from point 2 to point 3 (linear interpolation motion)–from point 3 to point 4 (circular interpolation motion)–from point 4 to point 5 (linear interpolation motion)–and so on through point number 10. Remember that since you are programming the work surface tool path, the radius of all circular motions (R-word) will be the actual workpiece radius, R3.0 (or R0.125 if programming in inch mode) for each circular motion in our example.

When the milling cutter reaches point number 10, it is finished milling the contour, but if you allow it to retract in Z while at point 10 in XY, it will leave a nasty witness mark on the workpiece (a gouge-line on the side of the milled surface). Most programmers will come up with one more work surface tool path motion to bring the milling cutter away from the last surface to mill. This is best done in a circular motion (commonly called an arc-out motion). The size of this arc must, of course, be larger than the radius of the largest planned cutter. It must also include enough clearance to allow the largest planned cutter to move away from the last surface being milled.

For our example, when using the planned cutter size (30.0 millimeters or 1.25 inches in diameter), the 25.0-mm (or 0.975-in) arc-out radius will provide 10.0-mm (or 0.350-in) motion away from the last surface milled (arc-out radius minus milling cutter radius). This distance provides clearance for the 6.0-mm (or 0.25-in) step around the contour. When the cutter reaches point number 11, it will be completely clear of the workpiece by 4.0-mm (or 0.1-in) when using a 30.0-mm (or 1.25-in) diameter cutter.

Here are a few rules that affect the way a work surface tool path must be programmed:

- The cutter radius compensation offset value must be equal to or smaller than the smallest inside radius in the tool path or an alarm will be generated– In our example, the offset must be equal to or smaller than 25.0-mm (or 0.975-in), which is the motion from point 10 to point 11. You must be especially concerned with this rule when machining pockets. And, by the way, this is also a limiter for maximum cutter size.
- Be careful with non-motion commands–Once cutter radius compensation is instated, the machine will be constantly looking ahead to see the motions coming up in a program. Some machines cannot look very far. If it cannot determine the direction of the next XY motion, the machine will generate an alarm.
- Be careful when machining recesses in X and Y–The cutter must be able to fit into the recess. (See the drawing below for an example.) This is yet another limiter for maximum cutter size. A small cutter may be able to fit into a recess, but a larger one may not. Again, an alarm will be generated if the cutter cannot fit into the recess. Be sure to specify a maximum cutter size in the setup documentation that considers this potential problem.
- Two movements in the same direction–Some (especially older) machines do not allow this. If two motions in the same direction are commanded, the machine will generate an alarm.

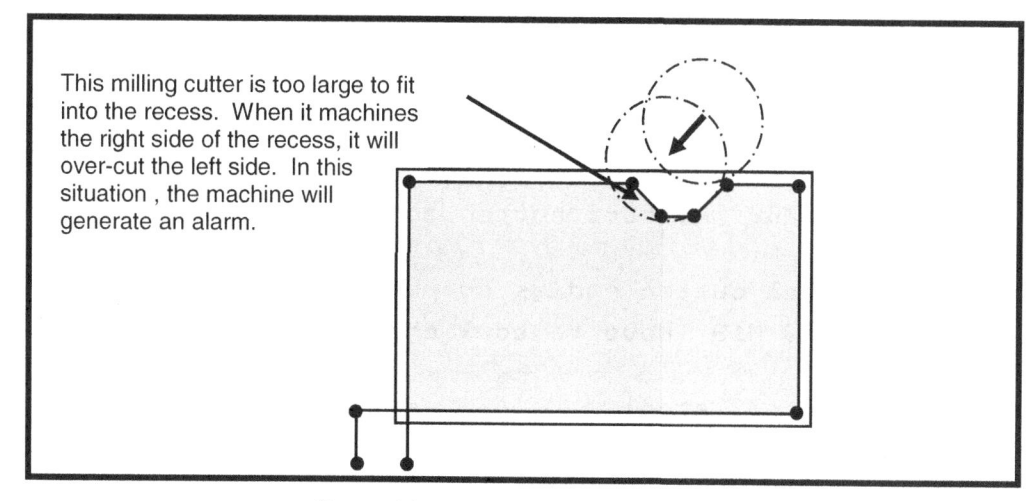

Figure 4.23: Tool must fit inside recesses

Step 3: Cancel Cutter Radius Compensation

Like many CNC features, cutter radius compensation is modal. The machine will continue to keep the milling cutter on the right or left side of all programmed motions until cutter radius compensation is cancelled. A G40 word is used to cancel cutter radius compensation. The very next X and/or Y coordinate (either within the G40 command or after it) will be a taken as a centerline position.

Though it is not always possible to do so, we recommend retracting the milling cutter in the Z-axis prior to canceling cutter radius compensation. This way, the cutter will be clear of the workpiece should an unexpected X and/or Y movement take place during cancellation.

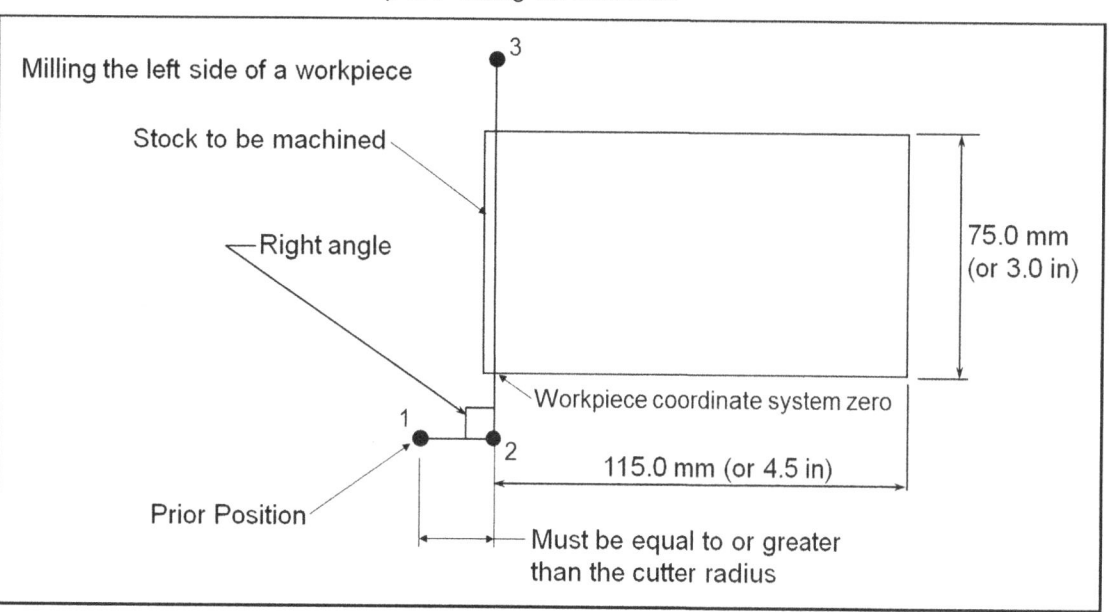

Here is the entire milling operation, including the cancellation of cutter radius compensation:

Metric-mode example:

```
%
O0016 (Program number)
N010 G21 G90 G94 G54
N020 T01 M06 (25.0 millimeter end mill)
N030 S350 M03 T02 (Start spindle fwd at 350 rpm, and get tool station one ready)
N040 G00 X-14.5 Y-14.5 (Rapid to prior position)
```

Key Concept 4: Know The Compensation Types

```
N050 G43 H01 Z2.0 (Instate tool length compensation)
N060 Z-27.0 M08 (End mill is now at Z-axis work surface and coolant is on)
N070 G41 D01 X0.0 (Instate cutter radius compensation)
N080 G01 Y89.5 F5.0 (Mill left side)
N090 G00 Z2.0 M09 (Retract cutter above work surface, turn off coolant)
N100 G40 (Cancel cutter radius compensation)
N100 G91 G28 Z0 M19 (Move to tool change position, orient spindle)
N110 M01 (Optional stop)
.
.
.
```

Inch-mode example:

```
%
O0016 (Program number)
N010 G20 G90 G94 G54
N020 T01 M06 (1.0 end mill)
N030 S350 M03 T02 (Start spindle fwd at 350 rpm, and get tool station one ready)
N040 G00 X-0.6 Y-0.6 (Rapid to prior position)
N050 G43 H01 Z0.1 (Instate tool length compensation)
N060 Z-1.1 M08 (End mill is now at Z-axis work surface and coolant is on)
N070 G41 D01 X0.0 (Instate cutter radius compensation)
N080 G01 Y3.6 F5.0 (Mill left side)
N090 G00 Z0.1 M09 (Retract cutter above work surface, turn off coolant)
N100 G40 (Cancel cutter radius compensation)
N100 G91 G28 Z0 M19 (Move to tool change position, orient spindle)
N110 M01 (Optional stop)
.
.
.
```

In block N090, we retract the tool to above the workpiece—then in block N100, we cancel cutter radius compensation. From this point on, any XY position specified in the program will be to the spindle centerline.

Don't Forget To Cancel!

You must remember to cancel cutter radius compensation. If you do not, the next tool will remain under the influence of cutter radius compensation—and will move in an unusual manner. Consider, for example, a center drilling operation that follows a contour milling operation. If you forget to cancel cutter radius compensation after milling, the center drill will move to XY positions that are incorrect by the amount of the cutter radius compensation offset value, even though specified hole-centerline coordinates in the program are correct. This problem can be quite difficult to diagnose.

What If I Have More Than One Contour To Mill?

If you have multiple contours to mill, you must handle them separately—Consider, for example, having to mill the right side of a workpiece and then the left side. This requires two separate work surfaces. First, instate, machine, and cancel for the right side. Then move on to the left side. Instate, machine, and cancel. If you must rapid the tool in XY at any time during a milling cutter's motions, you probably have more than one contour to machine—and again—you must handle them separately.

Examples

We offer two complete examples to help solidify your understanding of cutter radius compensation.

Example 1

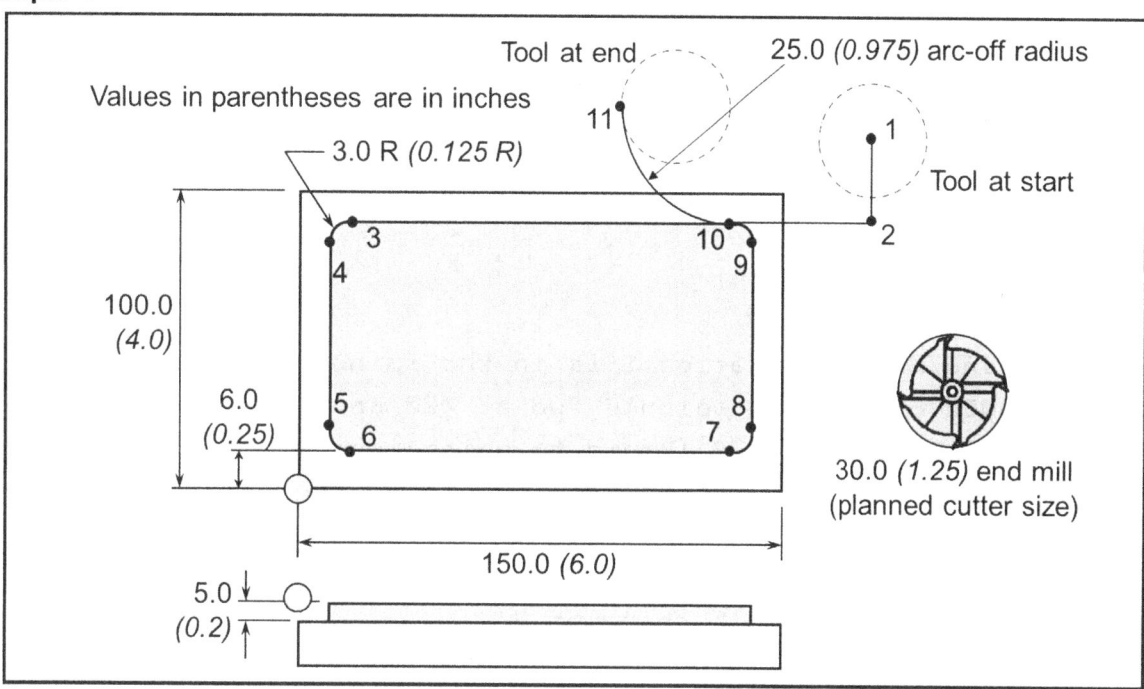

Figure 4.24: Complete example 1 using cutter radius compensation

The milling cutter will be milling the 0.25 inch step around the outside of this workpiece. The planned cutter size is a 1.25 diameter end mill (having a 0.625 radius). Based upon the distance from point one to point two in the program, the maximum cutter size for this program is 1.45 inches in diameter.

Program with comments:

Metric-mode example

```
%
O0017 (Program number)
N010 G21 G90 G94 G54
(30.0 end mill in station 1 is in the spindle)
N020 S400 M03 (Start spindle fwd at 400 rpm)
N030 G00 X167.0 Y111.0 (Rapid to prior position pt 1)
N040 G43 H01 Z2.0 (Instate tool length compensation)
N050 G01 Z-5.0 F750.0 (Fast feed to work surface)
N060 G42 D01 Y94.0 (Instate cutter radius compensation to pt 2)
N070 X9.0 F125.0 (Mill to point 3)
N080 G03 X6.0 Y91.0 R3.0 (Circular mill to point 4)
N090 G01 Y9.0 (Mill to point 5)
```

Key Concept 4: Know The Compensation Types

```
    N100 G03 X9.0 Y6.0 R3.0 (Circular mill to point 6)
    N110 G01 X141.0 (Mill to point 7)
    N120 G03 X144.0 Y9.0 R3.0 (Circular mill to point 8)
    N130 G01 Y91.0 (Mill to point 9)
    N140 G03 X141.0 Y94.0 R3.0 (Circular mill to point 10)
    N150 G02 X116.0 Y119.0 R25.0 (Arc-off the work surface to point 11)
    N160 G00 Z2.0 (Retract to above the work surface in Z)
    N170 G40 (Cancel cutter radius compensation)
    N180 G91 G28 Z0.0 (Rapid to Z-axis zero return position)
    N190 G28 X0.0 Y0.0 (Rapid XY to zero return position)
    N200 M30 (End of program)
    %
```

Inch-mode example

```
    %
    O0017 (Program number)
    N010 G20 G90 G94 G54
    (1.25 end mill in station 1 is in the spindle)
    N020 S400 M03 (Start spindle fwd at 400 rpm)
    N030 G00 X6.725 Y4.475 (Rapid to prior position pt 1)
    N040 G43 H01 Z0.1 (Instate tool length compensation)
    N050 G01 Z-0.2 F30.0 (Fast feed to work surface)
    N060 G42 D01 Y3.75 (Instate cutter radius compensation to pt 2)
    N070 X0.375 F5.0 (Mill to point 3)
    N080 G03 X0.25 Y3.625 R0.125 (Circular mill to point 4)
    N090 G01 Y0.375 (Mill to point 5)
    N100 G03 X0.375 Y0.25 R0.125 (Circular mill to point 6)
    N110 G01 X5.625 (Mill to point 7)
    N120 G03 X5.75 Y0.375 R0.125 (Circular mill to point 8)
    N130 G01 Y3.625 (Mill to point 9)
    N140 G03 X5.625 Y3.75 R0.125 (Circular mill to point 10)
    N150 G02 X4.65 Y4.725 R0.975 (Arc-off the work surface to point 11)
    N160 G00 Z0.1 (Retract to above the work surface in Z)
    N170 G40 (Cancel cutter radius compensation)
    N180 G91 G28 Z0.0 (Rapid to Z-axis zero return position)
    N190 G28 X0.0 Y0.0 (Rapid XY to zero return position)
    N200 M30 (End of program)
    %
```

Since the cutter is well clear of the right side of the workpiece in block N030, we're allowing the fast-feed approach in line N050. In block N070, when milling begins, we include the cutting feedrate.

Example 2

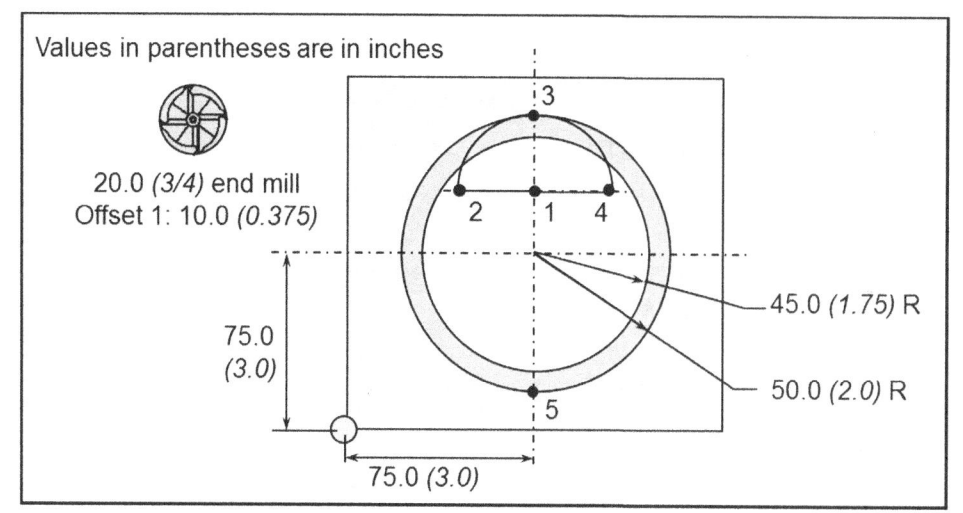

Figure 4.25: Complete example 2 using cutter radius compensation

This time, the milling cutter will be machining the inside of a counter-bored hole. The 3.5 diameter (1.75 radius) hole exists in the workpiece when this program is run (we're just milling the counter-bore).

Program with comments:

<u>Metric-mode</u> example:

```
%
O0018 (Program number)
N010 G21 G90 G94 G54
(20.0-mm end mill in station 1 is in the spindle)
N020 S500 M03 (Start spindle fwd at 500 rpm)
N030 G00 X75.0 Y108.0 (Rapid to prior position pt1)
N040 G43 H01 Z2.0 (Instate tool length compensation, move tool to just above workpiece)
N050 G01 Z-6.0 F750.0 (Fast feed to work surface)
N060 G42 D01 X58.0 F100.0 (Instate cutter radius compensation to pt 2)
N070 G02 X75.0 Y125.0 R17.0 (Arc-in to point 3)
N080 Y25.0 R50.0 (Circular mill right half of circle to pt 4)
N090 Y125.0 R50.0 (Circular mill left half of circle back to pt 3)
N100 X92.0 Y108.0 R17.0 (Arc-off to pt 5)
N110 G00 G40 X75.0 (Cancel cutter radius compensation during a movement back to the center of the approach circle pt 1)
N120 Z2.0 (Retract to just above the workpiece in Z)
N130 G91 G28 Z0.0 (Rapid to Z-axis zero return position)
N140 G28 X0.0 Y0.0 (Rapid XY to zero return position)
N150 M30 (End of program)
%
```

Key Concept 4: Know The Compensation Types

Inch-mode example:

```
%
O0018 (Program number)
N010 G20 G90 G94 G54
(3/4" end mill in station 1 is in the spindle)
N020 S500 M03 (Start spindle fwd at 500 rpm)
N030 G00 X3.0 Y4.275 (Rapid to prior position pt1)
N040 G43 H01 Z0.1 (Instate tool length compensation, move tool to just above workpiece)
N050 G01 Z-0.25 F40.0 (Fast feed to work surface)
N060 G42 D01 X2.275 F4.0 (Instate cutter radius compensation to pt 2)
N070 G02 X3.0 Y5.0 R.725 (Arc-in to point 3)
N080 Y1.0 R2.0 (Circular mill right half of circle to pt 4)
N090 Y5.0 R2.0 (Circular mill left half of circle back to pt 3)
N100 X3.725 Y4.275 R0.725 (Arc-off to pt 5)
N110 G00 G40 X3.0 (Cancel cutter radius compensation during a movement back to the center of the approach circle pt 1)
N120 Z0.1 (Retract to just above the workpiece in Z)
N130 G91 G28 Z0.0 (Rapid to Z-axis zero return position)
N140 G28 X0.0 Y0.0 (Rapid XY to zero return position)
N150 M30 (End of program)
%
```

Notice the arc-in and arc-out techniques. Arc-in is important when the cutter must come directly into contact with the first surface to mill. This is always the case when milling inside contours (like pockets). Cutter radius compensation is being instated during a movement from point one to point two, a straight line motion. Also notice that cutter radius compensation is being canceled in this program during an X motion–in block N0110. Remember, you are not allowed to instate or cancel cutter radius compensation during a circular motion.

What if I Use a CAM System to Prepare Programs?

Most CAM systems give you a choice between two possibilities when it comes to how they will generate CNC programs that include cutter radius compensation. One of your choices will be to have the CAM system create programs as we have just shown–the tool path is the work surface path and the cutter radius compensation offset value must be set equal to the cutter's radius. Since this method is quite logical to setup people and operators (they always enter the milling cutter's radius in the offset), this is the method we recommend–even if you use a CAM system.

But again, most CAM systems do provide a second choice. There are many CAM system programmers that prefer the second method. With the second method, the tool path is the milling cutter's centerline path (as is shown in lesson 3.1). The programmer specifies a planned cutter size to the CAM system. It is based on this planned size that the CAM system will generate the centerline tool path in the CNC program. The cutter radius compensation offset value must be set to the radial difference from the planned cutter size to the actual size of the cutter being used.

For example, say the programmer plans on using a 25.0-mm (or a 1.0-in) diameter end mill to mill a contour. They specify this planned cutter size to the CAM system–and the CAM system will generate a centerline tool path for this 25.0-mm (or 1.0-in) diameter cutter. When the job is to be run, if the setup person uses the planned end mill (25.0-mm or 1.0-in), they will enter a value of zero into the cutter radius compensation offset. In this case, it will be just as if cutter radius compensation is not being used.

If the setup person uses a 20.0-mm end mill (or 0.750-in end mill for the inch-mode example), they must place a value of -2.5 (or -0.125 for the inch-mode example) into the cutter radius compensation offset. Since this cutter is smaller than the planned cutter size, the offset value must be negative. The radial difference between a 25.0-mm diameter and a 20.0-mm diameter is 2.5 millimeters, calculated by subtracting 20.0 from 25.0 and dividing the result by two. For the inch-mode example, the radial difference between a 1.0-in diameter and a 0.75-in diameter is 0.125-in (1.0 minus 0.750, result divided by two).

If the setup person uses a 30.0-mm end mill (or a 1.25-in end mill), they must place a value of 2.5 (or 0.125 for the inch-mode example) into the cutter radius compensation offset. Since this cutter is larger than the planned cutter size, the offset value must be positive (again the plus sign is assumed). The radial difference between a 25.0-mm diameter and a 30.0-mm diameter is 2.5-mm, calculated by subtracting 25.0-mm from 30.0-mm and dividing the result by two. For the inch-mode example, the radial difference between a 1.0-in diameter and a 1.250-in diameter is 0.125-in (1.250 minus 1.000, divided by two).

While we prefer the first method shown, again, many CAM system programmers do use this second method. They would probably contend that if the setup person uses the planned cutter size (as they normally will), they can leave the offset value zero. They may feel this is easier for the setup person.

This may be related to how a programmer is initially taught to use cutter radius compensation. Many CAM system programmers have been taught to use the second method. If you use a CAM system to prepare your programs, this is really a matter of personal preference. It doesn't really matter which method is used, as long as everyone understands it.

Setup Person's Cutter Radius Compensation Responsibilities

In the setup documentation, the programmer must specify the milling cutters in a program that use cutter radius compensation. They should also specify the minimum and maximum diameter for each milling cutter as discussed earlier in this lesson. And they should specify which cutter radius compensation offset numbers are being used.

With this information, the setup person will assemble each cutting tool. They must also determine each milling cutter's cutter radius compensation offset value. If using our recommended method (offset is the cutter radius) the setup person will simply measure the milling cutter's diameter (with a micrometer, for example) and divide the result by two. This is the value that is entered into the cutter radius compensation offset register.

Rough and Finish Milling with the Same Set of Tool Path Coordinates

This is a very helpful technique for manual programmers. Though it has some limitations, it keeps you from having to calculate two sets of tool path coordinates—one for roughing and another for finishing. (But if you use a CAM system to prepare programs, the CAM system can create the roughing tool path just as easily as it can create the tool finishing path.)

The trick is to enter a larger cutter radius compensation offset value for the rough milling cutter. How much larger will determine the amount of stock that will be left for finishing. If, for example, you want to leave 0.5-mm (or 0.020-in if using the inch mode) for finishing, increase the value of the rough milling cutter's cutter radius compensation offset by 0.5 (or 0.020 for the inch-mode example). The figure below illustrates.

Key Concept 4: Know The Compensation Types

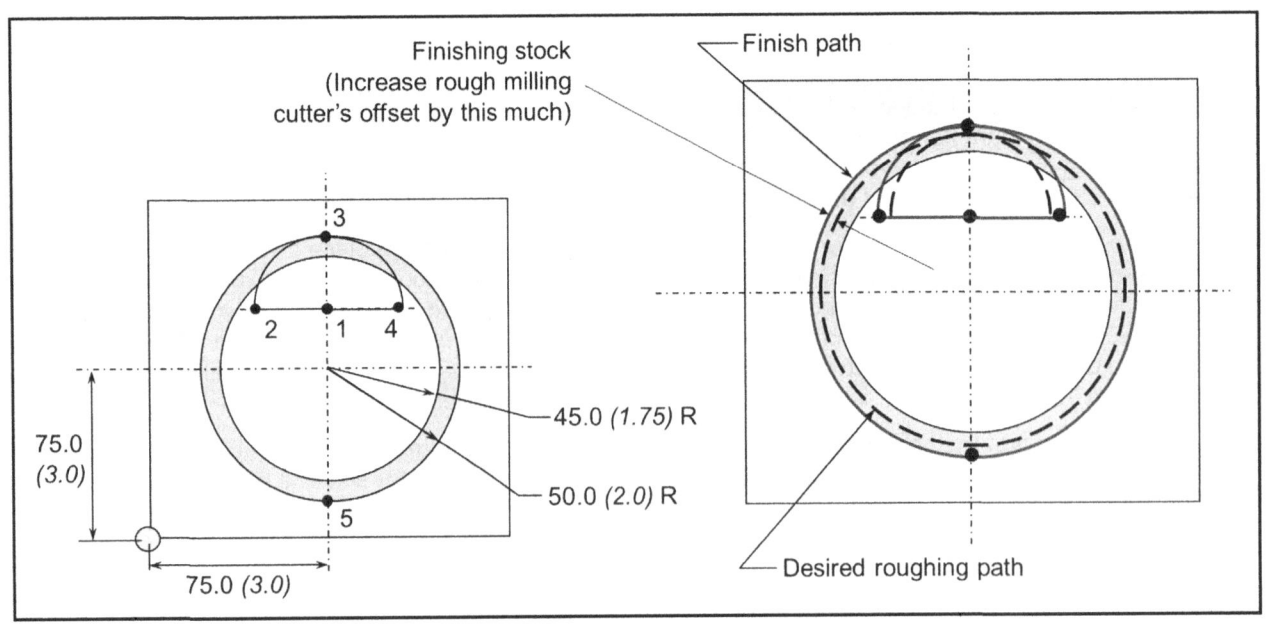

Figure 4.26: Rough mill with the finish milling cutter's tool path

Say you want to rough and finish mill the workpiece shown in the figure above with the same 20.0-mm (or 0.75-in) end mill. Two offsets will be used for cutter radius compensation. Offset number 1 is the roughing offset and must be set to a value that is larger than the actual cutter radius by the amount of finishing stock. We want to leave 0.5-mm finishing stock (or 0.020-in for the inch-mode example), so the setup person must enter a value of 10.5 (or 0.395 for the inch-mode example) in offset number 1 (10.0-mm actual cutter radius plus 0.5-mm finishing stock, or for the inch-mode example: 0.375-in actual cutter radius plus 0.020-in finishing stock). Offset number 31 is finishing offset and will be set to the milling cutter's actual size, 10.0 (or 0.375 for the inch-mode example). You must, of course, choose offsets that are not being used for another purpose. If adjustments must be made to the finished hole for diameter sizing purposes, changes will be made to offset number 31 for this example.

Program with comments:

Metric-mode example

```
%
O0019 (Program number)
N010 G21 G90 G94 G54
  (20.0-mm end mill in station 1 is in the spindle)
N020 S450 M03 (Start spindle fwd at 500 rpm)
N030 G00 X75.0 Y108.0 (Rapid to prior position pt1)
N040 G43 H01 Z2.0 (Instate tool length compensation, move tool
to just above workpiece)
N050 G01 Z-6.0 F750.0 (Fast feed to work surface)
N060 G42 D01 X58.0 F100.0 (Instate cutter radius compensation
using offset increased by 0.5 to pt 2)
N070 G02 X75.0 Y125.0 R17.0 (Arc-in to point 3)
N080 Y25.0 R50.0 (Circular mill right half of circle to pt 4)
N090 Y125.0 R50.0 (Circular mill left half of circle back to pt
3)
N100 X92.0 Y108.0 R17.0 (Arc-off to pt 5)
N110 G00 G40 X75.0 S500 (Cancel cutter radius compensation
during a movement back to the center of the approach circle pt
1, increase speed for finishing)
```

```
N120 G42 D31 X58.0 F75.0 (Instate cutter radius compensation
using offset with actual cutter radius to pt 2)
N130 G02 X75.0 Y125.0 R17.0 (Arc-in to point 3)
N140 Y25.0 R50.0 (Circular mill right half of circle to pt 4)
N150 Y125.0 R50.0 (Circular mill left half of circle back to pt
3)
N160 X92.0 Y108.0 R17.0 (Arc-off to pt 5)
N170 G00 G40 X75.0 (Cancel cutter radius compensation during a
movement back to the center of the approach circle)
N180 Z2.0 (Retract to just above the workpiece in Z)
N190 G91 G28 Z0 (Move to the Z-axis zero return position)
N200 M30 (End of program)
%
```

Inch-mode example

```
%
O0019 (Program number)
N010 G20 G90 G94 G54
  (0.75-in end mill in station 1 is in the spindle)
N020 S450 M03 (Start spindle fwd at 500 rpm)
N030 G00 X3.0 Y4.275 (Rapid to prior position pt1)
N040 G43 H01 Z0.1 (Instate tool length compensation, move tool
to just above workpiece)
N050 G01 Z-0.25 F40.0 (Fast feed to work surface)
N060 G42 D01 X2.275 F5.0 (Instate cutter radius compensation
using offset increased by 0.020 to pt 2)
N070 G02 X3.0 Y5.0 R.725 (Arc-in to point 3)
N080 Y1.0 R2.0 (Circular mill right half of circle to pt 5)
N090 Y5.0 R2.0 (Circular mill left half of circle back to pt 3)
N100 X3.725 Y4.275 R.725 (Arc-off to pt 4)
N110 G00 G40 X3.0 S500 (Cancel cutter radius compensation
during a movement back to the center of the approach circle
pt1)
N120 G42 D31 X2.275 F4.0 (Instate cutter radius compensation
using offset with actual cutter radius to pt 2)
N130 G02 X3.0 Y5.0 R.725 (Arc-in to point 3)
N140 Y1.0 R2.0 (Circular mill right half of circle to pt 5)
N150 Y5.0 R2.0 (Circular mill left half of circle back to pt 3)
N160 X3.725 Y4.275 R.725 (Arc-off to pt 4)
N170 G00 G40 X3.0 (Cancel cutter radius compensation during a
movement back to the center of the approach circle)
N180 Z0.1 (Retract to just above the workpiece in Z)
N190 G91 G28 Z0 (Move to the Z-axis zero return position)
N200 M30 (End of program)
%
```

Key Concept 4: Know The Compensation Types

Notice that coordinates in blocks N120 through N170 use exactly the same as the coordinates as are used in blocks N060 through N100 (though the offset number, and feedrate are different). Also notice that we used different cutting conditions for roughing and finishing. In lesson 6.2, we show a method using sub-programming that will keep you from having to write these commands twice.

Though we use the same milling cutter for roughing and finishing, you can, of course, use two different cutters and still use this technique. Simply increase the rough milling cutter's offset by the amount of stock you want to leave for the finishing cutter.

A Warning

Be careful not to break one of the rules of cutter radius compensation. The smallest inside radius in your tool path must be larger than the cutter radius compensation value stored in the roughing offset. In our example, we use a very large approach radius (17.0-mm or 0.725-in for the inch-mode example. The milling cutter can be up to 34.0-mm (or 1.450 inches) in diameter before this program will fail.

Trial Machining with Cutter Radius Compensation

Just as tool length compensation allows trial machining and sizing for depth (Z) dimensions, so cutter radius compensation allows trial machining for XY dimensions. Again, trial machining is required on the first workpiece when a dimension machined by a milling cutter has a very tight tolerance. This technique ensures that the cutter will not machine too much material on its very first try. If done for each milling cutter that has a tight tolerance, the first workpiece will pass inspection.

Like trial machining for depth dimensions, trial machining for XY dimensions involves five steps. We will be referring to the drawing shown in the figure below.

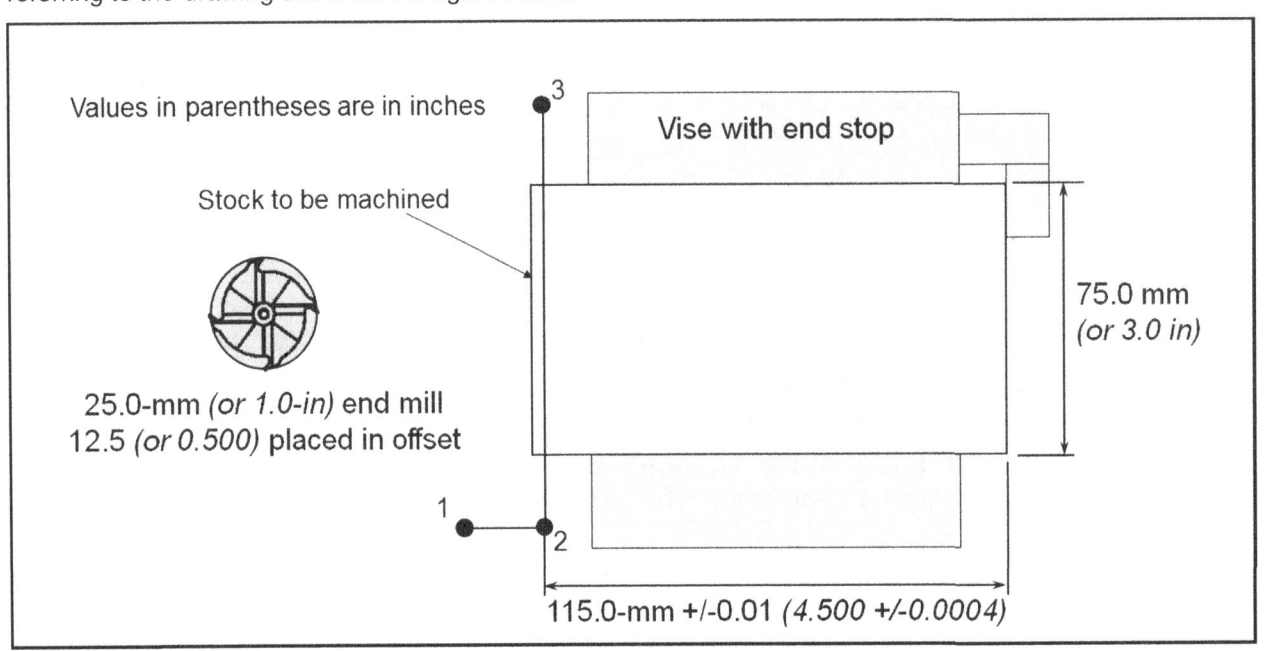

Figure 4.27: Drawing for trial machining discussions

1: Recognition of a tight tolerance that worries you–If you're worried that a dimension's tolerance is so small that your initial cutter radius compensation offset setting is not accurate enough, trial machining must be done. In our case, say we're using the program shown earlier that mills the left end of this workpiece. As you are running the very first workpiece, you notice the very tight plus or minus 0.01-mm (or 0.0004-in) tolerance for the overall length of the workpiece. You're worried that your initial cutter radius compensation offset setting is not accurate enough. If the cutter is running-out in its holder even a tiny amount, for instance, this dimension will be machined undersize, scrapping the workpiece.

2: Increase the value of the cutter radius compensation offset by about 0.25-mm (or 0.010-in)–Since we're using a 25.0-mm (or 1.0-in) diameter milling cutter, we'll increase the offset value to 25.25 (or 0.510 for the inch-mode example). This is 0.25-mm (0.010-in) larger than the true radius of this cutter.

3: Let the milling cutter machine under the influence of the trial machining offset and stop the cycle after the tool is finished–In our example, the milling cutter will be milling the left end of the workpiece. When it is finished, you stop the machine. Our trial machining offset will force the milling cutter to leave excess stock on the machined surface.

4: Measure the current dimension and reduce the cutter radius compensation offset accordingly–If everything is perfect, the current overall length of this workpiece should be exactly 115.25-mm (or 4.510-in). But when you measure, you find it to be 115.2-mm (or 4.508 in). It is currently only 0.2-mm (0.008-in) oversize–and it's a good thing you trial machined. Either you did not measure the diameter of the end mill correctly or the end mill is running out in its holder. Either way, you must reduce the cutter radius compensation offset by 0.2-mm (or 0.008-in), making the offset value 25.05 (or 0.502 for the inch-mode example).

5: Re-run the milling cutter under the influence of the adjusted offset–This time, the milling cutter will machine the overall length perfectly to size. You measure just to be sure. It comes out to 114.994-mm (or 4.9998-in for the inch-mode example). The tiny 0.006-mm (0.0002-in) deviation is caused by the difference in tool pressure from the first time the cutter machined (the normal amount of stock is being removed) to the smaller depth-of-cut it took after trial machining. The next workpiece (when the milling cutter takes its full depth-of-cut) will be perfect.

How Do You Know When To Trial Machine?

We've said that when tolerances are under about 0.05-mm (0.002-in) (overall), your initial tooling measurements may not be accurate enough to ensure that the first workpiece is machined correctly. But this is largely a matter of your own personal experiences – and some trial-and-error.

When in doubt, it's best to trial machine. And when you're trial machining, you can learn a lot. When you increase the initial offset to trial machine, by say 0.25-mm (0.010-in), the cutter *should* machine leaving *exactly* 0.25-mm *(0.010-in)* too much stock. *Did it?* If it did not, how close was it?

This difference is the amount of error in your initial setting – caused mostly by imperfect measuring – with possibly a little tool pressure or cutter run-out. This tells you whether you needed to trial machine in the first place.

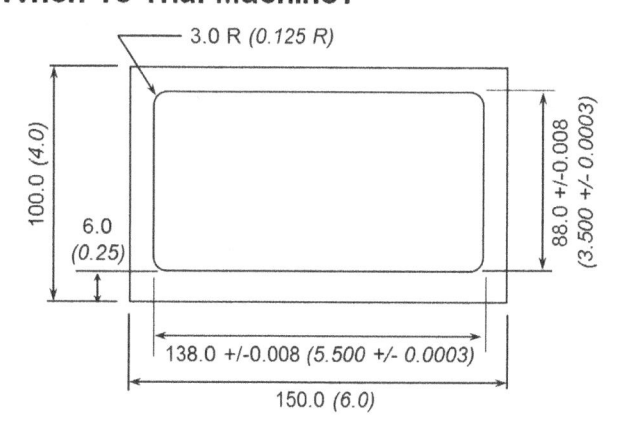

If the difference is well less than your tolerance, the workpiece *would have been machined properly* – without trial machining. If this happens consistently, you shouldn't have to trial machine for this tolerance with this tool and workpiece material in the future.

Dimensions: 3.0 R (0.125 R); 100.0 (4.0); 6.0 (0.25); 88.0 +/-0.008 (3.500 +/- 0.0003); 138.0 +/-0.008 (5.500 +/- 0.0003); 150.0 (6.0)

What If the Milling Cutter Machines On Both Sides of the Workpiece?

In our example for trial machining (milling the left side of a workpiece), the milling cutter is only machining one end of the workpiece–and it's pretty easy to make sizing adjustments. But keep in mind that milling cutters often machine on two sides of the workpiece. After machining the hole in figure 4.26, for example, you will be measuring between two surfaces machined by the milling cutter. This is also the case when machining a rectangular pocket, milling both ends of a workpiece, or milling the raised step shown in the first example program.

When measuring between two surfaces machined by the milling cutter, you must pay particular attention to cutter radius compensation offset value adjustments. Say for example, you measure the diameter of the hole in Figure 4.26 after finish machining. It should be 100.0-mm (or 4.0-in for the inch-mode example)–but you find it to be 99.99-mm (or 3.9996-in). The overall hole-size is 0.01-mm (0.0004-in) too small. If you want to machine the next workpiece perfectly (make it precisely 100.0-mm [4.0-in]), what do you do?

Again, this milling cutter is machining on <u>both</u> sides of the hole. You must reduce its cutter radius compensation offset value by 0.005 mm (0.0002-in). This will cause the milling cutter to machine 0.05-mm (0.0002-in) more on both sides, increasing the hole-diameter by 0.01-mm (0.0004-in).

Key Concept 4: Know The Compensation Types

A Tip for Remembering Which Way to Adjust the Offset

A common mistake made by beginning setup people and operators is adjusting the cutter radius compensation offset in the wrong direction. Here's a way to remember which way is which.

Using our recommended method (offset is cutter radius), think about what will happen if you enter the radius of a milling cutter as zero. The machine will think you have a milling cuter with zero diameter. It will make the spindle center follow the tool path you have programmed, which is right on the work surface. The milling cutter will machine way too much material (scrapping the workpiece, of course). So, <u>reducing the offset value will make the milling cutter machine more material</u>.

How Important Is It To Make Your First Workpiece a Good One?

We've been talking a great deal about trial machining to ensure that your first workpiece gets machined properly. There are companies, however, that are not overly concerned if setup people scrap the first workpiece–as long as they learn enough in the process to make the next workpiece properly. When procuring raw material, these companies actually purchase more than the quantity needed in the production run (manufacturing people refer to the extra material as setup pieces or practice pieces).

While these companies do exist, the vast majority of CNC using companies expect their setup people to make every workpiece a good one. It is a tenet of Lean Manufacturing. They know that if their setup people use trial machining techniques for dimensions with critical tolerances as they machine the first workpiece, the first workpiece will be an acceptable workpiece. While mistakes are bound to occur, there is really no excuse for scrapping the first workpiece if trial machining techniques are used (assuming the program and process are correct).

When Trial Machining Is Not Required

Again, only tight XY tolerances require trial machining with cutter radius compensation. If the milled surface has a large tolerance, you should be able to measure the cutter radius compensation value accurately enough. The cutter will machine somewhere within the tolerance band on its first try (the workpiece will be acceptable). But it is still important to adjust the offset in such a way that the next workpiece machined will have the dimension come out right at its target value.

In our example for milling the left side of a workpiece, say the overall length tolerance is plus or minus 0.125-mm (or 0.005-in). In this case trial machining is not required–so you let the milling cutter machine with the initial offset setting (12.5-mm [or 0.500-in]). Once the cutter is finished you measure the overall length and find it to be 100.05-mm (4.502-in). It is 0.05-mm (or 0.002-in) larger than its target value, but not out of tolerance (it is a good workpiece). Even so, you should still decrease the offset by 0.05-mm (or 0.002-in) to make the milling cutter machine 0.05-mm (0.002-in) more material on the next workpiece. Then, the overall length will come out right at its target dimension. If this is done with all cutting tools, the production run will begin with all workpiece attributes being machined at their target dimensions.

Sizing with Cutter Radius Compensation

Sizing is required when the wear a cutting tool experiences during its life affects the surfaces it machines. With very tight tolerances (say, under about 0.01-mm [0.0004-in] overall or so), you may need to perform several sizing adjustments during a milling cutter's life. The example shown in Figure 4.27 may require this kind of sizing.

After trial machining for the first workpiece, the setup person has the 115.0-mm (4.500-in) overall-length-dimension coming out perfectly to size. They turn the job over to a CNC operator to run production. After thirty workpieces or so, the operator measures the 115.0-mm (4.500-in) dimension and finds it to be 115.003-mm (or 4.5001-in). The cutter has worn a tiny amount. It is still cutting properly and does not have to be replaced–but it is showing signs of wear. After thirty more workpieces, the operator measures again and finds the overall length to be 115.006-mm (4.5002-in). It is still within the tolerance band, but if this trend continues for much longer, workpieces will be machined with this dimension out of its tolerance band.

At this point, the operator decides to make a sizing adjustment. They reduce the milling cutter's cutter radius compensation value by 0.006-mm (or 0.0002-in). The next workpiece to be machined will come out back at the 115.0-mm (4.500-in) target dimension.

This sizing adjustment may have to be repeated several times (based upon the size of the tolerance and how much the milling cutter wears) before the milling cutter is completely dull and must be replaced.

Lesson 4.3: Cutter Radius Compensation

What Do You Shoot For When Making Sizing Adjustments?
By the way, this is why many CNC people do not use the mean value as the target value. If you do, you're only working with half the tolerance band. The tighter the tolerance, the more of a problem this can be. In our example, if the setup person trial machines using a target value of 115.993-mm (4.4998-in) instead of 115.0-mm (4.500-in), which is closer to the low limit for this tolerance band, the milling cutter will last for twice as many workpieces before a sizing adjustment will be needed.

What If I Use a CAM System and the Offset Value Is the Deviation From the Planned Cutter Size?
Making offset adjustments for trial machining and sizing will be exactly the same. You will just be working with smaller offset values. When you want to machine more material, reduce the offset value. When you want to trial machine (leaving more material), increase the offset value.

Do I Have to Make All These Calculations When Adjusting Offsets?
As we said during our discussions of tool length compensation, most CNC controls allow you to modify offset values incrementally. FANUC calls this feature input plus (actually the [INPUT+] soft key on the display screen). With this feature, you need only know the amount of needed offset adjustment. If you need an offset to be reduced by 0.05-mm (or 0.002-in), simply type -0.05 (or -0.002 for the inch-mode example) and press the [INPUT+] soft key. The control will automatically calculate the new value for the offset and enter it.

When the CNC has 4 columns per offset, one on the columns will be for geometry (the nominal radius of the tool) and one for wear (fine tuning or wear amount). The [INPUT+] soft key applies to both the geometry and wear fields. One benefit of using the separate geometry and wear fields is that limits can be set on the maximum values allowed. For example, tool geometry may be limited between 0.0 and the maximum radius of a tool possible. The wear may be limited to ±0.05-mm (or ±0.020-in). This helps error proof data entry.

Why Can't I Just Change Programmed Coordinate/s In the Program To Make Sizing Adjustments?
You can find this discussion in lesson 4.2. The same points made for tool length compensation apply to cutter radius compensation–so we won't repeat the discussion here. But again, never change programmed coordinates to make sizing adjustments.

Key Points for Lesson 4.3:
- ✓ Cutter radius compensation is used only for milling cutters–and only when side milling.
- ✓ Cutter radius compensation allows the programmer to ignore the size of the milling cutter as the program is being written.
- ✓ Cutter radius compensation simplifies calculations for programming, allows a range of cutter sizes to be used, allows trial machining and sizing, and allows one tool path to be used for both rough milling and finish milling.
- ✓ There are three steps to programming cutter radius compensation: instate it, program the tool path, and cancel it.
- ✓ The setup person must store the cutter's radius in the cutter radius compensation offset (based upon our recommended method for manual programmers).
- ✓ For sizing, if you want the milling cutter to remove more material, *reduce* the offset value.

Key Concept 4: Know The Compensation Types

Exercise: Practice Programming Cutter Radius Compensation

In this exercise you will practice specifying cutter radius compensation commands. The workpiece engineering drawing has been marked up with the planned machining process, and we have included a tool-path drawing to show how to position cutting tools through their milling operations.

To mill the oval slot, for example, first plunge at pt14, then feed to pt15, instate cutter comp. to pt 16, mill around slot and back to point 17, arc-off to point 22, cancel comp on move to pt 23.

First fill in the coordinate sheet. Then complete the G-code program that follows, specifying the feeds and speeds required for each tool, filling in modal G-codes and other modal words, and only specifying moving axes.

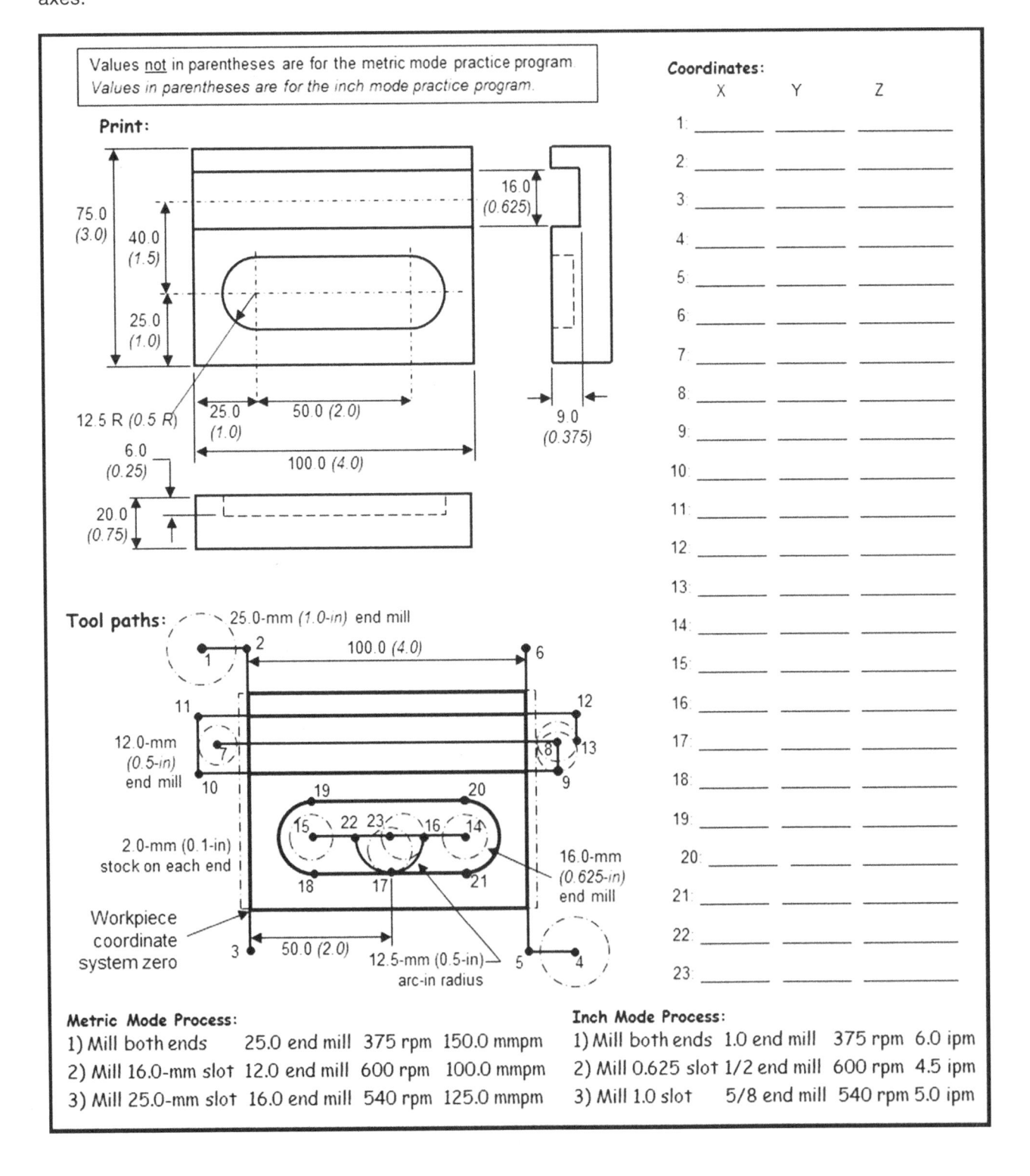

Values **not** in parentheses are for the metric mode practice program.
Values in parentheses are for the inch mode practice program.

Print:

Dimensions: 75.0 (3.0), 40.0 (1.5), 25.0 (1.0), 16.0 (0.625), 12.5 R (0.5 R), 25.0 (1.0), 50.0 (2.0), 100.0 (4.0), 9.0 (0.375), 6.0 (0.25), 20.0 (0.75)

Tool paths:
- 25.0-mm (1.0-in) end mill
- 12.0-mm (0.5-in) end mill
- 16.0-mm (0.625-in) end mill
- 2.0-mm (0.1-in) stock on each end
- 12.5-mm (0.5-in) arc-in radius
- 100.0 (4.0), 50.0 (2.0)
- Workpiece coordinate system zero

Coordinates:

#	X	Y	Z
1			
2			
3			
4			
5			
6			
7			
8			
9			
10			
11			
12			
13			
14			
15			
16			
17			
18			
19			
20			
21			
22			
23			

Metric Mode Process:
1) Mill both ends   25.0 end mill   375 rpm   150.0 mmpm
2) Mill 16.0-mm slot 12.0 end mill   600 rpm   100.0 mmpm
3) Mill 25.0-mm slot 16.0 end mill   540 rpm   125.0 mmpm

Inch Mode Process:
1) Mill both ends   1.0 end mill   375 rpm   6.0 ipm
2) Mill 0.625 slot 1/2 end mill   600 rpm   4.5 ipm
3) Mill 1.0 slot   5/8 end mill   540 rpm   5.0 ipm

Fill in the blanks for this program (works for both metric mode and inch modes):

```
%
O0020
N005 T01 M06 (25.0 millimeter end mill)
N010 G____ G54 G90 S____ M03 T02 (Select measurement system mode)
N015 ____ ____ ____ (pt 1)
N020 ____ ____ ____ M08
N023 G01 Z____ F____ (Fast feed to work surface)
N025 ____ ____ ____ (pt 2)
N030 ____ F____ (pt 3)
N035 ____ ____
N040 ____
N045 ____ ____ (pt 4)
N050 ____ ____ ____
N055 ____ ____ ____ (pt 5)
N060 ____ ____ (pt 6)
N065 ____ ____ M09
N070 ____
N075 G91 G20 Z0 M19
N080 M01
N085 T02 M06 (12.0 millimeter end mill)
N090 G54 G90 ____ M03 T03
N095 ____ ____ ____ (pt 7)
N100 ____ ____ ____ M08
N105 ____ ____ ____
N110 ____ ____ (pt 8)
N115 ____ ____ ____ ____ (pt 9)
N120 ____ ____ (pt 10)
N125 ____ ____ (pt 11)
N130 ____ ____ (pt 12)
N135 ____ ____ ____ (pt 13)
N140 ____ ____ M09
N145 G91 G20 Z0 M19
N150 M01
N155 T03 M06 (16.0 millimeter end mill)
N160 G54 G90 ____ M03 T01
N165 ____ ____ ____ (pt 14)
N170 ____ ____ ____ M08
N175 ____ ____ ____
N180 ____ (pt 15)
N185 ____ ____ ____ ____ (pt 16)
```

Lesson 4.3: Cutter Radius Compensation

FANUC CERT – Machining Center ©2017 CNC Concepts, Inc.

Key Concept 4: Know The Compensation Types

```
N190 ____ ____ ____ ____ ____ (pt 17)
N195 ____ ____ (pt 18)
N200 ____ ____ ____ (pt 19)
N205 ____ ____ (pt 20)
N210 ____ ____ ____ (pt 21)
N215 ____ ____ (pt 17)
N220 ____ ____ ____ ____ (pt 22)
N225 ____ ____ ____ (pt 23)
N230 ____ ____ M09
N235 G91 G28 Z0 M19
N240 M30
%
```

Metric mode answer program for cutter radius compensation exercise

```
%
O0020
N005 T01 M06 (25.0 millimeter end mill)
N010 G21 G54 G90 S375 M03 T02
N015 G00 X-14.5 Y89.5 (pt 1)
N020 G43 H01 Z2.0 M08
N023 G01 Z-22.0 F750.0
N025 G42 D01 X0 (pt 2)
N030 Y-14.5 F150.0 (pt 3)
N035 G00 Z2.0
N040 G40
N045 X114.5 Y-14.5 (pt 4)
N050 G01 Z-22.0 F750.0
N055 G42 D31 X100.0 (pt 5)
N060 Y89.5 F150.0 (pt 6)
N065 G00 Z2.0 M09
N070 G40
N075 G91 G28 Z0 M19
N080 M01

N085 T02 M06 (12.0 millimeter end mill)
N090 G54 G90 S600 M03 T03
N095 G00 X-8.0 Y65.0 (pt 7)
N100 G43 H02 Z2.0 M08
N105 G01 Z-9.0 F750.0
N110 X108.0 F100.0 (pt 8)
N115 G42 D02 Y57.0 F750.0 (pt 9)
N120 X-14.0 F100.0 (pt 10)
N125 Y73.0 F750.0 (pt 11)
N130 X114.0 F100.0 (pt 12)
N135 G40 Y65.0 F750.0 (pt 13)
N140 G00 Z2.0 M09
N145 G91 G28 Z0 M19
N150 M01

N155 T03 M06 (16.0 millimeter end mill)
N160 G54 G90 S540 M03 T01
N165 G00 X75.0 Y25.0 (pt 14)
N170 G43 H03 Z2.0 M08
N175 G01 Z-6.0 F125.0
N180 X25.0 (pt 15)
N185 G42 D03 X62.5 F750.0 (pt 16)
N190 G02 X50.0 Y12.5 R12.5 F125.0 (pt 17)
N195 G01 X25.0 (pt 18)
N200 G02 Y37.5 R12.5 (pt 19)
N205 G01 X75.0 (pt 20)
N210 G02 Y12.5 R12.5 (pt 21)
N215 G01 X50.0 (pt 17)
N220 G02 X37.5 Y25.0 R12.5 (pt 22)
N225 G40 G01 X50.0 (pt 23)
N235 G01 Z2.0 M09
N235 G91 G28 Z0 M19
N240 M30
%
```

Inch mode answer program for cutter radius compensation exercise

```
%
O0020
N005 T01 M06 (1.0 inch end mill)
N010 G20 G54 G90 S375 M03 T02
N015 G00 X-0.6 Y3.6 (pt 1)
N020 G43 H01 Z0.1 M08
N023 G01 Z-0.85 F40.0
N025 G42 D01 X0 (pt 2)
N030 Y-0.6 F6.0 (pt 3)
N035 G00 Z0.1
N040 G40
N045 X4.6 Y-0.6 (pt 4)
N050 G01 Z-1.1 F40.0
N055 G42 D31 X4.0 (pt 5)
N060 Y3.6 F6.0 (pt 6)
N065 G00 Z0.1 M09
N070 G40
N075 G91 G28 Z0 M19
N080 M01

N085 T02 M06 (1/2 end mill)
N090 G54 G90 S600 M03 T03
N095 G00 X-0.35 Y2.5 (pt 7)
N100 G43 H02 Z0.1 M08
N105 G01 Z-0.375 F40.0
N110 X4.35 F4.5 (pt 8)
N115 G42 D02 Y2.1875 F40.0 (pt 9)
N120 X-0.6 F4.5 (pt 10)
N125 Y2.8125 F40.0 (pt 11)
N130 X4.6 F4.5 (pt 12)
N135 G40 Y2.5 F40.0 (pt 13)
N140 G00 Z0.1 M09
N145 G91 G28 Z0 M19
N150 M01

N155 T03 M06 (5/8 end mill)
N160 G54 G90 S540 M03 T01
N165 G00 X3.0 Y1.0 (pt 14)
N170 G43 H03 Z0.1 M08
N175 G01 Z-0.250 F5.0
N180 X1.0 (pt 15)
N185 G42 D03 X2.5 F40.0 (pt 16)
N190 G02 X2.0 Y0.5 R0.5 F5.0 (pt 17)
N195 G01 X1.0 (pt 18)
N200 G02 Y1.5 R0.5 (pt 19)
N205 G01 X3.0 (pt 20)
N210 G02 Y0.5 R0.5 (pt 21)
N215 G01 X2.0 (pt 17)
N220 G02 X1.5 Y1.0 R0.5 (pt 22)
N225 G40 G01 X2.0 (pt 23)
N235 G01 Z0.1 M09
N235 G91 G28 Z0 M19
N240 M30
%
```

Lesson 4.4–Workpiece Coordinate System Offsets

Tool length and cutter radius compensation allow you to ignore certain attributes of cutting tools as you write your CNC programs. In like fashion, workpiece coordinate system setting offsets allow you to ignore the precise location of the work holding devices on the machine table as you write programs.

Objectives

After completing this lesson, students should be able to:

- ✓ Describe the applications for assigning multiple workpiece coordinate system zero points
- ✓ Describe how to program multiple workpiece coordinate system zero points
- ✓ Describe how to shift the workpiece coordinate system reference point for modular qualified setups using the external workpiece coordinate system offset
- ✓ Describe applications for the external workpiece coordinate system offset
- ✓ Describe how to set the workpiece coordinate system offsets in the program using the G10 command

Introduction

You know from lesson 1.6 that workpiece coordinate system setting offsets are used to assign workpiece coordinate system zero. Based upon our discussions so far, you know that the distances from the workpiece coordinate system zero point on the physical part to the zero return (reference) position must be determined (either measured or calculated). The polarity of these values is almost always negative since the zero return position is usually close to the plus over-travel limit of each axis. These values are entered into workpiece coordinate system offset #1 to assign workpiece coordinate system zero. (You might want to review lesson 1.6 before continuing with this lesson.)

We've only shown how to assign one workpiece coordinate system zero point, using workpiece coordinate system offset #1 (G54). In this lesson, we'll be showing when multiple workpiece coordinate system zeros are required as well as how they are assigned and programmed.

Also, with the method shown in lesson 1.6, the point-of-reference for workpiece coordinate system offset entries is the zero return position. There are situations when the zero return position doesn't make the best point of reference for work offset entries. In this lesson, we'll show how the point of reference can be shifted–and when you can benefit from shifting it.

And finally, there are a few tricks you can use with workpiece coordinate system offsets to enhance your machining center's performance. We'll include them in this lesson.

Do You Need to Learn More about Workpiece Coordinate System Offsets?

There are many applications that don't need any more from workpiece coordinate system offsets than we show in lesson 1.6. Many applications only have one workpiece coordinate system zero point per program. The setup person measures the workpiece coordinate system offset values during setup and enters them into work offset #1.

However, when it comes to becoming more efficient with workpiece coordinate system zero assignment (and in turn–reducing setup time), many applications can benefit by improved methods. You cannot become more efficient, of course, unless you know something better is possible.

While the material we present in this lesson may not be of immediate need to you, you will definitely want to understand its content. If you do decide to skip ahead, be sure to come back to this material once you've acquainted yourself with your machining center and its usage.

Key Concept 4: Know The Compensation Types

Assigning Multiple Workpiece Coordinate System Zero Points

Most programs for vertical machining centers require but one workpiece coordinate system zero point per program—but there are exceptions. Consider, for example, the setup shown below.

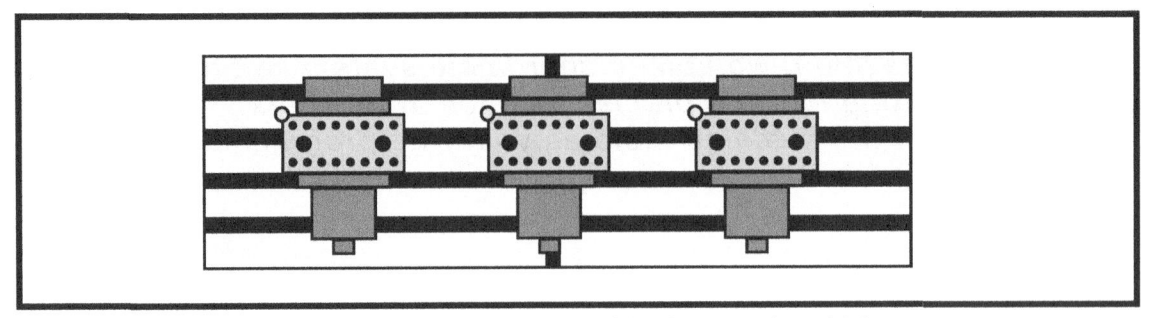

Figure 4.28: Three setups made on the table of a vertical machining center.

In this example, there are three identical workpieces being run on the table at the same time. With this application, the manufacturer will save on tool changing time (one tool will machine all three workpieces before the next tool).

If only one workpiece coordinate system zero point is assigned for all three workpieces, the programmer must, of course, know the exact distance between the workpieces. And of course, the setup person must position the workpieces precisely in the setup. This can be very cumbersome, especially for vise setups.

Workpiece coordinate system offsets will commonly allow you to assign up to six workpiece coordinate system zero points for use by a single program. (FANUC also provides an option for forty-eight additional work offsets programmed with G54.1, but most machine tool builders supply only six.) With multiple workpiece coordinate system setting offsets, you can run multiple identical workpieces—as in the figure above, different sides of the same workpiece (possibly utilizing an indexing device), or completely different workpieces. We show a rotary device application for multiple workpiece coordinate system offsets in lesson 6.4.

With multiple workpiece coordinate system zero points comes the need to determine multiples sets of workpiece coordinate system offset values. And one way to assign multiple workpiece coordinate system zeros is simply an extension of what you already know. This method of workpiece coordinate system zero assignment works very well when there are no relationships among the workpiece coordinate system zero points to be assigned—as is the case when workpieces are held in vises.

With this method, the distances between each workpiece coordinate system zero point and the zero return position in all axes is determined and entered into each workpiece coordinate system offset. The figure below shows the workpiece coordinate system offset values using this method.

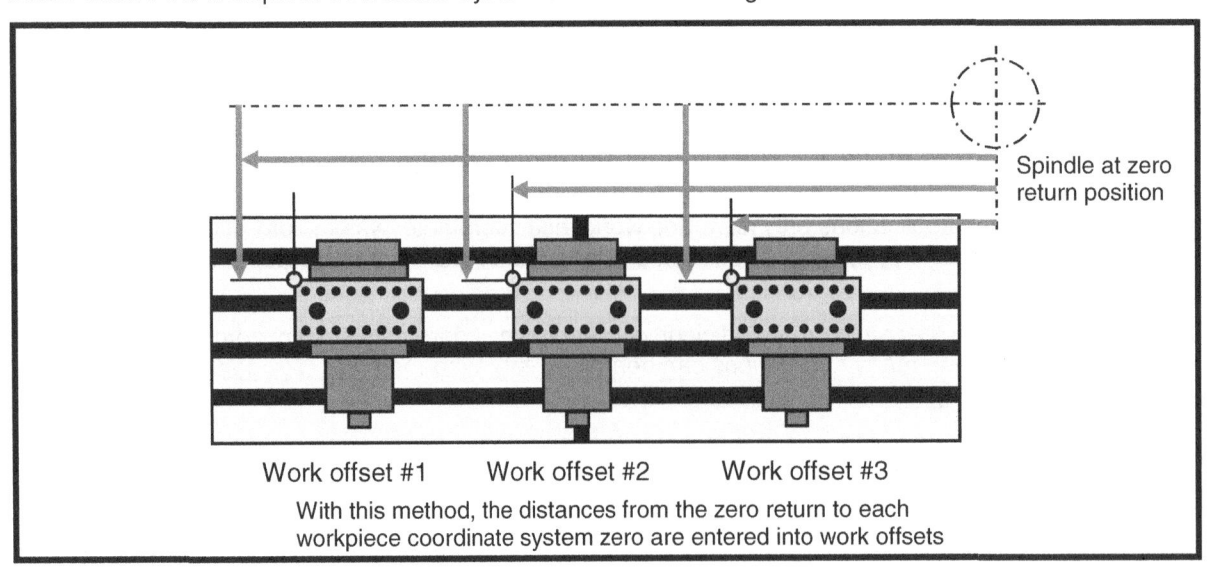

Figure 4.29: Workpiece coordinate system zero assignment values for multiple workpiece coordinate system zero points

Lesson 4.4: Workpiece Coordinate System Offsets

When using this method, all workpiece coordinate system offset values will be negative. For the example in the figure above, the workpiece coordinate system zero assignment values for the left-most workpiece will be placed in workpiece coordinate system offset #1 (G54). For the middle workpiece, they will be placed in workpiece coordinate system offset #2 (G55). And for the right-most workpiece, they will be placed in workpiece coordinate system offset #3 (G56).

The Z-axis value for each workpiece coordinate system setting offset must also be determined and entered into the appropriate workpiece coordinate system setting offset–as is shown in the figure below.

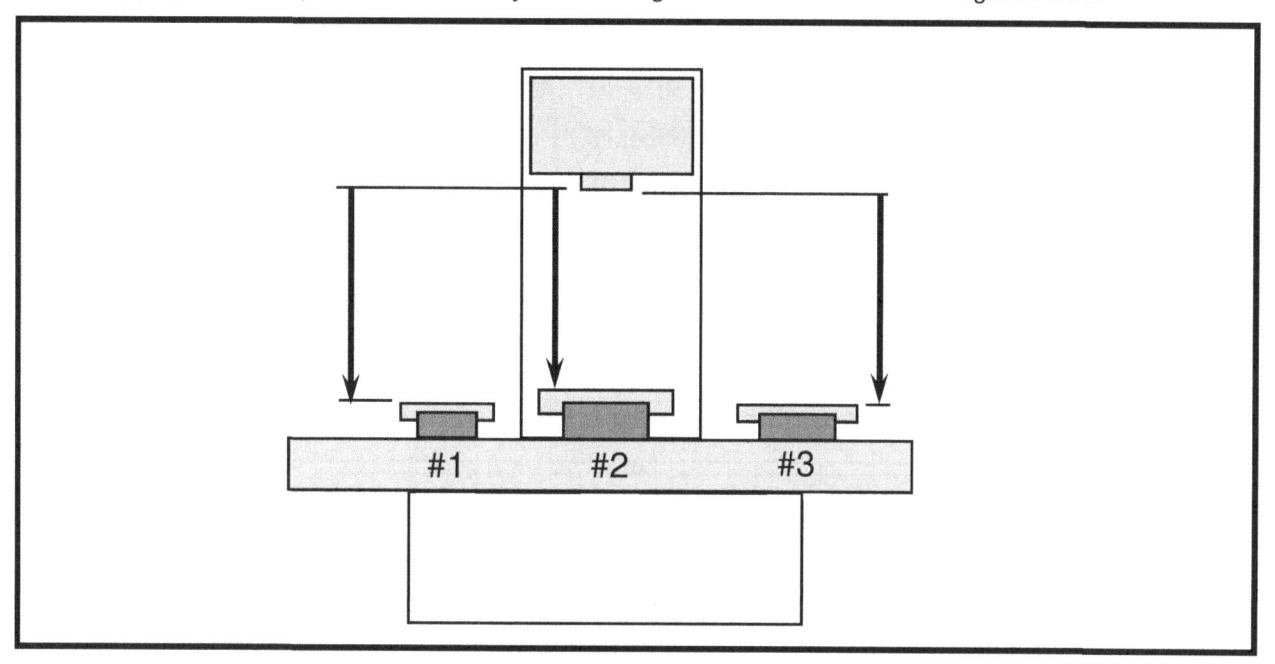

Figure 4.30: The workpiece coordinate system zero assignment values for the Z-axis

Here is an example of the workpiece coordinate system offset screen with values entered in the workpiece coordinate system offsets #1 through #3. Notice the large negative values. These are the distances from the zero return position to the workpiece coordinate system zero point in each axis (and for each offset).

		WORK COORDINATES			
(G54)					
NO.		DATA	NO.		DATA
000	X	0.0000	002	X	-10.3455
EXT	Y	0.0000	G55	Y	-9.2314
	Z	0.0000		Z	-10.8437
001	X	-15.3432	003	X	-3.3432
G54	Y	-9.2134	G56	Y	-5.2343
	Z	-11.1245		Z	-11.0343

Figure 4.31: coordinate system offsets for 3 workpiece (shown in inch mode)

Key Concept 4: Know The Compensation Types

Programming with Multiple Workpiece Coordinate System Zero Points

You know that G54 is used to instate workpiece coordinate system offset #1. When this word is executed, the machine knows the location of the workpiece coordinate system zero point. All motions commanded in the absolute mode will be specified from this location. Here are the G-codes related to all six of the workpiece coordinate system setting offsets.

- G54: workpiece coordinate system setting offset #1
- G55: workpiece coordinate system setting offset #2
- G56: workpiece coordinate system setting offset #3
- G57: workpiece coordinate system setting offset #4
- G58: workpiece coordinate system setting offset #5
- G59: workpiece coordinate system setting offset #6

Programming with multiple workpiece coordinate system zero points is very easy. You simply instate the workpiece coordinate system offset for the workpiece coordinate system zero point (workpiece) that you are working with prior to making any cutting tool motions. The figure below shows the setup to be used for our example program. While no dimensions are given, the comments in the program should make it easy to follow along.

Example program showing multiple workpiece coordinate system setting offsets

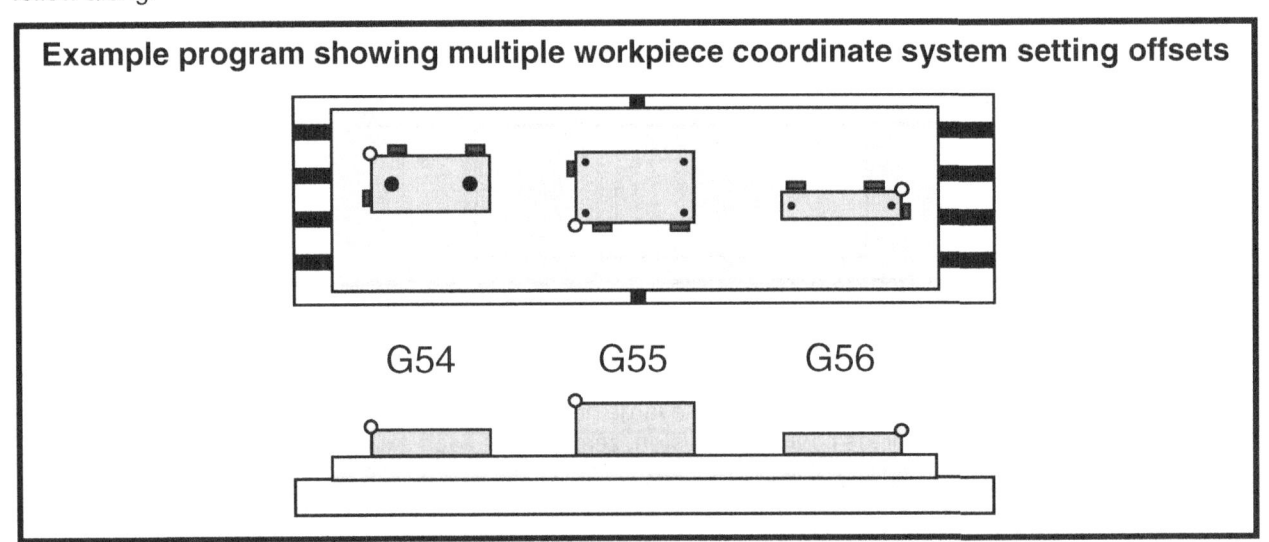

Figure 4.32: Setup for example program

Program with comments:

Metric-mode example

```
%
O0021 (Program number)
N010 G21 G90 G54 (Select workpiece coordinate system setting
offset #1)
N020 T01 M06 (Spot drill)
N030 S1000 M03 T02 (Start spindle fwd at 1000 rpm, get tool two
ready)
N040 G00 X12.0 Y-12.0 (Move to left hole of left workpiece)
N050 G43 H01 Z2.0 (Instate tool length compensation, move tool
down to just above work surface)
N060 G01 Z-4.0 F100.0 (Center drill left-most hole in left
workpiece)
N070 G00 Z2.0 (Retract from hole)
```

N080 X88.0 (Move to right hole in left workpiece)
N090 G01 Z-4.0 (Center drill hole)
N100 G00 Z2.0 (Retract from hole)
N110 **G55** Z2.0 (Select coordinate system number two, move to 0.1 above middle workpiece-NOTICE THAT MIDDLE WORKPIECE IS HIGHER THAN THE LEFT ONE!)
N120 X9.0 Y9.0 (Move to lower-left hole of middle workpiece)
N130 G01 Z-4.0 (Center drill hole)
N140 G00 Z2.0 (Retract from hole)
N150 Y66.0 (Move to upper-left hole of middle workpiece)
N160 G01 Z-4.0 (Center drill hole)
N170 G00 Z2.0 (Retract from hole)
N180 X116.0 (Move to upper-right hole of middle workpiece)
N190 G01 Z-4.0 (Center drill workpiece)
N200 G00 Z2.0 (Retract from hole)
N210 Y9.0 (Move to lower-right hole of middle workpiece)
N220 G01 Z-4.0 (Center drill workpiece)
N230 G00 Z2.0 (Retract from hole)
N240 G56 X-91.0 Y-12.0 (Select workpiece coordinate system setting offset #3, move to left-most hole of right workpiece)
N250 Z2.0 (Move to just above workpiece-NOTICE THAT RIGHT WORKPIECE IS LOWER THAN MIDDLE WORKPIECE!)
N260 G01 Z-4.0 (Center drill hole)
N270 G00 Z2.0 (Retract from hole)
N280 X-9.0 (Move to right-most hole of right workpiece)
N290 G01 Z-4.0 (Center drill hole)
N300 G00 Z2.0 (Retract from hole)
N310 G91 G28 Z0 M19 (Move to the Z-axis zero return position, orient spindle)
N320 M01 (Optional stop)
.
. (Program continues)
.

Inch-mode example

%
O0021 (Program number)
N010 G20 G90 G54 (Select workpiece coordinate system setting offset #1)
N020 T01 M06 (Spot drill)
N030 S1000 M03 T02 (Start spindle fwd at 1000 rpm, get tool two ready)
N040 G00 X0.5 Y-0.5 (Move to left hole of left workpiece)
N050 G43 H01 Z0.1 (Instate tool length compensation, move tool down to just above work surface)

Key Concept 4: Know The Compensation Types

```
N060 G01 Z-0.2 F4.0 (Center drill left-most hole in left
workpiece)
N070 G00 Z0.1 (Retract from hole)
N080 X3.5 (Move to right hole in left workpiece)
N090 G01 Z-0.2 (Center drill hole)
N100 G00 Z0.1 (Retract from hole)
N110 G55 Z0.1 (Select coordinate system number two, move to 0.1
above middle workpiece-NOTICE THAT MIDDLE WORKPIECE IS HIGHER
THAN THE LEFT ONE!)
N120 X0.375 Y0.375 (Move to lower-left hole of middle
workpiece)
N130 G01 Z-0.2 (Center drill hole)
N140 G00 Z0.1 (Retract from hole)
N150 Y2.625 (Move to upper-left hole of middle workpiece)
N160 G01 Z-0.2 (Center drill hole)
N170 G00 Z0.1 (Retract from hole)
N180 X4.625 (Move to upper-right hole of middle workpiece)
N190 G01 Z-0.2 (Center drill workpiece)
N200 G00 Z0.1 (Retract from hole)
N210 Y0.375 (Move to lower-right hole of middle workpiece)
N220 G01 Z-0.2 (Center drill workpiece)
N230 G00 Z0.1 (Retract from hole)
N240 G56 X-3.625 Y-0.5 (Select workpiece coordinate system
setting offset #3, move to left-most hole of right workpiece)
N250 Z0.1 (Move to just above workpiece-NOTICE THAT RIGHT
WORKPIECE IS LOWER THAN MIDDLE WORKPIECE!)
N260 G01 Z-0.2 (Center drill hole)
N270 G00 Z0.1 (Retract from hole)
N280 X-0.375 (Move to right-most hole of right workpiece)
N290 G01 Z-0.2 (Center drill hole)
N300 G00 Z0.1 (Retract from hole)
N310 G91 G28 Z0 M19 (Move to the Z-axis zero return position,
orient spindle)
N320 M01 (Optional stop)
.
. (Program continues)
.
```

Important! A workpiece coordinate system offset-selecting G-code (G54-G59) by itself will not cause any motion. It simply tells the machine the location of the current workpiece coordinate system zero point. In block N110, for example, as we switch from workpiece coordinate system setting offset #1 (G54) to workpiece coordinate system setting offset #2 (G55), we must include a motion to the new Z surface above the middle workpiece (Z2.0 in the metric-mode example or Z0.1 in the inch-mode example). If the tool is not moved to this position in Z prior to the movement to the X and Y position in block N120, it will crash into the middle workpiece. When using multiple workpiece coordinate system zeros in a program, you must always be concerned with this potential for collisions.

Lesson 4.4: Workpiece Coordinate System Offsets

The Potential Trade-Off with this Method

Again, programming multiple workpiece coordinate system offsets is very easy. But if the setup person must measure each individual set of workpiece coordinate system zero assignment values and enter them into workpiece coordinate system setting offsets every time the job is run, setting up the machine will be more difficult. If the setup is qualified (the work holding device can be placed on the machine table in exactly the same location every time the setup is made), the setup person will only have to measure these values once (the first time the setup is made). Workpiece coordinate system offset values can then be documented for the next time the job is run. Workpiece coordinate system offset entries for qualified setups can even be programmed–as we'll show a little later–to eliminate the need for the setup person to enter them each time the setup is made.

Reminder About Tool Length Compensation Values

Remember that workpiece coordinate system offset values in the Z-axis are directly tied to the way you use tool length compensation. If using our recommended method (tool length is the offset value), each workpiece coordinate system offset Z value will be specified from the point of reference for workpiece coordinate system offset entries, which is the spindle nose at the zero return position as we have shown so far.

If you use the second method shown (each tool length compensation value is the distance from the tool tip at the zero return position to the workpiece coordinate system zero surface in Z), one of your workpiece coordinate system setting offset Z values will be zero. This will be the workpiece you use to measure tool length compensation values. All other workpiece coordinate system setting offset Z values must be specified from this one. In essence, you're setting up a reference workpiece coordinate system setting offset from which others must be are entered.

Shifting the Workpiece Coordinate System Offsets Reference Point

There is another way to specify workpiece coordinate system offset values in work offsets. Programming remains exactly the same regardless of which method is used. With this second method, the zero return position is not the point of reference for each work offset entry. Instead, you can specify a more logical position as the point of reference workpiece coordinate system offset entries. This technique is most helpful when your work holding tooling is very predictable. It can make workpiece coordinate system zero assignment much easier, eliminating the need for any measurements. Consider the sub-plate shown in the figure below.

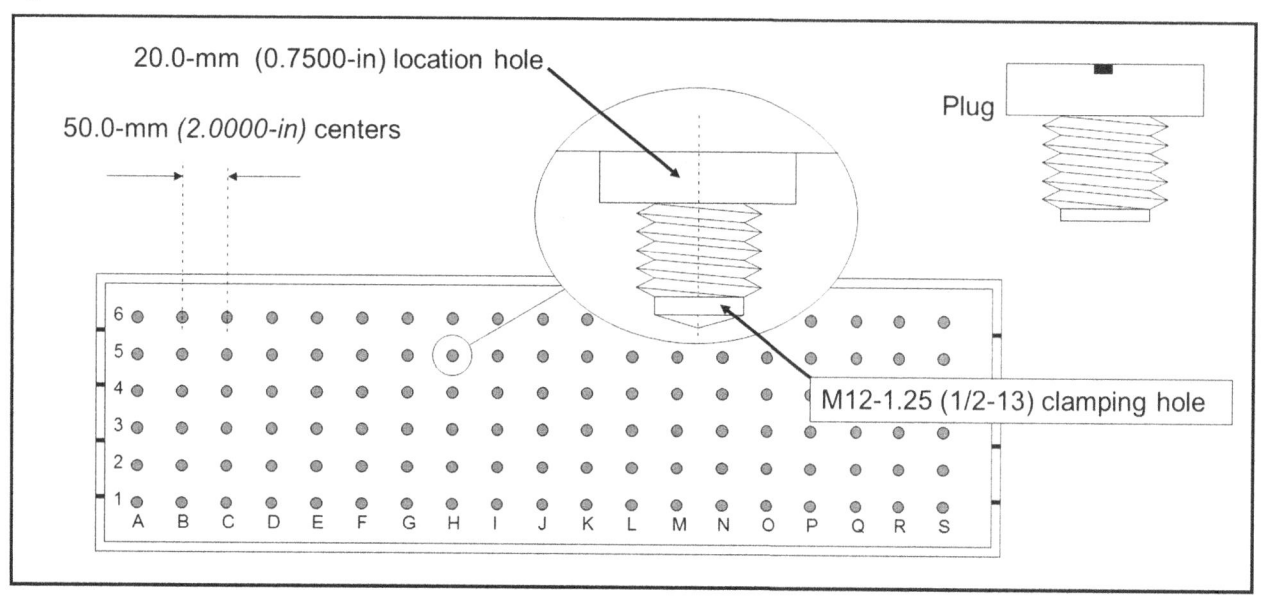

Figure 4.33: Sub-plate used on a vertical machining center

This kind of sub-plate is available from commercial suppliers–and many companies make their own. It has been semi-permanently mounted to the machine table (it could be removed if necessary, but is seldom taken off the machine table). There are a series of precise, evenly spaced location-and-mounting holes to locate and secure work holding devices. As the expanded view shows, each hole is a combination of both (location and mounting). And each hole is nicely labeled, so it can be easily referenced in setup documentation (the hole at H-5 is circled in the figure above).

Key Concept 4: Know The Compensation Types

The idea is to be able to quickly and easily mount work holding devices. Again, notice that each hole has the ability to locate the work holding device (with a 20.0-mm or 0.7500-in precision hole) or to clamp it (with an M12-1.25 or 1/2-13 tapped hole). Holes that are not currently being used will be plugged to keep chips and debris from getting into them.

This sub-plate is used with special component-type work holding tooling (think of the children's toys made by Lego). These components are of high precision, so when one is placed on the sub-plate, you'll know its precise location relative to all of the holes in the sub-plate. We'll pick the hole labeled A-1 (the lower left hole) as the reference hole–the one to which we'll shift the point of reference for workpiece coordinate system offset entries in the X and Y axes.

We show one type of locating component in the figure below–a corner locator. This is but one of many types of components made for the sub-plate. All component tools will locate and clamp in the holes in the sub-plate. In the figure below, the corner locator has been placed in holes F-4 & 5 and G-4 & 5.

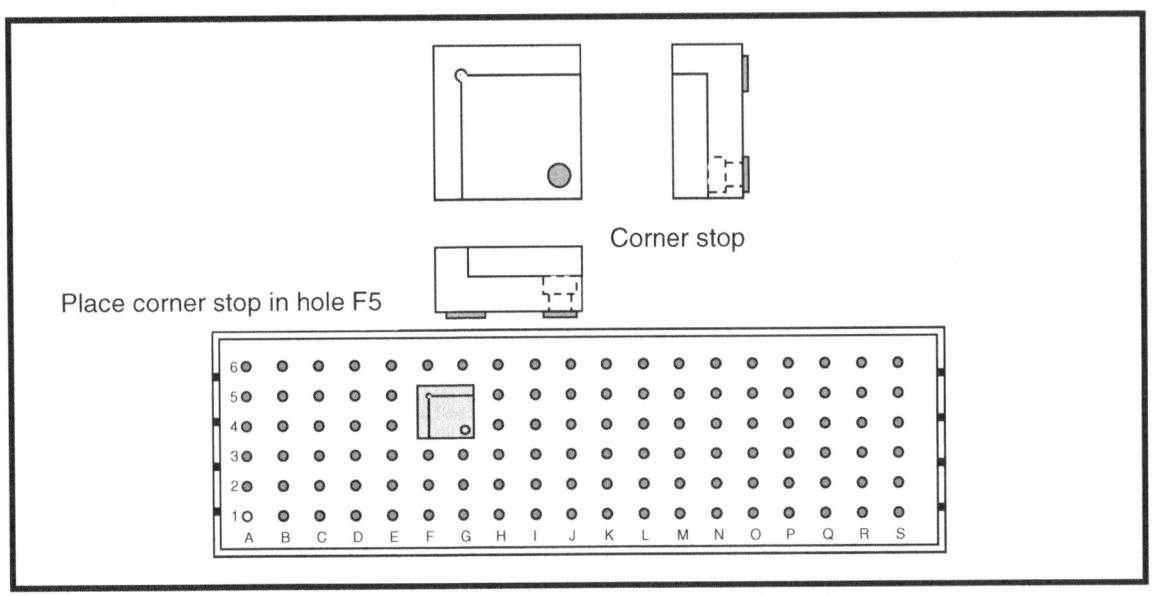

Figure 4.34: Component work holding tools used on sub-plate.

Again, there is high precision with these components and the sub-plate. The corner-locator's location edges are perfectly aligned with the upper-left location hole (F-5 in the figure above). Since the holes are on precise two-inch centers, the location edge in X is exactly ten inches from the hole A-1 in the X-axis and eight inches from hole A-1 in the Y-axis.

Again, workpiece coordinate system offsets will allow us to shift the X and Y-axis point-of-reference from the spindle center at the zero return position to the center of the hole at A-1. This is done with the external (also called the common) workpiece coordinate system offset (offset #0). Look at the figure below.

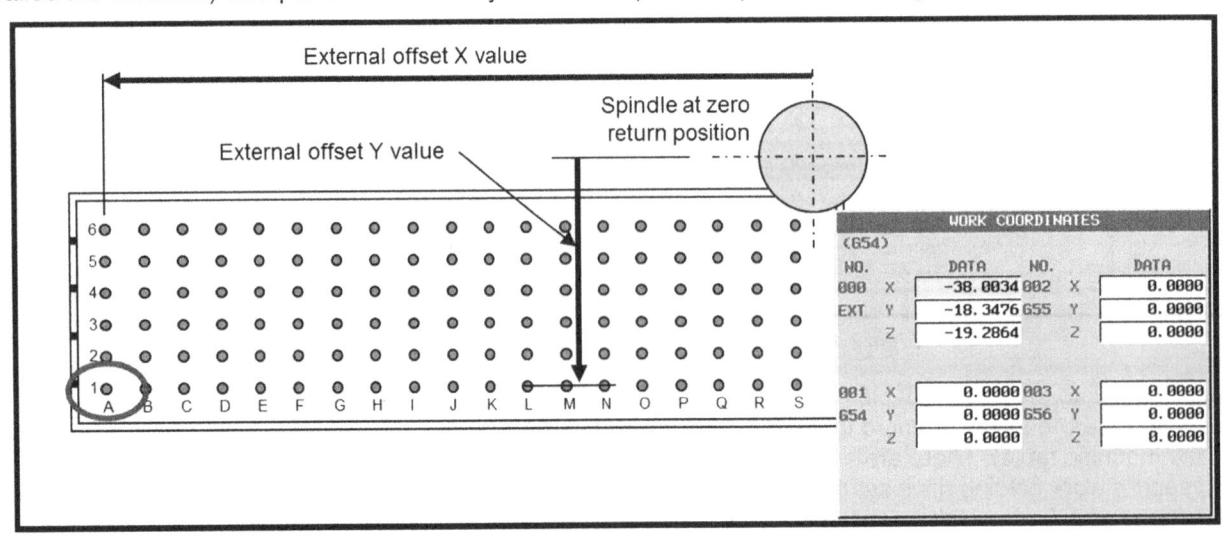

Figure 4.35: Shifting the point of reference with the external workpiece coordinate system offset (shown in inch mode)

The polarity for shifting is taken <u>from</u> the zero return position <u>to</u> the new reference point–meaning these values are negative. The setup person will measure the distances from the zero return position to the center of the hole at A-1 and enter them into the external workpiece coordinate system offset. This is only done once–it is not repeated as long as the sub-plate remains on the machine table.

While our drawings don't show it, this component tooling is also very precise in the Z-axis. So we will shift the point of reference in the Z-axis from the spindle nose at the zero return position to the top of the sub-plate. This distance is entered in the Z-register of the external offset. Again, the polarity for this entry will be negative (from the spindle nose to the top of the sub-plate). (You can only [feasibly] shift the Z-axis point of reference for workpiece coordinate system setting offset entries if you use our recommended method for tool length compensation [tool length is offset].)

Once the point of reference for workpiece coordinate offset entries has been shifted to hole A-1, all X and Y workpiece coordinate system setting offset entries will be specified from the hole at A-1, making them very easy to predict (no more measuring). Say the workpiece coordinate system zero point is the corner of the corner locator. In this case, the work offset X-value will be 10-inches. The Y-value will be 8-inches. See the figure below for an illustration.

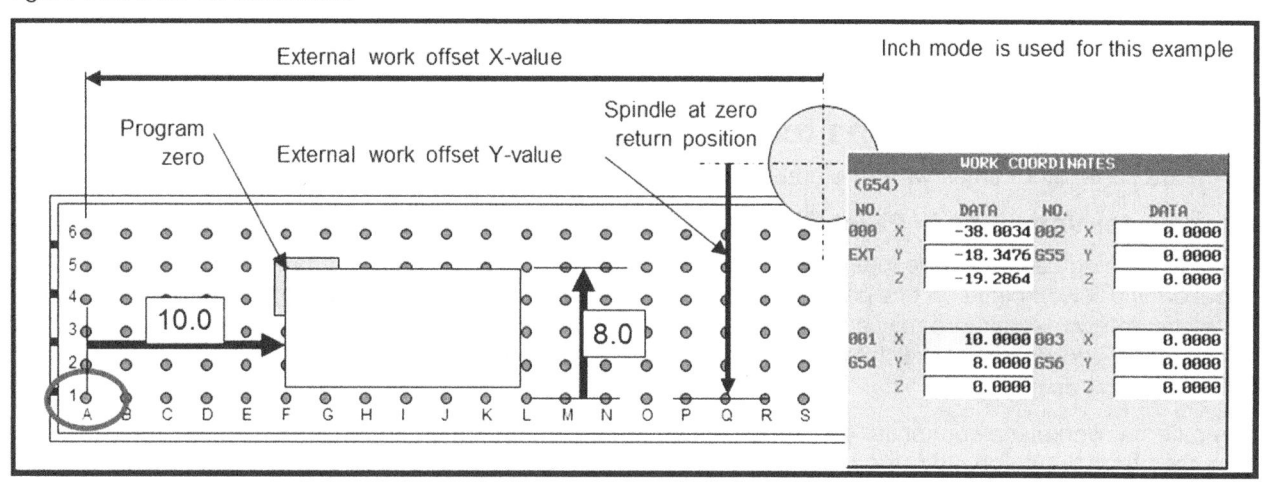

Figure 4.36: Workpiece coordinate system offset entries after point of reference has been shifted (shown in inch mode)

In Z, all Z-axis workpiece coordinate system setting offset entries will be specified from the top of the sub-plate. Say the corner locator shown in the figure above is 0.500 thick (to the bottom of the workpiece). The workpiece coordinate system setting offset Z-setting will be 0.5 plus the workpiece thickness–assuming workpiece coordinate system zero in Z is the top of the workpiece. If the workpiece is 1.0 inch thick, the workpiece coordinate system setting offset Z value will be 1.500-inches.

Again, after the point of reference has been shifted, polarity is determined from the new point of reference to the workpiece coordinate system zero points. For our application, the X, Y, and Z workpiece coordinate system offset entries will now be positive values.

Notice how easy this makes it for the setup person. They simply place the component tools on the sub-plate. If they must, they can easily determine the workpiece coordinate system zero assignment values without measuring anything.

Since the programmer specifies that the corner locator in Figure 4.36 will be placed in at hole F-5, they will know the workpiece coordinate system zero assignment values before the program is written. As we show next, workpiece coordinate system setting offset entries can be programmed. This will completely eliminate the task of workpiece coordinate system zero assignment during setup!

Programming Workpiece Coordinate System Offset Entries

FANUC CNCs provide the ability to program offset entries, including tool length and cutter radius compensation offset entries and workpiece coordinate system setting offset entries. The word used is G10. The format for G10 may vary among CNC models, so you must reference the manufacturer's programming

Key Concept 4: Know The Compensation Types

manual for the specific format for your machine. We'll show how G10 is programmed for workpiece coordinate offsets on many of the current FANUC models (the FANUC 16M, 18M, 21i/18i/16i, 30i/31i/32i, 0i).

```
G10 L2 P_ X_ Y_ Z_
```

The L-word specifies the kind of data to be entered. For workpiece coordinate system offsets the value is L2.

A P-word specifies the workpiece coordinate system offset number. P1, for example, specifies workpiece coordinate system offset #1.

X, Y, and Z-words specify the values to be placed in the corresponding workpiece coordinate system offset registers.

G10 is also affected by the current positioning mode, absolute (G90) or incremental (G91). In absolute mode, workpiece coordinate system offset registers will be over-written, which is what we want to do. In incremental mode, the current values in the workpiece coordinate system offset will be modified by the values included in the G10 command. Here's an example:

```
N010 G90 G10 L2 P1 X10.0 Y8.0 Z1.5 (Overwrite workpiece
coordinate system setting offset number one's current values)
```

This command can be included at the beginning of the program and will work nicely to keep the setup person from having to enter the workpiece coordinate system values for the setup shown in the previous illustration.

Some Other Applications for the External Work Offset

There are some other times when the common offset can be helpful.

Allowing for Variations in Pallet Changers

Many machining centers made today are equipped with pallet changers. A pallet changer allows the operator to be loading a workpiece on one pallet while the machine is running the other. Though these devices are quite accurately made, there may be a small deviation from one pallet to the next, meaning the distance from the machine's zero return position to the pallet vary from pallet to pallet. This of course, will cause problems with machining accuracy.

The external workpiece coordinate system offset can help you overcome this problem. You can measure the distances from the machine's zero return position to each pallet's center and top. This need be done only once. If there are deviations in pallets, each pallet will have its own set of external workpiece coordinate system setting offset values. A G10 command at the beginning of the program being run on each pallet can specify the correct common workpiece coordinate system setting offset values for the pallet being run. (P0 specifies workpiece coordinate system offset #0–the external work offset).

Allowing for Variations after a Mishap

Though you will do everything you can to avoid mishaps, it seems inevitable that someday you will have a crash. When a crash occurs, it is likely that one or more axes will "slip" by a small amount, meaning workpiece coordinate system offset values for all qualified setups previously made will be incorrect. If you make a lot of qualified setups, and if your workpiece coordinate system offset entries are included in your program with G10 commands, this can be a real problem. When the machine is repaired, the repair person should be able to get each axis back to its original position. But if they cannot, you can avoid having to change the workpiece coordinate system offset values in all programs by storing the variations in the external work offset.

Differences in Spindle Gap from One Machine to Another

If you are measuring tool lengths off-line for several machining centers, you may notice a small difference in the way the spindle taper is bored from one machine to the next. This, of course, will cause problems with your tool length compensation values. Instead of having to make a special measuring gauge for each machine, the deviation from machine to machine can be stored in the Z-axis value of the external work offset. From this point, the same tool length compensation value will work in every machine.

To Enhance Safety During Dry-Runs

As you begin verifying your CNC programs (methods will be shown in Key Concept 10), it can sometimes be helpful to run the program above the setup. If you increase the Z-value of the common workpiece coordinate system setting offset by say, 5-inches, all movements the machine makes will be 5-inches above their normal movements.

Key Points for Lesson 4.4:

- ✓ Workpiece coordinate system setting offsets allow you to ignore the positions of your work holding tools on the machine table as you write programs.
- ✓ You can assign up to six workpiece coordinate system zero points to be used by a program (up to forty-eight if you have purchased the additional workpiece coordinate system setting offset option).
- ✓ One way to assign multiple workpiece coordinate system setting offsets is to determine (measure) the distances from the zero return position to each individual workpiece coordinate system zero point in each axis and enter these workpiece coordinate system zero assignment values into the related workpiece coordinate system setting offsets.
- ✓ You can shift the point of reference for workpiece coordinate system setting offset entries from the zero return position to a more logical position. The common workpiece coordinate system setting offset is used for this purpose. This method works well when your work holding setups are very predictable.
- ✓ The common offset can be used to deal with variations among machines, to help re-align after a crash, and to help with program verification.

Talk to experienced people in your company…

… to learn more about work offsets.

1) Ask if any of the setups made on your machining centers use multiple workpiece coordinate system zeros.

2) Ask if repeated setups are qualified, and if they are, whether G10 is being used to program workpiece coordinate system setting offset settings.

3) Ask whether the point of reference for workpiece coordinate system setting offset entries has been shifted to a more local position. Does your company use sub-plates?

Key Concept 5: You Must Provide Structure to your CNC Programs

While there are many ways to write programs, you must ensure that your programs are safe and easy to use, yet as efficient as possible. This can be a real challenge since safety and efficiency usually conflict with one another.

Key Concept 5 contains two lessons:

    5.1: Introduction to program structure

    5.2: Structured program format

Key Concept 5 is you must structure your CNC programs using a strict format–while incorporating a design that accomplishes the objectives you intend. Though it is important to create efficient programs, safety and ease-of-use must take priority–at least until you gain proficiency. In Key Concept 5, we'll be showing techniques that stress safety as the top priority.

To this point in the course, we have been presenting the building blocks of CNC programming–providing the needed individual tools. Machine components, axes of motion, workpiece coordinate system zero, and programmable functions are presented in Key Concept 1. Preparation steps in Key Concept 2. Motion types in Key Concept 3. Compensation types in Key Concept 4. In Key Concept 5, we're going to draw all of these topics together, showing you what it takes to write CNC machining center programs completely on your own–making you a self-sufficient programmer.

Lesson 5.1 will introduce you to a CNC program's structure, showing you the reasons why programs must be strictly formatted. We'll also review some the program-structure-related points we've made to this point–and present a few new ones. And we'll address some variations related to how certain machine functions are handled.

In lesson 5.2, we'll show the four types of program structure as they are applied to both vertical and horizontal machining centers. These formats can be used as a crutch to help you write your first few programs.

You may be surprised at how much you already know about a program's structure, especially if you have been doing the exercises included in this text and/or those in the related exercises. We have introduced many of the CNC words used in programming, and we have been following the structure-related suggestions that we will be recommending here in Key Concept 5.

For example, do you recognize any of these words and commands? If so, write their meanings in the space provided.

    O0001: _____

    T01 M06: _____

    S1000 M03 T02: _____

    G91 G28 Z0 M19: _____

    M01: _____

Key Concept 5: You Must Provide Structure To Your CNC Programs

    G43 H01 Z0.1 M08: _____

    M30: _____

If you can't remember one or more of these CNC words or commands, don't worry. We'll be explaining them in detail during Key Concept 5. But if you do recognize most of them, you're already well on your way to understanding program structure.

Lesson 5.1–Introduction to Program Structure

Structuring a CNC program is the act of writing a program in a way that the CNC machine can recognize and execute safely, efficiently, and with a high-degree of operator-friendliness.

Objectives

After completing this lesson, students should be able to:

- ✓ Describe how to prioritize safety, ease-of-use, and efficiency in program structure
- ✓ Describe the reasons for structuring programs in a consistent manner
- ✓ Describe the machine variations that may affect program structure
- ✓ Describe the initialized G-codes and how to allow for them in your programs
- ✓ Describe the four basic program elements and how to structure them in your programs

Introduction

You know that CNC programs are made up of blocks, that each block is made up of words, and that each word is made up of a letter address and a numerical value.

You also know that programs are executed sequentially–command by command. The machine will read, interpret, and execute the first command in the program. Then it will move on to the next command. Read–interpret–execute. It will continue doing so until the entire program has been executed.

And you have seen several complete programs–you have even worked on a few if you have done the exercises in this text. You have probably noticed that there is quite a bit of consistency and structure in CNC programs.

Our focus in this lesson will be to help you understand more about the structure that is used in CNC programming.

Objectives of Your Chosen Program Structure

CNC machines have come a long way. In the early days of NC (before computers), a program had to be written just so. If anything was out-of-place, the machine would generate an alarm–failing to execute the program. While today's CNC machines are much more forgiving, you must still write CNC programs in a rather strict manner.

There are many ways to write a workable program–and the methods you use in structuring your programs will have an impact on three important objectives:

- Safety
- Efficiency
- Ease-of-use (operator friendliness)

It may be impossible to come up with a perfect balance among these objectives. Generally speaking, what you do to improve one objective will negatively affect the other two. When faced with a choice, a beginning programmer's priorities should always lean toward safety and ease-of-use. Our recommended programming structure stresses these two objectives. We will, however, show some of the efficiency-related short-comings of our recommended methods–so you can improve efficiency as you gain proficiency.

We're going to be assuming that you have control of the structure you use to write programs. A company may, however, already have a programmer that is writing programs with a different structure. As long as these programs are working–and satisfying the company's objectives–you're going to have to adapt to the established structure. If you understand the reasons for formatting, and if you understand one successful method for structuring programs, it shouldn't be too difficult to adapt.

Key Concept 5: You Must Provide Structure To Your CNC Programs

Reasons for Structuring Programs with Consistent Format

Let's begin by discussing the reasons why you must write your programs using a strict structure.

Familiarization

You must have some way to get familiar with CNC programming. You'll need some help writing your first few programs. The formats we show in the next lesson will provide you with this help. You'll be able to use our given formats as a crutch until you (eventually) have them memorized.

Actually, the formats we show in the next lesson will keep you from having to memorize anything. Instead, you will look at a word or command in the format and then you must remember its use. Compare this to recognizing the road signs you see as you drive an automobile. It is unlikely that you could recite every road sign from memory–but when you see one–you immediately know its meaning. Think of our given formats as like a set of road signs designed to help you write CNC programs.

As you'll see, you'll have road signs to help you when you begin writing a program (program start-up format), when you're finished with a tool (tool ending format), when you begin a new tool (tool start-up format), and when you end a program (program ending format).

Consistency

If you have been doing the exercises in this text, you've already worked on a few actual programs, filling in the blanks with needed CNC words. You have also seen several complete example programs in this course. You probably noticed that these programs are written in a very consistent manner. And the commands within each tool of each program are consistent with the other tools in the program.

Consistency within programs is important for three reasons. First, consistency helps you to become familiar with programming. Repeated commands soon become memorized.

Second, and more importantly, consistency among programs will help everyone that must work with your programs, including other programmers, setup people, and operators. Your programs will be easier to work with if you always structure them in the same manner.

Third, it is important to be able to repeat your past successes. If a program is running properly–achieving all of the objectives you intend–using its structure in your next program will ensure continued success.

Re-running Tools in the Program

This is the most important reason for structuring your programs using a strict format. As you know, trial machining involves five steps:

1. Recognizing a close tolerance
2. Adjusting an offset to leave additional stock
3. Machining under the influence of the trial machining offset
4. Measuring the machined dimension and adjusting the offset accordingly
5. Re-running the tool.

Trial machining is but one time when tools must be re-run. Say you're verifying a long program. You are fifteen tools into a twenty-tool program when you find a mistake. You must stop the cycle to correct the mistake. With the mistake corrected, you'll want to pick up where you left off–at the beginning of the fifteenth tool. You wouldn't want to re-run the entire program (from the beginning) just to get to tool number fifteen (doing so would be a waste of time).

Your ability to re-run tools is directly related to the structure you use to program. If you don't structure your programs properly, it will not be possible to re-run tools.

Here is a specific example that stresses why program formatting is so very important for re-running tools. Say you are programming two tools that run in sequence (say, tools one and two). Both tools happen to run at 400 rpm. Close to the beginning of the first tool, you have an S400 word in the command that starts the spindle. As you continue writing the commands for the second tool, you decide to leave out the S400 word, since the spindle speed is modal.

Everything will work just fine as long as the program runs in sequence–from beginning to end (the second tool immediately follows the first tool). But say the operator runs the entire program before they discover that the second tool has done something wrong (maybe it's a drill that has not gone deep enough). After correcting the problems, if the operator attempts to re-run the second tool by itself (as they should), it will run at the same speed as the last tool in the program–probably not 400 rpm! This is but one of many times when you must include redundant words (words that are already instated) in each tool—just to gain the ability to re-run tools.

In essence, you must make each tool in the program independent from the rest of the program. Think of the programming commands for each tool in a program as making a mini-program, self-sufficient and capable of activating all necessary machine functions.

You will often be tempted to leave out (necessary) redundant words because they are modal and remain in effect from tool to tool. To fight this temptation, think of each tool in the program as if it is the first tool. Everything you include in the program for the first tool (except the program number) will be needed in the program for each successive tool. This includes the tool change command (like T01 M06), the workpiece coordinate system setting offset designator (like G54), the absolute mode selection (G90), the spindle starting words (like S1000 M03), the next tool (like T02), movement to the first XY position in both axes (like G00 X2.0 Y2.0), the tool length compensation command (like G43 H01 Z0.1), the coolant activation (M08), and the feedrate word in the tool's first cutting command (like F5.0).

In the next lesson, we will be providing two complete sets of program format—one for vertical machining centers and the other for horizontal machining centers. These formats include all of the words needed in each tool—allowing any tool in the program to be re-run.

Efficiency Limitations

Again, the highest priority for our given formats is safety. Even so, beginners must exercise extreme caution when verifying their first few programs. Next we emphasize ease-of-use. Programs written with our recommended formats will be relatively easy to work with.

Our given formats do not place an emphasis on efficiency. While programs written with these formats will not be wasteful, they will not be as efficient as they could be. But remember, when you start emphasizing efficiency, safety and ease-of use will probably suffer. Here are some of the efficiency-related limitations of our given formats:

> **One activity per command**—For the most part, our given formats will cause the machine to do one thing at a time. In reality, the machine can sometimes be doing more. For example, our recommended format for tool start-up will have the spindle to start in one command and the movement to the first XY position in the next. While these two functions can be done together, it is helpful for beginners to concentrate on but one thing at a time.
>
> **Approach motions**—During each approach to the workpiece, we first move the cutting tool to its first XY position. Then we move it to its first Z position. While these commands can be combined to minimize program execution time, a three-axis approach movement tends to be scary for entry-level setup people and operators.
>
> **Rapid approach distance**—In our given formats, we recommend using an approach distance of 0.100 inch (about 2.5 mm) when approaching qualified (known, consistent) surfaces. While it is more efficient to reduce rapid approach distance, it is not as safe. As you gain confidence and proficiency, you should consider reducing rapid approach distance to improve efficiency.

Machine Variations That Affect Program Structure

We're providing two sets of format—one for vertical machining centers and another for horizontal machining centers. But even within a given machine type, there are lots variations among machine tool builders.

Our given formats are aimed at basic machining centers—those without a lot of bells and whistles. About the only programmable functions we address are automatic tool changer, spindle, coolant, and feedrate. If your machine has other accessories, you must consult your machine tool builders programming manual to learn how they are programmed. As is discussed in lesson 1.1, most accessories are handled with M-codes—so look for the machine's list of M-codes.

Even within our limited coverage of accessories, there are variations. Here we list them.

M-Code Differences

M-code number selection is left completely to the discretion of your machine tool builders, and no two builders seem to be able to agree on how all M-codes should be numbered. For very common machine functions like spindle on/off (M03, M04, M05) and flood coolant on/off (M08, M09), machine tool builders have standardized. But for less common functions like indexers, pallet changers, automatic clamping, chip

Key Concept 5: You Must Provide Structure To Your CNC Programs

conveyers, high pressure coolant systems, and through-the-tool coolant systems, you must find the list of M-codes in your machine tool builder's programming manual.

Here is a list of very common M-codes you will need for our program formats. Fortunately, most machine tool builders utilize these M-code numbers just as we show below.

- M00 - Program stop (halts the program's execution until the operator reactivates the cycle)
- M01 - Optional stop (used to cause the machine to stop between tools during program verification)
- M03 - Spindle on CW (for right hand tools)
- M04 - Spindle on CCW (for left hand tools)
- M05 - Spindle off (not normally needed with our formats since M19, M06, M30 will stop the spindle)
- M06 - Tool change (places the tool in the ready station into the spindle)
- M08 - Flood coolant on (cools and lubricates the machining operation)
- M09 - Coolant off (used to turn off coolant before each tool change)
- M19 - Spindle orientation (used to save a little time at each tool change–spindle begins orienting for a tool change on the way to the tool change position)
- M30 - End of program (on many machines, M02 can also be used)

Automatic Tool Changer Variations

All true CNC machining centers have automatic tool changers to automatically load tools into the machine's spindle. Automatic tool changers incorporate some kind of tool storage magazine to hold tools that are not in use. They also incorporate some kind of mechanism that exchanges them between the magazine and the spindle. Machining centers vary with regard to how many tools they can hold.

Even with the wide variety of automatic tool changing systems in use today, programming remains remarkably similar. There are but two basic programming styles.

T-Word Brings a Tool To the Ready Station, M06 Commands the Tool Change

The vast majority of automatic tool changers on quality production machines are programmed in this manner. We have been showing this method in all of the example programs in this text. With this method, a T-word is used to specify the tool to be brought to the ready or waiting station. The T-word by itself does not make the tool change. It simply rotates the magazine, bringing the specified tool to the ready position (the position that allows it to be placed into the spindle). An M06 word commands the tool change, placing the tool that is currently in the ready station into the spindle. M06 also places the tool that is currently in the spindle back into the magazine. The command

```
T07 M06
```

will first rotate the magazine, bringing tool number seven to the ready station. Then it will place tool number seven in the spindle. The tool that is currently in the spindle when this command is executed will be placed back into the magazine, probably in its original tool stations (though tool changer designs vary in this regard).

Almost all current model machining centers have random access tool changers. This means you can use tools in any order, regardless of tool station number. Though our example programs have been using cutting tools in sequential order (tool one, then tool two, then tool three, and so on), you can use tools in any order in your programs.

Do You Have a Double-Arm Tool Changer?
If the automatic tool changer has a double-arm changing arm (as many do), you'll have an added benefit. As the tool in the spindle is machining a workpiece, you can command that the next tool be placed in the ready station–so one tool can be machining while the next is getting ready. This will save tool changing time. We've been assuming this style of automatic tool changer is being used in all example programs shown in this course.

Example programs using T-word and M06
Metric-mode example

```
%
O0022 (Program number)
```

```
N010 G21
(6.0-mm drill)
N020 T01 M06 (Load tool number one in spindle)
N030 G90 G94 G17 G54 (Start-up G-codes)
N040 S1200 M03 T02 (Select workpiece coordinate system setting offset #1, absolute mode, start spindle fwd at 1200 rpm, get tool number two ready)
N050 G00 X25.0 Y25.0 (Rapid to hole location in X and Y)
N060 G43 H01 Z2.0 (Instate tool length compensation for tool one, approach in Z to just above work surface)
N070 M08 (Turn on the coolant)
N080 G01 Z-16.0 F100.0 (Drill hole)
N090 G00 Z2.0 M09 (Rapid out of hole, turn off coolant)
N100 G91 G28 Z0 M19 (Rapid to the tool change position, orient spindle)
N110 M01 (Optional stop)
N120 T02 M06 (Load tool number two in spindle)
(3/8 drill)
N130 G90 G94 G17 G54 (Restate start-up G-codes)
N140 S1000 M03 T03 (Start spindle fwd at 1000 rpm, get tool number three ready)
N150 G00 X50.0 Y25.0 (Rapid to hole position in X and Y).
```

Inch-mode example

```
%
O0022 (Program number)
N010 G20
(0.25-in drill)
N020 T01 M06 (Load tool number one in spindle)
N030 G90 G94 G17 G54 (Start-up G-codes)
N040 S1200 M03 T02 (Select workpiece coordinate system setting offset #1, absolute mode, start spindle fwd at 1200 rpm, get tool number two ready)
N050 G00 X1.0 Y1.0 (Rapid to hole location in X and Y)
N060 G43 H01 Z0.1 (Instate tool length compensation for tool one, approach in Z to just above work surface)
N070 M08 (Turn on the coolant)
N080 G01 Z-0.65 F4.0 (Drill hole)
N090 G00 Z0.1 M09 (Rapid out of hole, turn off coolant)
N100 G91 G28 Z0 M19 (Rapid to the tool change position, orient spindle)
N110 M01 (Optional stop)
N120 T02 M06 (Load tool number two in spindle)
(3/8 drill)
```

Key Concept 5: You Must Provide Structure To Your CNC Programs

```
N130 G90 G94 G17 G54 (Restate start-up G-codes)
N140 S1000 M03 T03 (Start spindle fwd at 1000 rpm, get tool
number three ready)
N150 G00 X2.0 Y1.0 (Rapid to hole position in X and Y).
```

In block N020, tool 1 is being placed in the spindle. While tool 1 begins machining, tool 2 is rotated to the ready station (in block N040). Later, in block N120, when the tool change is commanded, the magazine will probably have already completed its rotation to station 2, so the tool change can take place immediately. Tool 2 will be placed in the spindle (and tool 1 will be placed back in the magazine). In block N140, the magazine will begin rotating to tool 3 as tool 2 begins machining. This process is repeated for all tools used in the program.

Where Is the Tool Change Position?
With the vast majority of machining centers, the tool magazine remains stationary. That is, the spindle must move to a special position before a tool change can occur. This special position is called the tool change position. For older machines, if the axes are not at the tool change position when an M06 is commanded, an alarm will be generated. Programming an M06 on more contemporary machines will actually run a special macro program that cause a rapid motion to the tool change position and then make the tool change.

For many vertical machining centers, the tool change position is the Z-axis zero return position (G28). For most horizontal machining centers, the tool change position is the Y and Z-axis zero return positions. In the example above (for a vertical machining center), notice that block N100 sends the Z-axis to its zero return position and orients the spindle prior to the tool change command in block N120.

What Is the M19 Doing In Block N100?
M19 activates a machine function called a spindle orient. Before a tool change can occur, the spindle must rotate to align the keyway in the cutting tool holder with the key in the tool changer-arm. The spindle orient function performs this spindle rotation. While M06 will automatically perform a spindle orient, including M19 during the movement to the tool change position will save some time (about one-half to three seconds per tool change, depending upon the machine). The spindle will be oriented properly by the time the machine reaches the tool change position.

Why Is the T-Word Repeated In Each M06 Command?
You may be wondering why we repeat the tool station number in the M06 command. In block N120, for example, there is a T02 word, even though the T02 word has already been specified in block N040. This is another example of a seemingly redundant word that must be repeated for the purpose of being able to re-run tools. If we must re-run tool 2 and if tool 2 is not currently in the spindle, the restart block will be the M06 command (block N120 in the example above). Without a T-word in this block, how will the machine know which tool to place in the spindle when you re-run tool 2?

Single-Arm or Umbrella Tool Changers

A single-arm or umbrella tool changer is less efficient. It must replace the tool from the spindle back to the magazine before the next tool can be loaded into the spindle. For most machines with this kind of tool changer, programming is quite similar. A T-word still rotates the magazine to the ready station and an M06 still makes the tool change.

But with a single-arm or umbrella tool changer, you are not allowed to get the next tool ready while the tool in the spindle is machining the workpiece. Here is the same program shown above, modified for this kind of tool changer. Notice that the only difference is the T word has been removed from the command after the tool change (blocks N040 and N140).

This example is shown using inch mode.

```
%
O0023 (Program number)
N010 G20
(1/4 drill)
N020 T01 M06 (Load tool number one in spindle)
N030 G90 G94 G17 G54 (Start-up G-codes)
```

```
N040 S1200 M03(Select workpiece coordinate system setting
offset #1, absolute mode, start spindle fwd at 1200 rpm)
N050 G00 X1.0 Y1.0 (Rapid to hole location in X and Y)
N060 G43 H01 Z0.1 (Instate tool length compensation for tool
one, approach in Z to just above work surface)
N070 M08 (Turn on the coolant)
N080 G01 Z-0.65 F4.0 (Drill hole)
N090 G00 Z0.1 M09 (Rapid out of hole, turn off coolant)
N100 G91 G28 Z0 M19 (Rapid to the tool change position, orient
spindle)
N110 M01 (Optional stop)
N120 T02 M06 (Load tool number two in spindle)
(3/8 drill)
N130 G90 G94 G17 G54 (Restate start-up G-codes)
N140 S1000 M03 (Start spindle fwd at 1000 rpm)
N150 G00 X2.0 Y1.0 (Rapid to hole position in X and Y).
```

Certification Cart Tool Changer

The FANUC Certification Cart (Mill) has a unique style of automatic tool changer. See the reading materials for Lesson 1.1 for a description. It is programmed with an M06 and a T word, but does not have the ability to get the next tool ready while the spindle tool is being used. During a tool change, the tool in the spindle will first be placed back in its storage pot, then the spindle will pick up the next (commanded) tool. Additionally, and very importantly, word order in the tool change command is critical. The M06 must precede the T-word in a tool change command. Here is a command that will place tool number 7 in the spindle:

```
M06 T07
```

If the T-word precedes the M06, the Certification Cart will generate an alarm.

T-Word Does Everything

Programming for some single-arm and umbrella tool changers is easier. There is no M06. The T word does everything, including rotating the magazine to the ready station, and making the tool change. With this kind of machine, it is not possible to get the next tool station ready while the tool in the spindle machines the workpiece. Here is the program again, modified for this kind of tool changer.

This example is shown using inch mode.

```
%
O0024 (Program number)
N010 G20
(1/4 drill)
N020 T01 (Load tool number one in spindle)
N030 G90 G94 G17 G54 (Start-up G-codes)
N040 S1200 M03(Select workpiece coordinate system setting
offset #1, absolute mode, start spindle fwd at 1200 rpm)
N050 G00 X1.0 Y1.0 (Rapid to hole location in X and Y)
N060 G43 H01 Z0.1 (Instate tool length compensation for tool
one, approach in Z to just above work surface)
N070 M08 (Turn on the coolant)
N080 G01 Z-0.65 F4.0 (Drill hole)
N090 G00 Z0.1 M09 (Rapid out of hole, turn off coolant)
```

Key Concept 5: You Must Provide Structure To Your CNC Programs

```
N100 G91 G28 Z0 M19 (Rapid to the tool change position, orient spindle)
N110 M01 (Optional stop)
N120 T02 (Load tool number two in spindle)
(3/8 drill)
N130 G90 G94 G17 G54 (Restate start-up G-codes)
N140 S1000 M03 (Start spindle fwd at 1000 rpm)
N150 G00 X2.0 Y1.0 (Rapid to hole position in X and Y).
```

Tool Change at Beginning Or End?

Some debate exists regarding whether it is best to make the first tool change at the very beginning of the program–or to assume the first tool is in the spindle at the beginning of the program, making a tool change at the end of the program that places the first tool in the spindle.

We strongly recommend making the first tool change at the beginning of the program–as our example programs–and our given formats–show. With this method, you can rest assured that the correct tool will be in the spindle when machining begins.

If you assume that the first tool is in the spindle when the program begins, the results could be disastrous. Any number of things could cause the wrong tool to be in the spindle when the setup person or operator activates the program.

Does the Machine Even Have an Automatic Tool Changer?

Our given formats assume that your machine does have an automatic tool changer. If it does not (as is the case with some CNC milling machines) the operator must manually load each tool during the cycle. In this case, an M00 (program stop) is used to stop the cycle for the manual tool exchange. The programmer rapids the machine to a convenient tool-loading position and commands an M00. The machine will stop. A message in the program close to the M00 will specify which tool is to be loaded. Once the operator has loaded the tool, the cycle can be reactivated. Here is the program again.

This example is shown using <u>inch mode</u>.

```
%
O0025 (Program number)
N010 G20
(center drill)
N020 M00 (LOAD THE CENTERDRILL)
N030 G90 G94 G17 G54 (Start-up G-codes)
N040 S1200 M03 (Select workpiece coordinate system setting offset #1, absolute mode, start spindle fwd at 1200 rpm)
N050 G00 X1.0 Y1.0 (Rapid to hole location in X and Y)
N060 G43 H01 Z0.1 (Instate tool length compensation for tool one, approach in Z to just above work surface)
N070 M08 (Turn on the coolant)
N080 G01 Z-0.65 F4.0 (Drill hole)
N090 G00 Z0.1 M09 (Rapid out of hole, turn off coolant)
N100 G91 G28 Y0 Z0 M19 (Rapid to the tool change position, orient spindle)
N110 M01 (Optional stop)
N120 M00 (LOAD THE 3/8-INCH DRILL)
(3/8 drill)
N130 G90 G94 G17 G54 (Restate start-up G-codes)
```

```
N140 S1000 M03 (Start spindle fwd at 1000 rpm)
N150 G00 X2.0 Y1.0 (Rapid to hole position in X and Y).
```

In block N020, the machine stops and the operator is told to load the first tool. In block N100, the machine is sent to a convenient tool loading position (the Y and Z-axis zero return position). In block N120, the machine stops again and the operator is told to load the next tool.

Understanding the G28 Command

Again, the tool change position on most machining centers is the zero return position in at least one axis. Before a tool change can occur, the machine must be sent to this position. The most universal way to command the machine to move to the zero return position is to use G28. This method will work on all FANUC and FANUC-compatible machining centers.

G28 is a two-step command. Two things will happen whenever a G28 is commanded. First the machine will move (at rapid) the axes included in the G28 command to an intermediate position. Then the machine will rapid the axes to the zero return position. When an axis reaches its zero return position, a corresponding axis-origin-light may come on–indicating that the axis is at its zero return position.

The intermediate position confuses most people. In absolute mode, which most programmers prefer for general purpose programming, the intermediate position is specified relative to the workpiece coordinate system zero. In incremental mode, it is specified relative to the tool's current position. The best way to gain an understanding is to give a few examples, explaining after each.

Consider this command, which we use in all example programs in this text.

```
G91 G28 Z0
```

In step one of this G28 command, the tool will move to an intermediate position that is incrementally nothing (zero) from its current position in the Z-axis. In step two, it will go to the zero return position (in Z only). The X and Y axes will not move. For all intents and purposes, we're telling the machine to move straight to its zero return position in the Z-axis.

```
G91 G28 X0 Y0
```

In step one of this G28 command, the tool will move to an intermediate position that is incrementally nothing (zero) from its current position in X and Y. In step two, it will go to the zero return position in X and Y (simultaneously). We're telling the machine to move straight to its zero return position in X, and Y.

```
G91 G28 X0 Y0 Z3.0
```

In step one of this G28 command, the tool will move to an intermediate position that is incrementally nothing (zero) from its current position in X, Y. But in Z, it will move three inches in the positive direction. Maybe the tool is in a pocket you need to clear before moving in X and Y. In step two, it will go to the zero return position in X, Y, and Z (simultaneously).

Watch out! Here's what can happen in *absolute* mode. Consider this command.

```
G28 X0 Y0 Z0
```

Assuming the machine is currently in absolute mode (G90) when this G28 command is executed, step one tells the machine to move to the workpiece coordinate system zero point (probably a crash) in X, Y, and Z. Then, in step two, the machine will move to the zero return position in X, Y, and Z (if it still can).

This is the reason why we recommend temporarily switching to the incremental mode with G28. We're not really programming any coordinates incrementally, we're simply trying to send the machine straight to the zero return position.

What About G53?

Again, we use G28 in our given formats because it is universal. It will work on all FANUC and FANUC-compatible machining centers. But if you have a newer machine that is equipped with G53, there is another method that is much easier to use and understand. And you don't have to temporarily switch to the incremental mode.

Key Concept 5: You Must Provide Structure To Your CNC Programs

G53 is called temporary positioning relative to the machine coordinate system. The origin for the machine coordinate system is the zero return position. Like G28, G53 will automatically invoke the rapid mode. Also like G28, G53 is non-modal (it will only have effect in the command in which it is included). The command

    G53 X0 Y0 Z0

will send the machine (at rapid) straight to the zero return position in X, Y, and Z. And you don't have to be concerned with an intermediate position.

If all of your machining centers have the G53 command you may want to use it instead of G28. For all vertical machining center programs, simply replace the

    G91 G28 Z0 M19

commands with

    G53 Z0 M19

But if any of your machining centers do not have G53, we recommend that you use G28 for all programs to maintain compatibility from one machine to another (you never know when a program originally written for one machine must be run on another).

A Possible Problem with Initialized Modes

Many modal CNC words are initialized (automatically instated at power-up). If you work in metric mode, for example, your machine should be set to power-up in metric mode, meaning G21 is *initialized*. And if you work exclusively in the metric mode, you shouldn't need to include a G21 word in any of your programs. You may have noticed the example programs shown in this course include measurement system invoking G-codes (G21 for metric mode or G20 for inch mode).

> **What will happen if the machine is in the wrong measurement system?**
>
> The machine will incorrectly interpret all measurement-system-related words (coordinates, feedrates, work offset values, and tool length values). An X position of X25.0, for example, which is supposed to be 25.0 millimeters, will be interpreted as 25.0 inches (or vise-versa).

You should not assume that the machine is in your measurement system of choice.

Most programmers don't like to assume that the machine is still in all of its initialized modes when their programs are run. They include a series of commands in the program (commonly called safety commands) to re-instate all of the initialized modes—even though the related words may be redundant. Consider these commands:

    N010 G21 G17 G40 (Select metric mode, select XY plane, cancel
    cutter radius compensation)
    N020 G64 G80 (Select normal cutting mode, cancel canned cycle
    mode)

Remember that some CNCs allow only three compatible G-codes per command—so you may need multiple commands to specify all of the related modes. While we have not discussed some of these G-codes, rest assured that they involve modes that must be instated when your programs run.

Key Points for Lesson 5.1:

- ✓ Safety and ease-of-use should take priority over efficiency for entry-level programmers.
- ✓ You must use a strict structure in your programs to help you become familiar with programming, to ensure consistency among programs, and to gain the ability to re-run tools.
- ✓ When you gain proficiency, you can modify the given structured program formats to achieve more efficiency.
- ✓ Machining centers vary, meaning you may have to modify the given formats for your particular machine tools—especially related to the way your automatic tool changer is programmed.
- ✓ You must understand the way G28 works.
- ✓ Safety commands in your program can be used to ensure that the machine is still in all of its initialized modes.

Lesson 5.2–Structured Program Format

The program formats described in this lesson will keep you from having to memorize most of the words and commands needed in CNC programming. As you'll see, a large percentage of most programs is related to a common structure.

Objectives

After completing this lesson, students should be able to:

- ✓ Describe the four structures of the structured program format
- ✓ Describe the application of the structured programming format for a vertical machining center
- ✓ Describe the application of the structured programming format for a horizontal machining center
- ✓ Describe use of M01 optional stop command

Introduction

You know the reasons why programs must be formatted using a strict structure. In lesson 5.2, we're going to show a structured program format. We will show one program format for vertical machining centers and another for horizontal machining centers. We will also explain every word in each program format in detail. We will also show an example program that stresses the use of the structured program format.

There are four structure elements used in the program format:

1. Program-startup structure
2. Tool-startup structure
3. Tool-end structure
4. Program-end structure

Any time you begin writing a new program, start with the program-startup structure. You can copy this structure to begin your program. The actual values of some words will change based on what you wish to do in your own program, but the structure will remain the same every time you begin writing a new program.

After writing the program-startup structure comes the first tool-startup structure. This must be customized for the specific tool details. The tool-startup structure is followed by the motions for the first machining operation. Obviously, the motions to complete a particular machining operation will be somewhat unique, and you are on your own to write this section. When finished with the motions for a machining operation with a tool, you complete that section with the tool-end structure.

The pattern of tool-startup structure → machining operation motions → tool-end structure is repeated once for each tool operation required to machine the workpiece. Again, a tool-startup structure, followed by the motions for the machining operation and terminated by the tool-end structure.

When all the machining operations are programmed, you finish the program with the program-end structure.

One of the most important benefits of using the structured program format is that you do not have to memorize anything. You simply copy the structures and edit the appropriate words.

Key Concept 5: You Must Provide Structure To Your CNC Programs

The structured program formats assume that you are using workpiece coordinate system offsets to assign workpiece coordinate system zero.

Structured Program Format for Vertical Machining Centers

This structured program format is used for vertical machining centers that have workpiece coordinate system offsets. This particular format assumes that the machine has a double-arm automatic tool changer, that the tool change position is the Z-axis zero return position, and that the machine is resting at the tool change position in Z when the program begins. When this program ends, the machine is left at the X, Y and Z-axis zero return position. This makes a convenient workpiece loading position (with the table out toward the operator in the Y-axis).

For now you are to use the strict structure of these given formats. But the values that are shown in **bold** will change from program to program and from tool to tool.

Program Start-Up Structure:

```
%
O0001 (PROGRAM DESCIPTION HERE)
N010 G17 G21 G94 (SELECT XY-PLANE, METRIC MODE, PER-MINUTE FEEDRATE MODE)
N020 G40 G64 (CANCEL CUTTER COMP. AND SELECT NORMAL CUT. MODE)
N030 G69 G80 (CANCEL ROTATION AND CANNED CYCLES)
```

Tool Start-Up Structure:

```
(DESCRIPTION/NAME FOR THIS TOOL)
N100 T01 M06 (LOAD TOOL IN SPINDLE)
N110 G54 G90 S300 M03 T02 (SELECT WORK OFFSET/ABSOLUTE MODE, START SPINDLE, READY NEXT TOOL)
N120 G00 X125.0 Y125.0 (MOVE TO XY APPROACH POSITION)
N130 G43 H01 Z2.0 (INSTALL TOOL LEN. COMP., MOVE TO Z APPROACH PSN)
N140 M08 (COOLANT ON)
N150 G01 . . . .  F75.0 (CUTTING MOVES WITH FEEDRATE)
```

Tool Ending Structure:

```
N200 M09 (COOLANT OFF)
N210 G91 G28 Z0 M19 (TOOL CHG. PSN., PRE-ORIENT SPINDLE)
N220 M01 (OPTIONAL STOP)
```

Program Ending Structure:

```
N310 G28 X0 Y0 (RETURN TO XY ZERO RETURN - LOAD/UNLOAD POSITION)
N320 M30 (END OF PROGRAM)
%
```

Here are some notes to help you understand the format.

The Percent Signs That Begin and End the Program (%)

These are called file delimiters. In order to load programs, FANUC requires that the first and last character of the file to be loaded be a percent sign (%).

O-Word

FANUC programs can be named with a four-digit number. The O-word specifies the program number. Since several programs can reside in the CNC memory, this is what isolates one program in CNC memory from another.

N010
Blocks N010 through N030 are safety blocks to confirm that the machine is in a known initialized state. These commands will vary based upon the particular modal-optional features equipped on your machining center (like scaling, coordinate rotation, Custom Macro, and others). We recommend that you specify the canceling G-code (the initialized state) for all modal functions. Our example structure includes but a few of them.

G17 selects the XY-plane. If another plane (XZ or YZ) has been incorrectly instated with G18 or G19, circular motions, cutter radius compensation, and several other CNC features will not behave as expected.

The format shown above assumes you work in metric mode, and G21 selects metric mode. If you work in inch mode, of course, you must specify a G20 instead of G21.

G94 selects the feed-per-minute mode, so upcoming feedrates must be specified in either inches- or millimeters-per-minute, depending upon your measurement system choice. If you elect to work in feed-per-revolution mode, specify G95 instead of G94. If you do, feedrates must be specified in either inches- or millimeters-per-revolution.

N020
G40 ensures that cutter radius compensation is cancelled as your program begins.

G64 instates the normal cutting mode. It cancels something called exact stop check (which makes the machine come to a stop after every motion) and single direction positioning (which ensures that every motion approaches its end point from the same direction). These features are discussed in lesson 6.3.

N030
G69 cancels something called coordinate rotation, which can be used to rotate a series of programmed coordinates and is discussed in lesson 6.3.

G80 cancels any of the hole-machining canned cycles, which are discussed in lesson 6.1.

Important Note about Safety Commands
Our given safety commands assume that the machining center has certain features. While our three safety commands should work nicely for most machining centers, you must be prepared to modify them should your machine not have the related features.

Here are the safety commands that we recommend for the **Certification Cart**:

```
N010 G17 G20 G94
N020 G40 G64 G80
```

N100
Our tool startup structure shows tool station number one (T01) being placed in the spindle. If you wish to start with another tool, of course, you must specify the tool station number for your first tool.

> Remember, the Certification Cart (Mill) requires that the M06 precedes the T-word in tool changing command. Here is block N100 for the Certification Cart: **N100 M06 T01**

N110
Our tool startup structure also shows a specific workpiece coordinate system offset number, speed, and next tool station number. You will be specifying the workpiece coordinate system offset, speed, and next tool station number needed by your tool.

N120
This is the first rapid positioning movement in XY made by the tool. The actual position, of course, will be based upon the workpiece and first operation.

N130
This block includes the Z-axis motion to instate tool length compensation. The H-word value must match the tool station number at the beginning of the format. And the Z-position in this command must be appropriate to the work surface the tool is approaching. (If your machining center has a totally enclosed work area, you can include the M08 in this command to turn the coolant on as well.)

N140

Key Concept 5: You Must Provide Structure To Your CNC Programs

Assuming your workpiece and cutting tool require coolant, turn it on in this block. If your machining center has a totally enclosed work area, the M08 can be included in the previous command.

N150
Each cutting tool's feedrate is part of the tool startup format. Though the format shows a specific feedrate value, you will of course, specify the feedrate for your first tool in its first cutting command, and be sure it is specified using the appropriate unit (in millimeters/inches per minute/revolution). Also, your first cutting command may not be a G01 command.

N200
Though we're showing coolant being turned off in a block by itself, there are times when you can include the M09 in the last machining command.

N210
This block sends the machine to the Z-axis zero return position, which is the tool change position for most vertical machining centers. We provide a full description of the G28 command in lesson 5.1. This command also includes an M19 word to cause the spindle to begin rotating to its orient position on the way to the tool change position. While the up-coming tool change command (M06) will orient the spindle, placing the M19 in this command will save tool changing time, since the spindle rotation will be completed by the time the machine gets to its tool change position.

N220
An optional stop word (M01) will cause the control to look to the position of an on/off switch (commonly labeled optional stop) on the machine operator panel. If this switch is off, the control will continue with the program, ignoring the M01. If the switch is on, the control will stop the machine at this point. For most machines, the spindle, coolant, and axes are stopped until the operator reactivates the cycle. We recommend placing an M01 at the end of each tool to allow the setup person or operator to stop the machine after each tool. This is especially helpful during the program's verification, when it is necessary to check what one tool has done before moving on to the next tool.

N310 - N330
These commands should be quite familiar. The G28 command in block N310 sends the machine to a convenient workpiece loading position in the X and Y-axes. As you have seen in the programming activities, this may not be entirely necessary based upon the size of the machine.

Most machines use an M30 as the end-of-program word. This command will stop the machine. It will also turn off anything that is still running (like spindle and coolant). And M30 will rewind the program back to the beginning so that when the operator activates the next cycle, the machine will start from the beginning of the program.

A Note About Documentation Comments

As you know, you can include documenting comments in your program. They must be included in parentheses. Unlike we have been showing in the example programs shown in this course, most CNCs require that all characters in parentheses be upper case characters.

As we show in the tool startup and program startup format, you should include a comment naming each tool just above the tool change command. Also, it helps to skip a block just before the message to make it stand out (this helps to nicely separate the tools in your program).

You should also include documenting comments whenever you think it might be helpful to the setup person or operator. For a series of milling passes, for example, you could include the messages (PASS ONE), (PASS TWO), (PASS THREE), and so on to clarify where each pass begins.

If you include any program stop commands in your program (M00), you should include a message that tells the operator what it is they are supposed to do at the program stop–like

```
N070 M00 (CLEAR CHIPS FROM POCKET)
```

Though the topic is beyond the scope of this course, many FANUC CNCs also support a feature called Custom Macro. One of the capabilities of Custom Macro is to actually stop the cycle and display a message on the Operator Message screen.

Therefore programming

```
N070 #3006=1 (CLEAR CHIPS FROM POCKET)
```

acts similar to the M00, but the message is more proactively presented to the operator. The #3006=1 makes the text in the parenthesis appear on the operator message page.

While we do not show it in the example programs in this course, you should also include documenting comments at the very beginning of your program, to specify information like the part number, the last revision number, the process operation number, the date, the programmer's name, and any other general information that will help people understand what your program is doing.

Example Program for Vertical Machining Centers

Here is an example program that stresses the use of program structure. Although the program is quite simple (it is actually the same program shown earlier during our discussion of tool length compensation), it shows all of the principles of program formatting. Pay primary attention to the strict structure followed for each tool. The program can be broken into mini-programs, each making up one tool. Again, each tool is independent of the rest of the program.

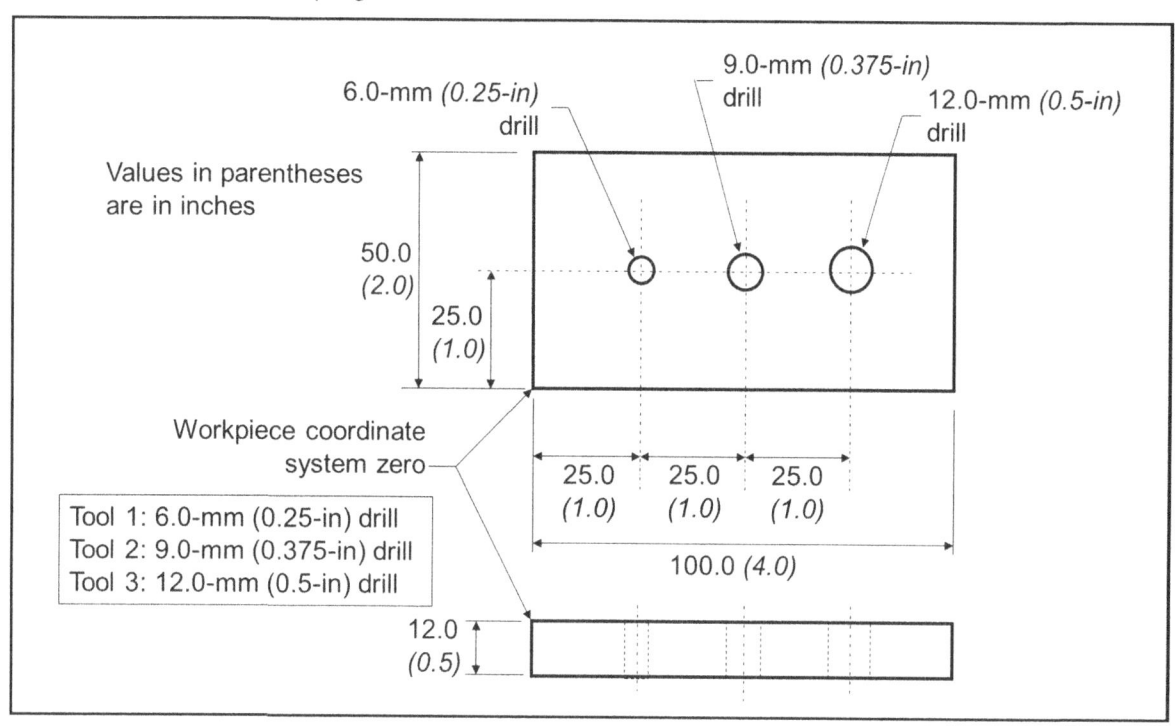

Figure 5.2: Example program to demonstrate the used of structured program format

Key Concept 5: You Must Provide Structure To Your CNC Programs

Metric-mode example

Program-startup structure:
```
%
O0026 (EXAMPLE PROGRAM FORMAT)
N010 G17 G21 G94 (SELECT XY-PLANE, METRIC MODE, SELECT FEED-PER-MINUTE MODE)
N020 G40 G64 (CANCEL CUTTER COMP. AND SELECT NORMAL CUT. MODE)
N030 G69 G80 (CANCEL ROTATION, AND CANNED CYCLES)
```

Tool-startup structure:
```
(6.0-MM DRILL)
N100 T01 M06 (LOAD TOOL IN SPINDLE)
N110 G54 G90 S1200 M03 T02 (START SPINDLE, READY NEXT TOOL)
N120 G00 X25.0 Y25.0 (MOVE TO XY APPROACH POSITION)
N130 G43 H01 Z2.0 (INSTALL TOOL LEN. COMP., MOVE TO Z APPROACH PSN)
N140 M08 (COOLANT ON)
N150 G01 Z-15.8 F100.0 (CUTTING MOVES WITH FEEDRATE)
N160 G00 Z2.0 (RETRACT FROM HOLE)
```

Tool-end structure:
```
N170 M09 (COOLANT OFF)
N180 G91 G28 Z0 M19 (TOOL CHG. PSN., PRE-ORIENT SPINDLE)
N190 M01 (OPTIONAL STOP)
```

Tool-startup structure:
```
(9.0-MM DRILL)
N200 T02 M06 (LOAD TOOL IN SPINDLE)
N210 G54 G90 S1000 M03 T03 (START SPINDLE, READY NEXT TOOL)
N220 G00 X50.0 Y25.0 (MOVE TO XY APPROACH POSITION)
N230 G43 H02 Z2.0 (INSTALL TOOL LEN. COMP., MOVE TO Z APPROACH PSN)
N240 M08 (COOLANT ON)
N250 G01 Z-16.7 F125.0 (CUTTING MOVES WITH FEEDRATE)
N260 G00 Z2.0 (RETRACT FROM HOLE)
```

Tool-end structure:
```
N270 M09 (COOLANT OFF)
N280 G91 G28 Z0 M19 (TOOL CHG. PSN., PRE-ORIENT SPINDLE)
N290 M01 (OPTIONAL STOP)
```

Tool-startup structure:
```
(12.0-MM DRILL)
N300 T03 M06 (LOAD TOOL IN SPINDLE)
N310 G54 G90 S800 M03 T01 (START SPINDLE, READY NEXT TOOL)
N320 G00 X75.0 Y25.0 (MOVE TO XY APPROACH POSITION)
N330 G43 H03 Z2.0 (INSTALL TOOL LEN. COMP., MOVE TO Z APPROACH PSN)
N340 M08 (COOLANT ON)
N350 G01 Z-17.6 F150.0 (CUTTING MOVES WITH FEEDRATE)
N360 G00 Z2.0 (RETRACT FROM HOLE)
```

Tool-end structure:
```
N370 M09 (COOLANT OFF)
N380 G91 G28 Z0 M19 (TOOL CHG. PSN., PRE-ORIENT SPINDLE)
N390 M01 (OPTIONAL STOP)
```

Program-end structure:
```
N400 G28 X0 Y0 (RETURN TO XY ZERO RETURN - LOAD/UNLOAD PSN)
N410 M30 (END OF PROGRAM)
%
```

Lesson 5.2: Structured Program Format

Inch-mode example

Program-startup structure:
```
%
O0026 (EXAMPLE PROGRAM FORMAT)
N010 G17 G20 G94 (SELECT XY-PLANE, INCH MODE, SELECT FEED-PER-MINUTE MODE)
N020 G40 G64 (CANCEL CUTTER COMP. AND SELECT NORMAL CUT. MODE)
N030 G69 G80 (CANCEL ROTATION AND CANNED CYCLES)
```

Tool-startup structure:
```
(0.25-IN DRILL)
N100 T01 M06 (LOAD TOOL IN SPINDLE)
N110 G54 G90 S1200 M03 T02 (START SPINDLE, READY NEXT TOOL)
N120 G00 X1.0 Y1.0 (MOVE TO XY APPROACH POSITION)
N130 G43 H01 Z0.1 (INSTALL TOOL LEN. COMP., MOVE TO Z APPROACH PSN)
N140 M08 (COOLANT ON)
N150 G01 Z-0.65 F4.0 (CUTTING MOVES WITH FEEDRATE)
N160 G00 Z0.1 (RETRACT FROM HOLE)
```

Tool-end structure:
```
N170 M09 (COOLANT OFF)
N180 G91 G28 Z0 M19 (TOOL CHG. PSN., PRE-ORIENT SPINDLE)
N190 M01 (OPTIONAL STOP)
```

Tool-startup structure:
```
(0.375-IN DRILL)
N200 T02 M06 (LOAD TOOL IN SPINDLE)
N210 G54 G90 S1000 M03 T03 (START SPINDLE, READY NEXT TOOL)
N220 G00 X2.0 Y1.0 (MOVE TO XY APPROACH POSITION)
N230 G43 H02 Z0.1 (INSTALL TOOL LEN. COMP., MOVE TO Z APPROACH PSN)
N240 M08 (COOLANT ON)
N250 G01 Z-0.7 F5.0 (CUTTING MOVES WITH FEEDRATE)
N260 G00 Z0.1 (RETRACT FROM HOLE)
```

Tool-end structure:
```
N270 M09 (COOLANT OFF)
N280 G91 G28 Z0 M19 (TOOL CHG. PSN., PRE-ORIENT SPINDLE)
N290 M01 (OPTIONAL STOP)
```

Tool-startup structure:
```
(0.5-IN DRILL)
N300 T03 M06 (LOAD TOOL IN SPINDLE)
N310 G54 G90 S800 M03 T01 (START SPINDLE, READY NEXT TOOL)
N320 G00 X3.0 Y1.0 (MOVE TO XY APPROACH POSITION)
N330 G43 H03 Z0.1 (INSTALL TOOL LEN. COMP., MOVE TO Z APPROACH PSN)
N340 M08 (COOLANT ON)
N350 G01 Z-0.75 F6.0 (CUTTING MOVES WITH FEEDRATE)
N360 G00 Z0.1 (RETRACT FROM HOLE)
```

Tool-end structure:
```
N370 M09 (COOLANT OFF)
N380 G91 G28 Z0 M19 (TOOL CHG. PSN., PRE-ORIENT SPINDLE)
N390 M01 (OPTIONAL STOP)
```

Program-end structure:
```
N400 G28 X0 Y0 (RETURN TO XY ZERO RETURN - LOAD/UNLOAD PSN)
N410 M30 (END OF PROGRAM)
%
```

Key Concept 5: You Must Provide Structure To Your CNC Programs

A Few Questions About the Program:
Let's see how well you understand the structure related to programming. Answers are below.

1) What is the purpose of blocks N010 through N030?
___

2) In block N110, T02 is specified to get the next tool ready. So is the T02 word absolutely mandatory in block N200? Why or why not?
___

3) All of the holes are at the same Y position (Y25.0 for the metric-mode example or Y1.0 for the inch-mode example). In block N120, Y25.0 or Y1.0 is specified. The Y-axis does not move between this command and N220 and N320. Is the Y-word specification absolutely mandatory in blocks N220 and N320? Why/why not?
___

Answers: 1) N010 through N030 are safety commands to ensure that initialized modes are still instated. 2) Yes – for the purpose of being able to re-run tool number two. Block N200 will be the restart command for tool number two (if tool number two is not currently in the spindle). If the T02 word is not in this command, the machine would not know which tool to place in the spindle. 3) Yes – again for the purpose of re-running tools. Notice that the program ends with the Y-axis at the zero return position (block N410). If you run the entire program before you discover something that requires tool number two or tool number three to be re-run, and if the Y-word is not repeated in blocks N220 and N320, the machine would drill the hole at the Y-axis zero return position, not Y25.0 (metric-mode example) or Y1.0 (inch-mode example).

More On the Optional Stop Word (M01)
In our example, the setup person should confirm that each drill has machined its hole in the correct location and that it has broken through the bottom of the workpiece. In other situations, they may have to check the surface finish machined by the tool – or they may have to measure a dimension machined by the tool to confirm that it has been machined within the specified tolerance band. Any number of things may have to be checked.

Where Is the Restart Command For Each Tool?
You know that setup people and operators must re-run tools on a regular basis. To do so, they must position the program's cursor to the beginning of the block that begins the tool. The restart block will be the tool change command (blocks N100, N200, and N300 in the example program above).

For example, say the setup person has run the entire program before finding something that requires tool number 2 to be re-run. In this case, the restart block will be block N200. So they turn on the optional stop switch to make the machine stop when tool number two is finished and they re-run tool number 2.

When the machine reaches block N290, the optional stop block will cause the machine to stop (because the optional stop switch is currently on). They check the machining operation and still something is wrong. They correct the problem and must re-run tool number 2 again.

Some machines will generate an alarm if a T-word is commanded for a tool when the tool is already in the spindle. In that case, restart at the block after to tool change. For tool number 2 the restart block will now be block N210.

Program Structure for the Certification Cart (Mill)

We have made a few comments about programming structure differences for the Certification Cart. Here is the structure we recommend.

If you are taking a class and the Certification Cart is being used for practice, use this format to structure your programs. Note that this is the format used for programs Certification Cart projects.

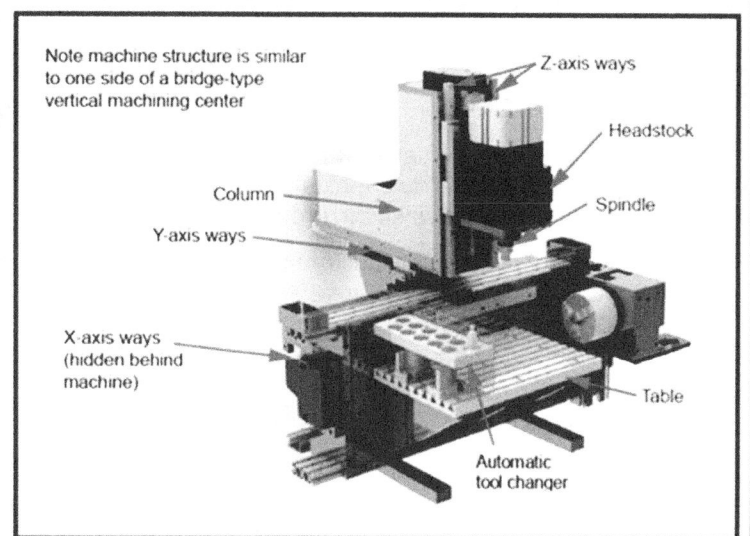

Note machine structure is similar to one side of a bridge-type vertical machining center

Program structure (shown in metric mode). Bolded values change from program and tool to tool:

Program Start-Up Structure:
```
%
  O0001 (PROGRAM DESCIPTION HERE)
  N010 G17 G21 G94 (SELECT XY-PLANE, METRIC MODE, PER-MINUTE FEEDRATE MODE)
  N020 G40 G64 G80 (CANCEL CUTTER COMP., SCALING, AND CANNED CYCLES)
```

Tool Start-Up Structure:
```
  (DESCRIPTION/NAME FOR THIS TOOL)
  N100 M06 T01 (LOAD TOOL IN SPINDLE)
  N110 G54 G90 S300 M03 (SELECT WORK OFFSET/ABSOLUTE MODE, START SPINDLE)
  N120 G00 X125.0 Y125.0 (MOVE TO XY APPROACH POSITION)
  N130 G43 H01 Z2.0 (INSTALL TOOL LEN. COMP., MOVE TO Z APPROACH PSN)
  N140 M08 (COOLANT ON)
  N150 G01 . . . . F75.0 (CUTTING MOVES WITH FEEDRATE)
```

Tool Ending Structure:
```
  N200 M09 (COOLANT OFF)
  N210 G91 G28 Z0 (TOOL CLEARANCE POSITION)
  N220 M01 (OPTIONAL STOP)
```

Program Ending Structure:
```
  N310 G28 X0 Y0 (RETURN TO XY ZERO RETURN - LOAD/UNLOAD POSITION)
  N320 M30 (END OF PROGRAM)
%
```

Key Concept 5: You Must Provide Structure To Your CNC Programs

Format for Horizontal Machining Centers

There are not many differences between the structured program format for vertical versus horizontal machining centers. The only real difference has to do with the tool change position. For most horizontal machining centers, the tool change position is the zero return position in the Y and Z axes (instead of just the Z-axis). Also, this format has a program end with the machine at the zero return position in all three axes (for machines with pallet changers, this is commonly the pallet change position).

This structured program format assumes that the machine has a double-arm automatic tool changer, that the tool change position is the Y and Z-axis zero return position, and that the machine is resting at the tool change position in Y and Z when the program begins.

Again, you are to use the strict structure of these given formats. But the actual values that are shown in **bold** will change from program to program and from tool to tool.

Program-Startup Structure:

```
%
O0001 (PROGRAM DESCIPTION HERE)
N010 G17 G21 G94 (SELECT XY-PLANE, METRIC MODE, SELECT PER-MINUTE-FEEDRATE MODE)
N020 G40 G64 (CANCEL CUTTER COMP. AND SELECT NORMAL CUT. MODE)
N030 G69 G80 (CANCEL ROTATION AND CANNED CYCLES)
```

Tool-Startup Structure:

```
(DESCRIPTION/NAME FOR THIS TOOL)
N100 T01 M06 (LOAD TOOL IN SPINDLE)
N110 G54 G90 S300 M03 T02 (SELECT WORK OFFSET/ABSOLUTE MODE, START SPINDLE, READY NEXT TOOL)
N120 G00 X125.0 Y125.0 (MOVE TO XY APPROACH POSITION)
N130 G43 H01 Z2.0 (INSTALL TOOL LEN. COMP., MOVE TO Z APPROACH PSN)
N140 M08 (COOLANT ON)
N150 G01 . . . . F75.0 (CUTTING MOVES WITH FEEDRATE)
```

Tool-End Structure:

```
N200 M09 (COOLANT OFF)
N210 G91 G28 Y0 Z0 M19 (TOOL CHG. PSN., PRE-ORIENT SPINDLE)
N220 M01 (OPTIONAL STOP)
```

Program-End Structure:

```
N310 G28 X0 (RETURN TO X ZERO RETURN - LOAD/UNLOAD PSN)
N320 M30 (END OF PROGRAM)
%
```

Lesson 5.2: Structured Program Format

Here is the example program, modified for a horizontal machining center. Only the inch-mode version of the program is shown.

Program-startup structure:
```
%
O0027 (EXAMPLE PROGRAM FORMAT)
N010 G17 G20 G94 (SELECT XY-PLANE, INCH MODE, SELECT PER-MINUTE-FEEDRATE MODE)
N020 G40 G64 (CANCEL CUTTER COMP. AND SELECT NORMAL CUT. MODE)
N030 G69 G80 (CANCEL ROTATION AND CANNED CYCLES)
```

Tool-startup structure:
```
(1/4 DRILL)
N100 T01 M06 (LOAD TOOL IN SPINDLE)
N110 G54 G90 S1200 M03 T02 (START SPINDLE, READY NEXT TOOL)
N120 G00 X1.0 Y1.0 (MOVE TO XY APPROACH POSITION)
N130 G43 H01 Z0.1 (INSTALL TOOL LEN. COMP., MOVE TO Z APPROACH PSN)
N140 M08 (COOLANT ON)
N150 G01 Z-0.65 F4.0 (CUTTING MOVES WITH FEEDRATE)
N160 G00 Z0.1 (RETRACT FROM HOLE)
```

Tool-end structure:
```
N170 M09 (COOLANT OFF)
N180 G91 G28 Y0 Z0 M19 (TOOL CHG. PSN., PRE-ORIENT SPINDLE)
N190 M01 (OPTIONAL STOP)
```

Tool-startup structure:
```
(3/8 DRILL)
N200 T02 M06 (LOAD TOOL IN SPINDLE)
N210 G54 G90 S1000 M03 T03 (START SPINDLE, READY NEXT TOOL)
N220 G00 X2.0 Y1.0 (MOVE TO XY APPROACH POSITION)
N230 G43 H02 Z0.1 (INSTALL TOOL LEN. COMP., MOVE TO Z APPROACH PSN)
N240 M08 (COOLANT ON)
N250 G01 Z-0.7 F5.0 (CUTTING MOVES WITH FEEDRATE)
N260 G00 Z0.1 (RETRACT FROM HOLE)
```

Tool-end structure:
```
N270 M09 (COOLANT OFF)
N280 G91 G28 Y0 Z0 M19 (TOOL CHG. PSN., PRE-ORIENT SPINDLE)
N290 M01 (OPTIONAL STOP)
```

Tool-startup structure:
```
(1/2 DRILL)
N300 T03 M06 (LOAD TOOL IN SPINDLE)
N310 G54 G90 S800 M03 T01 START SPINDLE, READY NEXT TOOL)
N320 G00 X3.0 Y1.0 (MOVE TO XY APPROACH POSITION)
N330 G43 H03 Z0.1 (INSTALL TOOL LEN. COMP., MOVE TO Z APPROACH PSN)
N340 M08 (COOLANT ON)
N350 G01 Z-0.75 F6.0 (CUTTING MOVES WITH FEEDRATE)
N360 G00 Z0.1 (RETRACT FROM HOLE)
```

Tool-end structure:
```
N370 M09 (COOLANT OFF)
N380 G91 G28 Y0 Z0 M19 (TOOL CHG. PSN., PRE-ORIENT SPINDLE)
N390 M01 (OPTIONAL STOP)
```

Program-end structure:
```
N400 G28 X0 RETURN TO X ZERO RETURN - LOAD/UNLOAD PSN)
N410 M30 (END OF PROGRAM)
%
```

Key Concept 5: You Must Provide Structure To Your CNC Programs

The workpiece for this example program is not really a good application for horizontal machining centers, but it works nicely to show the format. Most horizontal machining centers have a rotary device within the table that allows several sides of the workpiece to be exposed to the spindle for machining. We show how rotary devices are programmed in lesson 6.4.

Key Points for Lesson 5.2:

- ✓ There are four structures in the structured program format to help you write programs: program-startup structure, tool-startup structure, tool-end structure and program-end structure.
- ✓ You are only on your own to come up with the cutting motion commands for each tool–a large percentage of most programs is related to the overall program structure.
- ✓ The tool-startup structure includes all CNC words to make each tool independent of the rest of the program–so tools can be re-run.
- ✓ The restart block for a tool is the tool change command (with the M06).
- ✓ Tool-end structure includes an optional stop word (M01) so setup people and operators can easily stop the machine after each tool to inspect what the tool has done (by turning on the optional stop switch on the machine operator panel).

Talk to experience people in your company...

... to learn more about the structure used in your CNC programs.

1) Ask to see a few programs actually being used on your company's machining centers. Compare them to the example programs shown in this text. Question an experienced person about any CNC words you don't recognize. Did the programmer use safety commands? Where is the tool change position?

2) Ask to see the procedure used to re-run a tool.

3) Ask what commands are related to the automatic tool changer.

Exercise: Write Your First Program By Yourself

In this exercise you will practice formatting a CNC program. Study the workpiece tool path drawing and process. Fill in the coordinate sheet and, using the appropriate process, write a program (on a separate sheet of paper) that will machine the workpiece. Use the measurement system (metric mode or inch mode) used by your company or school.

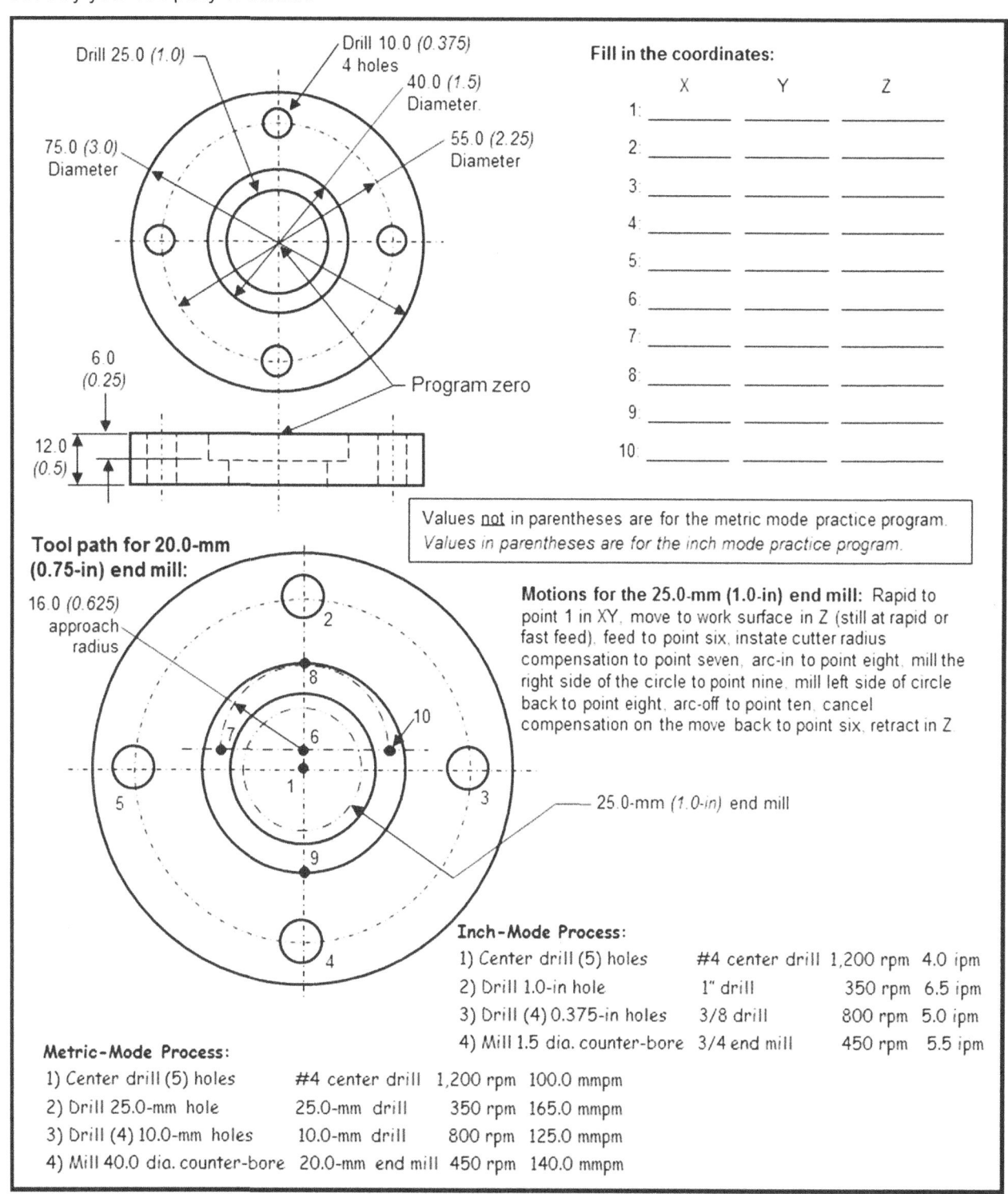

Fill in the coordinates:

	X	Y	Z
1:	___	___	___
2:	___	___	___
3:	___	___	___
4:	___	___	___
5:	___	___	___
6:	___	___	___
7:	___	___	___
8:	___	___	___
9:	___	___	___
10:	___	___	___

Drill 25.0 (1.0)
Drill 10.0 (0.375) 4 holes
40.0 (1.5) Diameter
75.0 (3.0) Diameter
55.0 (2.25) Diameter
6.0 (0.25)
12.0 (0.5)
Program zero

Values **not** in parentheses are for the metric mode practice program.
Values in parentheses are for the inch mode practice program.

Tool path for 20.0-mm (0.75-in) end mill:
16.0 (0.625) approach radius

Motions for the 25.0-mm (1.0-in) end mill: Rapid to point 1 in XY, move to work surface in Z (still at rapid or fast feed), feed to point six, instate cutter radius compensation to point seven, arc-in to point eight, mill the right side of the circle to point nine, mill left side of circle back to point eight, arc-off to point ten, cancel compensation on the move back to point six, retract in Z.

25.0-mm (1.0-in) end mill

Inch-Mode Process:
1) Center drill (5) holes   #4 center drill   1,200 rpm   4.0 ipm
2) Drill 1.0-in hole   1" drill   350 rpm   6.5 ipm
3) Drill (4) 0.375-in holes   3/8 drill   800 rpm   5.0 ipm
4) Mill 1.5 dia. counter-bore   3/4 end mill   450 rpm   5.5 ipm

Metric-Mode Process:
1) Center drill (5) holes   #4 center drill   1,200 rpm   100.0 mmpm
2) Drill 25.0-mm hole   25.0-mm drill   350 rpm   165.0 mmpm
3) Drill (4) 10.0-mm holes   10.0-mm drill   800 rpm   125.0 mmpm
4) Mill 40.0 dia. counter-bore   20.0-mm end mill   450 rpm   140.0 mmpm

Answer programs are on the next page.

Metric-mode answer program for program formatting exercise

```
%
O0028
N010 G17 G21 G94
N020 G40 G64
N030 G69 G80

(CENTER DRILL)
N100 T01 M06
N110 G54 G90 S1200 M03 T02
N120 G00 X0 Y0 (pt 1)
N130 G43 H01 Z2.0
N140 M08
N150 G01 Z-3.0 F100.0
N160 G00 Z2.0
N170 Y27.5 (pt 2)
N180 G01 Z-3.0
N190 G00 Z2.0
N200 X27.5 Y0 (pt 3)
N210 G01 Z-3.0
N220 G00 Z2.0
N230 X0 Y-27.5 (pt 4)
N240 G01 Z-3.0
N250 G00 Z2.0
N260 X-27.5 Y0 (pt 5)
N270 G01 Z-3.0
N280 G00 Z2.0
N290 M09
N300 G91 G28 Z0 M19
N310 M01

(25.0-MM DRILL)
N320 T02 M06
N330 G54 G90 S350 M03 T03
N340 G00 X0 Y0 (pt 1)
N350 G43 H02 Z2.0
N360 M08
N370 G01 Z-21.5 F165.0
N380 G00 Z2.0
N390 M09
N400 G91 G28 Z0 M19
N410 M01

(10.0-MM DRILL)
N420 T03 M06
N430 G54 G90 S800 M03 T04
N440 G00 X0 Y27.5 (pt 2)
N450 G43 H03 Z2.0
N460 M08
N470 G01 Z-17.0 F125.0
N480 G00 Z2.0
N490 X27.5 Y0 (pt 3)
N500 G01 Z-17.0
N510 G00 Z2.0
N520 X0 Y-27.5 (pt 4)
N530 G01 Z17.0
N540 G00 Z2.0
N550 X-27.5 Y0 (pt 5)
N560 G01 Z-17.0
N570 G00 Z2.0

N580 M09
N590 G91 G28 Z0 M19
N600 M01

(20.0-MM END MILL)
N610 T04 M06
N620 G54 G90 S450 M03 T01
N630 G00 X0 Y0 (pt 1)
N640 G43 H04 Z2.0
N650 M08
N660 G01 Z-6.0 F750.0
N670 Y4.0 F140.0 (pt 6)
N680 G42 D04 X-16.0 (pt 7)
N690 G02 X0 Y20.0 R16.0 (pt 8)
N700 Y-20.0 R20.0 (pt 9)
N710 Y20.0 R20.0 (pt 8)
N720 X16.0 Y4.0 R16.0 (pt 10)
N730 G01 G40 X0 (pt 6)
N740 G00 Z0.1
N750 M09
N760 G91 G28 Z0 M19
N770 M01

(PROGRAM END)
N780 G91 G28 Z0 M19
N790 G28 X0 Y0
N800 M30
%
```

Inch-mode answer program for program formatting exercise

```
%
O0028
N010 G17 G20 G94
N020 G40 G64
N030 G69 G80

(CENTER DRILL)
N100 T01 M06
N110 G54 G90 S1200 M03 T02
N120 G00 X0 Y0 (pt 1)
N130 G43 H01 Z0.1
N140 M08
N150 G01 Z-0.12 F4.0
N160 G00 Z0.1
N170 Y1.125 (pt 2)
N180 G01 Z-0.12
N190 G00 Z0.1
N200 X1.125 Y0 (pt 3)
N210 G01 Z-0.12
N220 G00 Z0.1
N230 X0 Y-1.125 (pt 4)
N240 G01 Z-0.1
N250 G00 Z0.1
N260 X-1.125 Y0 (pt 5)
N270 G01 Z-0.1
N280 G00 Z0.1
N290 M09
N300 G91 G28 Z0 M19
N310 M01

(1.0-IN DRILL)
N320 T02 M06
N330 G54 G90 S350 M03 T03
N340 G00 X0 Y0 (pt 1)
N350 G43 H02 Z0.1
N360 M08
N370 G01 Z-0.83 F6.5
N380 G00 Z0.1
N390 M09
N400 G91 G28 Z0 M19
N410 M01

(0.375-IN DRILL)
N420 T03 M06
N430 G54 G90 S800 M03 T04
N440 G00 X0 Y1.125 (pt 2)
N450 G43 H03 Z0.1
N460 M08
N470 G01 Z-0.64 F5.0
N480 G00 Z0.1
N490 X1.125 Y0 (pt 3)
N500 G01 Z-0.64
N510 G00 Z0.1
N520 X0 Y-1.125 (pt 4)
N530 G01 Z-0.64
N540 G00 Z0.1
N550 X-1.125 Y0 (pt 5)
N560 G01 Z-0.64
N570 G00 Z0.1

N580 M09
N590 G91 G28 Z0 M19
N600 M01

(0.75-IN END MILL)
N610 T04 M06
N620 G54 G90 S450 M03 T01
N630 G00 X0 Y0 (pt 1)
N640 G43 H04 Z0.1
N650 M08
N660 G01 Z-0.25 F30.0
N670 Y0.125 F5.5 (pt 6)
N680 G42 D04 X-0.625 (pt 7)
N690 G02 X0 Y0.75 R0.625 (pt 8)
N700 Y-0.75 R0.75 (pt 9)
N710 Y0.75 R0.75 (pt 8)
N720 X0.625 Y0.125 R0.625 (pt 10)
N730 G01 G40 X0 (pt 6)
N740 G00 Z0.1
N750 M09
N760 G91 G28 Z0 M19
N770 M01

(PROGRAM END)
N780 G91 G28 Z0 M19
N790 G28 X0 Y0
N800 M30
%
```

Key Concept 6: Features That Help Simplify Programming

You now have the basic tools you need to write CNC programs. However, writing programs with only the tools we have shown will be very tedious. In Key Concept 6, we'll be showing features that make programming easier, shorten the program's length, and in general, facilitate writing programs.

Key Concept 6 contains four lessons:

- 6.1: Canned cycles for drilling
- 6.2: Sub-programming techniques
- 6.3: Other special programming features
- 6.4: Programming rotary devices

Key Concept 6 will be of great interest to you. While you have learned how to write simple CNC programs, you may be wondering if there are easier ways to get things done.

With hole-machining, for example, we have shown only drilling operations. As you know, there are other kinds of hole-machining operations (like tapping, reaming, boring, and counter-boring). Using our current programming methods, drilling a hole requires three blocks per hole (rapid to XY position, drill hole, retract from hole). In lesson 6.1 we will show you a better way, using what FANUC calls *Canned Cycles for Drilling*.

In lesson 6.2, you'll learn how to use subprograms to make programming more efficient and flexible. In lesson 6.3, we'll present several special features that can help you develop programs. While they won't all be of immediate use, you'll surely find at least some to be very helpful. And in lesson 6.4, we'll show how rotary devices are programmed.

Almost every CNC function presented thus far is essential to your ability to program CNC machining centers. You will be using the features and techniques presented in Key Concepts 1 through 5 in every program you write (with the possible exception of cutter radius compensation). What you have learned so far are the rudimentary tools of CNC programming.

Key Concept 6 will introduce you to tools that will simplify many programming tasks. Though not every feature and technique discussed will be of use to every CNC programmer, you will pick up several techniques that will streamline common programming activities. They make programming easier, they make programs shorter, they reduce the potential for mistakes, and in general, they simplify the task of writing CNC programs.

You may be wondering why we waited so long to show these features, since they make programming so much easier. Do not underestimate the importance of what you have learned so far. Just as it helps to understand how arithmetic calculations are done manually before you use a calculator, so does it help to understand the basic way of programming machining operations before you move on to more advanced methods. At the very least, the programming activities you have learned so far should give you an appreciation for what is presented in Key Concept 6. A good foundation in CNC will let you grasp the value of the features we introduce.

Some of the features shown in Key Concept 6 are options on some FANUC CNCs. But most machine tool builders will include them with the machines they sell. If you come across one of these features that your machine does not have, remember that many of them are field-installable, meaning if you have a need for them, they can be added to your control at any time (for an additional price).

Key Concept 6: Features That Help Simplify Programming

Here is a list of the special features we will discuss during this Key Concept in the order we cover them:
- ✓ G73 - G89 Canned Cycles for Drilling - To simplify hole-machining operations (lesson 6.1)
- ✓ M98 - M99 Sub-programming - To minimize redundant commands (lesson 6.2)
- ✓ / - Block delete techniques - To give your operator a choice (lesson 6.3)
- ✓ Sequence number techniques (lesson 6.3)
- ✓ G02-G03 helical interpolation - For thread milling, machining external threads and when holes are too large to tap (lesson 6.3)
- ✓ Other G-codes not discussed to this point (lesson 6.3)
- ✓ M-codes not addressed to this point (lesson 6.3)
- ✓ Rotary devices (lesson 6.4)

As stated, you will not have immediate need for some of these features. In fact, some you may never need. But we urge you to study this information if for no other reason than to gain a better understanding of what is possible in CNC programming. You can always come back and review a needed feature when the need arises, if you know it is available.

Lesson 6.1–Canned Cycles For Drilling

Armed only with what you know so far, programming hole-machining operations is very tedious. Canned cycles for drilling will dramatically simplify the programming of these very common machining operations.

Objectives

After completing this lesson, students should be able to:

- ✓ Describe the meaning of canned
- ✓ Describe the various canned cycles and their application
- ✓ Describe when and why holes must be peck drilled and how to select the appropriate cycle
- ✓ Describe how to program canned cycles
- ✓ Describe how to use G98 and G99 (initial and rapid plane) to clear obstructions between holes

Introduction

Almost all programs have at least some hole-machining operations. If you have been doing the exercises in this text, you have seen how tedious, time consuming, and error-prone it can be to program hole-machining operations with G00 and G01. You know that with G00 and G01, each hole will require at least three blocks, making the program quite long. And we have only performed basic drilling operations. Peck drilling, tapping, boring, and counter-boring operations will require even more blocks per hole.

Canned Cycles for Drilling will simplify the programming of hole-machining operations. Only one block is required per hole, regardless of the machining style (drill, peck drill, tap, ream, bore, counter-bore, etc.). Additionally, canned cycles are modal, meaning once you instate a canned cycle, you can continue machining additional holes with the same geometry by simply programming the coordinates of the hole. This will dramatically shorten the program's length, make programming easier, less time-consuming, and less error-prone.

What Does "Canned" Mean?

A canned cycle is a series of preset movements that the CNC will execute based upon a limited amount of program information. The zero return command (G28) is a kind of canned cycle. G28 actually makes the machine do two things. First, the machine will move to the intermediate position. Second, it will move to the zero return position.

If you have the single block switch turned on (a function that makes the machine execute one block in the program at a time), you actually have to start the cycle twice to make the CNC complete the G28 command. Again, the first time you activate the cycle (by pressing the cycle start button), the machine executes the motion to the intermediate position. The second time, it moves to the zero return position.

Hole-machining canned cycles are much more elaborate. Even the standard drilling cycle will make the machine do at least three things per block. With the chip-breaking peck drilling cycle, the CNC may generate over one-hundred actions from a single program block.

Invoking Canned Cycles:

You invoke the complete canned cycle specification for the first hole to machine. With a standard drilling cycle, for example, the instating command includes the cycle type (G81), the first hole position in XY, the rapid plane (R-word), the hole's bottom position in Z (commonly its depth), and the machining feed rate. The instating command machines the first hole. To machine additional holes with the same geometry, you simply program the XY coordinates, one set per block. After the last hole, you must cancel the cycle with a G80 word.

Key Concept 6: Features That Help Simplify Programming

Here is a list of the most common hole-machining canned cycles as FANUC names them in approximate order of popularity:

G80 – Cancel any of the canned cycles

G81 – Drilling cycle

G73 – High-speed peck drilling cycle (breaks chips as the hole is machined)

G83 – Peck drilling cycle (causes the drill to retract between pecks)

G82 – Counter-boring cycle

G84 – Right hand tapping cycle (also used for rigid tapping)

G74 – Left hand tapping cycle (also used for rigid tapping)

G86 – Boring cycle (rapids out of the hole)

G89 – Boring cycle with dwell (pauses at hole bottom, rapids out)

G76 – Fine boring cycle (leaves no witness mark in hole)

G85 – Boring cycle (retracts from hole at the programmed feedrate)

Canned Cycle Commonalities

Canned cycles share two things in common.

First, they are all modal. Once a canned cycle is instated, the machine will continue machining one hole per command until the canned cycle is cancelled. G80 is the word used to cancel canned cycles. Once you instate a canned cycle, you must remember to cancel it when you're finished machining holes. If you don't, the control will continue machining (unwanted) holes even after you have finished listing hole-coordinates.

Second, all canned cycles will cause these basic actions:

1. A rapid positioning movement to the hole position in XY (if the tool is not already in this position).
2. A rapid positioning movement to the R-plane (referred to as the *rapid plane*). This approach position is just above the surface to machine.
3. Using the canned cycle's machining style, the hole will be machined.
4. With some canned cycles, an action will occur at the hole-bottom (like a spindle reversal for tapping).
5. The cutting tool will come out of the hole to the R-plane. (With some canned cycles this motion will be done at the machine's rapid rate. With others it will be done at a feedrate.)
6. Depending upon the current status of two G-codes, the tool may retract further in Z (at rapid) to clear obstructions between holes.

Description of Each Canned Cycle

You choose how the hole will be machined using one of a series of G-codes. For the most part, each hole-machining canned cycle is aptly named. Let's describe the machining operation performed by each cycle type in detail.

G80 – Cancel the Canned Cycle Mode

This is not actually a canned cycle commanding word. It is the word used to cancel the canned cycle mode when you finish machining holes with any canned cycle. Again, you <u>must</u> remember to cancel the canned cycle when you are finished machining holes.

G81 – Drilling Cycle

This canned cycle causes the tool to feed to the hole-bottom. Then it will retract at rapid. It is the most popular canned cycle – commonly used for center drilling, spot drilling, drilling, reaming, and rough boring.

G73 – High-Speed Peck Drilling Cycle

This canned cycle is used when drilling a material that has the tendency to produce long, stringy chips. If the chips do not break, they will have the tendency to gather around the drill (forming what machinists commonly call a rat's-nest). Eventually the bundle of chips will interfere with machining. This cycle is used to force the chips to break at regular intervals as the hole is drilled. The breaking of the chip is accomplished by a small retract movement (about 0.1-mm or 0.005-in).

For example, if the peck amount (specified by a Q-word in the canned cycle command) is set to 3.0-mm (or 0.100-in), the tool will plunge into the hole to a depth of 3.0-mm (or 0.100-in), then retract about 0.1-mm (or 0.005-in). The chip will break at this point. Then the tool will plunge another 3.0-mm (or 0.100-in) and retract. It will continue with this process until the hole-bottom is reached.

The retract amount is adjustable by a control parameter, meaning it can be changed if you do not agree with its current setting. We have seen some machine tool builders that set the retract amount to an excessive value. If, on your machine, the tool appears to retract more than about 0.1-mm (or 0.005-in) per peck, you will be wasting cycle time whenever you use this cycle. (The parameter number for retract amount is specified in the FANUC programming manual in the description of G73).

G83 – Peck Drilling Cycle (full retract between pecks)

If drilling a hole having a depth that is greater than about three times the drill diameter, the chips machined by the drill will tend to bind up in the drill's flutes. If the drill continues into the hole, this binding will get progressively worse, and will eventually cause the drill to break. The deep-hole peck drilling cycle allows you to easily specify a series of pecks (peck depth is specified with a Q-word in the G83 command). After each peck, the machine will retract the drill (at rapid) all the way out of the hole to clear chips. The tool will then rapid back into the hole to within a small distance (set by another control parameter) of where it left off. It will then feed in another peck amount. The machine will continue this process until the hole-bottom is reached. The tool then retracts from the hole a final time.

Figure 6.1 provides a summary of the motions performed by the three drilling cycles.

G81 Drilling cycle

Used for: center drilling, spot drilling, drilling, reaming, and rough boring.

1) Rapid to XY position
2) Rapid to R-plane
3) Machine hole as follows:
 a: feed to hole bottom (Z)
4) No bottom action is required
5) Rapid out of hole

G73 High-speed peck drilling cycle

Used for: drilling holes into gummy materials when chips will not break. This cycle will break chips at regular intervals.

1) Rapid to XY position
2) Rapid to R-plane
3) Machine hole as follows:
 a: feed into hole by Q amount
 b: retract a small amount
 c: feed in by another Q amount
 d: retract a small amount
 e: continue until hole-bottom (Z) is reached
4) No bottom action is required
5) Rapid out of hole

Note: 0.100 inch (2.5 mm) works nicely for the Q value

G83 Peck drilling cycle

Used for: drilling deep holes that are deeper than about 3-4 times the drill diameter. This cycle will retract the tool out of the hole between pecks to clear chips.

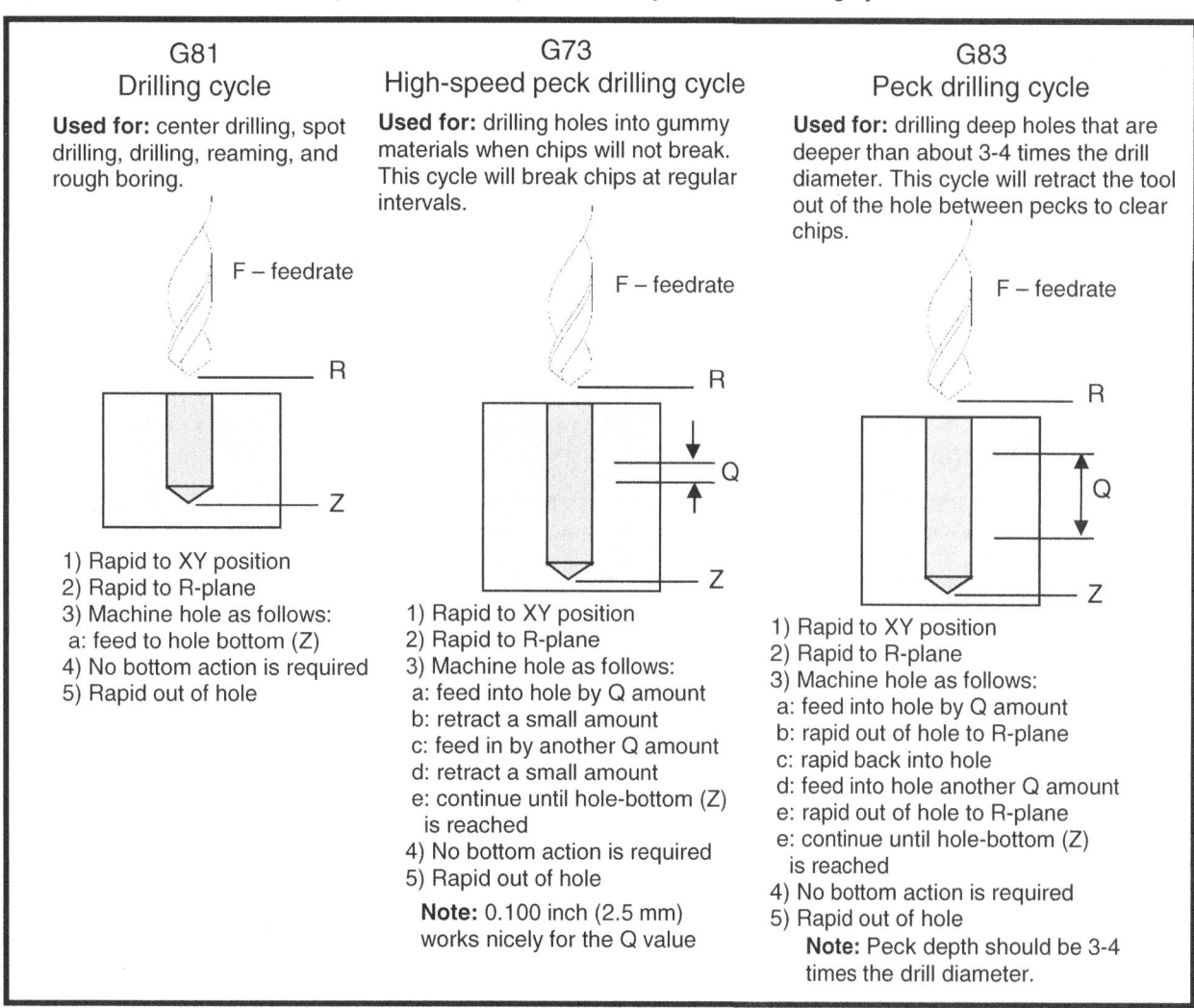

1) Rapid to XY position
2) Rapid to R-plane
3) Machine hole as follows:
 a: feed into hole by Q amount
 b: rapid out of hole to R-plane
 c: rapid back into hole
 d: feed into hole another Q amount
 e: rapid out of hole to R-plane
 e: continue until hole-bottom (Z) is reached
4) No bottom action is required
5) Rapid out of hole

Note: Peck depth should be 3-4 times the drill diameter.

Figure 6.1 – The three drilling canned cycles

Key Concept 6: Features That Help Simplify Programming

G84 – Right-Hand Tapping Cycle (also used for rigid tapping)

Since most tapped holes require right-hand threads, G84 is the most common tapping cycle. The spindle must be running in the forward direction (M03) when this cycle is commanded. Once the tap feeds to the hole-bottom, the spindle direction will change (to reverse – M04) and the tap will feed back out of the hole. Then the spindle will change direction again (back to forward) to get ready to tap the next hole.

With many (especially older) machining centers, the machine cannot perfectly synchronize the spindle reversal at the hole bottom with the retract motion out of the hole. These machines require a special tension/compression tap holder that allows the tap to float (up and down) in its holder. Figure 6.2 shows one.

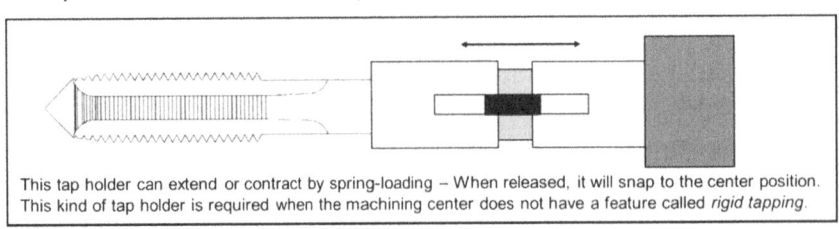

This tap holder can extend or contract by spring-loading – When released, it will snap to the center position. This kind of tap holder is required when the machining center does not have a feature called rigid tapping.

Figure 6.2 – A tension-compression tap holder

Rigid Tapping

There is a feature called rigid tapping (also called synchronous tapping) that allows perfect synchronization between the spindle reversal at the hole-bottom and the retract motion. Most current model machining centers have this feature. Machines that have rigid tapping can tap faster, provide better threads in the tapped hole, and eliminate the need for the special (and expensive) tension/compression tap holder.

The word M29 is used to select the rigid tapping mode. The M29 command must include an S word (to specify the spindle rpm for tapping) and it must be the command just prior to the G84 tapping command.

Feedrate for Tapping

As you know, older machining centers require that you program feedrate in per-minute fashion (specified with G94). To calculate the per-minute feedrate, you must multiply the pitch of the thread times the previously calculated spindle speed in rpm.

For threads specified using the Metric measurement system, pitch is part of the thread designation. For a M12.0-1.25 metric thread, for instance, the pitch is 1.25 millimeters. For threads specified using the Imperial measurement system, the thread specification designates the number of threads per inch. Pitch is equal to one divided by the number of threads per inch. For a 1/2-13 thread, for example, pitch is 0.0769 (1 divided by 13).

If tapping using a speed of 230 rpm, an M12-1.25 thread requires a per-minute feedrate of 287.5 millimeters per minute (230 times 1.25). At 230 rpm, a 1/2-13 thread requires a per-minute feedrate of 17.69 inches per minute (230 times 0.0769).

Newer machines provide the ability to specify feedate in per-revolution fashion (specified with G95). For these machines, feedrate specification for tapping is easier. Simply specify the pitch of the thread as the feedrate.

Rapid Plane for Tapping

If using a tension-compression holder (the machine does not have rigid tapping), the tap can actually float in the tap holder. That is, it can extend and contract along the Z-axis. If the tap holder extends as the hole is tapped, and if the rapid plane is only 2.0-mm (or 0.100-in) above the work surface, it is possible that the tap will still be in the hole after it retracts from the hole. If this happens, the tap will break during the machine's next XY movement. For this reason, most programmers will increase the approach distance to about 7.0-mm (or 0.25-in) when tapping with tension-compression holders.

If tapping with rigid tapping (again, specified with M29), the tap must not float in its holder, so most programmers maintain the 2.0-mm (or 0.1-in) approach distance (this is one of the reasons rigid tapping is faster – the tool doesn't need to move as far).

Tapping Can Be a Little Scary

Tapping canned cycles (G84 and G74) will cause the machine to disable certain control-panel functions, making it a little difficult to safely verify tapping operations. Under normal operation, when the machine executes a tapping canned cycle, it will disable feedrate override and feed hold. Feedrate override is a multi-position switch (like a rheostat) that provides control of programmed feedrate. But when tapping, this function must be disabled. If tapping is not performed at the programmed feedrate, the tap will break as the hole is machined. Feed hold is a push-button that, when pressed, will cause axis motion to stop. If motion is stopped when the tap is in a hole, of course, the tap will break.

Without these two program-verification functions, tapping is somewhat dangerous. You must be sure that the tap is programmed properly before allowing a tap to enter a hole.

Coolant for Tapping?
While some coolants work well for tapping certain materials, many companies use a special tapping compound. Some machining centers even have a special (automatic) function that applies tapping compound to the tap for each hole. Most machines do not. Instead, the program will stop (with an M00) command just before the tapping operation/s. During this program stop, the operator will clear the chips from holes to be tapped and apply the tapping compound to each hole.

When To Tap
If the tapping compound must be manually applied, most programmers will save all tapping until the very end of the program. This way, the operator will only have to clear chips and apply tapping compound once.

G74 – Left-Hand Tapping Cycle

This cycle is just like the G84 tapping cycle, except it is used for left-hand taps. The spindle must be running in the reverse direction (M04) when this cycle is commanded. Left-hand tapping is not often required – and of course – requires a special left-hand tap. G74 can also be used for rigid tapping. Figure 6.3 shows the two tapping cycles.

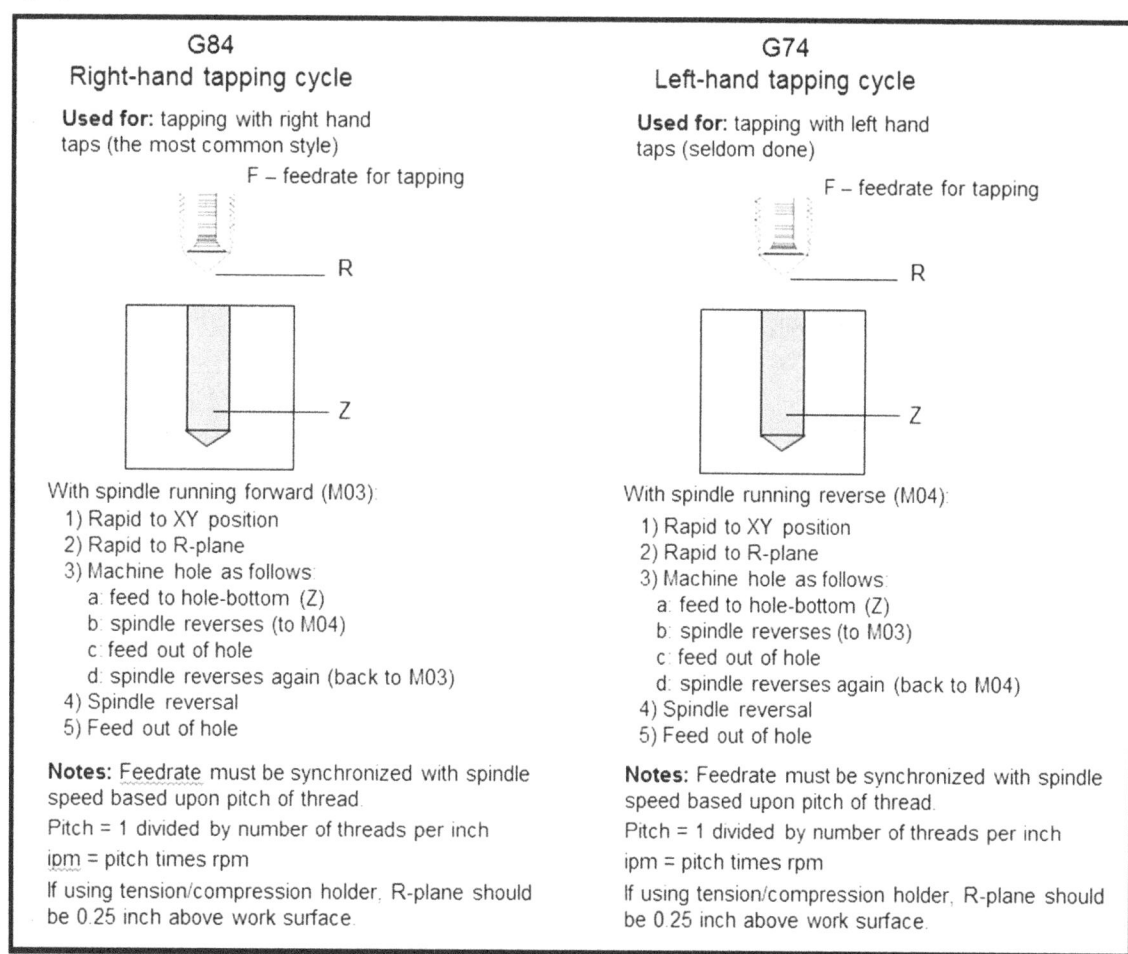

G84 Right-hand tapping cycle	G74 Left-hand tapping cycle
Used for: tapping with right hand taps (the most common style)	**Used for:** tapping with left hand taps (seldom done)
F – feedrate for tapping	F – feedrate for tapping

With spindle running forward (M03):
1) Rapid to XY position
2) Rapid to R-plane
3) Machine hole as follows:
   a: feed to hole-bottom (Z)
   b: spindle reverses (to M04)
   c: feed out of hole
   d: spindle reverses again (back to M03)
4) Spindle reversal
5) Feed out of hole
	With spindle running reverse (M04):
	1) Rapid to XY position
	2) Rapid to R-plane
	3) Machine hole as follows:
	   a: feed to hole-bottom (Z)
	   b: spindle reverses (to M03)
	   c: feed out of hole
	   d: spindle reverses again (back to M04)
	4) Spindle reversal
	5) Feed out of hole

Notes: Feedrate must be synchronized with spindle speed based upon pitch of thread. Pitch = 1 divided by number of threads per inch ipm = pitch times rpm If using tension/compression holder, R-plane should be 0.25 inch above work surface.	**Notes:** Feedrate must be synchronized with spindle speed based upon pitch of thread. Pitch = 1 divided by number of threads per inch ipm = pitch times rpm If using tension/compression holder, R-plane should be 0.25 inch above work surface.

Figure 6.3 – Motions and programming words or tapping cycles

G82 – Counter-Boring Cycle

This cycle causes the tool to feed to the hole bottom, pause for a specified length of time (pause time is specified with a P-word), and then retract from the hole at rapid. The pause at the hole-bottom allows the counter-boring tool (commonly an end mill) to relieve tool pressure, making the hole-bottom flat and precise.

Key Concept 6: Features That Help Simplify Programming

G86 – Boring Cycle (leaves drag line witness mark)
This cycle is used for rough or semi-finish boring, or for finish boring when it doesn't matter if a witness mark is left in the bored hole. The tool will feed to the hole-bottom, the spindle will stop, and the tool will retract from the hole (at rapid). It is during the retract motion (with the tool tip still in contact with the hole surface) that a line will be drawn on the inside of the hole.

G89 – Boring Cycle with Dwell (leaves drag line witness mark)
This cycle is like a combination of the G86 and G82 commands. This cycle is used when a boring bar is machining a blind hole to a precise depth. The boring bar will first feed to the hole-bottom and then pause for a specified time to relieve tool pressure. The spindle will then stop. Finally, the tool will rapid out of the hole. As with G86, a witness mark will be left in the hole.

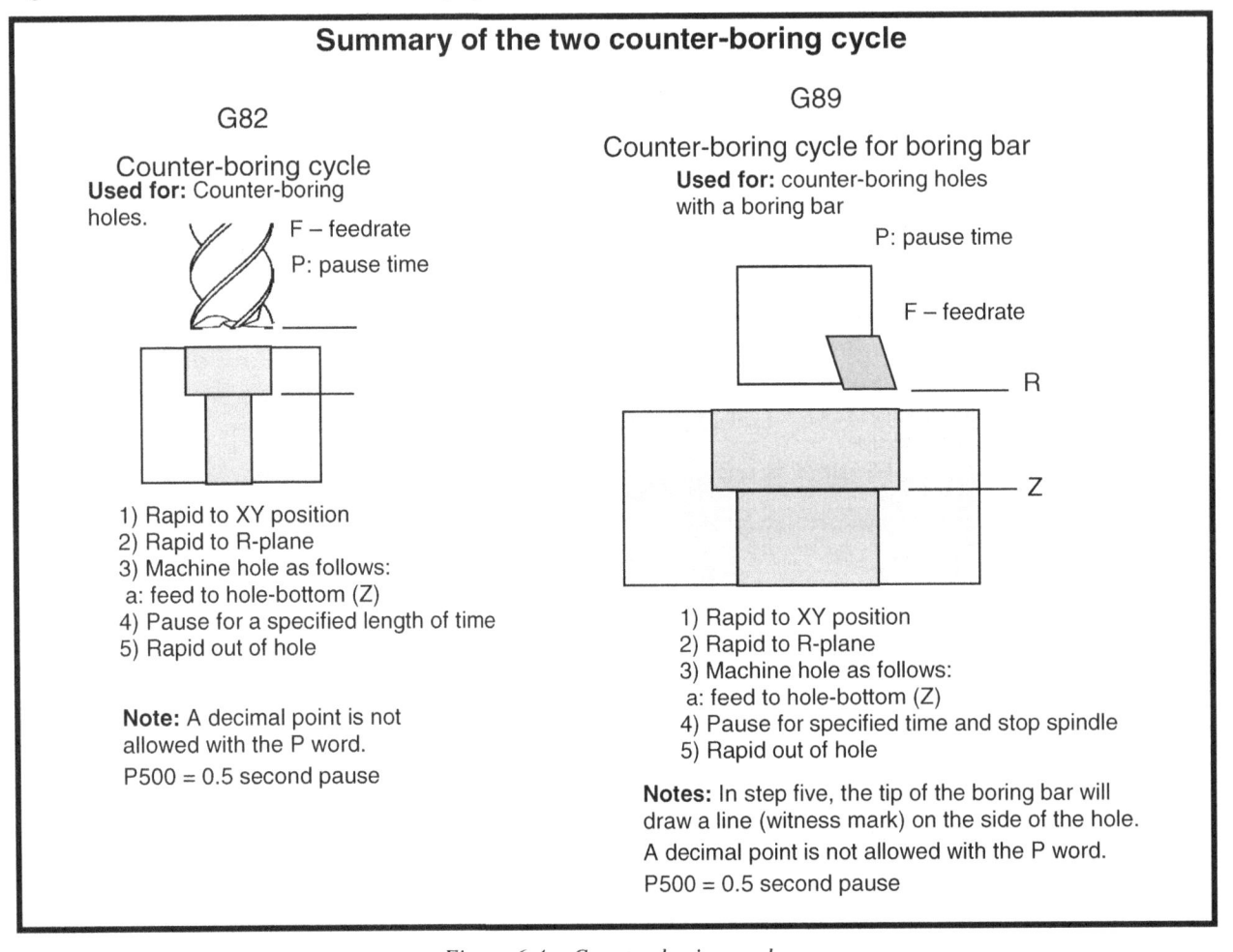

Figure 6.4 shows the two counter-boring cycles.

Summary of the two counter-boring cycle

G82
Counter-boring cycle
Used for: Counter-boring holes.
F – feedrate
P: pause time

1) Rapid to XY position
2) Rapid to R-plane
3) Machine hole as follows:
 a: feed to hole-bottom (Z)
4) Pause for a specified length of time
5) Rapid out of hole

Note: A decimal point is not allowed with the P word.
P500 = 0.5 second pause

G89
Counter-boring cycle for boring bar
Used for: counter-boring holes with a boring bar
P: pause time
F – feedrate
R
Z

1) Rapid to XY position
2) Rapid to R-plane
3) Machine hole as follows:
 a: feed to hole-bottom (Z)
4) Pause for specified time and stop spindle
5) Rapid out of hole

Notes: In step five, the tip of the boring bar will draw a line (witness mark) on the side of the hole.
A decimal point is not allowed with the P word.
P500 = 0.5 second pause

Figure 6.4 – Counter-boring cycles

G76 – Fine Boring Cycle (leaves no witness mark)
This cycle is used for finish boring when no witness marks can be left in the hole after boring. For each hole, the boring bar will feed to the hole-bottom. The spindle will then stop. The spindle will then rotate to its orient position (the M19 position). The tool's cutting edge will then be moved away from the hole-surface to allow some clearance between the tool tip and the hole (the direction and amount of the clearance motion is specified in the canned cycle command). The boring bar will then rapid out of the hole. Finally, the spindle will be restarted and the tool will move back to the hole's center line.

Controlling Move-Over At Hole-Bottom
Two words are used to control the move-over amount (I and J). An I-word in the G76 command specifies the amount and direction of move-over along the X-axis. A J-word in the G76 command specifies the amount and direction of move-over along the Y-axis. Say, for example, the boring bar tip is pointing in the X plus direction when the spindle is at its orient position. In this case, an I-0.02 (if using metric mode) or I-0.001 (if

using inch mode) can be included in the G76 command to tell the machine to move over 0.02-mm (or 0.001-in) in the X minus direction at the hole-bottom and after the spindle is oriented. We recommend specifying a very small move-over amount (no more than 0.02-mm or 0.001-in) in case the setup person makes a mistake when loading the boring bar and has the tip pointing in the wrong direction.

In order to specify these words correctly, you must know (when programming) which way the tip of the boring bar will be pointing when the spindle is at its orient position. This may be impossible, since the boring bar may be held in a collet holder. Unless the setup documentation is very specific about how the boring bar must be held in the holder, you won't know which way the boring bar is pointing when the spindle is oriented.

For this reason, many programmers will include the words I0 and J0 in the G76 command. The setup person will modify the program during setup, once they know which way the boring bar tip will be pointing.

A Tip for Boring Bar Tip Pointing

Most tool holders used on machining centers have a dimple in one of the keyways engaged by the tool changing arm. And most machining centers only allow you to load the tool holder into the spindle in one position (the two keyways in the tool holder are of different widths). One of the tool holder keyways has a dimple. Use the dimpled keyway as your target for the boring bar tip. If you're using a collet holder, have the setup person rotate the boring bar in the holder until its tip is pointing at the dimpled keyway. Then tighten the collet. When the spindle is at its orient position, it will always be pointing in the same direction. Figure 6.5 shows the two most popular boring cycles.

G86 Boring cycle

Used for: machining precise holes when a drag-line witness mark is permitted.

F – feedrate

1) Rapid to XY position
2) Rapid to R-plane
3) Machine hole as follows:
   a: feed to hole-bottom (Z)
4) Stop spindle
5) Rapid out of hole

Note: In step four, the tip of the boring bar will draw a line on the side of the hole.

G76 Fine boring cycle

Used for: machining precise holes when a drag-line witness mark is not permitted.

I: X move-over direction and amount
J: Y move-over direction and amount
F – feedrate

1) Rapid to XY position
2) Rapid to R-plane
3) Machine hole as follows:
   a: feed to hole-bottom (Z)
4) Stop spindle, orient spindle, move over
5) Rapid out of hole

Notes: Programmer must know which direction the tool tip will be pointing when the spindle is oriented.

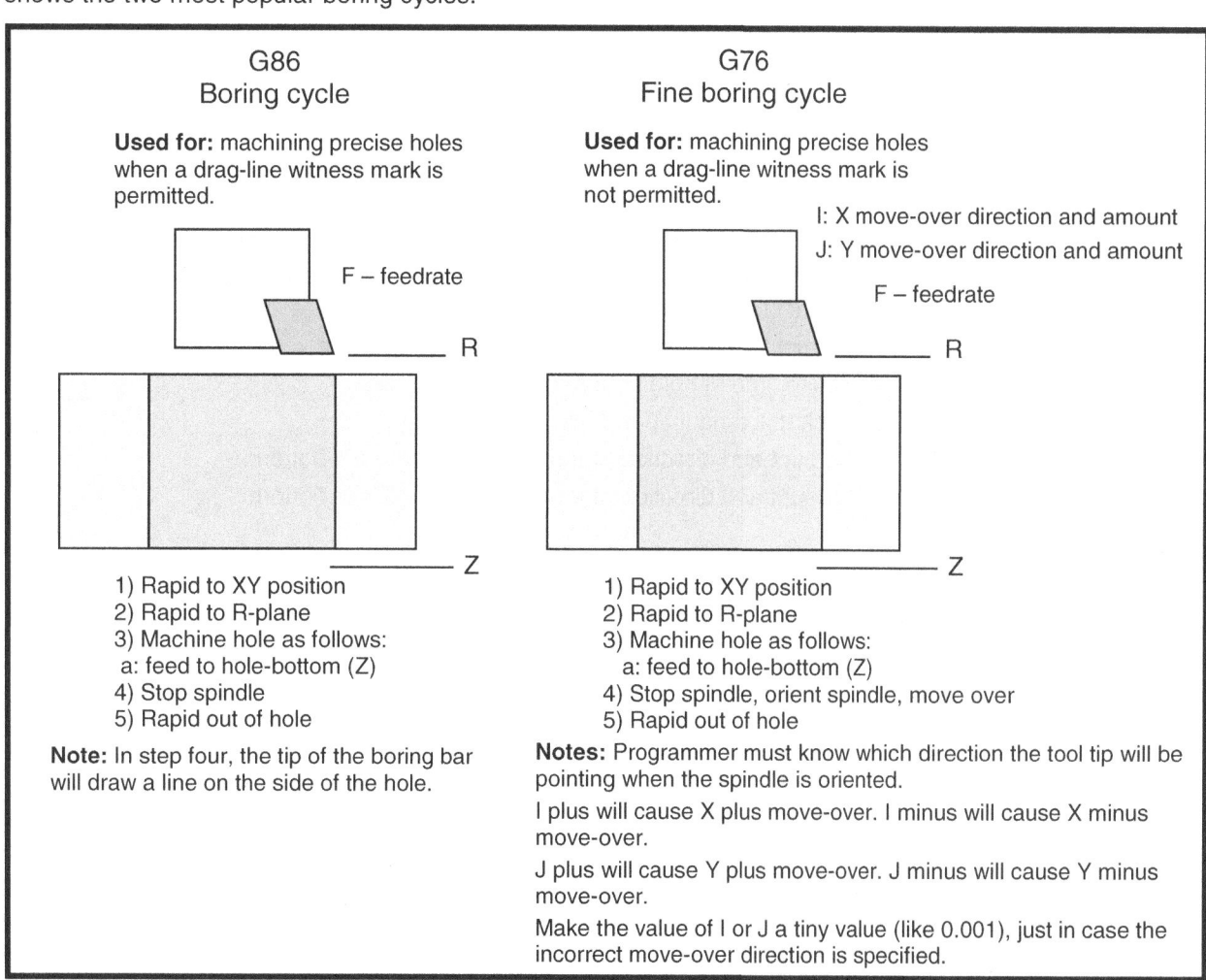

I plus will cause X plus move-over. I minus will cause X minus move-over.

J plus will cause Y plus move-over. J minus will cause Y minus move-over.

Make the value of I or J a tiny value (like 0.001), just in case the incorrect move-over direction is specified.

Figure 6.5 – Motions for the two most popular boring cycles

G85 – Boring Cycle (Feeds in and feeds out)

This cycle causes the tool to feed to the hole-bottom and retract from the hole at the same feedrate. While this cycle is sometimes used for reaming or finish boring, no machining will be done during the retract

Key Concept 6: Features That Help Simplify Programming

motion. For this reason, most programmers use the G81 standard drilling cycle for reaming and one of the boring cycles for boring.

G87 and G88 – Manual Cycles (not recommended)

G87 and G88 are manual cycles that we do not recommend. They perform differently, based on the settings of certain control parameters. Each requires manual intervention on the operator's part. For the most part, the machining operations they perform (back-boring that requires a pilot or cutting edge to be placed on and removed from the tool during the cycle) are just as easily programmed in long-hand fashion using G00 and G01. With G00 and G01, you'll have complete control of the motions being made by the back-boring tool.

Words Used In Canned Cycles

As you have probably noticed, canned cycles share many words in common. These words will have exactly the same meaning in all canned cycles. Although some canned cycles do not use some of these words, when you understand how the shared words work in one canned cycle, you will know how they work in all. Here is a list of all words used in canned cycles, which cycles they apply to, and a brief description of their use:

WORD:	STATUS:	DESCRIPTION:
N	All cycles	Sequence number
G73-G89	All cycles	Canned cycle type
X	All cycles	X coordinate of the hole-center
Y	All cycles	Y coordinate of the hole-center
R	All cycles	Rapid plane position above work surface (specified from workpiece coordinate system zero in Z)
Z	All cycles	Z position of hole-bottom (specified from program zero in the Z-axis)
F	All cycles	Feedrate for the machining operation
L	All cycles	Number of holes to be machined in the command (used with incremental mode only)
G98	All cycles	Retract to the initial plane (initialized)
G99	All cycles	Retract to the R-plane
P	G82, G89	Pause time at hole bottom (P500 = .5 second)
Q	G73, G83	Peck drill amount per peck
I	G76	Amount and direction of move-over in X at hole bottom
J	G76	Amount and direction of move-over in Y at hole bottom

A Simple Example

Here is a simple example program that stresses the points made so far. Figure 6.6 shows the workpiece. We're simply drilling four holes using the drilling cycle, G81.

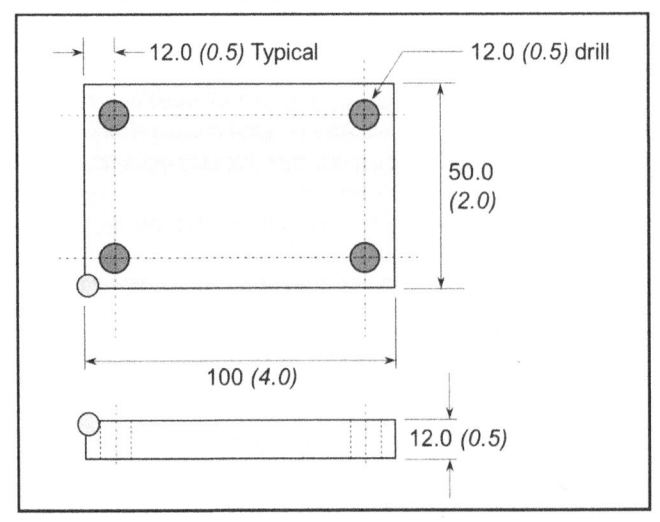

Figure 6.6 – A simple example

Lesson 6.1: Canned Cycles For Drilling

Program with comments:

Metric-mode example

```
%
O0029 (Program number)
(12.0-MM DRILL)
N005 G21 G54 G90 S611 M03 (Select workpiece coordinate system
setting offset #1, absolute mode, start spindle fwd at 611 rpm)
N010 G00 X12.0 Y12.0 (Rapid to first hole-location)
N015 G43 H01 Z2.0 M08 (Instate tool length compensation,
position tool to just above work surface, start coolant)
N020 G81 R2.0 Z-17.6 F125.0 (Drill lower-left hole)
N025 Y38.0 (Drill upper-left hole)
N030 X88.0 (Drill upper-right hole)
N035 Y12.0 (Drill lower-right hole)
N040 G80 M09 (Cancel canned cycle, turn off coolant)
N045 G91 G28 Z0 (Move to Z-axis zero return position)
N050 G28 Y0 (Move to Y-axis zero return position)
N055 M30 (End of program)
%
```

Inch-mode example

```
%
O0029 (Program number)
(0.5-IN DRILL)
N005 G20G54 G90 S611 M03 (Select workpiece coordinate system
setting offset #1, absolute mode, start spindle fwd at 611 rpm)
N010 G00 X0.5 Y0.5 (Rapid to first hole-location)
N015 G43 H01 Z0.1 M08 (Instate tool length compensation,
position tool to just above work surface, start coolant)
N020 G81 R0.1 Z-0.65 F5.0 (Drill lower-left hole)
N025 Y1.5 (Drill upper-left hole)
N030 X3.5 (Drill upper-right hole)
N035 Y0.5 (Drill lower-right hole)
N040 G80 M09 (Cancel canned cycle, turn off coolant)
N045 G91 G28 Z0 (Move to Z-axis zero return position)
N050 G28 Y0 (Move to Y-axis zero return position)
N055 M30 (End of program)
%
```

In block N010, the tool is positioned over the first hole. In block N015, the tool is brought to just above the work surface.

Block N020 instates the drilling cycle, telling the machine that the R-plane is 2.0-mm (or 0.1-in) above the workpiece (the tool is already in this position), the hole bottom position is -17.6-mm (or -0.65-in) below the top of the workpiece (this happens to be the hole depth since the top surface is workpiece coordinate system zero), and the feedrate for drilling is 125.0 mmpm (or 5.0 ipm). This command drills the first (lower-left) hole.

Key Concept 6: Features That Help Simplify Programming

Now we simply list hole-locations – and only the moving axes must be included in the command. While you can include both X and Y coordinates in every command, it will lengthen the program and may result in a typing mistake. As with all motion commands, we recommend that you only include moving axes in each command. Block N025 machines the upper-left hole. Block N030 machines the upper-right hole. And block N035 machines the lower-right hole.

Since there are no more holes to machine, the G80 in block N040 cancels the canned cycle.

This program uses G81 – the drilling cycle. But say the material being machined is very gummy, and chips are stringing up around the drill. Our next example program (shown only for the metric-mode example) machines the same workpiece, but uses G73 (the high-speed peck drilling cycle) instead of G81:

```
%
O0030 (Program number)
(12.0-MM DRILL)
N005 G21 G54 G90 S611 M03 (Select workpiece coordinate system
setting offset #1, absolute mode, start spindle fwd at 611 rpm)
N010 G00 X12.0 Y12.0 (Rapid to first hole-location)
N015 G43 H01 Z2.0 M08 (Instate tool length compensation,
position tool to just above work surface, start coolant)
N020 G73 R2.0 Z-17.6 Q3.0 F125.0 (Drill lower-left hole)
N025 Y38.0 (Drill upper-left hole)
N030 X88.0 (Drill upper-right hole)
N035 Y12.0 (Drill lower-right hole)
N040 G80 M09 (Cancel canned cycle, turn off coolant)
N045 G91 G28 Z0 (Move to Z-axis zero return position)
N050 G28 Y0 (Move to Y-axis zero return position)
N055 M30 (End of program)
%
```

With only two changes (changing G81 to G73 and adding the Q3.0 word), we've modified the way the four holes will be machined.

Understanding G98 and G99

FANUC and FANUC-compatible CNCs for machining centers provide the ability to easily avoid clamps and other obstructions between holes when using canned cycles – moving the tool over them in the Z-axis.

Remember the basic steps that all canned cycles will perform:

1) A rapid positioning movement to the hole position in XY (if the tool is not already in this position).

2) A rapid positioning movement to the R-plane (referred to as the *rapid plane*). This approach position is just above the surface to machine).

3) Using the canned cycle's machining style, the hole will be machined.

4) With some canned cycles a hole-bottom action will occur (like a spindle reversal when tapping).

5) The tool will come out of the hole to the R plane (at rapid or at a feedrate, based upon the canned cycle type).

6) Depending on the setting of a G code, the tool may retract further in Z.

In step number six, G98 and G99 specify to which of two possible Z positions the tool will be retracted. Your choices are:

- The R-plane specified in the canned cycle (which is right above the work surface)
- The *initial plane* (the last programmed Z position prior to the canned cycle command) – this will be a position that is above the obstruction between holes

G98 specifies the *initial plane*. The initial plane is the cutting tool's last Z position prior to the canned cycle command. To make use of obstruction clearing, the initial plane must be specified above all obstructions.

G99 specifies the R-plane, which is included in the canned cycle itself. The R-plane is right above the work surface. G98 and G99 are modal, meaning if you have a series of holes that require the same retract position, G98 or G99 need only be specified in the first command of the series.

Though the word order of your canned cycle command is unimportant, we recommend placing the G98 or G99 at the *end* of the command. This will help you remember what the tool will do *after* the hole is machined.

G98 (initial plane) is initialized when the machine power is turned on – and it is reinstated whenever the canned cycle is canceled (with G80). So if you do not include a G98 or G99 in the canned cycle command for the first hole, the control will automatically retract the tool to the initial plane. This is the safer of the two retract positions.

If you position the tool to the R-plane just prior to the canned cycle command (like we did in the example shown in Figure 6.6), the R-plane and the initial plane will be at the same Z-axis position. In this case, G98 and G99 will have no effect on the retract position – and you need not include either G-code in the canned cycle command. With either G-code, the tool will retract to the same position in Z.

By the way, G98 is initialized at power up, and it is reinstated whenever the CNC executes a G80, meaning if neither G98 nor G99 is included in the canned-cycle-instating command, G98 will be in effect. In this case the CNC will retract the tool to the (safer) initial plane.

Figure 6.7 shows an example of obstruction clearing. Notice the two clamps between the holes.

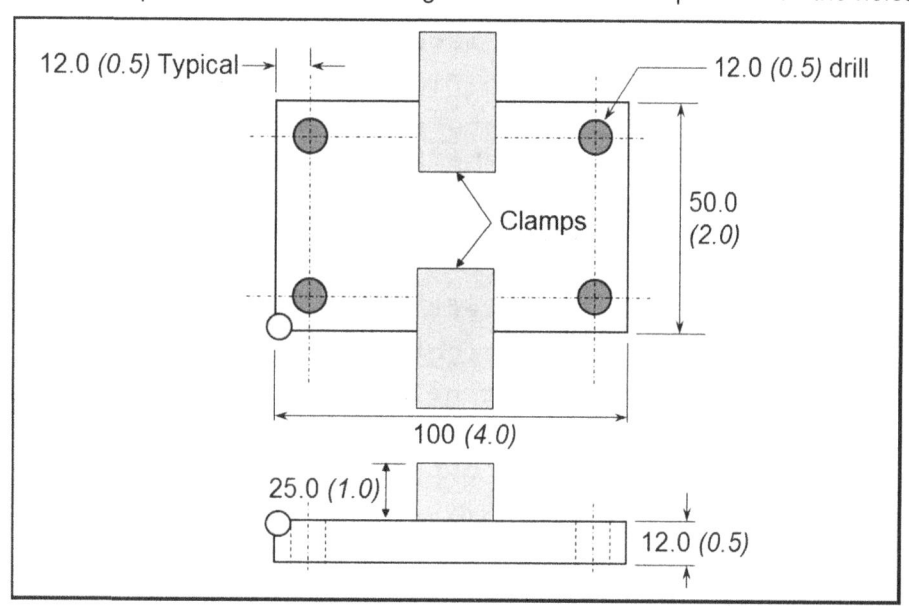

Figure 6.7 – Example showing obstruction clearing between holes

Program with comments:

<u>Metric-mode</u> example

```
%
O0031 (Program number)
(12.0-MM DRILL)
N005 G21 G54 G90 S611 M03 (Select workpiece coordinate system
 setting offset #1, absolute mode, start spindle fwd at 611 rpm)
N010 G00 X12.0 Y12.0 (Rapid to first hole-location)
N015 G43 H01 Z50.0 M08 (Instate tool length compensation,
 position tool to initial plane well above work surface, start
 coolant)
N020 G81 R2.0 Z-17.6 F125.0 G99 (Drill lower-left hole, retract
 to R-plane)
```

Key Concept 6: Features That Help Simplify Programming

```
N025 Y38.0 G98 (Drill upper-left hole, retract to initial plane)
N030 X88.0 G99 (Drill upper-right hole, retract to R-plane)
N035 Y12.0 (Drill lower-right hole, continue retracting to R-plane)
N040 G80 M09 (Cancel canned cycle, turn off coolant)
N045 G91 G28 Z0 (Move to Z-axis zero return position)
N050 G28 Y0 (Move to Y-axis zero return position)
N055 M30 (End of program)
%
```

Inch-mode example

```
%
O0031 (Program number)
(0.5-IN DRILL)
N005 G20 G54 G90 S611 M03 (Select workpiece coordinate system setting offset #1, absolute mode, start spindle fwd at 611 rpm)
N010 G00 X0.5 Y0.5 (Rapid to first hole-location)
N015 G43 H01 Z2.0 M08 (Instate tool length compensation, position tool to initial plane 2.0 inches above work surface, start coolant)
N020 G81 R0.1 Z-0.65 F5.0 G99 (Drill lower-left hole, retract to R-plane)
N025 Y1.5 G98 (Drill upper-left hole, retract to initial plane)
N030 X3.5 G99 (Drill upper-right hole, retract to R-plane)
N035 Y0.5 (Drill lower-right hole, continue retracting to R-plane)
N040 G80 M09 (Cancel canned cycle, turn off coolant)
N045 G91 G28 Z0 (Move to Z-axis zero return position)
N050 G28 Y0 (Move to Y-axis zero return position)
N055 M30 (End of program)
%
```

Notice the two clamps between the holes. In block N015, the tool is brought to the *initial plane, which is 50.0-mm (or 2.0-in) above the work surface.* This Z position is well above the clamps.

In block N020, the first hole is drilled. Since there is no clamp between this hole and the next one, a G99 is placed in this command to allow the tool to retract from this hole to the R-plane – saving a little time. Again, we recommend placing the G98 or G99 at the end of the command to help you remember what will happen *after* the hole is machined.

In block N025, the upper-left hole is machined. But there is a clamp between this hole and the next one, so a G98 is included in this command to cause the tool to retract to the *initial plane*, 50.0-mm (or 2.0-in) above the workpiece. When the XY positioning movement to the next hole occurs in block N030, the tool will be well above the clamp.

In block N030, the upper-right hole is machined, and since there is no clamp before the next hole, a G99 is placed in this command. After the last hole is machined (in block N035) the tool will again retract to the R-plane.

Canned Cycles and the Z-Axis

The initial plane, R-plane, and Z hole-bottom position values are absolute coordinates, taken from the Z-axis workpiece coordinate system zero point. The Z value in the canned cycle should not be thought of to as the hole-depth. The Z-word is the hole-depth only when the hole is being machined into the Z-axis workpiece coordinate system zero surface (which is often the case). It is more correct to say that the Z value of the canned cycle command is the hole-bottom position. There will be times when you must machine holes into Z surfaces other than the workpiece coordinate system zero surface in Z. Figure 6.8 is an example showing a time when you will not be machining into the Z zero surface.

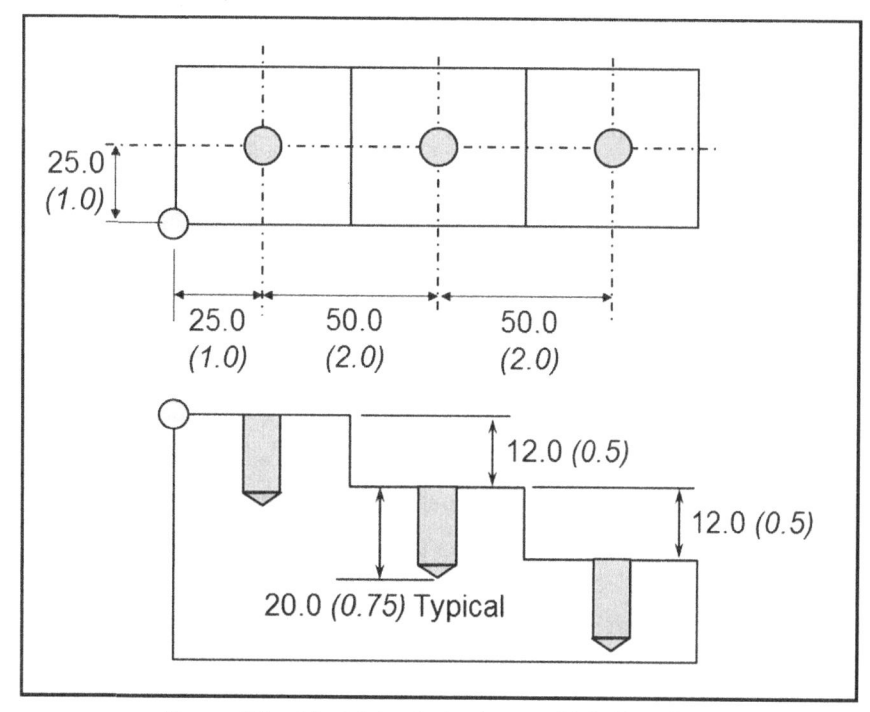

Figure 6.8 – Machining holes into multiple Z surfaces

As you can see, only one of the holes is being machined into the top (Z zero) surface of this workpiece. For the other two holes, you must manipulate the R-plane and Z hole-bottom position. When you have multiple surfaces to machine with a tool, we recommend working from top surface to bottom surface.

Program with comments:

Metric-mode example

```
%
O0032 (Program number)
N010 G54 G90 S600 M03 (12.0-mm drill)
N020 G00 X25.0 Y25.0
N030 G43 H01 Z2.0 M08 (Above highest surface)
N040 G81 R2.0 Z-20.0 F75.0 G99 (Just like shown so far)
N050 X75.0 R-10.0 Z-32.0 (Note new R and Z values)
N060 X125.0 R-22.0 Z-44.0 (Note new R and Z values)
N070 G80 M09 (Cancel cycle)
N080 G91 G28 Z0
N090 G28 Y0
N100 M30
%
```

Key Concept 6: Features That Help Simplify Programming

Inch-mode example

```
%
O0032 (Program number)
N010 G54 G90 S600 M03 (0.5-in drill)
N020 G00 X1.0 Y1.0
N030 G43 H01 Z0.1 M08 (Above highest surface)
N040 G81 R0.1 Z-0.75 F3.0 G99 (Just like shown so far)
N050 X3.0 R-0.4 Z-1.25 (Note new R and Z values)
N060 X5.0 R-0.9 Z-1.75 (Note new R and Z values)
N070 G80 M09 (Cancel cycle)
N080 G91 G28 Z0
N090 G28 Y0
N100 M30
%
```

Notice how easy it is to manipulate Z-axis positions by modifying R and Z values. You simply specify the current R-plane and Z-bottom-position for the holes that require surface changes. However, you must be careful! If you work from bottom to top, you may be in for a nasty surprise. Remember the basic sequence of any canned cycle:

1) A rapid positioning movement to the hole position in XY (if the tool is not already in this position).

2) A rapid positioning movement to the R-plane (referred to as the *rapid plane*). This approach position is just above the surface to machine).

3) Using the canned cycle's machining style, the hole will be machined.

4) Possibly an action will occur at the hole bottom.

5) The cutting tool will retract from the hole.

6) Possibly the tool will retract further in Z.

Consider this ***incorrect program*** shown using metric mode:

```
%
O0033 (Program number)
N010 G21 G54 G90 S600 M03 (12.0-mm drill)
N020 G00 X125.0 Y25.0 (Right-most hole on lowest surface)
N030 G43 H01 Z2.0 (Above highest surface)
N040 M08
N050 G81 R-22.0 Z-44.0 F75.0 G99 (Okay so far)
N060 X75.0 R-10.0 Z-32.0 (This command causes a crash)
N070 X25.0 R2.0 Z-20.0 (This command causes another crash)
N080 G80 M09 (Cancel cycle)
N090 G91 G28 Z0
N100 G28 X0 Y0
N110 M30
%
```

This program causes a crash (actually two crashes). In block N050 the first hole will be machined just fine. The tool will rapid in XY, then it will rapid to the R-plane, plunge the first hole and rapid back out to the R-plane (notice the G99 in N050. But in block N060, the tool will *first* move in X to the new position (crash) *before* moving to the new R-plane.

If you must machine from bottom to top, you could, of course, simply change the G99 in block N050 to a G98. This will cause the tool to come up to the initial plane (0.1 above the upper-most surface) before moving in XY between holes. While this will cause a little wasted (rapid motion) time, the tool will not crash into the workpiece.

Extended Example Showing Canned Cycle Usage

Here is a full example showing the use of several types of canned cycles (G81 – drilling, G84 - right-hand tapping, G82 – counter-boring, G73 – high speed peck drilling, and G76 – fine boring).

Figure 6.9 shows a drawing of the workpiece to be machined. Notice the two clamps that must be avoided. We'll use an initial plane of two inches above the work surface to clear these clamps.

This program is going to be using the fine boring cycle (G76) so no witness mark drag line will be left in the hole. We must know which direction the boring bar tip will be pointing when the spindle is at its orient position. We'll say it is pointing in the X plus direction, meaning an X minus direction move of (or 0.02-mm or 0.001-in) must be specified at the hole-bottom after the spindle orient (this is specified with I-0.02 using metric mode or I-0.001 using inch mode).

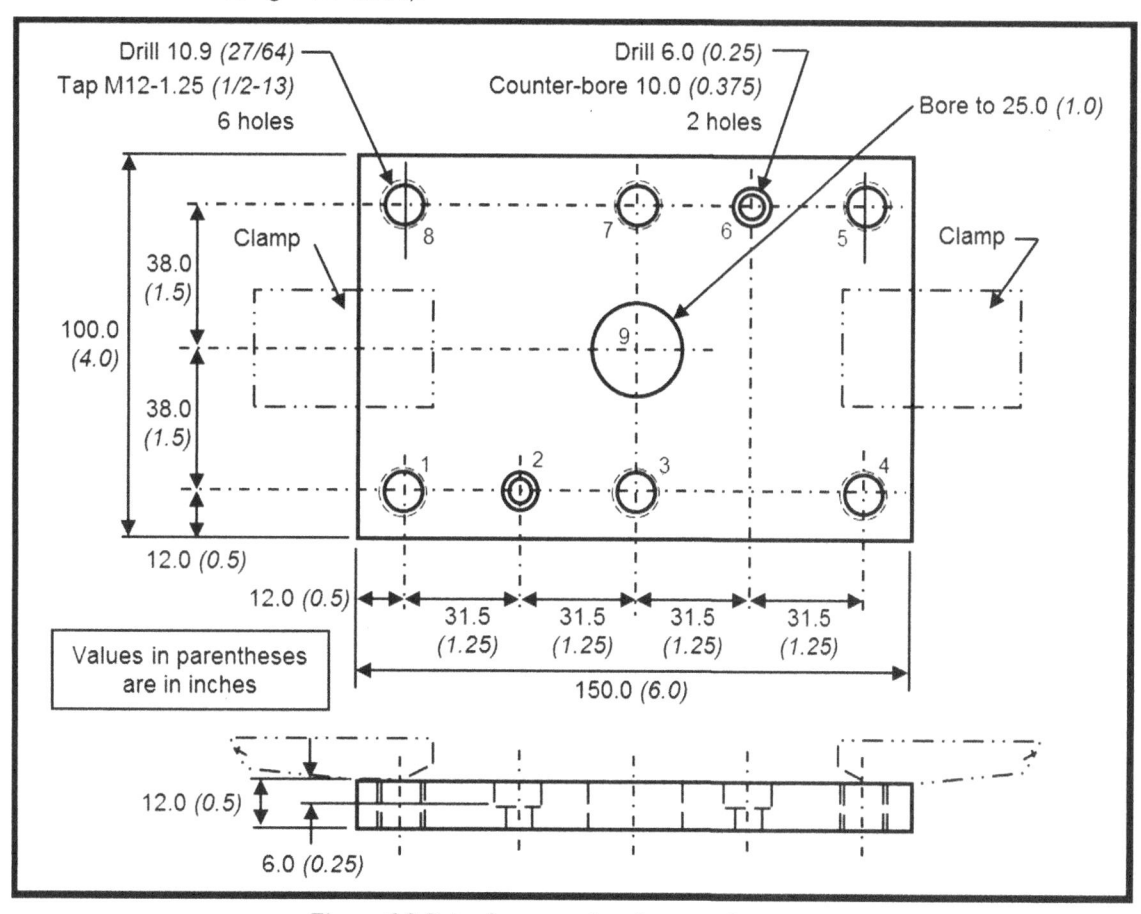

Figure 6.9 Print for canned cycle example program

Metric-mode process to machine the workpiece:

Seq.	Description	Cycle	Tool	Feed	Speed
1	Center drill all holes	G81	#4 center drill	75.0 mmpm	1200 rpm
2	Peck drill 10.9-mm holes (6)	G73	10.9 drill	125.0 mmpm	611 rpm
3	Tap M12-1.25 holes (6)	G84	M12-1.25 tap	287.5 mmpm	230 rpm
4	Drill 25.0 hole to 24.0-mm	G81	24.0 drill	125.0 mmpm	340 rpm
5	Bore 25.0-mm diameter hole	G76	25.0 boring bar	50.0 mmpm	500 rpm
6	Peck drill 6.0-mm holes (2)	G83	6.0 drill	60.0 mmpm	1000 rpm
7	Counter bore 10.0-mm holes	G82	10.0 end mill	75.0 mmpm	600 rpm

Key Concept 6: Features That Help Simplify Programming

Inch-mode process to machine the workpiece:

Seq.	Description	Cycle	Tool	Feed	Speed
1	Center drill all holes	G81	#4 center drill	3.0 ipm	1200 rpm
2	Peck drill 27/64 holes (6)	G73	27/64 drill	5.0 ipm	611 rpm
3	Tap 1/2-13 holes (6)	G84	1/2-13 tap	17.692 ipm	230 rpm
4	Drill 1.000 hole to 31/32	G81	31/32 drill	5.0 ipm	340 rpm
5	Bore 1.000 diameter hole	G76	1.000 boring bar	2.0 ipm	500 rpm
6	Peck drill 0.25 holes (2)	G83	0.25 drill	2.5 ipm	1000 rpm
7	Counter bore 0.375 holes	G82	0.375 end mill	3.0 ipm	600 rpm

Each process specifies the intended G-code type for the canned cycle to be used for the related operation. This program stresses the use of as many different canned cycles as possible. You may not agree with the necessity for some of the canned cycle types chosen.

Metric-mode example program:

```
%
O0034
N001 G17 G21 G94
N002 G40 G64
N003 G69 G80

(CENTER DRILL)
N005 T01 M06
N010 G54 G90 S1200 M03 T02
N015 G00 X12.0 Y12.0
N020 G43 H01 Z50.0 (Note two inch high initial plane)
N025 M08
N030 G81 R2.0 Z-3.0 F75.0 G99 (Center drill hole number one)
N035 X43.5 (#2)
N040 X75.0 (#3)
N045 X138.0 G98 (Note clamp after hole #4!)
N050 Y88.0 G99 (Back to R-plane! hole #5)
N055 X106.5 (#6)
N060 X75.0 (#7)
N065 X12.0 G98 (Come up above clamp just in case #8)
N070 X75.0 Y50.0 (#9)
N075 G80 M09 (Cancel cycle)
N080 G91 G28 Z0 M19
N085 M01

(10.9-MM DRILL)
N090 T02 M06
N095 G54 G90 S611 M03 T03
N100 G00 X12.0 Y12.0
N105 G43 H02 Z50.0 (Note two inch high initial plane)
N110 M08
N115 G73 R2.0 Z-16.27 Q3.0 F125.0 G99 (#1)
```

```
N120 X75.0 (#3)
N125 X138.0 G98 (Clear clamp! #4)
N130 Y88.0 G99 (#5)
N135 X75.0 (#7)
N140 X12.0 (#8)
N145 G80 M09 (Cancel cycle)
N150 G91 G28 Z0 M19
N155 M01

(M12-1.25 TAP - USES RIGID TAPPING)
N160 T03 M06
N165 G54 G90 S230 M03 T04
N170 G00 X12.0 Y12.0
N175 G43 H03 Z50.0 (Note two inch high initial plane)
N180 M08
N183 M29 S230 (Instates rigid tapping)
N185 G84 R2.0 Z-16.0 F287.5 G99 (#1)
N190 X75.0 (#3)
N195 X138.0 G98 (Note clamp! #4)
N200 Y88.0 G99 (#5)
N205 X75.0 (#7)
N210 X12.0 (#8)
N215 G80 M09 (Cancel cycle)
N220 G91 G28 Z0 M19
N225 M01

(24.0-MM DRILL)
N230 T04 M06
N235 G54 G90 S340 M03 T05
N240 G00 X75.0 Y50.0 (#9)
N245 G43 H04 Z2.0 (No need to clear clamps!)
N250 M08
N255 G81 R2.0 Z-20.2 F125.0 (Since R-plane and initial plane are the same, there is no need for either a G98 or G99)
N260 G80 M09 (Cancel cycle)
N265 G91 G28 Z0 M19
N270 M01

(25.0-MM BORING BAR)
N275 T05 M06
N280 G54 G90 S500 M03 T06
N285 G00 X75.0 Y50.0 (#9)
N290 G43 H05 Z2.0 (No need to clear clamps!)
N295 M08
```

Key Concept 6: Features That Help Simplify Programming

```
N300 G76 R2.0 Z-13.0 F50.0 I-0.02 (Since R-plane and initial plane are the same, there is no need for a G98 or G99)
N305 G80 M09 (Cancel cycle)
N310 G91 G28 Z0 M19
N315 M01
N320 T06 M06

(6.0-MM DRILL)
N325 G54 G90 S1000 M03 T07
N330 G00 X43.5 Y12.0 (#2)
N335 G43 H06 Z2.0 (No need to clear clamps)
N340 M08
N345 G83 R2.0 Z-14.8 Q6.0 F60.0 (Since the R-plane and the initial plane are the same, there is no need for G98 or G99)
N350 X106.0 Y88.0 (#6)
N355 G80 M09 (Cancel cycle)
N360 G91 G28 Z0 M19
N365 M01

(10.0-MM END MILL)
N370 T07 M06
N375 G54 G90 S600 M03 T01
N380 G00 X43.5 Y12.0 (#2)
N385 G43 H07 Z2.0 (No need to clear clamps)
N390 M08
N395 G82 R2.0 Z-6.0 F75.0 P500 (No need to clear clamps, so no need to use G98 or G99)
N400 X106.0 Y88.0 (#6)
N405 G80 M09 (Cancel cycle)
N410 G91 G28 Z0 M19
N415 G28 X0 Y0
N420 M01
N425 M30
%
```

Inch-mode example program:
```
%
O0034
N001 G17 G20 G94
N002 G40 G64
N003 G69 G80

(CENTER DRILL)
N005 T01 M06
N010 G54 G90 S1200 M03 T02
N015 G00 X0.5 Y0.5
```

```
N020 G43 H01 Z2.0 (Note two inch high initial plane)
N025 M08
N030 G81 R0.1 Z-0.12 F3.0 G99 (Center drill hole number one)
N035 X1.75 (#2)
N040 X3.0 (#3)
N045 X5.5 G98 (Note clamp after hole #4!)
N050 Y3.5 G99 (Back to R-plane! hole #5)
N055 X4.25 (#6)
N060 X3.0 (#7)
N065 X0.5 G98 (Come up above clamp just in case #8)
N070 X3.0 Y2.0 (#9)
N075 G80 M09 (Cancel cycle)
N080 G91 G28 Z0 M19
N085 M01

(27/64 DRILL)
N090 T02 M06
N095 G54 G90 S611 M03 T03
N100 G00 X0.5 Y0.5
N105 G43 H02 Z2.0 (Note two inch high initial plane)
N110 M08
N115 G73 R0.1 Z-0.75 Q0.1 F5.0 G99 (#1)
N120 X3.0 (#3)
N125 X5.5 G98 (Clear clamp! #4)
N130 Y3.5 G99 (#5)
N135 X3.0 (#7)
N140 X0.5 (#8)
N145 G80 M09 (Cancel cycle)
N150 G91 G28 Z0 M19
N155 M01

(1/2-13 TAP - USES RIGID TAPPING)
N160 T03 M06
N165 G54 G90 S230 M03 T04
N170 G00 X0.5 Y0.5
N175 G43 H03 Z2.0 (Note two inch high initial plane)
N180 M08
N183 M29 S230 (Instates rigid tapping)
N185 G84 R0.1 Z-0.65 F17.692 G99 (#1)
N190 X3.0 (#3)
N195 X5.5 G98 (Note clamp! #4)
N200 Y3.5 G99 (#5)
N205 X3.0 (#7)
```

Key Concept 6: Features That Help Simplify Programming

```
N210 X0.5 (#8)
N215 G80 M09 (Cancel cycle)
N220 G91 G28 Z0 M19
N225 M01

(31/32 DRILL)
N230 T04 M06
N235 G54 G90 S340 M03 T05
N240 G00 X3.0 Y2.0 (#9)
N245 G43 H04 Z0.1 (No need to clear clamps!)
N250 M08
N255 G81 R0.1 Z-0.85 F5.0 (Since R-plane and initial plane are the same, there is no need for either a G98 or G99)
N260 G80 M09 (Cancel cycle)
N265 G91 G28 Z0 M19
N270 M01

(1.0000 BORING BAR)
N275 T05 M06
N280 G54 G90 S500 M03 T06
N285 G00 X3.0 Y2.0 (#9)
N290 G43 H05 Z0.1 (No need to clear clamps!)
N295 M08
N300 G76 R0.1 Z-0.6 F2.0 I-0.001 (Since R-plane and initial plane are the same, there is no need for a G98 or G99)
N305 G80 M09 (Cancel cycle)
N310 G91 G28 Z0 M19
N315 M01
N320 T06 M06

(0.25-IN DRILL)
N325 G54 G90 S1000 M03 T07
N330 G00 X1.75 Y0.5 (#2)
N335 G43 H06 Z0.1 (No need to clear clamps)
N340 M08
N345 G83 R0.1 Z-0.65 Q0.4 F2.5 (Since the R-plane and the initial plane are the same, there is no need for G98 or G99)
N350 X4.25 Y3.5 (#6)
N355 G80 M09 (Cancel cycle)
N360 G91 G28 Z0 M19
N365 M01

(0.375-IN END MILL)
N370 T07 M06
N375 G54 G90 S600 M03 T01
```

Lesson 6.1: Canned Cycles For Drilling

```
N380 G00 X1.75 Y0.5 (#2)
N385 G43 H07 Z0.1 (No need to clear clamps)
N390 M08
N395 G82 R0.1 Z-0.25 F3.0 P500 (No need to clear clamps, so no need to use G98 or G99)
N400 X4.25 Y3.5 (#6)
N405 G80 M09 (Cancel cycle)
N410 G91 G28 Z0 M19
N415 G28 X0 Y0
N420 M01
N425 M30
%
```

Notes About the Program:

N020, N105, N175
The first three tools need to clear the clamps. Notice that the G43 command brings the tool up to the initial plane in Z (50.0-mm or 2.0-in above the work surface). This is the tool's last position just prior to the canned cycle command.

N030, N115, N185, N255, N300, N345, N395
Notice that the tool is already at the first hole-position when the canned cycle is instated (commanded in each tool's first XY motion). While the X and Y coordinates can be repeated in the canned cycle instating command, there is no need for them – and we recommend leaving them out.

N245, N290, N335
Since only one hole is being machined, there will be no need to clear obstructions, meaning the initial plane can be set the same as the R-plane.

N255, N300, N345
Since the R-plane and initial plane are the same for these tools, you can leave out the G98/G99 completely. Though the initial plane (G98) will be in effect (it is reinstated at each G80), the tool will still come out of the hole to a Z position of Z0.1.

N045, N050, N125, N130, N195, N200
Again, G98 and G99 are modal. They are only required when you want to switch retract planes. Since they control what happens after the hole is machined, it helps to place them at the end of the command.

N075, N145, N215, N260, N305, N355, N405
G80 is used to cancel each canned cycle.

N255
You may be questioning why a standard drilling cycle (G81) is used when only one hole is being machined. You may feel that it is just as easy to program one hole with G00 and G01. The need may arise during the program's verification for another machining method, like peck drilling. We recommend using canned cycles to machine all holes, even when there is only one hole to machine.

N300
We are using the G76 fine boring cycle. We have said we know that the boring bar will be pointing in the X plus direction when the spindle is at its orient position, so we can include the I-0.02 word (for the metric-mode example) or I-0.001 word (for the inch-mode example) in the G76 command to make the boring bar move over in the X minus direction at the hole-bottom. If you don't know which way the boring bar will be pointing at the spindle orient position (maybe you are using a collet holder and a straight-shank boring bar), we recommend including I0 and J0 in the G76 command. Once the setup is made and the setup person knows which way the boring bar tip is pointing at the spindle orient position (by testing it), they must modify the G76 command.

N183

Key Concept 6: Features That Help Simplify Programming

This program is for a machining center that has rigid tapping. M29 instates the rigid tapping mode and includes the speed for tapping. This command must be just prior to the tapping command.

N185
Again, we're using the rigid tapping feature, so the rapid plane is kept at 0.1 above the work surface. If you are using a tension/compression tap holder (your machine does not have rigid tapping), you must keep the rapid plane for tapping well above the work surface (6.0-mm or 0.25-in is sufficient). When the tap enters the hole, and especially during the spindle reversal at the hole-bottom, the tap will probably extend in the tension/compression holder. If it does, the tap may still be in the hole at the completion of the tapping cycle if a smaller R-plane value is used.

Using Canned Cycles In the Incremental Positioning Mode

You now know the function of canned cycles as they are used in the absolute positioning mode (G90). While we have been stressing the absolute positioning mode throughout this course, there is a benefit to working in the incremental mode (G91) with canned cycles when machining evenly spaced holes. All of the evenly spaced holes along a line can be machined in one command. By comparison, the absolute positioning mode allows but one hole per command to be machined. Consider the grid pattern of holes shown in figure 6.10. In the absolute mode, these one hundred holes require one hundred commands. In the incremental mode, this is reduced to about twenty commands.

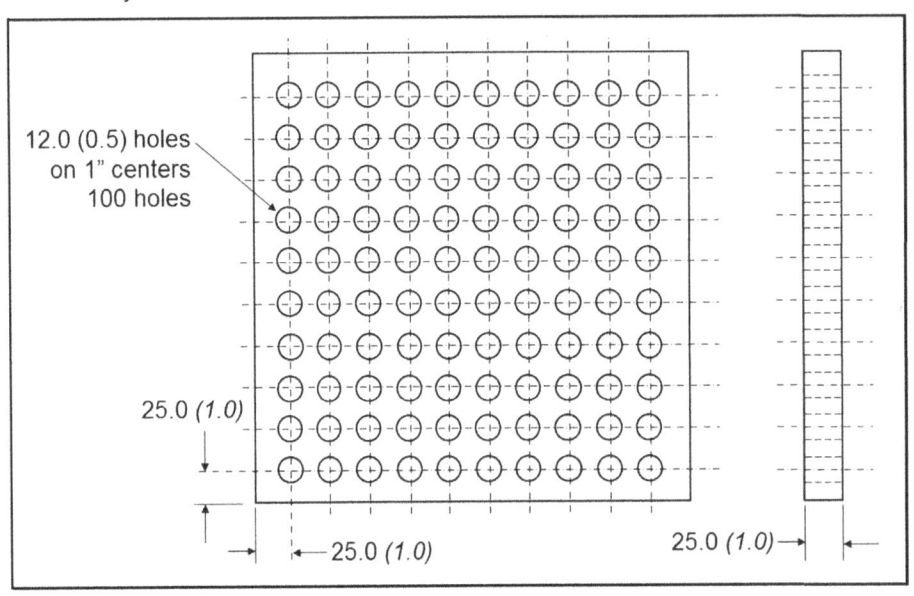

Figure 6.10 – An application for the incremental mode

A K-word in the canned cycle command is used to specify the number of evenly spaced holes to be machined. If the K-word is left out of the command (as is the case when you work in the absolute mode), the control will assume that one hole is to be machined.

> **NOTE:** With older FANUC controls, an L-word is used instead of K.

In order to use canned cycles in the incremental mode, you must understand the meanings of canned cycle-related words in the incremental mode. Here are the canned cycle words that change in the incremental mode:

    X is now the distance from the current X-axis position to the center of the hole in X.

    Y is now the distance from the current Y-axis position to the center of the hole in Y.

    R is now the distance from the current Z-axis position of the tool to the R-plane position.

    Z is a little different. Z is the distance from the R-plane to the bottom of the hole. If the R-plane is 0.100 in above the workpiece, the Z value will be the depth of the hole plus 0.100 in. With incrementally commanded canned cycles, *Z is always minus*.

Lesson 6.1: Canned Cycles For Drilling

The rest of the canned cycle words (G98, G99, I, J, F, P, and Q) mean exactly the same thing as they do in the absolute mode.

Here is an example for the workpiece shown in Figure 6.10. You should be impressed with how few commands this grid pattern requires.

Metric-mode example program:

```
%
O0035
N001 G17 G21 G94
N002 G40 G64
N003 G69 G80
N005 G54 G90 S600 M03 (12.0-mm drill)
N010 G00 X25.0 Y25.0 (Position of the lower-left hole)
N015 G43 H01 Z50.0 (Sets initial plane)
N020 M08
N025 G91 G81 R-48.0 Z-29.6 F125.0 G99 (Machine lower left hole in incremental mode- note R and Z values)
N030 X25.0 K9 (Machine first row of holes)
N035 Y25.0 (Machines first hole in second row)
N040 X-25.0 K9 (Machines nine holes in second row)
N045 Y25.0 (Machine first hole in third row)
N050 X25.0 K9 (Machine third row)
N055 Y25.0
N060 X-25.0 K9 (Fourth row)
N065 Y25.0
N070 X25.0 K9 (Fifth row)
N075 Y25.0
N080 X-25.0 K9 (Sixth row)
N085 Y25.0
N090 X25.0 K9 (Seventh row)
N095 Y25.0
N100 X-25.0 K9 (Eighth row)
N105 Y25.0
N110 X25.0 K9 (Ninth row)
N115 Y25.0
N120 X-25.0 K9 (Tenth row)
N125 G80 M09
N130 G91 G28 X0 Y0 Z0 M19
N135 M30
%
```

Inch-mode example program:

```
%
O0035
N001 G17 G20 G94
```

Key Concept 6: Features That Help Simplify Programming

```
N002 G40 G64
N003 G69 G80
N005 G54 G90 S600 M03 (0.5-in drill)
N010 G00 X1.0 Y1.0 (Position of the lower-left hole)
N015 G43 H01 Z2.0 (Sets initial plane)
N020 M08
N025 G91 G81 R-1.9 Z-1.18 F5.0 G99 (Machine lower left hole in
incremental mode- note R and Z values)
N030 X1.0 K9 (Machine first row of holes)
N035 Y1.0 (Machines first hole in second row)
N040 X-1.0 K9 (Machines nine holes in second row)
N045 Y1.0 (Machine first hole in third row)
N050 X1.0 K9 (Machine third row)
N055 Y1.0
N060 X-1.0 K9 (Fourth row)
N065 Y1.0
N070 X1.0 K9 (Fifth row)
N075 Y1.0
N080 X-1.0 K9 (Sixth row)
N085 Y1.0
N090 X1.0 K9 (Seventh row)
N095 Y1.0
N100 X-1.0 K9 (Eighth row)
N105 Y1.0
N110 X1.0 K9 (Ninth row)
N115 Y1.0
N120 X-1.0 K9 (Tenth row)
N125 G80 M09
N130 G91 G28 X0 Y0 Z0 M19
N135 M30
%
```

This technique dramatically reduces the number of commands needed to machine the holes (again, from one hundred to about twenty). In block N010, we position the drill to the first hole to machine (the lower left hole). In block N025, the first hole is machined – notice the incremental specifications for R and Z. And as with the absolute mode, if the drill is already at the hole-position in XY, you need not include the XY specifications in this command (if you do, they must be X0 and Y0). Block N030 machines the rest of the holes in the first row (notice the L9). Block N035 machines the first hole in the second row. Block N040 machines the rest of the holes in the second row. This is repeated for the balance of the holes in the pattern.

Key Points for Lesson 6.1:
- ✓ Hole-machining canned cycles dramatically simplify the programming of hole-machining operations.
- ✓ With canned cycles, you instate the cycle in a command that machines the first hole. You then list the coordinates for the rest of the holes to be machined. Finally, you cancel the cycle.
- ✓ There are a variety of hole-machining styles that are selected by G-codes.

Lesson 6.1: Canned Cycles For Drilling

✓ Canned cycles provide a way to clear obstructions between holes using two planes – the R-plane (specified by G99) and the initial plane (specified by G98).
✓ The initial plane is the tool's last Z position prior to the canned cycle command.
✓ The R and Z-words are absolute positions that can be changed for different Z surfaces.
✓ When you must machine a number of evenly spaced holes, you can use the incremental mode to minimize the number of required commands.

Exercise: Practice Programming Canned Cycles

Study the workpiece drawing and process. Fill in the coordinates and, using the appropriate process, write a program (on a separate sheet of paper) that will machine the workpiece. Use the measurement system (metric mode or inch mode) used by your company or school. To mill the 10.0-mm (0.375-in) radius slots: Plunge the 20.0-mm (0.75-in) end mill at one end, mill to the other, and then retract.

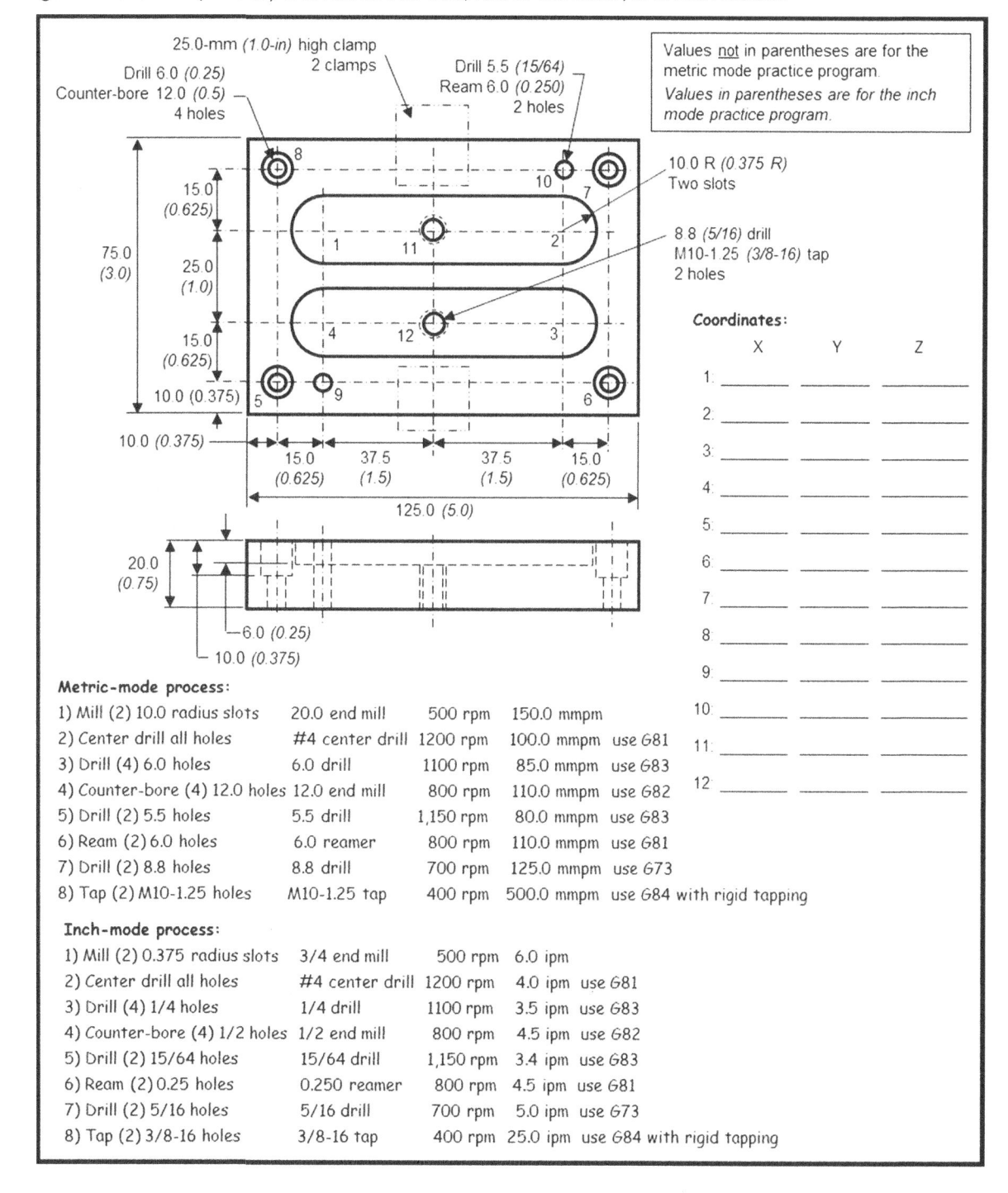

Metric-mode process:
1) Mill (2) 10.0 radius slots   20.0 end mill   500 rpm   150.0 mmpm
2) Center drill all holes   #4 center drill   1200 rpm   100.0 mmpm   use G81
3) Drill (4) 6.0 holes   6.0 drill   1100 rpm   85.0 mmpm   use G83
4) Counter-bore (4) 12.0 holes   12.0 end mill   800 rpm   110.0 mmpm   use G82
5) Drill (2) 5.5 holes   5.5 drill   1,150 rpm   80.0 mmpm   use G83
6) Ream (2) 6.0 holes   6.0 reamer   800 rpm   110.0 mmpm   use G81
7) Drill (2) 8.8 holes   8.8 drill   700 rpm   125.0 mmpm   use G73
8) Tap (2) M10-1.25 holes   M10-1.25 tap   400 rpm   500.0 mmpm   use G84 with rigid tapping

Inch-mode process:
1) Mill (2) 0.375 radius slots   3/4 end mill   500 rpm   6.0 ipm
2) Center drill all holes   #4 center drill   1200 rpm   4.0 ipm   use G81
3) Drill (4) 1/4 holes   1/4 drill   1100 rpm   3.5 ipm   use G83
4) Counter-bore (4) 1/2 holes   1/2 end mill   800 rpm   4.5 ipm   use G82
5) Drill (2) 15/64 holes   15/64 drill   1,150 rpm   3.4 ipm   use G83
6) Ream (2) 0.25 holes   0.250 reamer   800 rpm   4.5 ipm   use G81
7) Drill (2) 5/16 holes   5/16 drill   700 rpm   5.0 ipm   use G73
8) Tap (2) 3/8-16 holes   3/8-16 tap   400 rpm   25.0 ipm   use G84 with rigid tapping

Coordinates:

#	X	Y	Z
1			
2			
3			
4			
5			
6			
7			
8			
9			
10			
11			
12			

Key Concept 6: Features That Help Simplify Programming

Metric-mode answer program for canned cycles:

```
%
O0036 (Canned cycles practice)
N001 G17 G21 G23
N002 G40 G64
N003 G69 G80

(20.0-MM END MILL)
N005 T01 M06
N010 G54 G90 S500 M03 T02
N015 G00 X25.0 Y50.0 (pt 1)
N020 G43 H01 Z2.0
N025 M08
N030 G01 Z-6.0 F150.0
N035 X100.0 (pt 2)
N040 G00 Z2.0
N045 Y25.0 (pt 3)
N050 G01 Z-6.0
N055 X25.0 (pt 4)
N050 G00 Z2.0 M09
N055 G91 G28 Z0 M19
N060 M01

(CENTER DRILL)
N065 T02 M06
N070 G54 G90 S1200 M03 T03
N075 G00 X10.0 Y10.0 (pt 5)
N080 G43 H02 Z50.0
N085 M08
N090 G81 R2.0 Z-3.0 F100.0 G99
N095 X25.0 G98 (pt 9)
N100 X115.0 G99 (pt 6)
N105 Y65.0 (pt 7)
N110 X100.0 G98 (pt 10)
N115 X10.0 G99 (pt 8)
N120 X62.5 Y50.0 R-4.0 Z-9.0 G98 (11)
N125 Y25.0
N130 G80 M09
N135 G91 G28 Z0 M19
N140 M01

(6.0-MM DRILL)
N145 T03 M06
N150 G54 G90 S1100 M03 T04
N155 G00 X10.0 Y10.0 (pt 5)
N155 G43 H03 Z50.0
N160 M08
N165 G83 R2.0 Z-22.8 Q12.0 F85.0 G98
N170 X115.0 G99 (pt 6)
N175 Y65.0 G98 (pt 7)
N180 X10.0
N185 G80 M09
N190 G91 G28 Z0 M19
N195 M01

(12.0-MM END MILL)
N200 T04 M06
N005 G54 G90 S800 M03 T05
N210 G00 X10.0 Y10.0 (pt 5)
N215 G43 H04 Z50.0
N220 M08
N225 G82 R2.0 Z-6.0 P500 F110.0 G98
N230 X115.0 G99 (pt 6)
N235 Y65.0 G98 (pt 7)
N240 X10.0 (pt 8)
N245 G80 M09
N250 G91 G28 Z0 M19
N255 M01

(5.50-MM DRILL)
N260 T05 M06
N265 G54 G90 S1150 M03 T06
N270 G00 X25.0 Y10.0 (pt 9)
N275 G43 H05 Z50.0
N280 M08
N285 G83 R2.0 Z-22.65 Q12.0 F80. G98
N290 X100.0 Y65.0 (10)
N295 G80 M09
N300 G91 G28 Z0 M19
N305 M01

(0.6-MM REAMER)
N310 T06 M06
N315 G54 G90 S800 M03 T07
N320 G00 X25.0 Y10.0 (pt 9)
N325 G43 H06 Z50.0
N330 M08
N335 G81 R2.0 Z-21.0 F110.0 G98
N340 X100.0 Y65.0 (pt 10)
N345 G80 M09
N350 G91 G28 Z0 M19
N355 M01

(8.8-MM DRILL)
N360 T07 M06
N365 G54 G90 S700 M03 T08
N370 G00 X62.5 Y25.0 (pt 12)
N375 G43 H01 Z2.0
N380 M08
N385 G81 R-4.0 Z-23.64 F125.0 G98
N390 Y50.0 (pt 11)
N395 G80 M09
N400 M01

(M10-1.25 TAP)
N405 T08 M06
N410 G54 G90 S400 M03 T01
N415 G00 X2.5 Y1.0 (pt 12)
N420 G43 H08 Z50.0
N425 M08
N428 M29 S400
N430 G84 R-4.0 Z-24.0 F500.0 G98
N435 Y50.0 (pt 11)
N440 G80 M09
N445 G91 G28 Z0 M19
N450 G28 Y0
N455 M30
%
```

Inch-mode answer program for canned cycles:

```
%
O0036 (Canned cycles practice)
N001 G17 G20 G23
N002 G40 G64
N003 G69 G80

(0.75-IN END MILL)
N005 T01 M06
N010 G54 G90 S500 M03 T02
N015 G00 X1.0 Y2.0 (pt 1)
N020 G43 H01 Z0.1
N025 M08
N030 G01 Z-0.25 F6.0
N035 X4.0 (pt 2)
N040 G00 Z0.1
N045 Y1.0 (pt 3)
N050 G01 Z-0.25
N055 X1.0 (pt 4)
N050 G00 Z0.1 M09
N055 G91 G28 Z0 M19
N060 M01

(CENTER DRILL)
N065 T02 M06
N070 G54 G90 S1200 M03 T03
N075 G00 X0.375 Y0.375 (pt 5)
N080 G43 H02 Z2.0
N085 M08
N090 G81 R0.1 Z-0.12 F4.0 G99
N095 X1.0 G98 (pt 9)
N100 X4.625 G99 (pt 6)
N105 Y2.625 (pt 7)
N110 X4.0 G98 (pt 10)
N115 X0.375 G99 (pt 8)
N120 X2.5 Y2.0 R-0.15 Z-0.37 G98 (11)
N125 Y1.0
N130 G80 M09
N135 G91 G28 Z0 M19
N140 M01

(0.25-IN DRILL)
N145 T03 M06
N150 G54 G90 S1100 M03 T04
N155 G00 X0.375 Y0.375 (pt 5)
N155 G43 H03 Z2.0
N160 M08
N165 G83 R0.1 Z-0.855 Q0.5 F3.5 G98
N170 X4.625 G99 (pt 6)
N175 Y2.625 G98 (pt 7)
N180 X0.375
N185 G80 M09
N190 G91 G28 Z0 M19
N195 M01

(0.5-IN END MILL)
N200 T04 M06
N005 G54 G90 S800 M03 T05
N210 G00 X0.375 Y0.375 (pt 5)
N215 G43 H04 Z2.0
N220 M08
N225 G82 R0.1 Z-0.375 P500 F4.5 G98
N230 X4.625 G99 (pt 6)
N235 Y2.625 G98 (pt 7)
N240 X0.375 (pt 8)
N245 G80 M09
N250 G91 G28 Z0 M19
N255 M01

(15/64-IN DRILL)
N260 T05 M06
N265 G54 G90 S1150 M03 T06
N270 G00 X1.0 Y0.375 (pt 9)
N275 G43 H05 Z2.0
N280 M08
N285 G83 R0.1 Z-0.855 Q0.5 F3.4 G98
N290 X4.0 Y2.625 (10)
N295 G80 M09
N300 G91 G28 Z0 M19
N305 M01

(0.250-IN REAMER)
N310 T06 M06
N315 G54 G90 S800 M03 T07
N320 G00 X1.0 Y0.375 (pt 9)
N325 G43 H06 Z2.0
N330 M08
N335 G81 R0.1 Z-0.8 F4.5 G98
N340 X4.0 Y2.625 (pt 10)
N345 G80 M09
N350 G91 G28 Z0 M19
N355 M01

(5/16-IN DRILL)
N360 T07 M06
N365 G54 G90 S700 M03 T08
N370 G00 X2.5 Y1.0 (pt 12)
N375 G43 H01 Z0.1
N380 M08
N385 G81 R-0.15 Z-0.87 F5.0 G98
N390 Y2.0 (pt 11)
N395 G80 M09
N400 M01

(3/8-16 TAP)
N405 T08 M06
N410 G54 G90 S400 M03 T01
N415 G00 X2.5 Y1.0 (pt 12)
N420 G43 H08 Z2.0
N425 M08
N428 M29 S400
N430 G84 R-0.15 Z-1.0 F25.0 G98
N435 Y2.0 (pt 11)
N440 G80 M09
N445 G91 G28 Z0 M19
N450 G28 Y0
N455 M30
%
```

Lesson 6.2–Subprogram Techniques

There are times when a series of CNC commands must be repeated–within one program–and sometimes among several programs. Whenever you find yourself writing a series of commands a second time–you should consider using a subprogram. The longer the series of commands and the more often they must be repeated, the more a subprogram can help.

Objectives

After completing this lesson, students should be able to:

- ✓ Describe the benefits of using subprograms
- ✓ Describe the common applications for subprograms
- ✓ Describe the benefits of Custom Macro programming

Introduction

You know that a CNC machining center will execute a program in sequential order. It will start with the first command in the program: read it–interpret it–and execute it. Then it will move on to the next command–read, interpret, execute. It will continue this process for the entire program.

You also know that there are times when a series of commands in your program must be repeated. We've shown two examples so far.

- When using cutter radius compensation, the rough and finish mill can use the same set of coordinates. The example program (shown in lesson 4.3) includes the tool path coordinates twice, once for the rough milling cutter (using an offset value that is larger than the actual cutter size) and once for the finish milling cutter.

- When machining holes that require multiple machining operations (like center drilling, drilling, and tapping a series of holes)–the more holes that must be machined, the more blocks must be repeated.

In this lesson, you will learn about a way to change the order of program execution to some extent–and this will be especially helpful when commands must be repeated.

The Difference Between a Main Program and a Subprogram

A main program is the program that a setup person or operator will select and execute when they activate a cycle. Every program shown to this point in the course has been a main program. Main programs almost always end with M30 (or M02 with some machines).

A subprogram is a program that is called by a main program or by another subprogram. A subprogram always ends with M99.

In a main program, when you have a series of commands that you know will be repeated, you can call a subprogram that contains the repeated commands (you place the repeated commands in the subprogram instead of in the main program). Each time the commands must be repeated, you will call the same subprogram.

M98 is the word used to call a subprogram. A P-word in the M98 block specifies the program number of the subprogram to be executed. When the machine is finished executing the subprogram, it will come back to the main program to the block that follows the M98 command.

The figure below shows a simple application for a subprogram using metric mode.

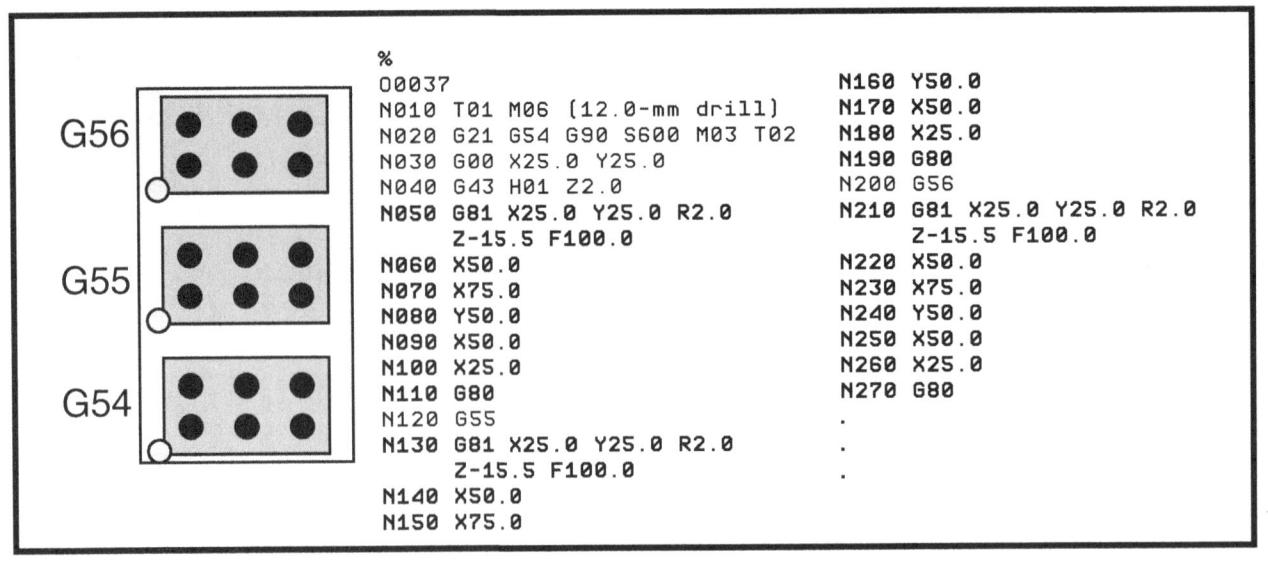

Key Concept 6: Features That Help Simplify Programming

```
%
O0037
N010 T01 M06 (12.0-mm drill)
N020 G21 G54 G90 S600 M03 T02
N030 G00 X25.0 Y25.0
N040 G43 H01 Z2.0
N050 G81 X25.0 Y25.0 R2.0
     Z-15.5 F100.0
N060 X50.0
N070 X75.0
N080 Y50.0
N090 X50.0
N100 X25.0
N110 G80
N120 G55
N130 G81 X25.0 Y25.0 R2.0
     Z-15.5 F100.0
N140 X50.0
N150 X75.0
N160 Y50.0
N170 X50.0
N180 X25.0
N190 G80
N200 G56
N210 G81 X25.0 Y25.0 R2.0
     Z-15.5 F100.0
N220 X50.0
N230 X75.0
N240 Y50.0
N250 X50.0
N260 X25.0
N270 G80
```

Figure 6.11–Application for subprogram usage (shown using metric mode)

In the example, three identical workpieces are being machined. The lower workpiece uses workpiece coordinate system offset #1 (G54), the middle workpiece uses workpiece coordinate system offset #2 (G55), and the upper workpiece uses offset #3 (G56). In the program, after selecting the workpiece coordinate system offset (in blocks N020, N120, and N200), the commands for the machining operation are specified (drilling six holes in this example). Notice that these are identical commands for each workpiece. So the words in blocks N050 through N110 are specified three times–once for each workpiece.

Now look at the figure below.

Main program:
```
%
O0038
N010 T01 M06
N020 G21 G54 G90
 S600 M03 T02
N030 G00 X25.0 Y25.0
N040 G43 H01 Z2.0
N050 M98 P1000
N060 G55
N070 M98 P1000
N080 G56
N090 M98 P1000
N100 G91 G28 Z0 M19
N110 M01
N120 T02 M06
.
.
.
```

Subprogram:
```
%
O1000
N010 G81 X25.0 Y25.0 R2.0
     Z-15.5 F100.0
N020 X50.0
N030 X75.0
N040 Y50.0
N050 X50.0
N060 X25.0
N070 G80
N080 M99
%
```

The subprogram is being executed three times (from lines N050, N070, and N090). After the subprogram is executed, the machine continues executing from the command after the calling M98 block.

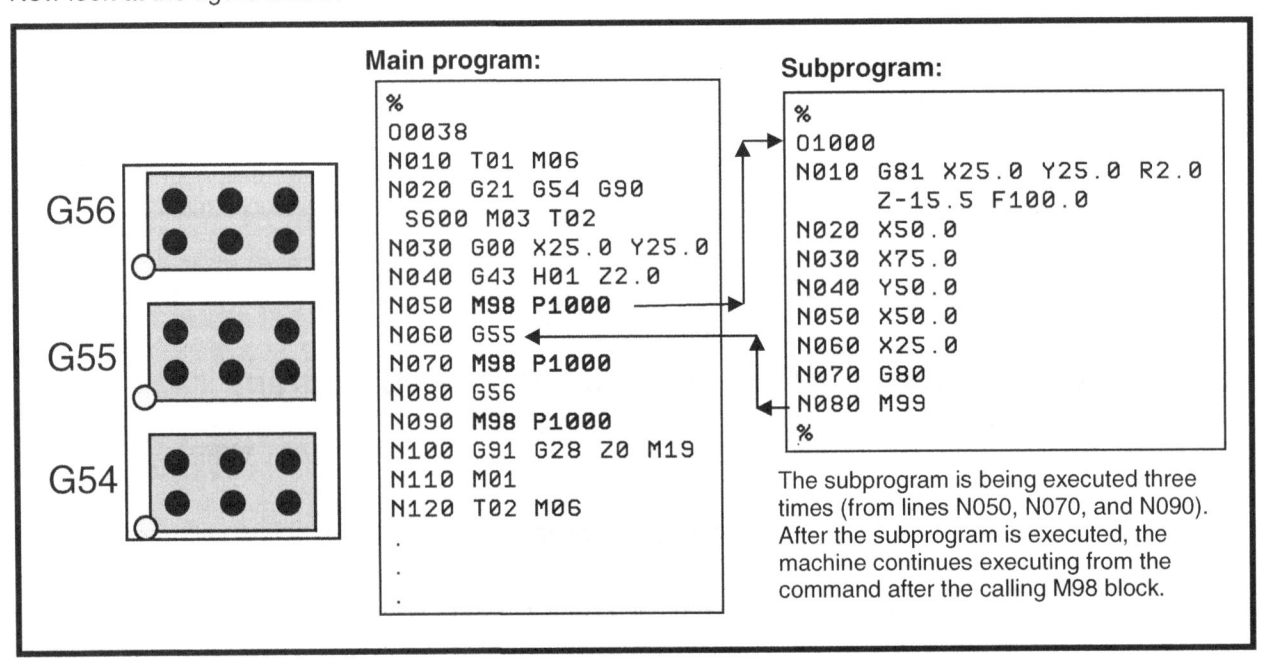

Figure 6.12–Now a subprogram is being used (shown using metric mode)

This time we have placed all of the repeated blocks in a subprogram. When these blocks must be executed, an M98 command is programed with a P-word specifying the program number for the subprogram. While this may not be the most demanding application for a subprogram (there aren't many commands being repeated), it illustrates how subprograms are used.

In block N050, the machine will execute the subprogram call for the first time. The holes in the lower workpiece will be drilled. When the machine is finished executing the subprogram (notice the M99 in block N080 of the subprogram), it returns to the main program–to the block that follows the calling M98 (block N060). Block N070 selects workpiece coordinate system offset #2 (G56). Block N070 calls the same subprogram again, causing the holes in the middle workpiece to be drilled. When finished, the machine will return (this time) to block N080. Workpiece coordinate system offset #3 (G56) is selected in block N090 and the subprogram is executed a third time—machining the holes in the upper workpiece.

A note about the P-word: Look at the P-word in blocks N050, N070, and N090 in the figure above. It is specified as P1000 (P one, zero, zero, zero) to call the program number O1000. A common mistake is to incorrectly specify the P-word as PO1000 (P oh, one, zero, zero, zero). This mistake will generate an alarm. The O letter address is the designator to specify a programmer number at the top of a program and 1000 is the actual program number.

Subprograms provide two important benefits—one rather obvious one and the other not so obvious.

Minimize number of blocks—subprograms minimize the number of blocks that must be specified in order to machine a workpiece. So programming becomes simpler and programs become shorter.

Simplifies verification and editing—If the subprogram is correct the first time it is executed, it will be correct every time it is executed. Consider the program in figure 6.11. The blocks to machine the six holes are specified three different times. A mistake could be made in any of these program sections—either writing or typing the blocks. When verifying or editing the program, you must be very careful with each of them—during the machining of all three workpieces. While you must still be careful when verifying and editing subprograms, if the holes are machined properly in the first workpiece, they will also be machined properly in the second and third workpieces—the same commands are used all three times.

Loading the Main Program and Subprograms

When you use subprograms, more than one program will be involved with a given job. It is not uncommon for a single job to require several programs. All programs related to the job must be loaded into the machine's memory before the job can be run. Your setup documentation must specify all program numbers related to the job—as well as where they can be found—so the setup person can quickly and easily load programs.

There is a way to load all programs that are related to a job from a single computer file. For FANUC controls, you specify a percent sign (%) before the first program and after the last program to be loaded. Several programs can be included between the percent signs. This shortens the time it takes to load programs, but it will not help when saving programs. With most FANUC controls, you can either save one program or all the programs in the CNC.

Be aware that using subprograms may require more time and effort from the setup person. Any programming time gained by using subprograms may be lost when setups are made. And of course, setup time (machine time) is much more valuable than programming time—especially if the machine is running production while programs are written.

Words Used with Subprograms

Let's summarize the words used with subprograms.

M98—used to call a subprogram from a main program—or another subprogram

M99—specifies the end a subprogram—the CNC will return to the calling program to the block that follows the calling M98

P-word— specifies the subprogram to be called and the number of times it will be executed

NOTE: With older FANUC CNCs, The P-word only specifies the sub-program number. An L-word is used to specify the number of executions.

We have introduced all these words except for the extended use of the P-word. The P-word specifies the program number and how many times the subprogram will be executed before returning to the main program. The first three digits of the P-word, if included, specify the number of executions. The last four digits specify the program number. Consider these commands:

```
N050 M98 P0051000 (Execute subprogram 1000 5 times)
N050 M98 P0101000 (Execute subprogram 1000 10 times)
```

Key Concept 6: Features That Help Simplify Programming

```
N050 M98 P1000 (Execute subprogram 1000 just once)
```

Nesting Subprograms

You can call a subprogram from a main program or from another subprogram. When you call a subprogram from a subprogram, it is called nesting subprograms. There is a limitation to how deep you can nest subprograms. With most CNCs, you can only nest four deep, as the Figure 6.13 shows.

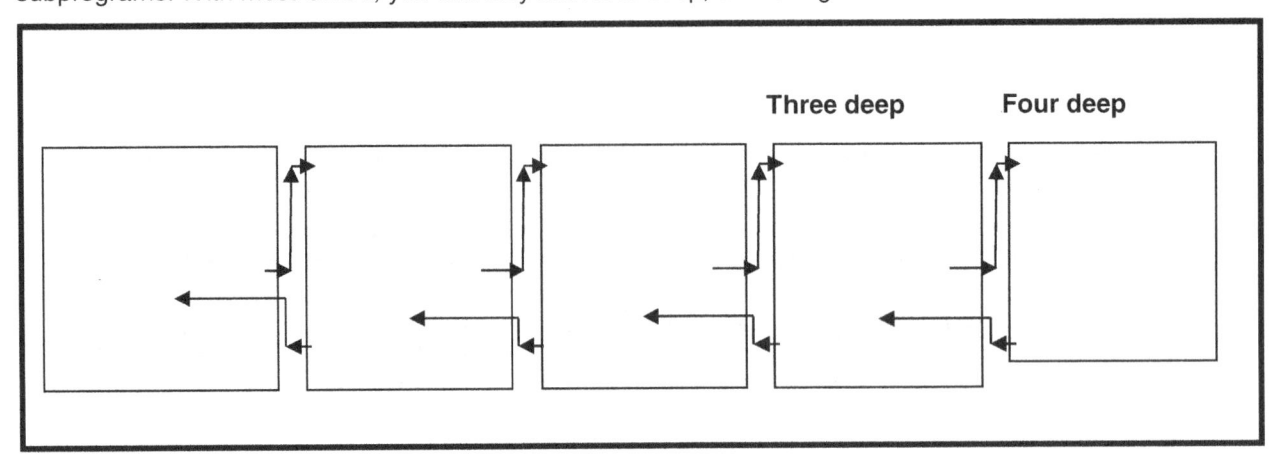

Figure 6.13–Nesting subprograms four deep

Machining Multiple Identical Pockets

The most popular application for subprograms of is performing multiple identical machining operations as we described above. Six identical holes have to be machined in three workpieces.

Before we show some more applications, you must understand a limitation of subprograms. If something must be different in the series of blocks from one execution of the subprogram to the next, a subprogram may not work. Subprograms will be executed in exactly the same manner every time the subprogram is called.

One limitation has to do with specifying motion commands in the absolute positioning mode where all coordinates are specified relative to the active workpiece coordinate system zero. When absolute positioning commands are repeated, the cutting tool will move through the same coordinates a second time. While this is acceptable for many applications, there are also times when it is not.

Consider, for example, machining the four identical pockets in the figure.

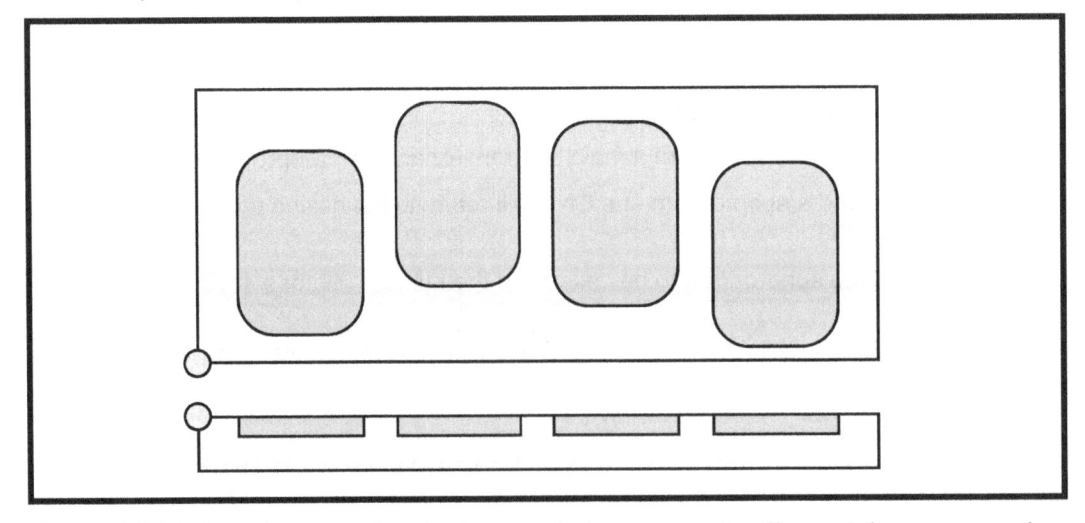

Figure 6.14–A time when repeating absolute positioning commands will repeat the same operation

If a subprogram was written for this application with the coordinates for the first pocket in the absolute positioning mode, all the coordinates are relative to the active workpiece coordinate system zero point in the lower left corner. When the subprogram is called without changing the workpiece coordinate system offset, the same pocket would be machined a second (and third, and fourth) time.

One solution is to program the pocket in the incremental positioning mode. In the main program, you position the end mill to the center of the pocket (or some other consistent position relative to each pocket) in the absolute positioning mode and then call the subprogram. Since the subprogram contains a series of incremental motions—each coordinate is specified from the tools current position. The figure shows this application.

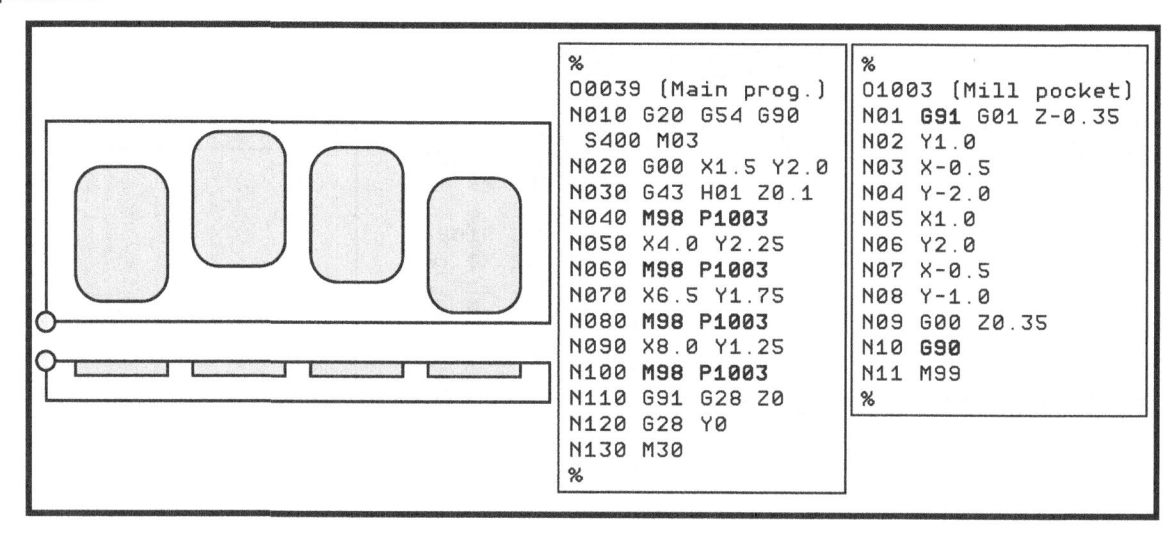

```
%
O0039 (Main prog.)
N010 G20 G54 G90
 S400 M03
N020 G00 X1.5 Y2.0
N030 G43 H01 Z0.1
N040 M98 P1003
N050 X4.0 Y2.25
N060 M98 P1003
N070 X6.5 Y1.75
N080 M98 P1003
N090 X8.0 Y1.25
N100 M98 P1003
N110 G91 G28 Z0
N120 G28 Y0
N130 M30
%
```

```
%
O1003 (Mill pocket)
N01 G91 G01 Z-0.35
N02 Y1.0
N03 X-0.5
N04 Y-2.0
N05 X1.0
N06 Y2.0
N07 X-0.5
N08 Y-1.0
N09 G00 Z0.35
N10 G90
N11 M99
%
```

Figure 6.15–Handling multiple identical machining operations in the incremental mode (shown using inch-mode)

While we haven't specified dimensions on the print, you should be able to follow along. In block N020 of the main program, we position the end mill to the center of the left-most pocket. In block N040, we call the subprogram. Block N01 of the subprogram selects the incremental positioning mode—and all of the motions in the subprogram are specified incrementally from the tool's current position. In block N10 of the subprogram, we reselect the absolute positioning mode. The subprogram ends in block N11, and the machine returns to block N050 of the main program. This block positions the end mill to the center of the second pocket from the left. Block N060 calls the subprogram again, which machines the second pocket. This process is repeated for the last two pockets.

Programming in the incremental mode is cumbersome. If your machining center is equipped with G52 (as newer machines are), there is a much easier way.

Understanding G52–Local Coordinate System

There is another way to handle this problem that defines a local coordinate system relative to the workpiece coordinate system. The G52 command includes axis designators (X, Y, and/or Z) that specify the distance from the active workpiece coordinate system to the local coordinate system in each axis. For example, the block

```
N050 G52 X1.5 Y2.0
```

(shown using inch mode) will temporarily shift the workpiece coordinate system zero point from workpiece coordinate system zero at the lower-left corner (in figure above) to the local coordinate system zero at the center of the left-most pocket. The subprogram can now be written in the absolute mode, using the local coordinate system zero at the center of the pocket as the reference point. See the figure below.

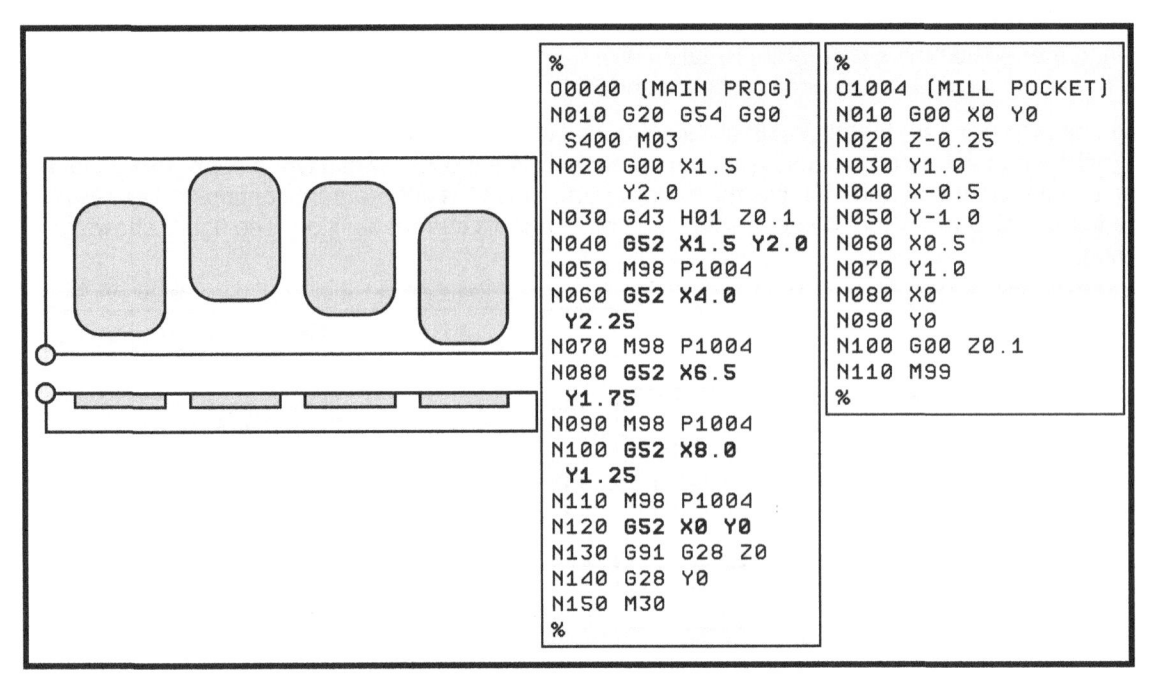

Key Concept 6: Features That Help Simplify Programming

```
                                  %                      %
                                  O0040 (MAIN PROG)      O1004 (MILL POCKET)
                                  N010 G20 G54 G90       N010 G00 X0 Y0
                                   S400 M03              N020 Z-0.25
                                  N020 G00 X1.5          N030 Y1.0
                                   Y2.0                  N040 X-0.5
                                  N030 G43 H01 Z0.1      N050 Y-1.0
                                  N040 G52 X1.5 Y2.0     N060 X0.5
                                  N050 M98 P1004         N070 Y1.0
                                  N060 G52 X4.0          N080 X0
                                   Y2.25                 N090 Y0
                                  N070 M98 P1004         N100 G00 Z0.1
                                  N080 G52 X6.5          N110 M99
                                   Y1.75                 %
                                  N090 M98 P1004
                                  N100 G52 X8.0
                                   Y1.25
                                  N110 M98 P1004
                                  N120 G52 X0 Y0
                                  N130 G91 G28 Z0
                                  N140 G28 Y0
                                  N150 M30
                                  %
```

Figure 6.15–Using G52 with subprograms (shown using inch mode)

Each time you shift the workpiece coordinate system zero point with G52, you specify the distance from the workpiece coordinate system zero point (lower-left corner in the figure above) to its new location (center of each pocket in our example). When you're finished, you must specify one more G52 command to shift the workpiece coordinate system zero back to its original location with the command G52 X0 Y0, as is done in block N120 of the main program above.

Multiple Hole-machining Operations on a Series of Holes

Another application for a subprogram is when you must perform multiple machining operations on holes. The more holes that must be machined, the more the subprogram will help. But note that if you have but a few holes to machine, you may not want to use a subprogram. The time and effort required to load the subprograms may not be worth the trouble.

But say for example, you have fifty holes that must be center drilled, drilled, and tapped. Without using a subprogram, this will require one-hundred-fifty blocks, even if you use canned cycles. If you use a subprogram, the number of blocks can be reduced to about fifty.

The next figure shows an example.

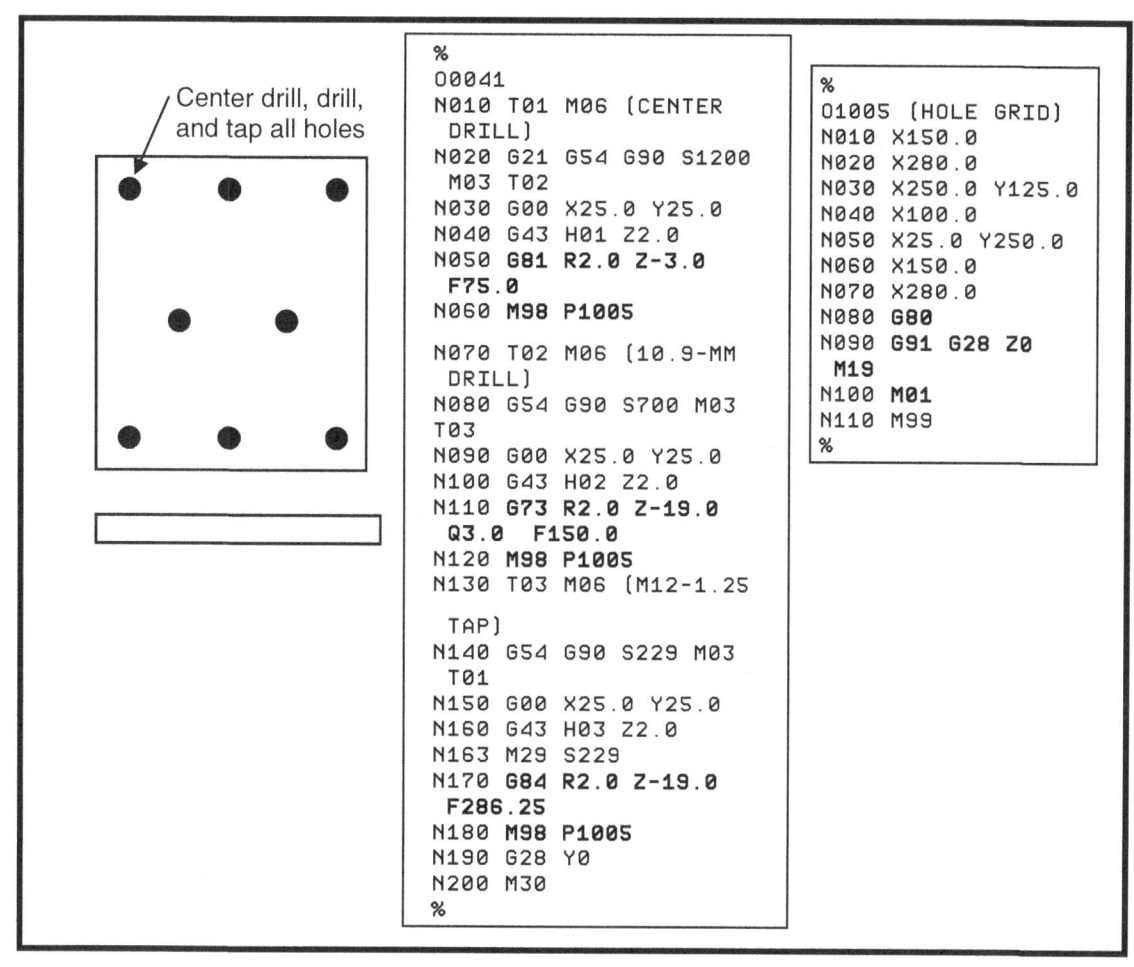

```
                    %
                    O0041
  Center drill, drill,    N010 T01 M06 (CENTER      %
  and tap all holes        DRILL)                   O1005 (HOLE GRID)
                    N020 G21 G54 G90 S1200          N010 X150.0
                     M03 T02                        N020 X280.0
                    N030 G00 X25.0 Y25.0            N030 X250.0 Y125.0
                    N040 G43 H01 Z2.0               N040 X100.0
                    N050 G81 R2.0 Z-3.0             N050 X25.0 Y250.0
                     F75.0                          N060 X150.0
                    N060 M98 P1005                  N070 X280.0
                                                    N080 G80
                    N070 T02 M06 (10.9-MM           N090 G91 G28 Z0
                     DRILL)                          M19
                    N080 G54 G90 S700 M03           N100 M01
                     T03                            N110 M99
                    N090 G00 X25.0 Y25.0            %
                    N100 G43 H02 Z2.0
                    N110 G73 R2.0 Z-19.0
                     Q3.0 F150.0
                    N120 M98 P1005
                    N130 T03 M06 (M12-1.25
                     TAP)
                    N140 G54 G90 S229 M03
                     T01
                    N150 G00 X25.0 Y25.0
                    N160 G43 H03 Z2.0
                    N163 M29 S229
                    N170 G84 R2.0 Z-19.0
                     F286.25
                    N180 M98 P1005
                    N190 G28 Y0
                    N200 M30
                    %
```

Figure 6.16–Using a subprogram with multiple operations on holes (shown using metric mode)

In block N050 of the main program, the first hole is machined using the appropriate canned cycle (G81, G82, G83, etc.). The subprogram is then called that machines the balance of holes. Since the canned cycle specified in the main program is modal, it is also active for the coordinates specified in the subprogram. Notice that the tool ending structure is also included in the subprogram (starting at block N090). When the machine returns to the main program (block N070), the next block is the tool change. This is repeated for each hole-machining tool.

Want to Include All of the Hole-Locations in the Subprogram?

If it makes more sense to you to specify all of the hole-locations in the subprogram, you can include an L0 word in the canned cycle block in the main program. As you know, the L-word specifies the number of holes to be machined by the canned cycle command. If you specify L0, you're telling the CNC not to machine a hole. Yet the canned cycle is being properly set up (with cycle type, Z-bottom position, R-plane, feedrate, etc.). The next motion block that is in the subprogram machines the first hole. Here are the two programs that show this technique (shown using metric mode).

```
%
O0042
N010 T01 M06 (CENTER DRILL)
N020 G21 G54 G90 S1200 M03 T02
N030 G00 X25.0 Y25.0
N040 G43 H01 Z2.0
N050 G81 R2.0 Z-3.0 F75.0 K0 (NO HOLE MACHINED)
N060 M98 P1006
N070 T02 M06 (10.9-MM DRILL)
```

```
N080 G54 G90 S700 M03 T03
N090 G00 X25.0 Y25.0
N100 G43 H02 Z2.0
N110 G73 R2.0 Z-19.0 Q3.0 F150.0 K0 (NO HOLE MACHINED)
N120 M98 P1006
N130 T03 M06 (M12-1.25 TAP)
N140 G54 G90 S229 M03 T01
N150 G00 X25.0 Y25.0
N160 G43 H03 Z2.0
N163 M29 S229
N170 G84 R2.0 Z-19.0 F286.25 K0 (NO HOLE MACHINED)
N180 M98 P1006
N190 G28 Y0
N200 M30
%


%
O1006 (HOLE GRID)
N01 X25.0 Y25.0 (FIRST HOLE MACHINED)
N02 X150.0
N03 X280.0
N04 X250.0 Y125.0
N05 X100.0
N06 X25.0 Y250.0
N07 X150.0
N08 X280.0
N090 G80
N09 G91 G28 Z0 M19 (The tool ending format is repeated too)
N10 M01
N11 M99
%
```

Rough and Finish Side Milling

During lesson 4.4, we provide an example of rough and finish milling. This requires that you specify the tool path coordinates twice–once for the roughing cutting and once for the finishing cutter. Here is the application again, but this time we'll use a subprogram in which to specify the (repeated) finishing tool path. Once again, this is a borderline application for a subprogram because the tool path is quite short, requiring but a few blocks, but illustrates the use of a subprogram to handle the application. Figure 6.17 shows the counter-bore that must be rough and finish milled.

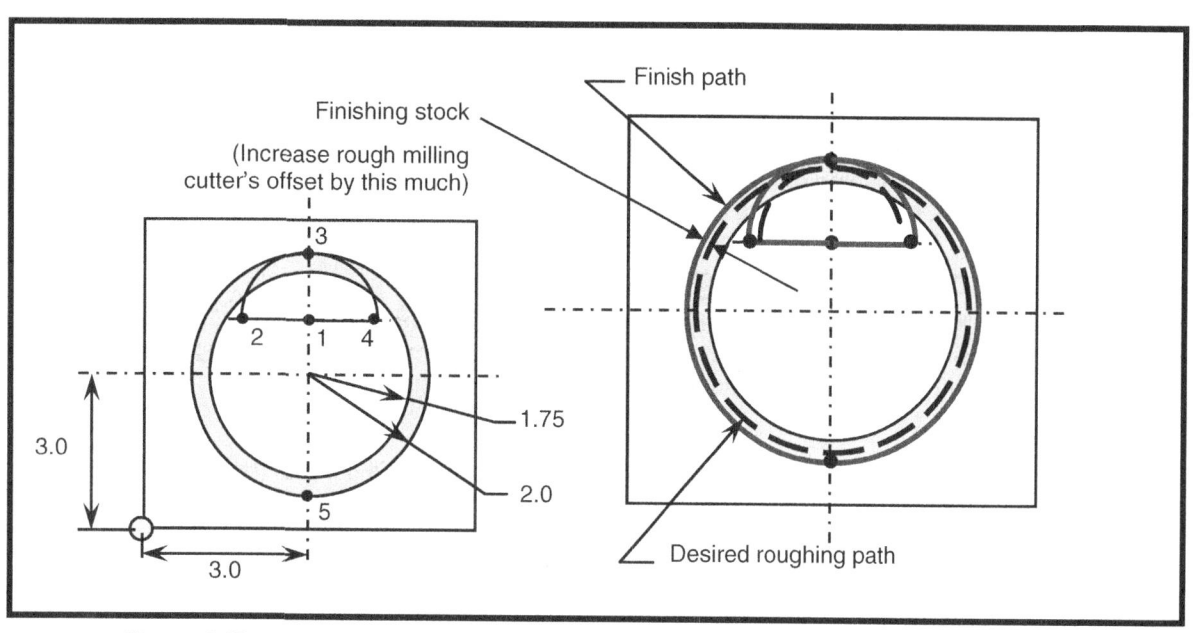

Figure 6.17–Rough and finish milling application for a subprogram (shown using inch mode)

```
%
O0028 (MAIN PROGRAM)
(3/4-IN END MILL IN STATION #1)
N010 G20 G54 G90 S450 M03 (Select workpiece coordinate system
setting offset #1, absolute mode, start spindle fwd at 500 rpm)
N020 G00 X3.0 Y4.275 (Rapid to prior position pt1)
N030 G43 H01 Z0.1 (Instate tool length compensation, move tool
to just above workpiece)
N040 G01 Z-0.25 F40.0 (Fast feed to work surface)
N050 G42 D01 X2.275 F5.0 (Instate cutter radius compensation
using offset increased by 0.020-in to pt 2)
N060 M98 P1001 (Rough mill counter-bore)
N070 G42 D31 X2.275 F4.0 (Instate cutter radius compensation
using offset with actual cutter radius to pt 2)
N080 M98 P1001 (Finish mill counter-bore)
N090 Z0.1 (Retract to just above the workpiece in Z)
N100 G91 G28 Z0 (Move to the Z-axis zero return position)
N110 M30 (End of program
%


%
O1001 (SUBPROGRAM TO COUNTERBORE)
```

Key Concept 6: Features That Help Simplify Programming

```
N010 G02 X3.0 Y5.0 R.725 (ARC-IN TO PT. 3)
N020 Y1.0 R2.0 (Circular mill right half of circle to pt 4)
N030 Y5.0 R2.0 (Circular mill left half of circle back to pt 3)
N040 X3.725 Y4.275 R.725 (Arc-off to pt 5)
N050 G00 G40 X3.0 (Cancel cutter radius compensation during a movement back to the center of the approach circle pt1)
N060 M99 (END OF SUBPROGRAM)
%
```

In blocks N050 and N070 of the main program, we instate cutter radius compensation using the appropriate offset (remember, the rough milling cutter's offset must be set to a value that is larger than its actual size to leave the desired amount of finishing stock). In blocks N060 and N080, the machine executes the subprogram that machines the counter-bore.

Utility Applications for Subprograms

Subprograms can also help improve the utilization of your CNC machining center. We call any application that does so a utility application. Utility applications can reduce setup time, they can make the machine easier to work with, and in general, they can enhance the way the machine performs.

Control Programs

Subprograms are often helpful when it is necessary to control the program's general flow. Consider, for example, a machining center that has a two-pallet pallet changer. This kind of machine allows the operator to be loading one pallet while the machine is working on the other. This, of course, reduces workpiece loading time.

Pallet changers allow a great deal of flexibility. It is possible to run first and second operation on the same workpiece, two identical workpieces, or two completely different workpieces. This flexibility can cause problems when it comes to program execution—especially when you're running two different workpieces.

If workpieces currently being run are unrelated to one another, you won't want the programs for each workpiece to be attached to one another. Instead, you'll want to have two completely independent programs. With subprograms, you can easily separate the two programs. We call this application a control program application. The figure below shows an example.

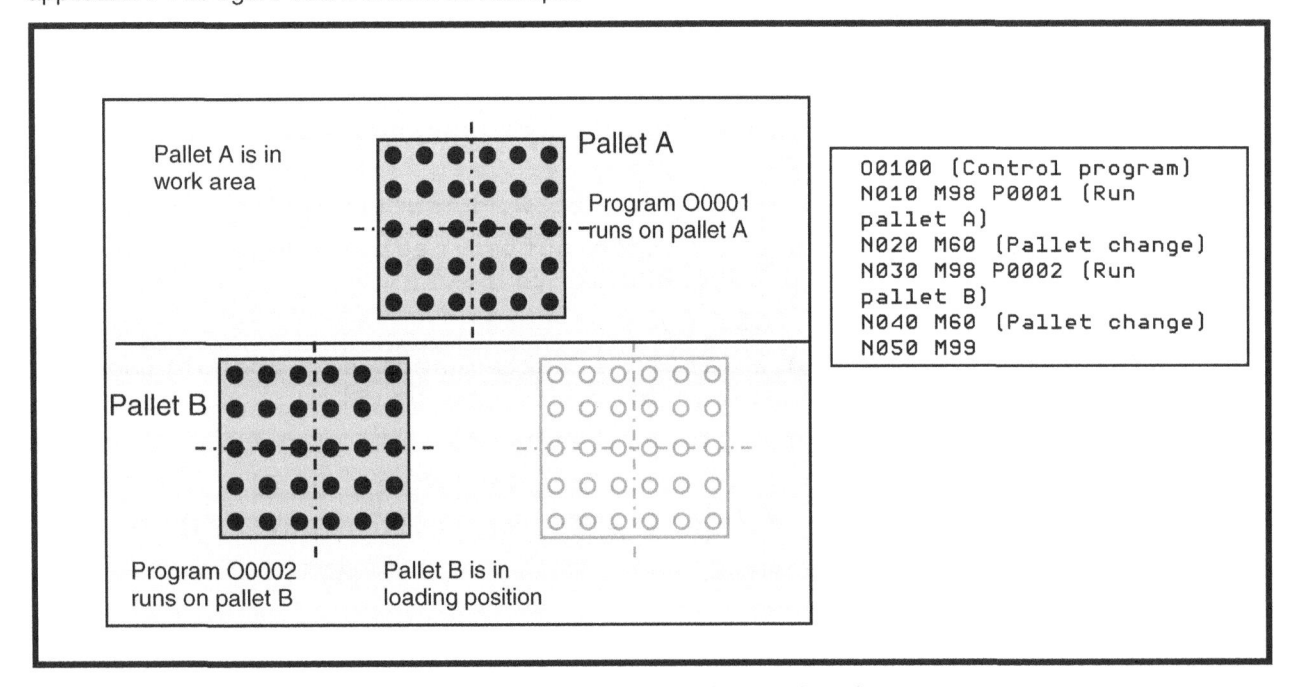

Figure 6.18–A control program application for subprograms

When program O0100 is run (the control program), it will first execute the program for pallet A (currently O0001). Program O0001 is the machining program that machines the entire workpiece on pallet A–and it must end with an M99. When this program is finished, block N020 will make the pallet change, placing pallet B in the work area. Then the machine will execute program O0002, machining the workpiece on pallet B. In block N040, the machine will make another pallet change, placing pallet A back in the work area.

Notice that this main program ends with an M99. When a main program ends with an M99, the machine will return to the beginning of the program and execute the program again–without stopping.

By the way, this might be an example of nesting subprograms (calling one from another). Programs O0001 and O0002 in Figure 6.18 are subprograms. If you need any of the other subprogram applications shown in this lesson in conjunction with them (like multiple hole-machining operations on a series of holes), you will be calling a subprogram from another subprogram. At this point, you'll be nested two deep. Remember, the limitation for most controls is nesting four deep.

What is Parametric Programming (Custom Macro)?

As you have seen, subprograms dramatically expand what can be done with manual programming functions. But as you know, they have one major limitation. If anything changes from one execution to the next, subprograms cannot be used.

Parametric programming gives you the ability to overcome this limitation. (FANUC parametric programming is called Custom Macro or Custom Macro B on older CNCs). In essence, parametric programming gives you the ability to write general-purpose subprograms.

While parametric programming is beyond our scope, you should at least be able to recognize applications for this very powerful programming tool. If your company has applications that fall in to one of these five categories, you'll want to learn more about parametric programming.

Part Families

Many companies machine workpieces that fit into a close family. While the workpieces are different, they have very similar attributes and are machined with the same process (sequence of machining operations). For example, one company machines air cylinders in a variety of sizes. Each component making up the air cylinder (end caps, piston, cylinder, etc.) falls into a close part family. The program used to machine one of the workpieces in the family will be very similar to programs for other workpieces in the family.

With conventional programming methods, the programmer will commonly modify one program to create another program for a different workpiece in the family. When finished, they will have one hard-and-fixed CNC program for each workpiece in the family. If a process or engineering change is made that affects all workpieces in the family, all of the CNC programs must be modified. Think of a part family that contains over one hundred workpieces. This could involve a great deal of work.

With parametric programming, one program can be developed to machine all workpieces in the family. In essence, a general purpose program is created that uses variables to represent the changing attributes for each workpiece in the family. The machine will behave differently based upon the current settings of these variables.

User Defined Canned Cycles

As you already know, CNC machining centers have hole-machining canned cycles. However, you may find yourself wishing that your machining center had other kinds of canned cycles. You may, for example, have to mill counter-bored holes similar to one shown in Figure 6.17 on a regular basis. They may be of differing sizes and require different end mills to be used.

If faced with this application, wouldn't it be nice to have a counter-boring milling cycle? With parametric programming, you can actually create your own canned cycles. And just about any application can be handled. A few classic examples include bolt pattern, pocket milling, thread milling, and face milling. But again, any time you find yourself having to perform the same kind of machining operation–and especially when the operation is required on a variety of different workpieces–is probably a good application for parametric programming.

Utilities

While utility applications for subprograms are relatively limited, utility applications for parametric programming are plentiful. You can do many things to enhance the way your CNC machining center

Key Concept 6: Features That Help Simplify Programming

performs. Examples include part counters, tool life managers, workpiece coordinate system zero assignment helpers, and tool length compensation value measuring helpers. Frankly speaking, just about any time you see a setup person or operator struggling with a machine-related task is probably a good time to consider developing a utility parametric program.

Complex Motions and Shapes

Since arithmetic calculations can be done within parametric programs, any shape that can be defined with an arithmetic formula can be machined by a parametric program. Consider, for example, the kind of motion that is necessary for milling taper threads. While helical interpolation (discussed in lesson 6.3) will help with the milling of straight threads, it will not help with milling tapered threads. When taper thread milling, the XY motion will not be circular. Instead, it will be in the form of a spiral motion. Parametric programming will allow you to create this kind of motion.

Driving Accessory Devices

Probes, post process gauging systems, and digitizing systems are among the accessory devices that require the use of parametric programming. By the way, if your machining center has a spindle probe, it is likely that it also has the parametric programming feature.

Talk to experienced people in your company...

... to learn more about how your company uses subprograms.

1) Ask an experienced person if your company uses subprograms. If the answer is yes, ask to see the applications they're used for.

2) If subprograms are used, ask how program numbers are organized. How quickly can the setup person load them (and remove them when the job is finished)?

Key Points for Lesson 6.2:

- ✓ Subprograms are used to repeat commands. Just about any time you find yourself repeating commands in a program, you should consider using a subprogram. The more commands that must be repeated, the fewer commands you'll need to for the job.
- ✓ One benefit of subprograms is obvious–shortened program length. But the other, easier program verification is not so obvious. When verifying your program, if subprogram runs properly the first time it is executed, it will run properly every time it is executed.
- ✓ M98 is the word used to call a subprogram. A P-word in the M98 command specifies the subprogram number. M99 is the word used to end the subprogram.
- ✓ Applications for subprograms fall into two basic categories–multiple identical machining operations and utilities.
- ✓ G52, temporary shift of workpiece coordinate system zero, can really help with multiple identical machining applications.
- ✓ Parametric programming, which is an option on most controls, gives you the ability to write multi-purpose subprograms.

Lesson 6.3–Other Special Programming Features

Current model CNC machining centers come with many features to help with special applications. While some of these features will be of little need to you in the immediate future, it is important to know they exist. You cannot begin to apply any feature that you don't know about.

Objectives

After completing this lesson, students should be able to:

- ✓ Describe the application for block delete
- ✓ Describe special techniques with sequence numbers
- ✓ Program thread milling operations using helical interpolation
- ✓ Describe applications for the dwell command
- ✓ Describe the use of G10 data setting

Introduction

CNC control manufacturers strive to equip their controls with as many helpful programming features as possible. Those mentioned so far (canned cycles and subprograms) are used on a very regular basis — and you should strive to master them. However, there are some special programming features that are not used nearly as regularly. Indeed there are some features that are extremely important to one company but of no value to another.

As you study this lesson, you need to consider your own company's CNC applications. If you are in doubt about the value of a given feature, ask your instructor or an experienced person in your company about its value to your company. You can minimize your studies about those features your company does not currently need. You can always come back and study this lesson in greater detail should the need arise.

As you study this lesson, remember that *your ingenuity is based predominantly upon your knowledge of what is possible*. You cannot apply a feature of which you are unaware. At the very least, this lesson will acquaint you with what is possible with special CNC programming features.

The organization of this lesson is not as tutorial as previous lessons. While we will explain each feature in detail and in tutorial format, we don't present them in a special order. Here are the topics contained in this lesson:

- Block delete (also called optional block skip)
- Special techniques with sequence numbers
- G-codes that have not yet been introduced (in numerical order)

Block Delete (also called optional block skip)

The block delete function is used to give the CNC operator a choice between one of two possibilities. An on/off switch on the control panel (commonly labeled block delete or optional block skip) is used to actually make the choice. Since applications for block delete vary, the programmer must make each use of block delete *very clear* to the operator. This should be done in the setup- and/or production-run-documentation.

A slash code (/) in the program tells the control to look to the position of the block delete switch. If the switch is on, the control will skip any words to the right of the slash code. If the switch is off, the control will execute these words.

Here is a simple example. Say your machining center does not have an adequate coolant switch. Your setup person has no way to turn off the coolant when programs are being verified (whenever an M08 is executed, coolant will come on). To solve this problem, you can place a slash code at the beginning of every coolant command. Here is one way to do so.

Key Concept 6: Features That Help Simplify Programming

```
/ N015 M08 (If the block delete switch is on, the M08 will be
skipped and coolant will stay off)
```

During program verification, the setup person can turn *on* the block delete switch. This will force the control to *skip* commands that turn coolant on (leaving the coolant off).

The slash code does not have to be placed at the beginning of a command. If for example, you have a totally enclosed work area, you may want the coolant to come on during each cutting tool's approach to the workpiece. Consider this command.

```
N015 G43 H01 Z2.0 / M08 (Only M08 is affected by slash code)
```

In this case, only the M08 is influenced by the slash code. The rest of the command will be executed regardless of the position of the block delete switch.

Applications for Block Delete

As stated, block delete can help whenever you wish to give the operator a choice between one of two possibilities. In the coolant example, either the setup person wants coolant or they do not (again, the choice is between one of two possibilities). There are *many* times when a programmer wants to give the operator a choice between one of two possibilities.

Another Optional Stop

As you know, the optional stop function (M01) lets the operator stop the machine at key times in the program. In the format shown in lesson 5.2, we have you place an M01 at the end of every tool (except the last one) to give the setup person and operator the ability to check and see what each tool has done prior to going on to the next tool. This is very important during the program's verification – and is especially important when trial machining. If you follow our recommended programming format, you cannot (feasibly) use the optional stop function for any *other* purpose that might come up.

Say you want your operator to perform sampling inspections on the workpiece at a critical point during the machining cycle. It might be a critical finish boring operation. After this operation in *every tenth workpiece*, you want the operator to take a measurement. This is an excellent application for optional stop - but if you have already placed an M01 at the end of every tool, you cannot use M01 to additionally make the machine stop only after the critical milling operation. If the operator turns on the optional stop switch, the machine will stop *at the end of every tool* and at the sampling measurement. This will be distracting and will not provide the desired result.

Consider this program stop command.

```
N060 M00
```

M00 is a *program stop*. It causes the machine to stop until the operator reactivates the cycle. There is no option to it – *the machine will stop*. This command can be used if you want your operator to measure *every* workpiece. Still, this is not the desired result. But if used in conjunction with the slash code, you can use the *block delete switch* as a kind of second optional stop function. Consider this command.

```
/ N060 M00
```

If the block delete switch is turned on, the control will *skip* the M00 command and the machine will not stop. For this application, *on* will be the normal setting for the block delete switch. When the operator turns off the block delete switch (at every tenth workpiece), the machine will stop after the finish boring operation to allow the measurement.

A warning about block delete: In this application, the block delete switch will function in just the opposite fashion as the *optional stop switch* (when the optional stop switch is turned on, the machine will stop). This can cause some confusion. Remember that when the optional block skip switch is turned *on*, words to the right of the slash code will be skipped. We're accustomed to something extra happening when a switch is turned on. With block delete, something *doesn't* happen when the switch is turned on.

Trial Machining

Throughout this text we have been stressing the need for trial machining. Remember the five steps required to trial machine:

   1: Recognition of a tight tolerance that worries you

2: Make an initial adjustment that causes additional stock to be left on the workpiece

3: Let the tool machine under the influence of the trial machining adjustment and stop the cycle after the tool is finished

4: Measure the machined dimension and adjust accordingly

5: Re-run the tool under the influence of the new adjustment

As shown so far, trial machining requires a great deal of manual intervention. After the initial adjustment and trial machining, for example, the setup person or operator must manually restart the cycle.

Think about step one. If a setup person or operator can recognize a tight tolerance that requires trial machining, so can a programmer. There are things a programmer can do in a program to simplify the trial machining operation – and the block delete function can help.

Whenever using block delete for trial machining, the block delete switch will be on during normal production (skipping the trial machining commands). When the setup person or operator wants to trial machine, they'll turn *off* the block delete switch.

Trial Boring

Our example will be for finish boring. But *any time* you notice your setup person struggling to hold size on the first workpiece is probably a good time to use block delete and *program the trial machining operation*.

Finish boring on a machining center requires that the boring bar be precisely set to machine the hole-diameter. The hole-tolerance, the quality of the boring bar, and the skill of the setup person determine how many attempts it will take to get the boring bar to machine the hole to size.

With conventional methods (not using block delete), the setup person will:

1: Recognize the tight tolerance for the hole-diameter the boring bar machines.

2: Intentionally set the boring bar undersize to avoid machining the hole too big on the very first try.

3: Run the tool until the boring bar enters the hole by a small amount and then (manually) stop the cycle and retract the boring bar from the hole far enough to take a measurement in the hole (again, this is done manually, using the machine's hand-wheel or jog function).

4: Based upon the measurement just taken they will manually adjust the boring bar.

5: The tool will be re-run

Depending upon the quality of the boring bar and the skill of the setup person, they may not be able to get the boring bar to machine perfectly to its intended size on the first try. They may have to repeat this process several times to get the boring bar machining properly.

Again, note the amount of manual intervention required for trial machining. Anything you can do in your program to help will minimize the related effort – and will reduce trial machining (setup) time. And by the way, if the boring bar dulls during the production run, the CNC operator must repeat the trial machining operation after replacing the boring bar.

By combining block delete with a subprogram, you can dramatically simplify the task of trial boring. In fact, if the subprogram is written properly, it will work for every trial boring operation, regardless of hole-location, hole-size, or boring bar style. This subprogram can be kept in the machine's memory on a permanent basis.

In order for the subprogram to work for any hole and any boring bar, it must be written in the incremental positioning mode. Our example subprogram assumes the tool has already been positioned to the XY location for the hole and is currently 0.1 in above the hole in Z.

Here is the universal trial boring subprogram (shown using metric mode):

```
%
O1000 (Subprogram for trial boring)
N1 G91 G86 R0 Z-8.0 (Trial machine the hole to 0.2 in deep)
N2 G80 (Cancel cycle)
N3 Z50.0 (Rapid away in Z)
N4 X75.0 Y75.0 (Move to a convenient measuring position)
N5 M00 (Stop for measurement)
```

Key Concept 6: Features That Help Simplify Programming

```
N6 G00 X-75.0 Y-75.0 (Move back to hole in XY)
N7 Z-50.0 (Move back to 0.1 above hole)
N8 G90 (Re-select absolute mode)
N9 M99 (End of trial boring subprogram)
```

Again, this subprogram is only executed when trial boring is required. It probably doesn't make much sense by itself. Here are the commands for the boring bar in the main program:

```
%
O0001 (Main program that includes finish boring)
 .
 .
N075 T05 M06 (Place boring bar in spindle)
N080 G54 G90 S500 M03 T06 (Select workpiece coordinate system
setting offset #1, absolute mode, start spindle, get next tool
ready)
N085 G00 X50.0 Y65.0 (Move boring bar to hole-center in XY)
N090 G43 H05 Z2.0 (Instate tool length compensation, bring
boring bar tip to within 0.1 inch of work surface)
N095 F50.0 M08 (Instate feedrate for boring, turn coolant on)
/ N100 M98 P1000 (First try)
/ N105 M98 P1000 (Second try)
/ N110 M98 P1000 (Third try)
/ N115 M98 P1000 (Fourth try)
N120 G86 R2.0 Z-28.0 F50.0 (Bore the hole to depth)
N125 X125.0 (Machine second hole)
N130 G80 (Cancel cycle)
N135 . . .
 .
 .
```

In regular production, the block delete switch will be turned on. So normally, the commands that execute the subprogram (blocks N100 through N115) will be skipped.

When machining the first workpiece (and whenever the boring bar is replaced), the setup person will intentionally set the boring bar undersize (so it cannot machine the hole too big) and turn *off* the block delete switch, telling the machine to execute the trial boring subprogram.

Block N1 of the subprogram will machine the hole to a depth of 6.0-mm. This is just deep enough to take a measurement. Block N2 cancels the cycle and block N3 and N4 position the boring bar out of the way so the setup person can take the measurement of the current hole-size. Block N5 stops the machine so the setup person can take the measurement.

Since the setup person intentionally set the boring bar undersize before running the program, the hole will be undersize. They will adjust the boring bar to machine a larger diameter based upon how more material must be removed from the hole. Leaving the block delete switch turned off, they simply reactivate the cycle.

Since the block delete switch is still off, block N105 of the main program will be executed – and the trial machining subprogram will be executed again. At block N5, the setup person measures again. If the hole is still too small, they will adjust the boring bar again and activate the cycle. The trial machining subprogram will run again, giving the setup person yet another try.

When the hole is to size, the setup person will turn *on* the block delete switch – and the balance of the slash-coded commands (calls to the trial machining subprogram) will be skipped. In block N120 of the main program, the hole will be bored to its complete depth.

This particular main program gives the setup person four tries to get the hole to size, but of course, you could easily add more with additional M98 commands.

Notice the feedrate in block N095 of the main program. Since we want the subprogram to work for any hole in any location with any boring bar, the feedrate (which changes based upon boring bar and hole-size) cannot be part of the trial boring subprogram.

A Warning About Block Delete Applications

You must always consider what will happen if the operator has the block delete switch in the *wrong* position when they run the program. In the case of the coolant example, the operator will get wet. In the case of the trial machining operation, trial machining may be ignored when it is required. While these are not pleasant situations, at least the operator is not placed in extreme danger. Consider this application when the operator (and/or machine) may be in danger if the block delete switch is not correctly positioned.

Say you have some castings that are varying substantially from one workpiece to the next. A face milling cutter is being used to machine a large portion of the workpiece, and the amount of material to be removed in this area is varying as much as 13.0-mm (or 0.5-in) from one casting to another (there is 13.0-mm or 0.5-in too much stock on some workpieces). In this case, you could make a series of milling passes to rough the excess stock under the influence of the slash code. Then you could program the last roughing pass in the normal manner without optional block skip.

If the workpiece does *not* have excess stock, the operator turns *on* the block delete switch so the machine will ignore the extra passes (saving time). If the workpiece has excess stock, the operator must turn the block delete switch *off*. If they do not do so, the milling cutter will be machining much more stock than it is intended to take. This could cause damage to the tool, workpiece, and machine.

When faced with this problem, most programmers will either have the workpieces separated (those that have excess stock from those that do not) or the program will be written to make the extra passes, whether they are needed or not. While cycle time may be compromised, safety is enhanced.

Sequence Number (N-word) Techniques

As you know, sequence numbers allow you to number each block in the program in an organized manner. Most programs in this text, for example, skip five or ten numbers between each sequence number (N005, N010, N015, etc.). This allows everyone to find important commands, and allows room between commands should you need to add more.

Eliminating Sequence Numbers

While sequence numbers are extremely helpful, especially to beginning programmers, they do require space in the machine's memory. FANUC controls have the reputation of having rather small memory capacities. The time may come when you want to load a lengthy program into the machine, but there is insufficient storage capacity. Either some (possibly important) programs must be deleted from memory to make room, or your programs must be made shorter. The first technique most programmers apply to reduce program length is to eliminate sequence numbers. While this makes it more difficult to reference important commands in each program, more and longer programs can be loaded into the machine.

Using Special Sequence Numbers In Program Restart Commands

Though you may consider it rather easy to find the beginning of each tool by scanning T words (with our format for a double-arm tool changer, the *second* T word is in the command that places it into the spindle), you can help your operators more easily find the restart command for each tool if you use a special sequence number to begin each tool. Place this special sequence number in the M06 command which places the tool into the spindle (commonly the restart command). Consider these commands.

```
N1001 T01 M06 (Place tool one in spindle)
   .
   .
   .
N1002 T02 M06 (Place tool two in spindle)
```

```
N1003 T03 M06 (Place tool three in spindle)
```

If this technique is used consistently, everyone will know how to easily find the beginning of each tool. Even if you eliminate sequence numbers for the purpose of conserving memory space, we recommend retaining these special restart sequence numbers.

Using Sequence Numbers As Statement Labels

Statement labels are destination points in your program. Though they are more commonly used in computer programming languages, they do have applications with CNC programming. You can actually change the execution order of your CNC program. An *unconditional branching command* (like the *GOTO statement* in a computer programming language) will cause the machine to go to a specific command designated with a statement label. In CNC programs, sequence numbers can be used as statement labels.

In a *main program*, the unconditional branching command is M99. A *P-word* in the M99 command specifies the statement label (sequence number). The command

M99 P055

tells the control to go to sequence number N055 in the main program and continue. Most controls require that the P-word value precisely matches the sequence number (P50 will not correctly specify a branch to N050 with most controls – the P-word must be P050). Additionally, *unconditional branching will only work in a main program.* As you know, when the machine executes an M99 in a *subprogram*, it will return to the main program.

Using Block Delete To Exit a Series of Commands

There are two helpful applications for statement labels and unconditional branching. The first is related to block delete.

During our discussion of block delete, we show an example related to trial boring. As you know from this discussion, the commands

```
/ N100 M98 P1000 (First try)
/ N105 M98 P1000 (Second try)
/ N110 M98 P1000 (Third try)
/ N115 M98 P1000 (Fourth try)
N120 G86 R2.0 Z-28.0 F50.0 (Bore the hole to depth)
```

give the operator *four tries* to get the boring bar precisely adjusted. If after the fourth try the boring bar is not correctly adjusted, the machine will bore the hole to depth. With unconditional branching, you can give the setup person an *unlimited number* of trial boring attempts. Consider these commands.

```
/ N100 M98 P1000 (Branch to trial boring subprogram)
/ N105 M99 P100 (Go back to N100)
N110 G86 R2.0 Z-28.0 F50.0 (Bore the hole to depth)
```

The machine will continue to execute the trial boring subprogram as long as the block delete switch is turned off.

Other G-codes of Interest

If we have not yet discussed a given G-code, it probably has a very special application. And frankly speaking, it is not be required by all CNC users. While some of these G-codes will be very important to you and your company's needs, others will not. Talk with experienced people in your company to see which of these G-codes your company uses.

Thread Milling, G02 & G03 (helical interpolation)

As you know, G02 and G03 are used to specify circular motion (introduced in lesson 3.1). They are also used to specify *helical motion* – as is required for *thread milling*.

Thread milling is an operation performed when holes are too large to be tapped and when male (external diameter) threads must be machined.

Thread milling cutters vary from one tooling manufacturer to another, and *you must match your programming methods to the style of thread milling cutter you are using*. Figure 6.20 shows three popular types of thread milling cutters.

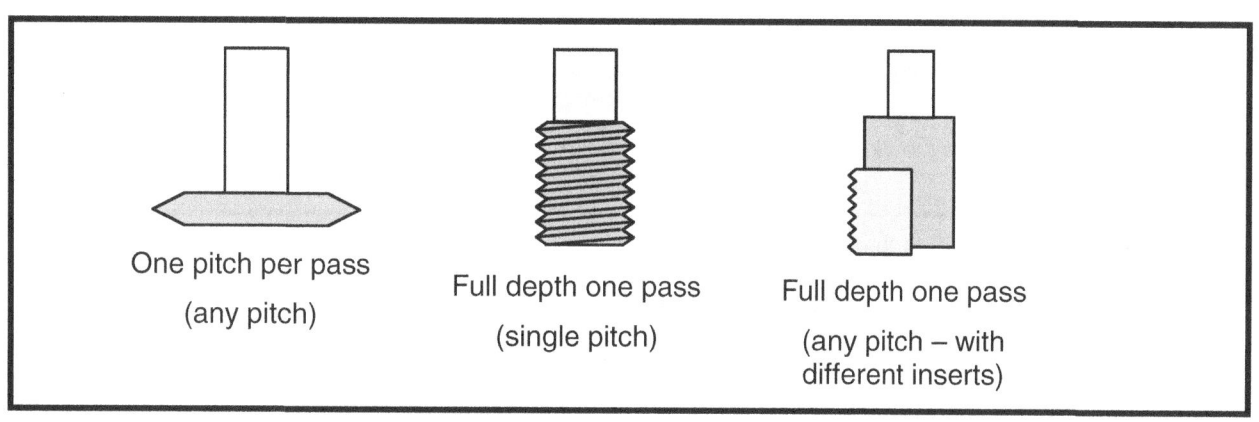

One pitch per pass (any pitch)

Full depth one pass (single pitch)

Full depth one pass (any pitch – with different inserts)

Figure 6.20 – Three types of thread milling cutters

The threading cutter on the left resembles a slotting cutter. The thread form is machined on its outside diameter. This cutter is inexpensive, but it isn't very efficient. It can only machine one pitch (crest) per pass around the diameter being threaded. A one inch deep, eight-threads-per-inch thread (1/8 in pitch) will require at least eight passes around the thread. See the right-most drawing in Figure 6.21. This kind of thread milling cutter can machine any thread pitch (eight-threads-per-inch, twelve-threads-per-inch, sixteen-threads-per-inch, etc.).

The middle thread milling cutter resembles a combination of a rough-milling cutter and a tap. It has several pitches (crests) machined in its outside diameter and can usually machine the entire thread in one pass around the threaded diameter. It will machine only one pitch. A different thread milling cutter is required for every thread pitch you must machine.

The right-most thread milling cutter is becoming the most popular style. One shank can hold inserts having any thread pitch. To change pitch, you simply change inserts. It can also machine most threads in one pass around the threaded diameter.

The motion required for thread milling is *helical motion*. With helical motion, two of the axes (X and Y) will be moving in a circular manner. The third axis (Z) will be moving in a linear manner. The motion resembles a spiral, but the radius of the spiral will remain constant. Figure 6.21 shows the motion.

Key Concept 6: Features That Help Simplify Programming

Two axes, X and Y, will be moving in a circular manner. The third axis, Z, will be moving in a linear manner.

Notice the arc-in and arc-out movements also required for approaching to and escaping from the diameter being thread milled.

Motion required for the middle and right-most thread milling cutters in Figure 6.20

Motion required for the left-most thread milling cutter in Figure 6.20

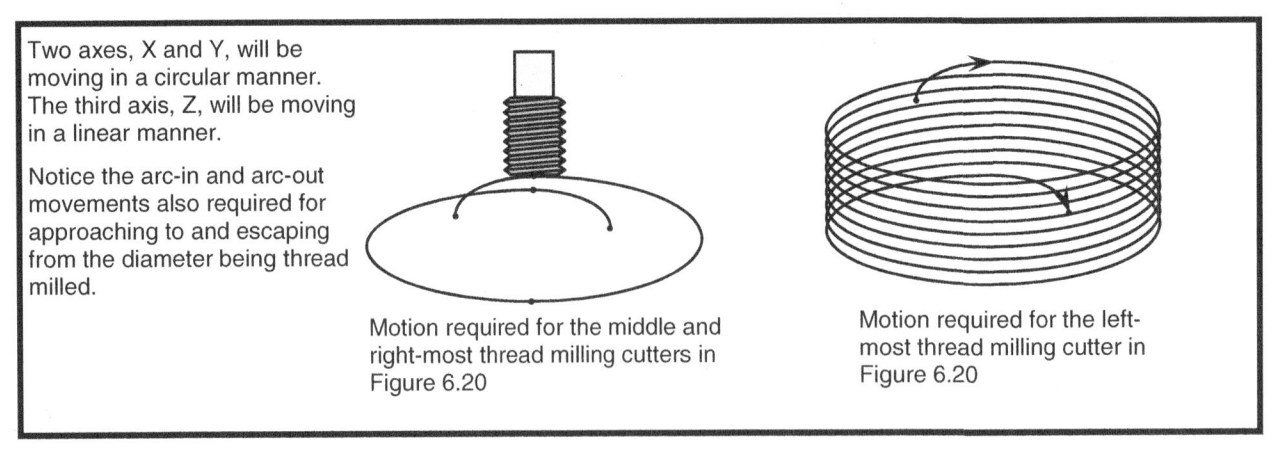

Figure 6.21 – Helical motion required for thread milling

The key to successful thread mill programming is in *matching the Z-axis departure in each helical motion to the portion of a full circle being machined*. If making a full circle motion in XY (all the way around the thread), the Z-axis must depart *one full thread pitch*. If making a one-quarter circle motion in XY (common with arc-in and arc-out motions), the Z-axis must depart *one quarter of the pitch*.

Figure 6.22 shows an example workpiece to be thread milled. In this example, an external (male) thread is being machined.

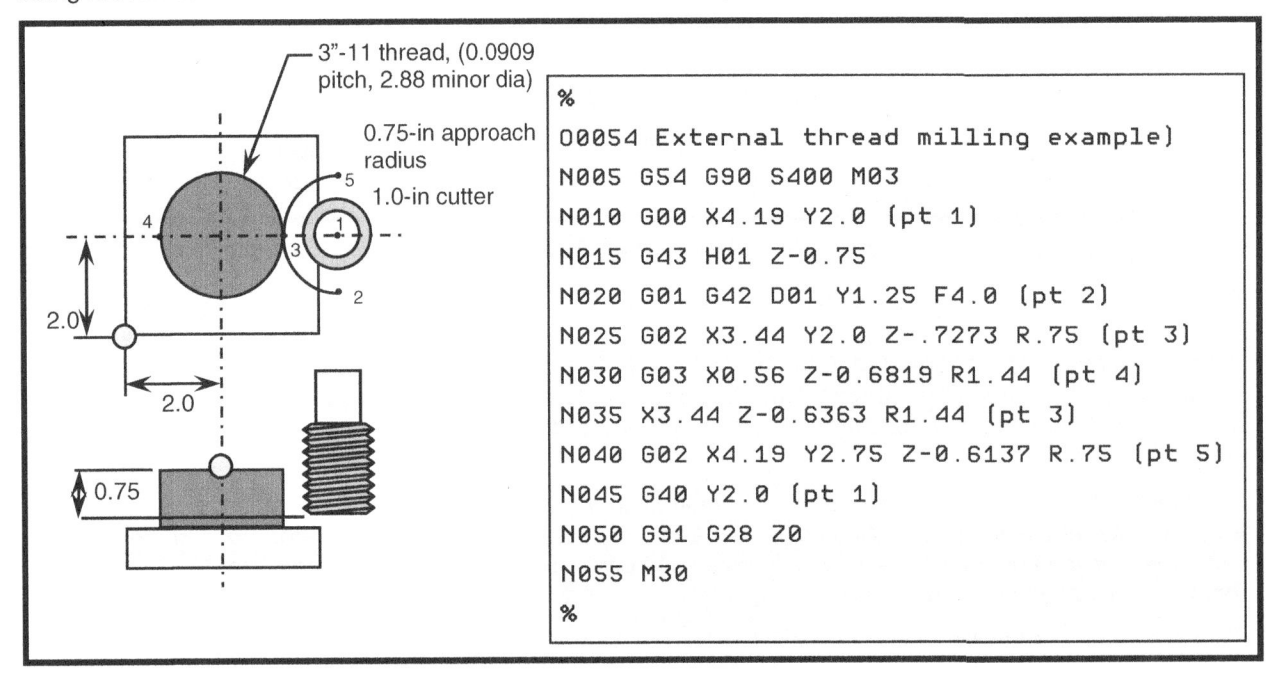

```
%
O0054 External thread milling example)
N005 G54 G90 S400 M03
N010 G00 X4.19 Y2.0 (pt 1)
N015 G43 H01 Z-0.75
N020 G01 G42 D01 Y1.25 F4.0 (pt 2)
N025 G02 X3.44 Y2.0 Z-.7273 R.75 (pt 3)
N030 G03 X0.56 Z-0.6819 R1.44 (pt 4)
N035 X3.44 Z-0.6363 R1.44 (pt 3)
N040 G02 X4.19 Y2.75 Z-0.6137 R.75 (pt 5)
N045 G40 Y2.0 (pt 1)
N050 G91 G28 Z0
N055 M30
%
```

Figure 6.22 – Drawing for external thread milling example program (shown using inch mode)

An eleven-threads-per-inch thread is being machined. The pitch is 0.0909 (1/11). The initial Z position when thread milling begins is Z-0.75 (in block N015). In block N025, the thread mill makes a one-quarter circle approach in XY (arc-in to point three), so the cutter must depart one-quarter of the pitch in Z during this motion (upward, in a Z plus direction if making a right-hand thread). One quarter of 0.0909 is 0.0227. When 0.0227 is added to -0.75, the result is -0.7273. So the ending Z position for point three is Z-0.7273 (in block N025). From point three, the cutter makes a half circle in XY to point four. So the cutter must depart by half the pitch in Z (0.0454). -0.7273 plus 0.0454 is -0.6819, which is the Z end point in block N030. Another half circle movement is next, meaning another 0.0454 Z departure in block N035. And finally, a one-quarter circle arc-out motion means a Z departure of 0.0227 in block N040.

Notice how cutter radius compensation can be used just as it can when machining XY contours. This is especially helpful for thread sizing. If the thread diameter is coming out too big or too small, the cutter radius compensation offset value can be adjusted.

Figure 6.23 shows another thread milling example, this time for an internal thread. Notice once again how the Z-axis departure for each helical motion is related to the percentage of a full circle being made

```
O0044
N005 G54 G90 S400 M03
N010 G00 X2.0 Y2.5
N015 G43 H01 Z0.1
N020 G01 Z-0.85 F30.
N025 G42 D01 X1.0
N030 G02 X2.0 Y3.5 Z-0.8812
  R1.0 F4.5
N035  X2.0 Y.5 Z-0.9437
  R1.5
N040 X2.0 Y3.5 Z-1.0062
  R1.5
N045 X3.0 Y2.5 Z-1.0374
  R1.0
N050 G00 G40 X2.0
N055 G00 Z0.1
N060 G91 G28 Z0
N065 G28 Y0
N070 M30
```

R1.0 Arc-in and out

Note that every circular move requires a Z axis departure

3"-8 Thread 3.0 major diameter

Program zero

1.0 tool dia.

0.125 pitch

This time the thread milling cutter must move in the Z minus direction while thread milling to machine a right-hand thread.

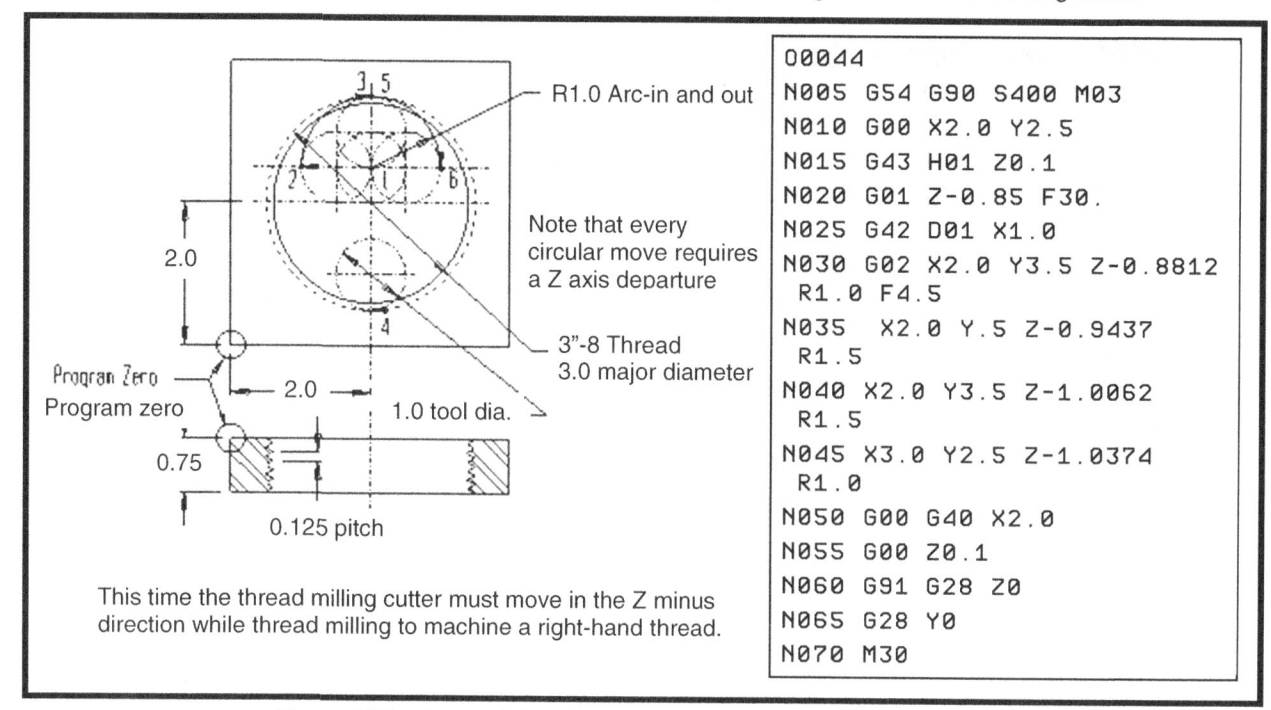

Figure 6.23 – Internal thread milling example (shown using inch mode)

As with the external thread milling example, cutter radius compensation (in block N025) can be used to help with thread sizing. If the thread does not come out to size, a cutter compensation offset (number 01 in our example) can be changed to adjust thread size.

Our example program shows the machining of a through-hole (a hole that goes all the way through the workpiece). If you must machine a thread in a *blind* hole, thread milling should be performed from bottom to top (G03 motions instead of G02) to keep chips from interfering with the machining operation.

G04 - Dwell Command

A *dwell command* causes axis motion to pause for a specified length of time. The most common application for the dwell command is to allow time for tool pressure to be relieved.

Two of the canned cycles (G82 and G89 for counter-boring) include a dwell for this purpose. When counter-boring to a precise depth, these cycles will cause the tool to pause at the hole bottom long enough to relieve tool pressure. This makes the hole-bottom flatter, and leaves a better finish. In the case of canned cycles, of course, the dwell is automatic. You have no need to use a G04.

But there are other times when it is necessary to relieve tool pressure. For example, if you plunge a center cutting end mill into solid stock prior to machining a pocket, you should allow the milling cutter to pause for a moment when it reaches the bottom of the pocket in Z. After tool pressure is relieved, the milling cutter can start moving in X and/or Y. A 0.5 second pause is usually sufficient to allow tool pressure to be relieved.

An *X-word* can be used in the G04 command to specify the length of time the axes will pause (in seconds). *All* axes will pause, *not just the X-axis*.

Here is an example that plunges a center cutting end mill into a pocket.

```
N055 G01 Z-12.0 F75.0 (Plunge into pocket in Z)
N060 G04 X0.5 (Pause all axis motion for 0.5 second)
N066 X75.0 F150.0 (Start milling the pocket)
```

Key Concept 6: Features That Help Simplify Programming

G04 is a *non-modal* G-code. It only takes effect in the command in which G04 is specified. Additionally, G04 does not affect current motion type (G01 in our example), meaning the current motion type is retained after the G04 command.

G09 and G61 - Exact Stop Check

Though not often required, G09 and G61 can be used when you must force a sharp outside corner during a contour milling operation. When executing any series of motion commands, the control will be *looking ahead* during one motion to determine what is coming up in the next command (this is the function the *look-ahead buffer*). One reason for this look-ahead function is to keep the machine from coming to a complete stop between commands.

However, under normal operation, the CNC machine will round corners to some extent. The faster the feedrate and the larger the machine, the greater the corner rounding will be. Figure 6.24 shows the effect of corner rounding.

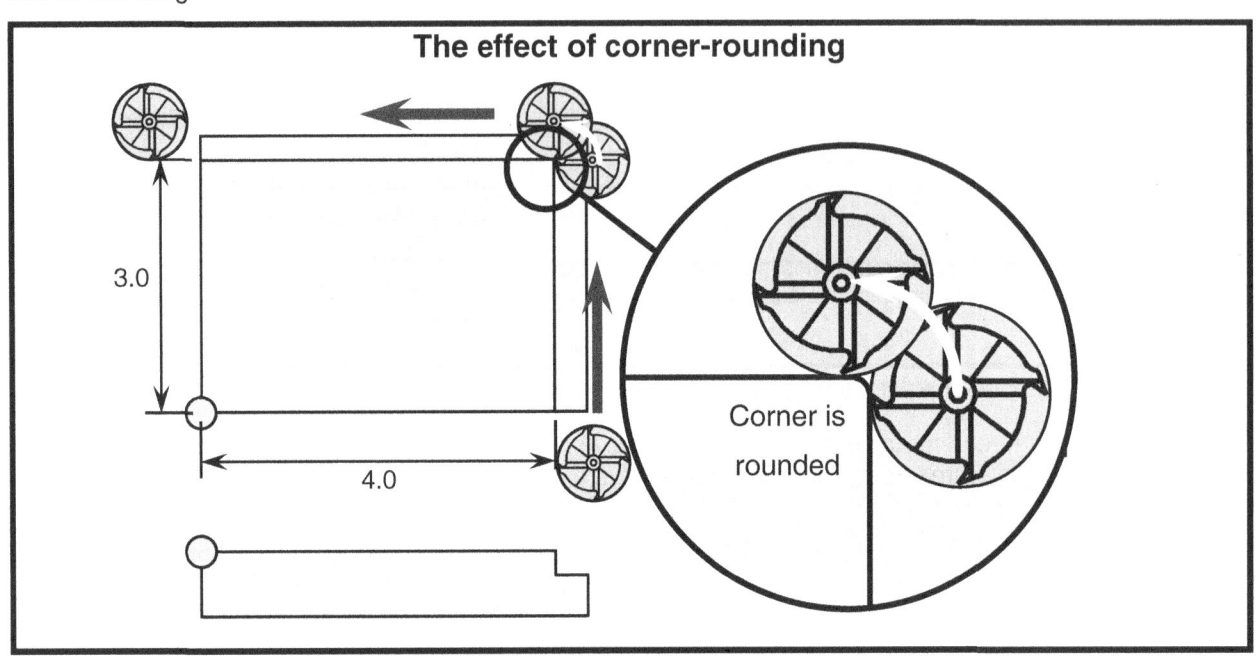

Figure 6.24 – Corner rounding caused by the look-ahead buffer

Say the corner shown in Figure 6.24 must be perfectly sharp. In this case, you cannot allow the machine to perform in its normal manner. The *exact stop check* function will cause the axes to come to a stop between commands (just a dwell command will). There are two G-codes for exact stop check. G09 is the non-modal exact stop check word and G61 is the modal word. G61 will cause the machine to continue coming to a stop after each command. If you use G61, you must remember to place the machine back into *normal cutting mode* by specifying G64 when you're finished.

Here is a program (shown using inch mode) for the workpiece in Figure 6.24. Since there is only one sharp corner, we're using the G09 non-modal G-code for exact stop check.

```
%
O0048 (Example for exact stop check)
N005 G20 G54 G90 S300 M03
N010 G00 X4.6 Y-0.6
N015 G43 H01 Z-0.25
N020 G42 D01 X4.0
N025 G09 Y3.0 F4.0
N030 X-0.6
N035 G00 Z0.1
N040 G40 M09
```

```
N045 G91 G28 Z0
N050 M30
%
```

G10 - Offset Setting By Programmed Command

G10 allows you to *program* offset entries. *Any* offset value can be entered or modified, including tool length compensation offset values, cutter radius compensation offset values, and workpiece coordinate system offset values. An *L-word* within the G10 command tells the control which type of offset is being changed. L-word values vary from one control model to another, so you must check in your control manufacturer's programming manual to find the L-word values for your particular control model

Here are the L word specifications for FANUC controls that have four offset registers per offset number (H GEOM, H WEAR, D GEOM, D WEAR):

- L2: Workpiece coordinate system offsets
- L10: H GEOM (Tool length compensation geometry offset)
- L11: H WEAR (Tool length compensation wear offset)
- L12: D GEOM (Cutter radius compensation geometry offset)
- L13: D WEAR (Cutter radius compensation wear offset)
- L52: Parameters

Again, you must confirm the values for L-words for your own control/s.

A *P-word* within the G10 command tells the control the offset number being entered. For offsets other than workpiece coordinate system setting offsets, an *R-word* specifies the value of the offset entry. When entering workpiece coordinate system setting offsets, X, Y, and Z specify the offset value entries.

G10 is affected by whether the machine is in absolute or incremental positioning mode. In absolute mode, G10 will *overwrite* current offset values with the values specified in the G10 command. In incremental mode, G10 *modifies* the current value of the offset by the amount specified in the G10 command. Here are some examples (shown using metric mode).

```
N005 G90 G10 L10 P1 R152.256 (Overwrite H GEOM offset number
one with 152.256-mm)
N010 G90 G10 L12 P1 R12.5 (Overwrite D GEOM offset number one
with 12.5-mm)
N015 G91 G10 L13 P2 R0.1 (Add 0.1-mm to D WEAR offset number
two's current value)
N020 G90 G10 L2 P1 X-305.378 Y-295.332 Z-269.226 (Overwrite
workpiece coordinate system setting offset number one with
these values)
```

Applications for G10

An application is shown in lesson 4.4 – retaining the workpiece coordinate system zero assignment values for qualified setups from one time the job is run to the next.

Any time you know the value of an offset before a setup is made is a good time to use G10. Here's another great application.

Programming Tool Length and Cutter Radius Compensation Offset Entries

In any setup time reduction program, one goal is to move tasks that are done while the machine is down between production runs *off line*. If setup-related tasks can be performed while the machine is running production, you can reduce the amount of time the machine is down between production runs (these tasks won't have to be done while the machine is down).

During our discussion of tool length compensation, for example, we describe how cutting tools for up-coming jobs can be assembled and measured while the machine is running workpieces for the current job. This

Key Concept 6: Features That Help Simplify Programming

keeps the assembly and measurement of cutting tools *off line*. But tool length and cutter radius compensation offsets must still be *entered* into the machine's offset registers. Since the tool setter knows these critical values long before the setup is made, why wait until the machine is down between production runs to type them?

While it will take a bit more work on the tool setter's part, they can create an offset entering program that can be used during setup to enter all offset values, keeping the setup person from having to do so while the machine is down. The time required for entering offsets will be the time it takes to load the offset-entering program and execute it once. With a good distributive numerical control (DNC) system, this task can be done in under thirty seconds.

Consider this template program (shown using metric mode) that the tool setter can use. While it allows up to twenty tools, you can, of course, modify this program to handle the number of tools your machining center/s can hold.

```
%
O5000 (Offset entering program)
G90 G10 L10 P1 R500.0
G10 L10 P2 R500.0
G10 L10 P3 R500.0
G10 L10 P4 R500.0
G10 L10 P5 R500.0
G10 L10 P6 R500.0
G10 L10 P7 R500.0
G10 L10 P8 R500.0
G10 L10 P9 R500.0
G10 L10 P10 R500.0
G10 L10 P11 R500.0
G10 L10 P12 R500.0
G10 L10 P13 R500.0
G10 L10 P14 R500.0
G10 L10 P15 R500.0
G10 L10 P16 R500.0
G10 L10 P17 R500.0
G10 L10 P18 R500.0
G10 L10 P19 R500.0
G10 L10 P20 R500.0
M30
%
```

To use this technique, the tool setter must have a computer close to the tool assembly and measuring station. After they measure each tool's length, they will modify the appropriate R500.0 word. The P-word corresponds to the tool station number. When finished measuring tool lengths, they will delete any unused commands from the template program. They can also, of course, add commands to specify cutter radius compensation offset values they have measured, like this command

```
G10 L12 P2 R12.5
```

which enters the value of D GEOM offset two as 12.5. When finished, this program will be saved in the DNC system and will be available to the setup person when the setup must be made.

Notice that the default value for each tool length is *500 millimeters*. This must be a value that is larger than your longest cutting tool. If the tool setter happens across a tool that cannot be assembled (or measured), they will leave the value for the tool set to 500-mm. After running this program, if the setup person sees a tool length of 500-mm, they'll know that the tool has not been measured and they'll have to measure it. And if they make a mistake (forgetting to do so), running the program with a tool's length set to twenty inches, the tool will stay far from the workpiece being machined.

Handling Differences Among Pallets

If you have machining centers equipped with pallet changers, you must be prepared for minor differences among your pallets. Even if you are running exactly the same setup on each pallet, you may still have to assign a *different* set of workpiece coordinate system setting offsets for each pallet.

Say for example, you have a horizontal machining center that has two pallets. Each pallet requires four workpiece coordinate system setting offsets. As you know, only six workpiece coordinate system setting offsets are available unless you purchase the option for more. If you use the *control program* shown in lesson 6.2, your main program will look something like this.

```
%
O0001 (Control program)
N005 M98 P1001 (Run pallet A program)
N010 M60 (Change pallets)
N015 M98 P1002 (Run pallet B program)
N020 M60 (Change pallets)
N025 M99 (Return to beginning and continue)
%
```

Programs O1001 and O1002 can each begin with four G10 commands to assign the four coordinate systems as they are needed with the current pallet. Here is an example of the beginning of each program (shown using inch mode).

```
%
O1001 (Program to machine workpiece on pallet A)
N005 G90 G10 L2 P1 X-12.8347 Y-12.1477 Z-10.3847
N010 G10 L2 P2 X-9.8347 Y-12.1477 Z-10.3847
N015 G10 L2 P3 X-6.8347 Y-12.1477 Z-10.3847
N020 G10 L2 P4 X-3.8347 Y-12.1477 Z-10.3847
  .
  .
  .
%
O1002 (Program to machine workpiece on pallet B)
N005 G90 G10 L2 P1 X-12.2214 Y-12.1465 Z-10.3847
N010 G10 L2 P2 X-9.2214 Y-12.1465 Z-10.3847
N015 G10 L2 P3 X-6.2214 Y-12.1465 Z-10.3847
N020 G10 L2 P4 X-3.2214 Y-12.1465 Z-10.3847
  .
  .
  .
```

Polar Coordinates (G15 and G16)

You know how to specify positions in the rectangular coordinate system using X, Y, and Z-axis designators. And *all* movements you need a tool to make can be specified with rectangular coordinates. However, there are times when you must perform trigonometry in order to determine rectangular coordinate values. Consider, for example, the bolt pattern shown in Figure 6.25.

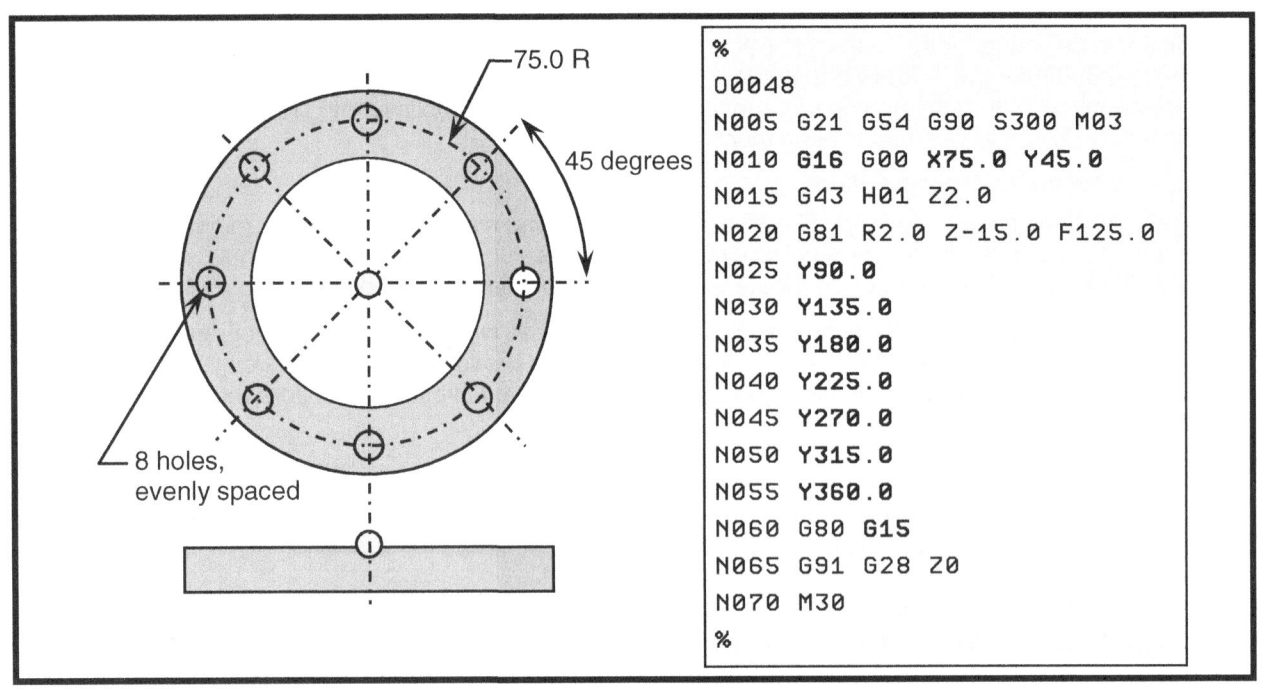

Key Concept 6: Features That Help Simplify Programming

```
%
O0048
N005 G21 G54 G90 S300 M03
N010 G16 G00 X75.0 Y45.0
N015 G43 H01 Z2.0
N020 G81 R2.0 Z-15.0 F125.0
N025 Y90.0
N030 Y135.0
N035 Y180.0
N040 Y225.0
N045 Y270.0
N050 Y315.0
N055 Y360.0
N060 G80 G15
N065 G91 G28 Z0
N070 M30
%
```

Figure 6.25 – Bolt hole-pattern dimensioned with polar coordinates (shown using metric mode)

Polar coordinates allow you to specify positions with a completely different coordinate system called the *polar coordinate system*. Yet with FANUC controls, X and Y-words are still used to specify coordinate values. In the polar coordinate system, X specifies the radius coordinate and Y specifies the angle coordinate.

The origin (workpiece coordinate system zero) of the polar coordinate system must be the same position as the origin in the rectangular coordinate system, the center of the ring in Figure 6.25. As discussed in lesson 6.2, you can use G52 to shift the origin if you need to. The origin for the angle (zero degrees) is the three o'clock position (the right most hole in Figure 6.25) and counter-clockwise is plus. Ninety degrees (Y90.0) is the upper-most hole (at the twelve o'clock position).

Look at the program in Figure 6.25. Block N010 selects polar coordinates mode with G16 and moves the drill to the two o'clock (45 degree) position. Again, the X-word in this command is specifying the *radius* of the bolt pattern. Y is specifying the angle.

In block N020, notice that canned cycles can be used with polar coordinates. This command drills the hole at the two o'clock position. Blocks N025 through N055 machine the rest of the holes. Since the radius remains consistent, the X-word is omitted from these commands. In block N055, note that three-hundred-sixty degrees is the same as zero degrees.

In block N060, we cancel the canned cycle mode and polar coordinates. From this point, all X and Y values will be taken as rectangular coordinates.

Plane Selection Commands (G17, G18, and G19)
Almost all machining is done in the XY plane. All examples shown in this text so far have needed only the XY plane. G17 instates the XY plane and is *initialized* at power-up. Since so much is done in the XY plane, and since it is initialized, we have barely mentioned this command (though it is included as one of the *safety command* words shown in lesson 5.2). If you work exclusively in the XY plane (as most programmers do) you will have no need for plane selection commands.

However, you must be aware of the times when you are not working the XY plane. Say for example, your vertical machining center is equipped with a *right-angle-head* which can point the tool at ninety degrees to the spindle (pointing the tool toward the X minus direction, for example). Any operation performed with this tool will *not* be in the XY plane. If the tool is pointing in the X direction (plus or minus), it will be working in the *YZ plane* (selected by G19). If the tool is pointing in the Y direction, it will be working in the *XZ plane* (G18).

If all you need to do is perform simple motions (G00 and G01), you still have no need for the plane selection commands. But if the tool will use circular commands, cutter radius compensation, or canned cycles, you must select the proper plane prior to giving these commands. Of course, you must also re-select the XY plane (G17) when you are finished.

Figure 6.26 shows an application for plane selection. A ball end mill is being plunged into a rounded pocket.

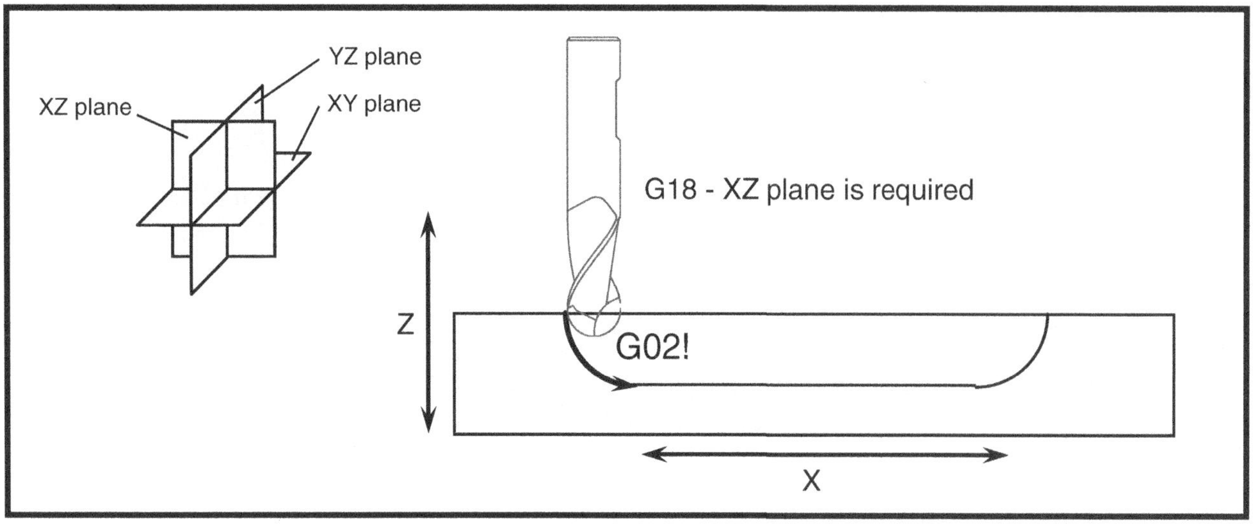

Figure 6.26 – Plane selection is required when making circular motions

Since the ball end mill is making a circular motion in the XZ plane, G18 must be specified prior to the circular motion. When finished making circular motions in the XZ plane, G17 must be specified to reselect the XY plane.

You may be questioning the motion direction shown in Figure 6.26 (G02). From your perspective (above the page) this looks like a counter-clockwise motion. You may feel this should be a G03 command. But remember that you must view circular motion direction (clockwise versus counter-clockwise) *from the plus side of the uninvolved axis*. For an XZ circular motion, Y is the uninvolved axis. For a vertical machining center, the plus side of the Y-axis is the column side of the machine (the opposite side of this page). When viewed from this perspective, the motion shown in Figure 6.26 is clockwise.

Inch/metric Mode Selection G20 and G21

Most programmers work *exclusively* in the measurement system mode that the majority of their engineering drawings use. If the majority of prints are dimensioned in inch, they work exclusively in the inch mode (G20). If they happen across a metric print from time to time, they will convert the print to inch (inches = millimeters divided by 25.4) and still program the job in the inch mode. If the majority of prints are dimensioned in metric, they work exclusively in the metric mode. If they happen across an inch print, they'll convert it to metric (millimeters = inches times 25.4) and program the job in the metric mode.

As you know, all current model CNC machines can be set to power-up in either measurement system mode. If you work exclusively in one measurement system mode or the other, you need not command G20 or G21 in your programs (though we recommend doing so as one of your safety command words).

If, of course, your company has some programs written in metric mode and others written in inch, it is very important that you include the measurement system mode selection G-code (G20 or G21) in your program startup format for *all* of your programs.

For the most part, you will experience no problems by choosing measurement system mode based upon the way your prints are dimensioned. But there is one important advantage of using the metric mode. This advantage has to do with the machine's *least input increment*.

In inch mode, the least input increment for most machining centers is 0.0001 inch. If your tolerances are quite open (not hard to hold), this least input increment value will be acceptable. However, if you are trying to hold closer tolerances of under 0.001 inch, it may be somewhat difficult to make adjustment (with offsets) to hold the tolerance based on the rather large least input increment of 0.0001 inch.

In metric mode, the least input increment is 0.001 millimeter. *0.001 millimeter is less than half of 0.0001 inch* (0.001 mm is actually 0.000039 inch, or about 40 millionths of an inch).

Think of it this way. A ten inch long linear axis has 100,000 positions in inch mode. In metric mode, the same ten inch long linear axis has 254,000 positions. This finer resolution makes it much easier to hold size when you must deal with tighter tolerances.

Key Concept 6: Features That Help Simplify Programming

Secondary Reference Position, G30

Though not a very common feature, some machining centers do make use of a *secondary reference position*. The *primary* reference position is the zero return (reference) position, commanded by G28. Most machine tool builders use this point of reference for tool changing and, if so equipped, for pallet changing. However, there are some machine tool builders that use the secondary reference position for these functions. Ask an experienced person in your company if any of your machines use this feature. Again, most do not.

G30 is programmed in *exactly* the same manner as G28. The only difference is that the machine will go to the secondary reference position. For example, the command

```
N075 G91 G30 X0 Y0 Z0
```

sends all axes to the secondary reference position.

Scaling Commands, G50 and G51

Scaling allows you to manipulate the *size* (scale factor) of your programmed movements. Though not commonly needed with conventional (manual) programming, these commands can be quite helpful in companies that machine with three dimensional programs, as is common when machining molds.

With scaling, one program can be used to machine different size molds. Since this function has such limited application, it is commonly an option that must be purchased for an additional price. Here are the words involved with scaling.

    G50 - Cancel scaling mode
    G51 – Activate scaling mode
    I - Scale center in X
    J - Scale center in Y
    K - Scale center in Z
    P - Scale factor

This example command will reduce all programmed motions to half size (scale factor of fifty percent).

```
G51 I0 J0 K0 P0.5
```

After making the motions under the influence of scaling, G50 must be commanded to cancel it.

G50.1 and G51.1 - Mirror Image Commands

As the name implies, mirror image is used to generate a series of movements that represent the mirror of the programmed path. All that really happens when mirror image is activated, however, is that the machine *reverses the sign* (plus to minus or vice versa) for the mirrored axis. An X-axis position of X50.0 before mirror image will be executed as X-50.0 after X-axis mirror image is turned on.

Applications for Mirror Image

Mirror image applies best to hole-machining operations. Say you have a workpiece that must be machined in a right-hand and a left-hand version. Mirror image allows you to use the same program that machines the left-hand workpiece to machine the right-hand workpiece.

Unfortunately, contour milling operations present a problem for mirror image. While mirror image will function properly, the problem is related to basic machining practice. A climb milling operation will be converted to a conventional milling operation when mirror image is activated (or vice versa). In most applications, this is unacceptable, since workpiece finish – and sometimes its size – will be substantially different from one style of milling to the other. Figure 6.27 illustrates the points made so far.

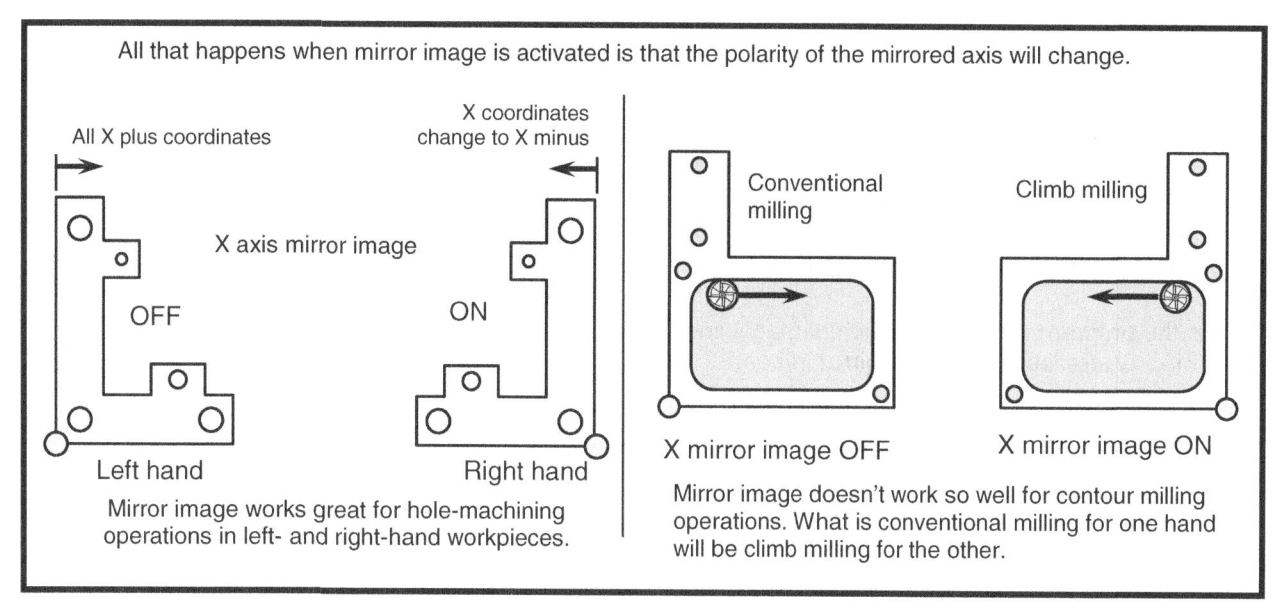

Figure 6.27 – Applications for mirror image

The Two Ways To Activate Mirror Image

Mirror image can be turned on manually (through the *setting page* of the display screen) or it can be activated from within a program. Your application determines whether you need to turn mirror image on manually or in your program.

Manually Activating Mirror Image
Say for example, you have five-hundred left-hand workpieces and five-hundred right-hand workpieces to run. You intend to run all of the left-hand workpieces in one setup, then tear down the setup and run the right-hand workpieces. In this case, there is no need to activate mirror image from within the program. Simply run one hand of the workpiece with mirror image turned off, and then turn on mirror image (manually) before you run the opposite hand.

In order for mirror image to work properly in this application, *you must also reverse the polarity of the workpiece coordinate system offset register for the axis being mirrored*. As always, the plus sign is assumed. You need only specify the sign if it's negative (-).

Activating Mirror Image From Within a Program
Some applications for mirror image require that you activate it from within in a program. In the previous left- and right-hand workpiece example, say you intend to run one set (left- *and* right-hand) of workpieces during each CNC cycle. If using mirror image to machine one of the workpieces, you will need to activate mirror image during the program's execution.

Most FANUC and FANUC compatible controls use a G51.1 word to activate mirror image. The letter address for the axis to be mirrored (X or Y) as well as the center position of mirror is included within this command. During setup, the workpiece coordinate system zero assignment values will be measured in the normal manner for each workpiece.

Here is an example program that combines a subprogram with mirror image. The center of mirror is right between the two workpieces that are spaced ten inches apart on the table (five inches to the right of the left workpiece).

Main Program (shown using inch mode):

```
%
O0001
N005 G20 G54 (½ drill)
N010 G50.1 (Cancels mirror image in X)
N015 M98 P1000 (Run entire left hand workpiece)
```

Key Concept 6: Features That Help Simplify Programming

```
N020 G51.1 X-5.0 (Turn on X-axis mirror image, center of mirror
is five inches from the workpiece coordinate system zero point)
N025 M98 P1000 (Run entire right hand workpiece)
N030 G50.1 (Cancel mirror image)
N035 G91 G28 X0 Y0 Z0 (Go to zero return position)
N040 M30
%
```

Here is the program that does the actual machining. Though it is quite simple (just drilling two holes with one tool), it nicely stresses the use of mirror image.

```
%
O1000 (Program written for left hand workpiece)
N005 G90 S500 M03
N010 G00 X1.0 Y1.0
N015 G43 H01 Z0.1 M08
N020 G81 X1.0 Y1.0 R0.1 Z-0.5 F6.0
N025 X2.0
N030 G80 M09
N035 G91 G28 Z0
N040 M99
%
```

Motion Relative to Zero Return Position (machine coordinate system), G53

G53 allows motion to be programmed relative to the machine's zero return (reference) position. The command

```
G53 X0 Y0 Z0
```

will send the machine (at rapid) to its zero return position in all three axes. As you know from lesson 5.1, this command is much easier to use than G28. However, if one or more of your (older) machines does not have G53, you should use G28 to maintain program compatibility among your machines.

G53 can also be helpful with certain accessory devices. If for example, you have a *manual pallet changer* (most popular on *vertical* machining centers), it is likely that the X and Y axes have to be moved to a specific position before the manual pallet change can be performed. The pallet change position relative to the workpiece coordinate system zero point in the program will vary from one program to the next. But the pallet change position relative to the zero return position will remain consistent for all programs. If your pallet change position is 380-mm in X and 300-mm in Y from the zero return position, the command

```
G53 X-380.0 Y-300.0
```

will rapid the machine directly to the pallet change position.

Single Direction Positioning Mode, G60

This feature can be used when hole-location is extremely critical. If more than one hole must be finish bored, it will, of course, be necessary to move the boring bar from hole to hole. Say the series of motions from one hole to another requires a reversal in axis direction. In this case, *backlash* (play within the axis) will affect positioning accuracy. While all machining centers have a feature called *backlash compensation*, if it is not perfectly adjusted, the holes will not be machined in their correct positions relative to one another.

Single direction positioning mode (G60) will cause the machine to approach each XY position from the same direction, eliminating potential backlash problems. G60 is modal, so it must only be programmed at the very beginning of the boring operation. Additionally, it only affects XY positioning movements and can be used even if you are using a canned cycle to bore the holes. G60 is canceled by the *normal cutting mode* command (G64).

Figure 6.28 is a drawing that shows what will happen when the machine is under the influence of the G60 command.

```
%
O0049
.
N095 T03 M06 (Boring bar)
N100 G54 G90 S800 M03 T04
N105 G60 G00 X25.0 Y25.0
N110 G43 H01 Z2.0
N115 G86 R2.0 Z-28.0 F50.0
N120 X100.0
N125 X50.0 Y50.0
N130 X100.0 Y100.0
N135 X25.0
N140 G80 G64
N145 G91 G28 Z0 M19
N150 M01
.
```

Position tolerance +/- 0.007-mm

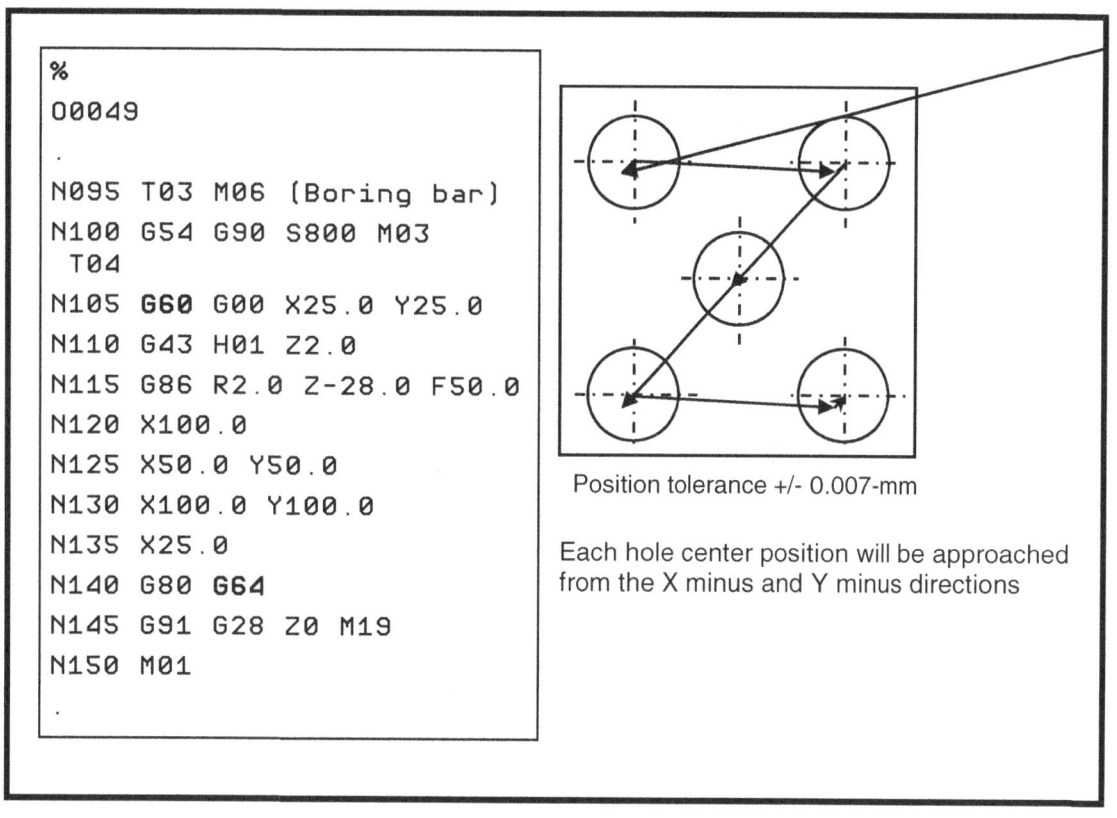

Each hole center position will be approached from the X minus and Y minus directions

Figure 6.28 – An application for single direction positioning

The tool will always approach from the same direction when under the influence of the G60 command. The direction of approach (usually from the X minus and Y minus direction) is set by a control parameter.

Coordinate Rotation G68 and G69

Coordinate rotation allows you to rotate a series of coordinates about a specified position and at a specified angle. At first glance this may appear to be just what the feature *polar coordinates* does (G15 and G16), but coordinate rotation is much more powerful. Polar coordinates gives you the ability to rotate just *one coordinate*, but coordinate rotation can rotate *a whole series of coordinates*, while maintaining the relationship among the coordinates. Figure 6.29 shows an application.

This shape must be machined in eight locations

Pick a convenient location and program the tool path

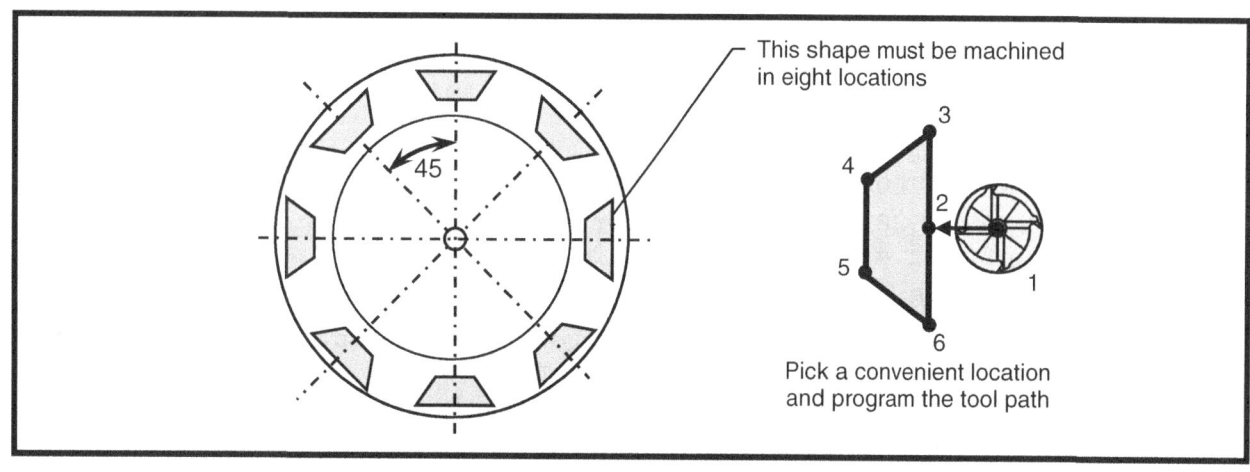

Figure 6.29 – Shape must be milled eight times

Key Concept 6: Features That Help Simplify Programming

The shape in Figure 6.29 is an outside raised shape that must be milled in eight locations. Coordinate rotation allows you to pick the most convenient location (the one that is best dimensioned on the print) and program the tool path in this location. We recommend placing this tool path in a subprogram:

Subprogram (shown using inch mode):

```
%
O2001 (Raised shape that must be rotated)
N1 G00 X3.5 Y0 (pt 1)
N2 G01 Z-0.25 F20.0
N3 G42 D01 X3.0 F3.0 (pt 2)
N4 Y1.0 (pt 3)
N5 X2.25 Y0.5 (pt4)
N6 Y-0.5 (pt 5)
N7 X3.0 Y1.0 (pt 6)
N8 Y0 (pt 2)
N9 G00 Z0.1
N10 G40
N11 M99
%
```

For our example, the shape is programmed for the right-most position (the three o'clock position). As with polar coordinates, this is the zero-degree position – and positive is counter-clockwise. The shape in the twelve o'clock position is at ninety degrees.

Here are the words involved with coordinate rotation:

- G68 - Instate rotation
- G69 - Cancel
- X, Y - Specify rotation center
- R - Rotation angle

Coordinate rotation allows you to specify the center of rotation right in the coordinate rotation command with X and Y. The command

N080 G68 X0 Y0 R45.0

will cause all upcoming programmed movements to be rotated about the workpiece coordinate system zero point at an angle of forty-five degrees.

G68 is modal and must be canceled when you're finished using it. G69 is the word to cancel it.

Here is a main program that uses the subprogram just shown to machine the shape in eight positions.

Program (shown using inch mode):

```
%
O0001 (Main program)
N005 G20 G54 G90 S800 M03 (Select workpiece coordinate system
setting offset #1, start spindle)
N010 G00 X0 Y0 (Rapid to center of workpiece in XY)
N015 G43 H01 Z0.1 (Instate tool length compensation, move tool
to just above work surface)
N020 M98 P2001 (Machine workpiece at three o'clock position –
subprogram is written for this location, so there is no need
for coordinate rotation yet)
```

```
N025 G68 X0 Y0 R45.0 (Instate rotation for forty-five degree angle)
N030 M98 P2001 (Machine workpiece at two o'clock position)
N035 G68 X0 Y0 R90.0 (Instate rotation for ninety degree angle)
N040 M98 P2001 (Machine workpiece at twelve o'clock position)
N045 G68 X0 Y0 R135.0 (Instate rotation for one-hundred-thirty-five degree angle)
N050 M98 P2001 (Machine workpiece at ten o'clock position)
N055 G68 X0 Y0 R180.0 (Instate rotation for one-hundred-eighty degree angle)
N060 M98 P2001 (Machine workpiece at nine o'clock position)
N065 G68 X0 Y0 R225.0 (Instate rotation for two-hundred-twenty-five degree angle)
N070 M98 P2001 (Machine workpiece at eight o'clock position)
N075 G68 X0 Y0 R270.0 (Instate rotation for two-hundred-seventy degree angle)
N080 M98 P2001 (Machine workpiece at six o'clock position)
N085 G68 X0 Y0 R315.0 (Instate rotation for three-hundred-fifteen degree angle)
N090 M98 P2001 (Machine workpiece at four o'clock position)
N095 G69
N100 G91 G28 Z0
N105 M30
%
```

In block N020, the first shape is machined. Since the subprogram is written for the shape in this location, there is no need for coordinate rotation yet.

Block N025 instates coordinate rotation, rotating about the workpiece coordinate system zero point at forty-five degrees. All upcoming motions will be rotated. Block N030 machines the shape at the two o'clock position (forty-five degrees).

Block N035 sets coordinate rotation for the ninety-degree location. Notice that coordinate rotation is *not accumulative* in the absolute mode. This angle (ninety-degrees) is an absolute position relative to zero degrees (three o'clock). Block N040 machines the shape at the twelve o'clock position. This process is repeated for the balance of shapes to mill.

In block N095, coordinate rotation is canceled. All motions from this point will be done in the normal manner.

Adjusting For Work Holding Devices That Are Not Square with the Axes

Coordinate rotation can also be helpful with setup imperfections. Large fixtures may be difficult to perfectly align with machine axes. If you have a spindle probe, it can be used to measure the angle of imperfection, along with the workpiece coordinate system zero location. Not only will the probe set workpiece coordinate system offset values, it will also specify the angle of imperfection with a G68 command.

Key Concept 6: Features That Help Simplify Programming

Key Points for Lesson 6.3:

- Control manufacturers provide a number of special features to help with programming. While some won't be of immediate need, it is important to know what is available.
- Block delete can be used to give the operator a choice between one of two possibilities.
- Block delete can help you program trial machining operations.
- Sequence numbers can be used as statement labels to change the order of program execution.
- G02 and G03 have a second function – they can be used to command helical motion, which is needed for thread milling.
- A dwell command can be used to make axis motions pause for a specified length of time – as is necessary when tool pressure must be relieved.
- Exact stop check (G09 and G61) can be used to force sharp corners when contour milling.
- G10 can be used to program offset entries.
- Polar coordinates (G15 and G16) can be used to rotate a coordinate about the current workpiece coordinate system zero point.
- Plane selection commands (G17, G18, and G19) must be used when machining occurs in the XZ or YZ plane.
- Mirror image (G50.1 and G51.1) can be used to machine a left-hand workpiece from a right-hand program.
- Single direction (G60) positioning will help machine bored holes in more accurate locations.
- Coordinate rotation (G68 and G69) will allow you to program a shape in one angular orientation and machine it in another.

Lesson 6.4–Programming Rotary Devices

Many CNC machining centers, and especially horizontal machining centers, are equipped with a rotary device that allow more than one workpiece surface to be exposed to the spindle for machining during the CNC cycle. This lesson will explain how they are programmed.

Objectives

After completing this lesson, students should be able to:

- ✓ Describe the difference between an indexer and a rotary axis
- ✓ Program rotation of indexers and rotary axes
- ✓ Describe the use of a rotary axis with absolute and incremental positioning
- ✓ Describe cutting and rapid rotary axis motion

Introduction

This lesson will only apply to you if your machining center has some kind of rotary device that allows the workpiece to be rotated during the CNC cycle. For *vertical* machining centers, you may have an indexer or rotary table (with full rotary axis) that rests *on top of the table.* For horizontal machining centers, your machine is likely to have an indexer or rotary axis built into the table of the machine.

The Difference Between an Indexer and a Rotary Axis

The major difference between an indexer and a rotary axis has to do with whether machining can occur *during rotation.* With an indexer, it cannot. Only a true rotary axis will allow machining during rotation. Frankly speaking, the vast majority of applications for rotary devices do not require machining during rotation. Instead, the rotary device is simply used to expose different surfaces of the workpiece to the spindle for machining. Machining is done *after rotation* – which either type of rotary device can do. *A rotary axis can be used as an indexer but an indexer cannot be used as a rotary axis.*

A Note to Horizontal Machining Center Programmers

All example programs to this point have been in the format for a vertical machining center. And we have tried to keep our examples as easy to understand as possible. Though our emphasis has been for vertical machining centers, note that every presentation to this point will also apply to horizontal machining centers – and – if you do have a horizontal machining center, it is mandatory that you understand the points made thus far.

The major usage difference between vertical and horizontal machining centers is that most horizontal machining centers are equipped with some form of rotary device within the machine's table. And by the way, if a horizontal machining center does not have a rotary device (some horizontal boring mills fit into this category), it is programmed in *exactly* the same manner as a vertical machining center.

Any of the programs shown in this text so far can be run on a horizontal machining center without major modifications. Only the minimal differences in program formatting shown in lesson 5.2 need to be done (changing movements to the tool change position). When you think about it, a horizontal machining center is nothing more than a vertical machining center that has been placed on its back.

When a rotary device is used, either on a vertical or horizontal machining center, programs get longer. They don't necessarily become more difficult to write, they just get longer. Again, several surfaces of the workpiece can be machined in one program, so more tools can be used by a program. (And by the way, the automatic tool changer magazines on horizontal machining centers tend to have a much larger capacities than those used on vertical machining centers. Indeed, some can hold well over one-hundred tools.)

Key Concept 6: Features That Help Simplify Programming

Additionally, most horizontal machining centers are equipped with *pallet changers*, meaning two or more jobs can be running on the machine at a given time (one workpiece is machined in the work area while the operator loads another). This means at least two programs will be needed in the machine at any given time.

The most difficult part of programming for a machine that has a rotary device is related to preparation and organization. Since more than one workpiece surface can be machined – and more cutting tools will be needed – the programmer must carefully plan the process. Developing a sequence of machining operations is relatively easy when machining but one surface on a vertical machining center. But it can be quite challenging when several surfaces must be machined. Just keeping track of which side of the workpiece is currently facing the spindle can be difficult. But with a workable process completed, programming for a horizontal machining center remains just as easy as it is for a vertical machining center.

Benefits of Rotary Devices

As stated, the most basic purpose of any rotary device is to expose several sides of the workpiece to the spindle for machining during one setup. This provides three benefits:

> 1) **The number of setups can be reduced.** The total time it takes to produce workpieces can be minimized and less handling is required. Fewer programs are needed, less machine down time, and faster through-put of workpieces in the shop are among the advantages of using rotary devices.
>
> 2) **It is easier to hold accuracy from one surface of the workpiece to another.** If multiple setups must be made, accuracy suffers since it may be impossible to locate the part perfectly in each setup. With an indexer or rotary table, the part is not removed for the entire machining cycle. The overall accuracy from surface to surface on the workpiece will be better.
>
> 3) **Chips fall easily from the work area and do not interfere with machining.** This is only true for rotary devices used on horizontal machining centers.

Indexers

An indexer is a device that allows the workpiece to be quickly rotated to a specified angular position. The rotation rate is *uncontrollable* and very fast, making it impossible to machine a workpiece while the indexer is rotating. Some indexers can only rotate in one direction, meaning the rotation will not always be the in most efficient direction.

There are many kinds of indexers. Usually the name for the indexer is related to the smallest angular increment of rotation it can perform. This increment usually determines the method by which the indexer is programmed.

Common indexers include:

- 90 degree indexers
- 45 degree indexers
- 5 degree indexers
- 1 degree indexers

Programming Indexer Rotation

Let's discuss how the most common indexers are programmed.

90 Degree and 45 Degree Indexers

These indexers are usually activated by a single *M-code*. The M-code number is determined by the machine tool builder or by the company that supplies the indexer. It should be listed in the series of M-codes that comes with the machine or indexer. You must, of course, know this M-code in order to program a 90 or 45 degree indexer.

When you want an index to occur, you simply specify the M-code. One index of the indexer's angle will occur. This can be somewhat cumbersome. Say for example, you have a 45 degree indexer and you want to index 180 degrees. If the word M13 is used to cause an index, you must program four consecutive M13 words (one per command) to total the desired 180 degrees.

Also note that most 90 and 45 degree indexers can only rotate in one direction (though there are exceptions). With this kind of indexer, rotation will not always be in the most efficient direction.

Lesson 6.4: Programming Rotary Devices

Five Degree Indexers

Five degree indexers are handled differently. Most manufacturers provide a *series* of M-codes for their five degree indexer, based on the most common angles of index. For example, M71 may specify a five degree index, M72 a fifteen degree index, M73 a forty five degree index, and M74 a ninety degree index. This will minimize the number of M-codes you will have to string together for an odd index angle.

Some 5 degree indexers force you to set (mechanically) the angle of index desired and only one M-code is used to activate the indexer (just like 45 and 90 degree indexers). This kind of indexer can be just as cumbersome to work with as a 45 or 90 degree indexer.

Like 45 and 90 degree indexers, many 5 degree indexers allow only one rotation direction. Those that do allow bi-directional rotation usually use two more M-codes to control rotation direction, one for clockwise and another for counter clockwise. Again, you must reference the indexer manufacturer's or machine tool builder's manual to learn more about how your indexer is programmed.

One Degree Indexer

This is the most popular type of indexer used on machining centers today. One degree indexers are usually the easiest indexers to work with. Most one degree indexers are programmed with a special letter address word (commonly A, B, or C). With this word, you can specify the exact angle of index desired. Say the letter address B is used for indexer rotation. If you want a twenty-seven degree index, you simply specify B27 (without a decimal point). Most one degree indexers have two M-codes to specify the direction of rotation.

As you can see, if your machine has an indexer, there are several possibilities for how it is programmed. We cannot be more specific. You must ask your instructor or an experienced programmer to explain how your indexer is programmed. It should, at least, be handled with one of the methods just presented.

Being able to program your indexer's rotation is but part of the challenge of programming rotary devices. After introducing rotary axes, we will discuss the overall approach to programming a machining center having either kind of rotary device.

Rotary Axes

Like an indexer, a rotary axis can be used with either a vertical or horizontal machining center. And as with an indexer, it is mounted on a vertical machining center's table. With a horizontal machining center, it is mounted within the machine as an integral part of the table mechanism.

A rotary table is much more flexible than an indexer. It incorporates a full axis of motion (again, a *rotary axis*). The method by which rotation is commanded is also much better. You can more precisely control the rotation angle as well as the direction of rotation. You can even control the motion *rate* (in degrees per minute), meaning you can machine during rotation. And rotary axes allow you to program rotation to positions specified with portions of a degree. The least input increment with most machines is 0.001 of a degree. So a rotary axis is like an indexer that has 360,000 positions.

As stated, a rotary axis is considered to be a true axis of the machine. Like any linear axis, it is designated with a letter address. *Which* letter address is used depends on the machine tool builder – and how the rotary axis is mounted on/in the machine tool. With horizontal machining centers (the rotary device is rotating the table itself), the designation for the rotary axis is always letter address B, and all machine tool builders adhere to this standard.

With vertical machining centers, when the rotary axis is mounted on the table in such a way that it is parallel to the X-axis (as it normally is), it is designated as the *A-axis*. By *parallel to an axis*, we mean the centerline of the rotary device is parallel to the axis. If the rotary axis is mounted parallel to the *Y-axis* (again, as it is with horizontal machining centers), it is called the *B-axis*. If it is mounted parallel to the Z-axis, it is called the *C axis*. Unfortunately, not all machine tool builders adhere to these naming standards. You must confirm your machining center's rotary axis name in the machine tool builder's programming manual.

For the purpose of this lesson, we will call the rotary axis the *B-axis* for horizontal machining centers and the *A axis* for vertical machining centers. If you understand our presentations, you will be able to easily apply what you have learned to your particular machine/s, regardless of what your rotary axis is named.

Key Concept 6: Features That Help Simplify Programming

How To Program a Rotary Axis Departure

A rotary axis allows a decimal point to be specified within the rotary axis designation. All rotary commands are given directly in *degrees of rotation*. For example, say you want to designate a rotary axis position of forty-five degrees for a horizontal machining center. Here is the command:

```
N050 G00 B45.0
```

A rotary axis can only be programmed to *three places* after the decimal point (not four, like X, Y, and Z in the inch mode). Also, if you are designating an angular departure that includes a value less than one degree, you must specify a *decimal portion of a degree*. Some design engineers designate portions of a degree in *minutes and seconds*. Minutes and seconds must be converted to decimal format for rotary axis designation. Here is a formula to convert minutes and seconds to decimal format:

> Degrees (in decimal format) equals Degrees plus (minutes divided by 60) plus (seconds divided by 3600)

Say you have the angular value specified as 13 degrees, 27 minutes, 37 seconds. And this angle must be specified in a rotary axis positioning motion. To make the conversion, you first divide 27 by 60. The result is 0.45 degrees. Then you divide 37 by 3600. The result is 0.010 degrees. Finally, you add 13 plus 0.45 plus 0.010 and the result is 13.46 degrees. If your rotary axis is designated with the B letter address, you will specify this command as:

```
N045 B13.46
```

There is one more important point about programming angular departure for a rotary table. All rotary axes can be used just like indexers. You can quickly rotate to an angular position to expose a surface for machining. Almost all rotary axes allow you to *clamp the rotary axis* after rotation and prior to performing powerful machining operations. This reduces the stress on the rotary axis drive system. Two M-codes are used for this purpose – one to clamp and one to unclamp. You must check your machine tool builder's manual to find the related M-codes. Forgetting to clamp your rotary axis prior to machining can result in undue wear and tear on the rotary axis.

Comparison To Other Axes

Just about everything you know about X, Y, or Z applies to a rotary axis. The methods by which you command motion in X, Y, and Z can still be used to control any rotary axis. Here is a list of the commonalities:

1) Zero Return Position
2) Polarity
3) Designation of Workpiece coordinate system zero
4) Absolute Mode (G90) and Incremental Mode (G91)
5) Rapid (G00) and Straight Line Cutting (G01)
6) Usage in Canned Cycles

Zero Return Position

Just like X, Y, and Z, a rotary axis will have a zero return position. And as with X, Y, and Z, part of powering up the machine may be to send the rotary axis to its zero return position. The zero return position is just as much a point of reference for the rotary axis as it is for any linear axes.

To command the rotary axis to go to the zero return position is identical to doing so for linear axes. If the rotary axis is designated with letter address B, here is a command that will sent the machine to zero return position in all axes:

```
N055 G91 G28 X0 Y0 Z0 B0
```

In this case, all four axes will move to zero return position at the same time. If you want to zero return only the B axis, here is the command:

```
N055 G91 G28 B0
```

As with X, Y, and Z, the zero return position for the rotary axis will make an excellent reference point for your program. Your rotary axis workpiece coordinate system zero point designation for the rotary axis is every bit as important as it is for X, Y, and Z, and will be specified from the rotary axis zero return position. The work offset page will have a rotary axis register (labeled as A, B, or C, depending upon your machine's rotary axis name). You will place the rotary axis workpiece coordinate system zero assignment value in this register, just like you do for any linear axis.

This next point may be a little difficult to visualize. Motions along X, Y, and Z axes are *linear*. There are definite ends (over-travel limits) to each linear axis of motion. X, Y, and Z axes will stop when their limits are reached (and an alarm will be sounded). Also, whenever you send a linear axis to its zero return position, the axis will move in the positive direction. Remember, the zero return position for any linear axis is close to its plus over-travel limit. The machine cannot help but move in the positive direction to get to the zero return position.

But with a rotary axis, rotation can continue indefinitely in either direction. That is, *a rotary axis cannot over-travel*. And as with a linear axis, the motion to the zero return position will be done in the positive direction. This presents a bit of a problem when commanding the rotary axis to go to its zero return position.

Let's illustrate the problem with an example. Say the rotary axis is resting at the zero return position when these commands are given:

```
N060 G00 G91 B1.0
N065 G00 G91 B-1.0
```

The rotary axis will move (incrementally) one degree in the plus direction and then rotate back one degree in the minus direction. Though the rotary axis ends up back at the same angular position as zero return, the zero return position has not truly been commanded (the B axis origin light will *not* come on, if the machine has one). Remember, the G28 command *sends* each axis to its zero return position. The commands above just happened to leave the rotary axis at the same angular position as zero return.

Additionally, when a rotary axis zero return is commanded, the control will make the axis approach the zero return position *from the minus direction* (just as it does with X, Y, and Z). Given this knowledge, what do you think will happen if (after running the commands above) this zero return command is given?

```
N070 G91 G28 B0
```

Though you may not like it, the control will actually rotate the axis a full revolution (360 degrees). The rotary axis ends up right where it starts (and the B axis origin light *will* come on). Again, the zero return position is reached from the minus side of each axis. And in this case, a full revolution is required to get there.

There will probably be many times when you are surprised by an unexpected rotary axis departure when commanding a zero return. When you find them, they are usually quite easy to correct. For example, given the scenario above in which the machine is making a full revolution to get to the zero return position, you can give the command:

```
N050 G91 G28 B-5.0
```

This will cause the rotary axis to move just five degrees in the negative direction (the intermediate position), and then rotate in the positive direction to the zero return position. Only short angular motions will be required.

Polarity

Like linear axes X, Y, and Z, a rotary axis has polarity (plus versus minus direction). As viewed from above the rotary axis, plus is the clockwise direction and minus is the counter-clockwise direction. However, not all machine tool builders adhere to this standard.

Designation of Workpiece Coordinate System Zero

If you are going to be programming the rotary axis in the absolute mode (G90), you must also designate an angular workpiece coordinate system zero point for the rotary axis. Like X, Y, and Z, the machine must know the location of origin point from which all coordinates will be specified. And as with any linear axis, this point of reference is called the workpiece coordinate system zero position. Also like X, Y, and Z, you will be including the B designation value in the workpiece coordinate system zero assignment (in the work offset).

Key Concept 6: Features That Help Simplify Programming

The method by which you determine polarity for the B axis workpiece coordinate system zero assignment value will be the same as for X, Y, and Z. For work offsets, the angular distance is taken *from* the zero return position *to* the workpiece coordinate system zero point.

Since very few drawings have angular datum surfaces, there is no need to be overly concerned with the workpiece coordinate system zero point location for a rotary axis. It is not nearly as important as it is for X, Y, and Z. If workpiece coordinate system zero is chosen wisely for X, Y, and Z, of course, many program coordinates will be taken right from the print. But with the rotary axis, no such help is available. For this reason, most programmers will simply designate that the side of the rotary table facing the spindle at the start of the program (commonly the zero return position) as the workpiece coordinate system zero side. In this case, the workpiece coordinate system zero assignment value for the rotary axis (in the work offset) will be *zero*.

As we progress in this presentation, you will see that it is sometimes wiser to program the rotary axis *incrementally* (especially when the rotary axis is used as an indexer). If the rotary axis is programmed exclusively in the incremental mode, *there is no need to assign a workpiece coordinate system zero point for the rotary axis*.

Absolute Positioning Mode

As you know, when you specify coordinates relative to the workpiece coordinate system zero point, you are working in the absolute positioning mode. And we have pointed out several times during our discussion of the linear axes (X, Y, and Z) that the absolute mode is best for specifying coordinates in the X, Y, and Z axes.

When it comes to the rotary axis, however, working in the absolute mode may not be the best way to specify axis rotation. Some unexpected things can happen if you do not fully understand how the absolute mode affects a rotary axis. In the absolute mode, the rotary axis is harder to visualize than X, Y, and Z.

Figure 6.30 shows a horizontal machining center that has a rotary axis internal to the table as viewed from above.

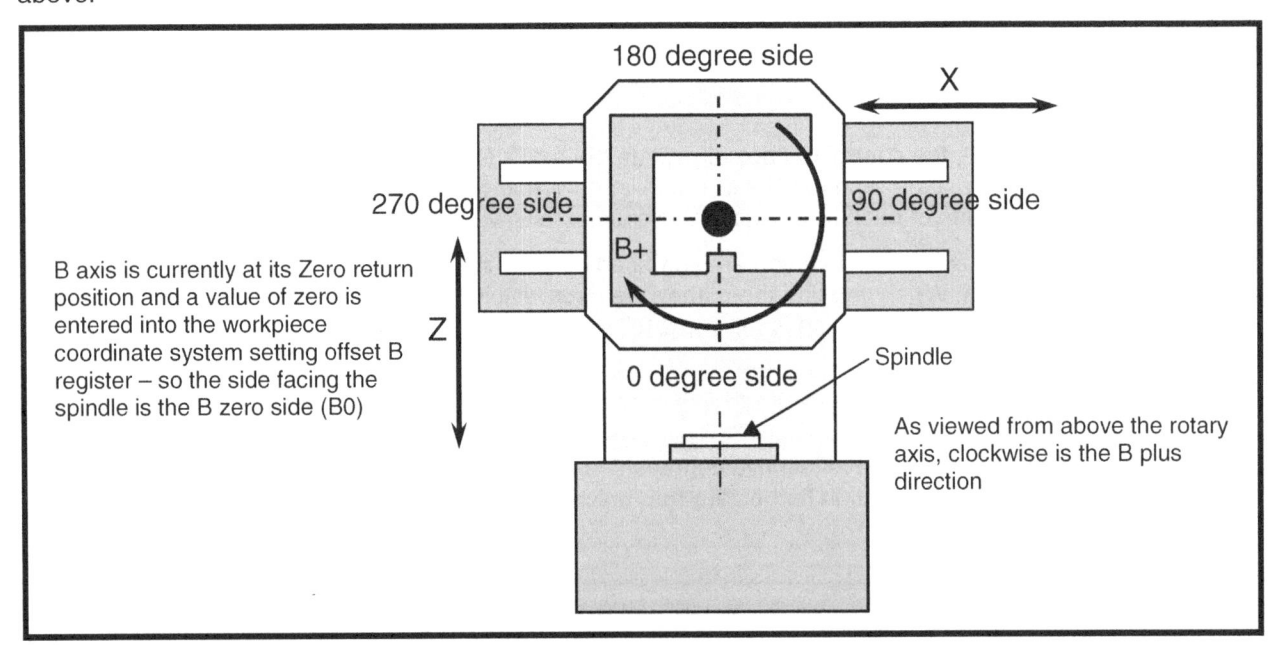

Figure 6.30 – Rotary axis as viewed from above a horizontal machining center – the column of the machine is at the bottom of this illustration

As you can see, we have depicted the B axis at its zero return position. And the entry in the work offset B register is zero. This means that the table side facing the spindle while the B axis is at zero return is the zero degree side. In the absolute positioning mode, whenever the control executes a B word of B0, the rotary axis will end its rotation with this side facing the spindle.

The ninety degree side (B90.0) is ninety degrees counter clockwise (plus) from the zero degree side. The one-eighty degree side (B180.0) is ninety degrees further from the ninety degree side. The two-hundred-seventy degree side (B270.0) is ninety degrees further. And the three-hundred-sixty degree side (B360.0) is ninety degrees further yet.

Here's where it gets a little complicated. You must remember that the specified table side will end up facing the spindle at the completion of the rotation command (when the machine is in the absolute positioning mode). The direction the table rotates (clockwise or counter-clockwise) is based upon the rotary axis position prior to the command.

If rotating from a small B coordinate value to a larger one (plus), the rotation direction (as viewed from above the table) will be clockwise. If rotating from a large B coordinate value to a smaller one (minus), the rotation direction will be counter-clockwise. This can lead to some undesirable and sometimes unexpected motions. Say for example, the two-hundred-seventy degree side (B270.0) is currently facing the spindle. In the absolute mode, you specify G00 B0 to rotate the axis at rapid to the zero degree side. The machine will rotate the table *counter-clockwise* (again, rotating from a large B coordinate value to a smaller one). So the axis will take the *long way* to get to the zero degree side.

To make the axis take the shorter direction (clockwise), you can, of course, specify G00 B360.0. The three-hundred-sixty degree side is the same as the zero degree side – and since three-hundred-sixty is greater than two-hundred-seventy, the rotation will be in a clockwise direction. And for continued rotations, you can just keep making the B word bigger.

But this will (eventually) lead to another problem. At some point (probably when the cycle is run the next time), the rotary axis will receive a command of B0 in the absolute positioning mode. When it does, the rotary axis will make a very long rotating motion to get to the commanded position. If, for example, the B axis had eventually accumulated up to one-thousand-eighty degrees (B1080.0) in the previous cycle, it will make three full clockwise rotations when B0 is commanded.

Most machines have two M-codes to help you override the natural motion direction related to the rotary axis. If you reference your machine tool builder's programming manual, you may will likely find M-codes to specify rotation direction. M12, for example, may be used to cause clockwise rotation while M13 may be used to cause counter clockwise rotation. With this kind of machine, if the rotary axis is currently at the two-hundred-seventy degree side when you give the command G00 B0 M12, the rotary axis will rotate clockwise (the short way) to get to the zero degree side. The two direction-specifying M-codes are very convenient. If your machine does not have them, it may be best to program the rotary axis with incremental positioning.

Study Figure 6.31 to help you understand a rotary table when it is programmed in the absolute positioning mode. These illustrations assume that your machine has no overriding M-codes to force rotation direction. Keep in mind that these examples also apply to a rotary axis that is placed on the table of a vertical machining center.

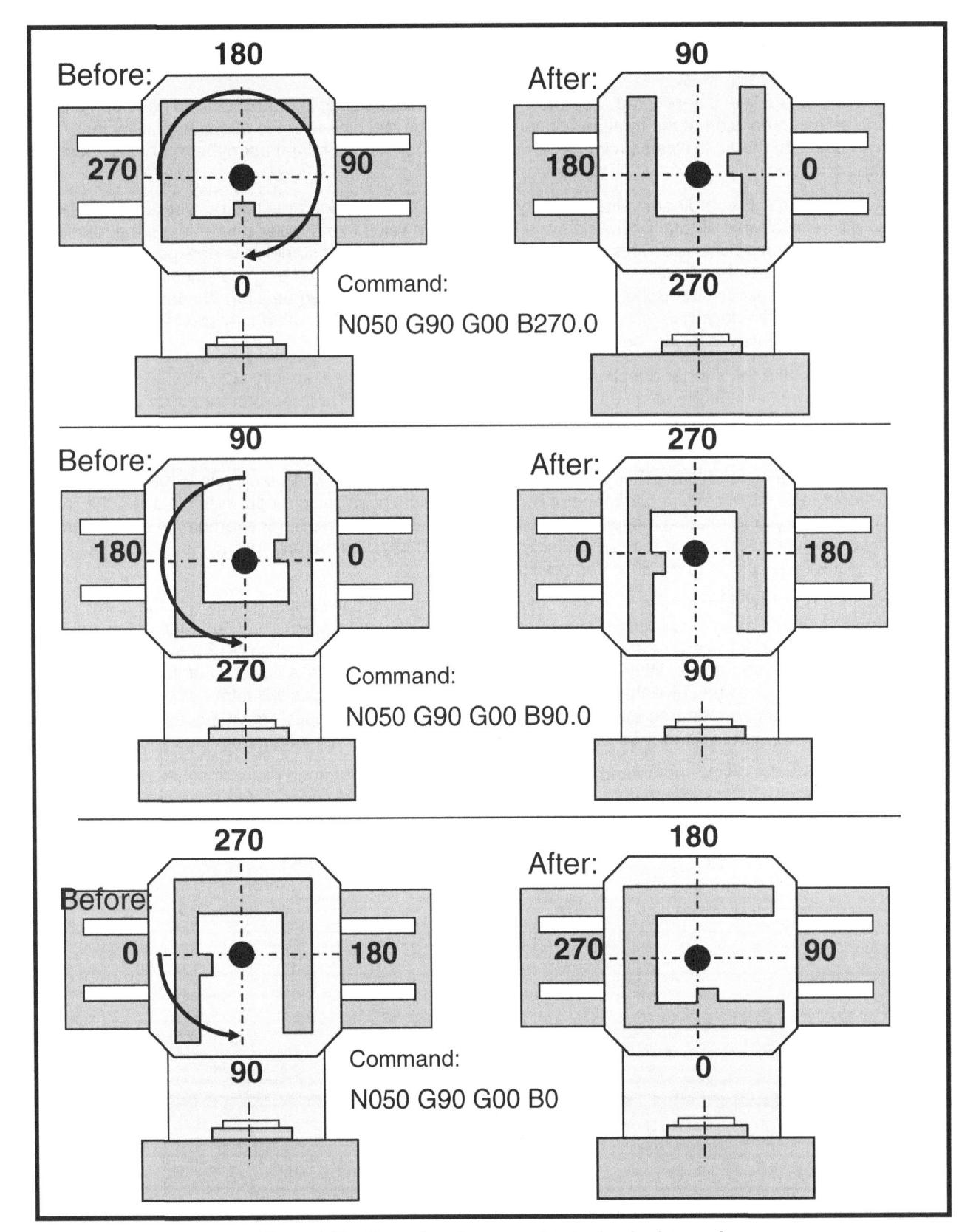

Key Concept 6: Features That Help Simplify Programming

Before: 180, 270, 90, 0

Command:
N050 G90 G00 B270.0

After: 90, 180, 0, 270

Before: 90, 180, 0, 270

Command:
N050 G90 G00 B90.0

After: 270, 0, 180, 90

Before: 270, 0, 180, 90

Command:
N050 G90 G00 B0

After: 180, 270, 90, 0

Figure 6.31 – Examples of rotary motion in the absolute mode

If you are using a rotary axis as simple indexer, the rotary axis departures you program will always be under three-hundred-sixty degrees. Indeed, they will usually be no more than one-hundred-eighty degrees. But a true rotary axis is not limited when it comes to the size of a departure. If you are machining a workpiece during rotation (a turbine rotor would require this kind of machining), you may need to command several rotations of the rotary axis in a single command. If the rotary axis is currently at the zero degree side (B0) when you give the command G90 G01 B3600.0 F60.0, the rotary axis will make ten rotations in the clockwise direction.

Incremental Positioning Mode

If your machining center does not have M-codes to override rotation direction in the absolute positioning mode, the incremental positioning mode may be best for most rotary axis applications. There is even a way to specify absolute positioning in linear axes at the same time you make an incremental rotary axis motion.

As with any linear axis motion, the incremental mode allows you to specify the rotary axis motion *from the current position*. In the incremental mode, clockwise rotation is still the positive direction and counter-clockwise is still the negative direction. If you want the rotary axis to rotate ninety degrees clockwise, you can specify this command:

```
N050 G91 G00 B90.0
```

G91 is modal, so you must remember to include a G90 in the next linear axis motion. If you want the rotary axis to rotate ninety degrees counter clockwise, you will specify this command:

```
N050 G91 G00 B-90.0
```

Most machines even allow you to position the rotary axis in the incremental mode *while making other axes move in the absolute mode*. Consider this command:

```
N050 G00 G91 B-90.0 G90 X1.5 Y4.0
```

The B axis will move incrementally *while* the X and Y axes will move to a position relative to workpiece coordinate system zero.

Important point: If you use this technique (incrementally rotating while positioning linear axes in absolute mode), *get in the habit of specifying the rotary axis motion first* – as we have done in the example. Again, G90 and G91 are modal. Whichever is last in the command (G90 in our case) will be the instated mode when the command is finished. Since subsequent commands in your program will require the absolute mode, you'll want this mode instated at the completion of the rotation command.

Study Figure 6.32 to learn more about rotary axis departures in the incremental mode.

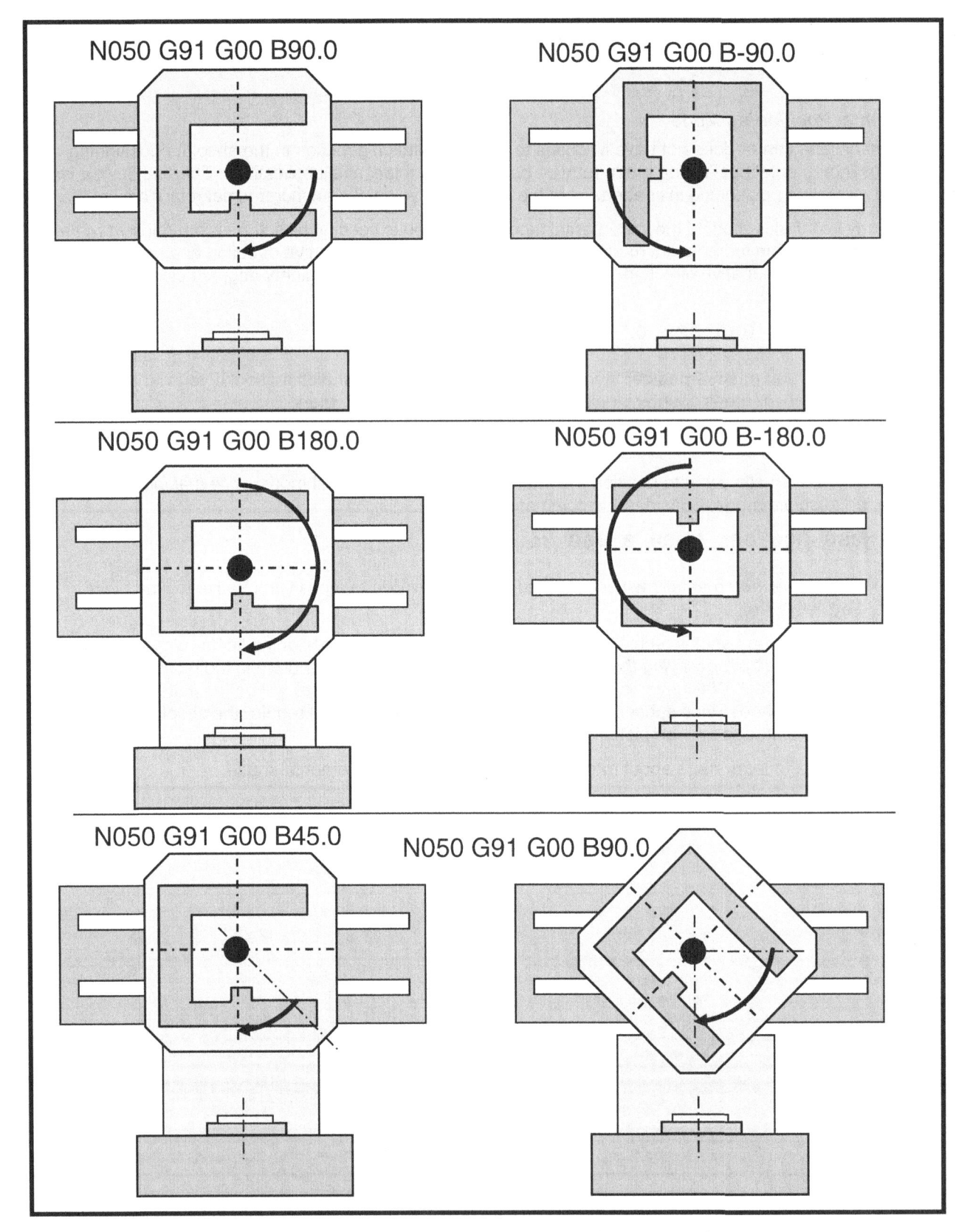

Key Concept 6: Features That Help Simplify Programming

Figure 6.32 – Examples of rotary axis motion in the incremental mode

As you can see, it is much easier to specify rotary axis departures in the incremental positioning mode.

Clamping the Rotary Axis for Machining After Rotation

Most rotary axes include a clamping system that can be used to lock the rotary axis after an index. This is important if you will be performing powerful machining operations after rotation. Two M-codes are used to control this clamping system, one to clamp, the other to unclamp. You must reference the set of M-codes for your machining center (commonly in the machine tool builder's programming manual) to find the related M-codes.

Rapid and Straight Line (linear) Motion

With one exception, the examples given so far have shown rapid motions for the rotary axis. In rapid mode, the rotary axis turns as quickly as it can. And again, when the rotary axis is rotated in the rapid mode, it is being used as a simple indexer. Machining will occur *after* rotation. Rapid rates for rotary axes vary from one manufacturer to another. Rotation rate is always specified in *degrees per minute*. And like the rapid rate for linear axes, it is usually very fast. If the rapid rate for your rotary axis is 3,600 degrees per minute, for example, your rotary axis can rotate at ten times in one minute.

A rotary axis can be programmed with G01 (straight line motion). But it cannot be programmed with G02 or G03 (circular motion), unless the machine is equipped with a special feature called *cylindrical interpolation*.

When a rotary axis departure is included in a G01 motion command, the motion rate (feedrate) must be specified in degrees per minute. This is the case even if the rotary motion also includes a linear axis motion. And frankly speaking, it is difficult to calculate a degrees-per-minute feedrate based upon the inches-per-minute feedrate you need for machining.

The method we show for calculating degrees per minute involves determining the amount of time the motion will take in minutes. First you must determine the desired feedrate in millimeters or inches per minute, based upon your measurement system choice (just as you will for any linear axis motion). Based upon the distance the cutting tool will be moving during the motion and the feedrate in millimeters (or inches) per minute, you can calculate the time required for the motion. By knowing the time required for the motion and the incremental angular departure for the motion in degrees, you can calculate the degrees-per-minute feedrate. Here are the related formulae:

- Time (in minutes) = Length of motion divided by desired millimeters (or inches) per-minute feedrate

- Degrees per minute (DPM) = Incremental rotation amount in degrees divided by time (in minutes)

Unfortunately, even determining the length of the motion can be difficult when the rotary axis is involved. While you may have to approximate, you can be pretty accurate. Say the tool tip is 100.0-mm (or 4.0-in) from the center of rotation and you need to make a ninety degree rotary axis departure. The circumference of a 100.0-mm (or 4.0-in) radius circle is 628.0-mm (or 25.12-in). This is calculated by multiplying pi times the diameter – 3.14 times 200.0-mm (or 8.0-in). Since the rotary motion is only one-quarter of the circle (ninety degrees), the motion distance will be one-quarter of 628.0-mm (or 25.12-in), resulting in 157.0-mm (or 6.28-in).

Let's say we've calculated the desired per-minute feedrate and it is 150.0 millimeters per minute (or 6.0 inches per minute). Based upon the first formula above, the time in minutes required for this motion will be 1.0466 minutes (157.0-mm divided by 150.0-mmpm or 6.28-in divided by 6.0-ipm).

Based upon the second formula above, the degrees-per-minute feedrate for this motion will be 85.99 dpm (90 degree rotation divided by 1.0466 minutes). This is specified with a feedrate word of F85.99.

Shown using metric mode, the commands to make this motion will be:

```
N045 G54 G90 S500 M03 T03 (Select work offset #1, absolute
mode, start spindle, get next tool ready)
N050 G00 X0 Y125.0 (Rapid to center of rotation in X and plunge
position in Y)
N055 G43 H01 Z117.0 (Rapid to just above work surface)
N060 G01 Z100.0 F150.0 (Plunge tool into workpiece)
```

```
N065 G91 B90.0 F85.99 (Machine slot in outside or round
workpiece)
N070 G90 Y175.0 F150.0 (Continue milling slot in Y-axis)
```

In block N045, notice that we've selected the absolute mode for linear axis motions. In block N060, we plunge into the slot at the desired feedrate (F150.0).

In block N065, the rotary axis is used to machine the slot for the incremental distance of ninety degrees. F85.99 is the degrees-per-minute feedrate.

In block N070, notice that we reselect the absolute mode and we re-specify the millimeters-per-minute feedrate.

The specification of the millimeters-per-minute feedrate in block N070 is extremely important. Without it, the machine will use the last programmed feedrate (F85.99) as a millimeters-per-minute feedrate in block N070, which will, of course, cause inappropriate machining - in this case way too slow, but in other cases may be much too fast.

Because degrees-per-minute feedrate is difficult to calculate, some programmers *cheat*. They approximate the degrees per minute feedrate and wait until the workpiece is actually being machined. Then they fine tune the feedrate by adjusting the *feedrate override switch*. After overriding the feedrate to the desired level, they change the programmed feedrate accordingly.

For example, say you program 100 degrees per minute for all rotary axis departures. At the machine, when verifying the program, you find that you must increase the feedrate override switch to 140% during the rotary axis motion before you are satisfied with the machining operation. In this case, the feedrate in the program must be changed to 140 degrees per minute.

While this technique works as long as you are a good judge of cutting conditions, you will never really know the actual cutting feedrate in millimeters or inches per minute.

Canned Cycle Usage

When used with canned cycles, the rotary axis can be used (just like X and Y) as a hole center-line coordinate. Every point made about X and Y during lesson 6.1 applies to the rotary axis. This allows you to easily machine a series of holes around the outside of the round workpiece.

Measurement system mode comment

Since rotary axis departures are specified in degrees per minute, and since the measurement system choice does not affect degree values (as it is linear motion distances), rotary axes are not affected by whether you work in metric mode or inch mode. You might consider the method by which you determine the degrees-per-minute feedrate to be an exception since it is based upon your desired cutting feedrate in millimeters or inches per minute. But once rotary axis feedrate is determined, it is specified, of course, in degrees per minute.

Approaching Rotary Device Applications

We now present a method for approaching rotary device programs. The basic approach is the same for both indexers and rotary axes. And it's also the same for horizontal and vertical machining centers. As you'll see, the real work lies in getting ready to write the program. The more sides of a workpiece that must be machined and the more operations that must be performed, the more difficult it will be to develop a sequence of machining operations.

Workpiece Coordinate System Zero Selection

The careful selection of the workpiece coordinate system zero point will make working with a rotary device much easier. Some odd things can happen when a *single* workpiece coordinate system zero is used for an entire program. A workpiece coordinate system zero that makes sense for one side of the workpiece will not make sense for another – after the rotary device rotates.

Look at Figure 6.33, an illustration that shows a front view of a vertical machining center that incorporates a rotary device. The rotary device is mounted parallel to the X-axis, meaning if it is a true rotary axis, it will be named the *A axis*.

Lesson 6.4: Programming Rotary Devices

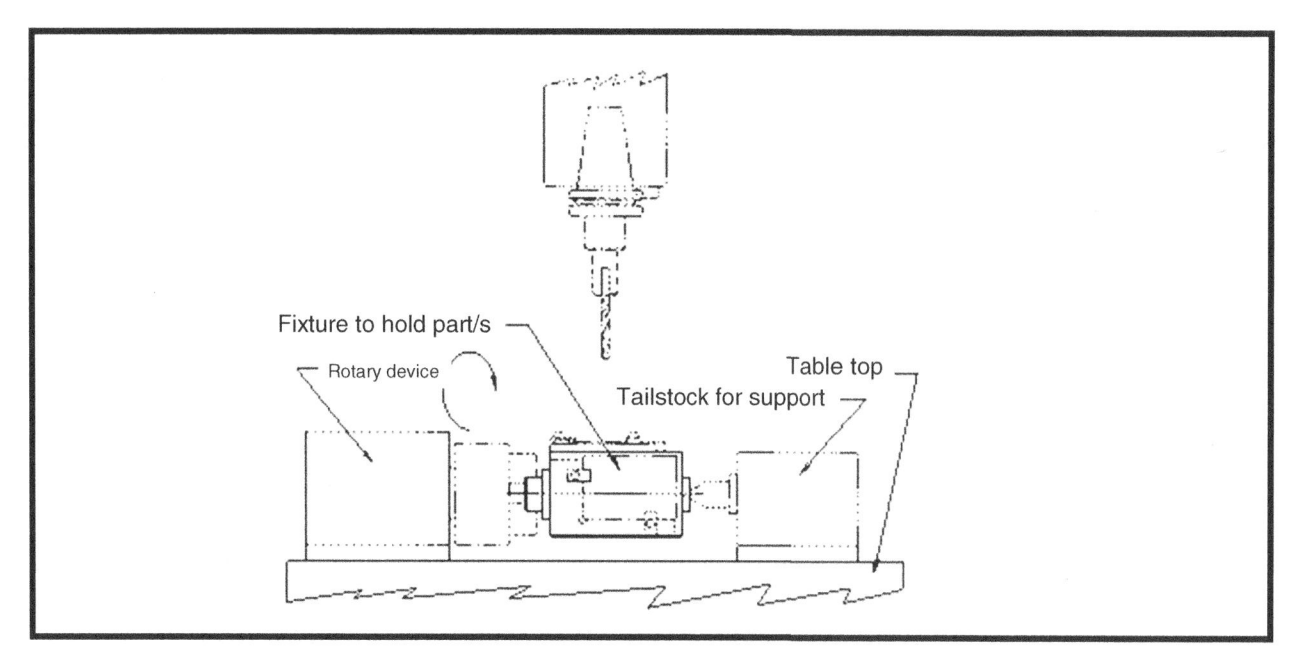

Figure 6.33 – Front view of vertical machining center with rotary device

As you can see, the rotary device is holding a fixture. This particular fixture is being used to hold two different workpieces. The fixture incorporates a turned diameter to accommodate the three-jaw chuck mounted on the rotary device. On the right side of the fixture is a hole to accommodate a tailstock for support.

Figure 6.34 is the assembly drawing for the fixture itself.

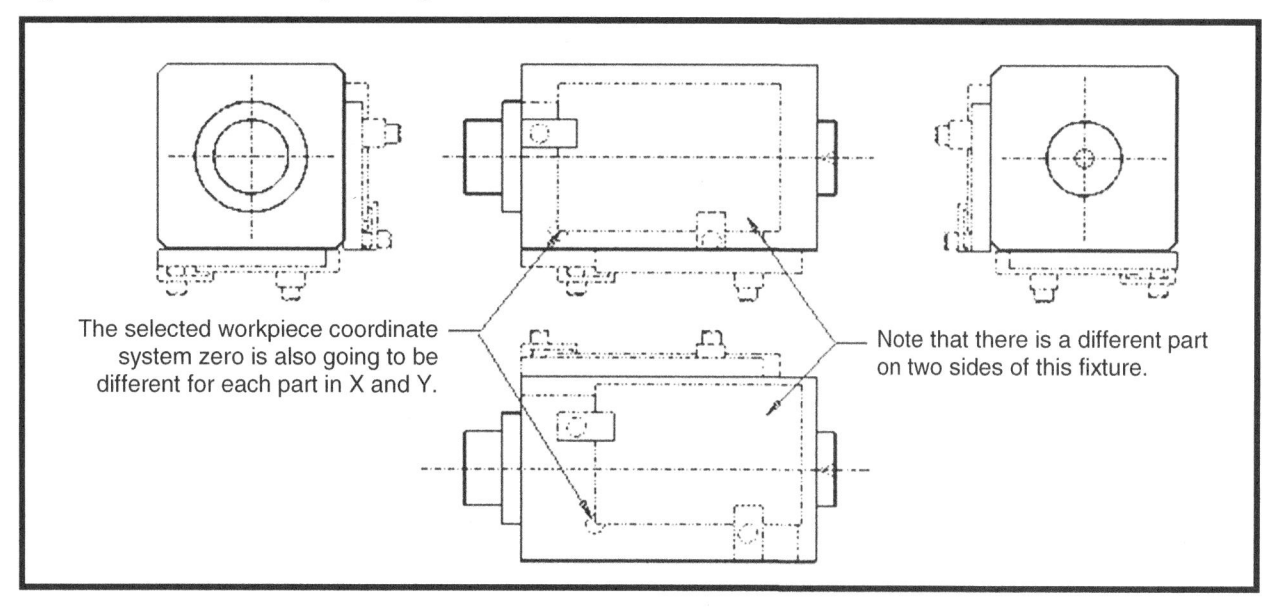

Figure 6.34 – Assembly drawing of fixture

Figure 6.34 better shows that two different parts are to be machined. The top view shows one of the workpieces and the front view shows the other.

Though it may be difficult to visualize, it will be next to impossible to assign *one workpiece coordinate system zero point* on the fixture that will allow programmed coordinates to be specified from the print for both workpieces. Figure 6.35 better illustrates that the two workpieces do not share the same location surfaces on the fixture. In this illustration, the two workpieces are super-imposed – one on top of the other.

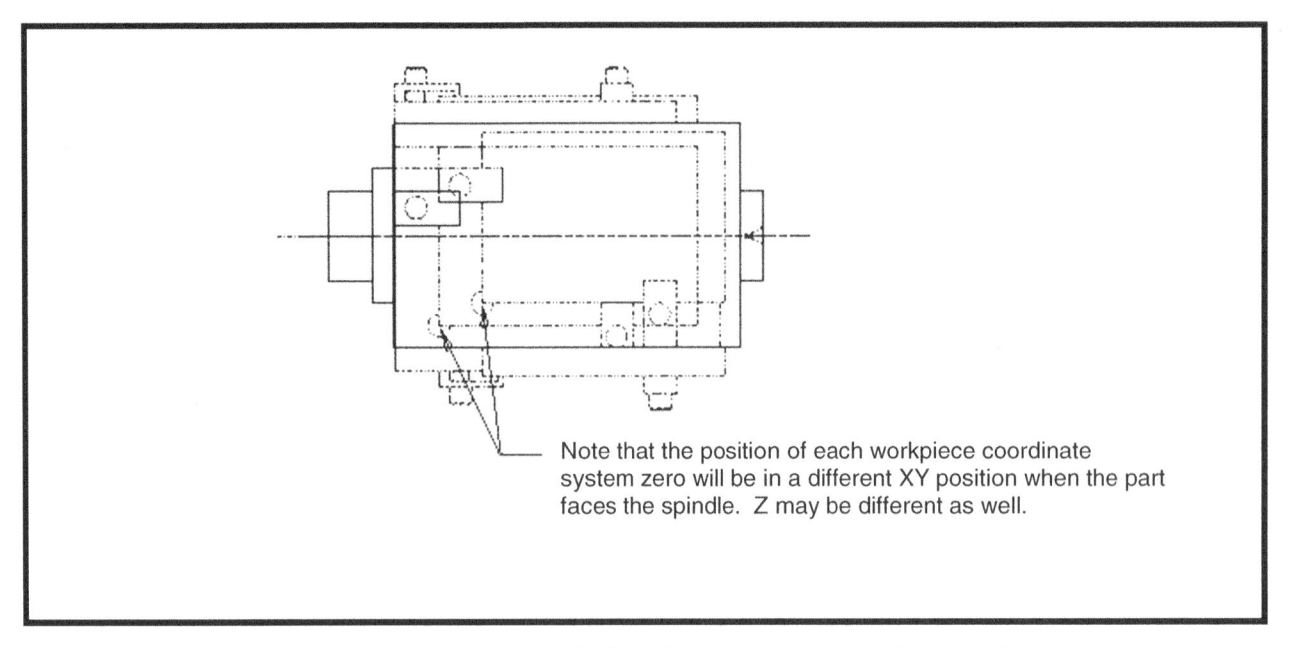

Key Concept 6: Features That Help Simplify Programming

Note that the position of each workpiece coordinate system zero will be in a different XY position when the part faces the spindle. Z may be different as well.

Figure 6.35 – Drawing shows both workpieces (superimposed) in one view

Figure 6.35 lets you look at each side of the fixture *as the spindle will see them*. It becomes clear that after an index, the workpiece coordinate system zero point for one workpiece will not work for the other.

This is a very common situation when working with rotary devices. How you handle this problem is based on whether or not the setup is qualified. For a truly qualified setup, you can assign one workpiece coordinate system zero point (commonly the center of rotation) and use it for all sides of the workpiece. While this makes it more difficult to calculate coordinates, only one workpiece coordinate system zero is needed – meaning only one work offset is needed. This makes determining workpiece coordinate system zero assignment values as easy as possible for the setup person. Even if more than one work offset is used (to make programming simpler), it will still be possible to use G10 commands to enter work offset values – again – keeping it simple for the setup person.

If the setup isn't qualified, multiple workpiece coordinate system zero points (and multiple work offsets) must be used – and the setup person must determine the workpiece coordinate system zero assignment values for each work offset during setup.

Assigning One Workpiece Coordinate System Zero Point Per Side

From a programmer's standpoint, this is the easiest way to handle the multiple sides being machined. You will simply pick the most logical location on each side of the indexing device. If the setup *is not* qualified, the setup person will measure the workpiece coordinate system zero assignment values during setup. If the setup *is* qualified, you can include G10 commands in your program to enter the workpiece coordinate system zero assignment values into work offsets.

Using One Central Workpiece Coordinate System Zero Point

This tends to be the *old school* of thought. In the past, this was the only way of handling the workpiece coordinate system zero point assignment (before work offsets were available).

Even though this technique will only require one workpiece coordinate system zero point for the entire program, you will still use a work offset (like work offset number one, G54) to assign it.

With this technique, the programmer will choose one central location to place the workpiece coordinate system zero point. While this makes it somewhat difficult to calculate coordinates for the program, at least all calculations will be made in the same way for each side of the fixture, making calculating coordinates *consistent*.

The workpiece coordinate system zero point will be in the *center of rotation* for two of the axes. The third axis workpiece coordinate system zero can be placed in a convenient position for that axis. For vertical machining centers, when the rotary table is mounted parallel to the X-axis (as shown in Figure 6.33), the workpiece coordinate system zero will be the center of rotation in the Y and Z axes. The workpiece coordinate system zero for the X-axis can be placed in a convenient location.

For horizontal machining centers, the workpiece coordinate system zero point will be the center of rotation for the X and Z axes. In Y, it can be placed in a convenient position. Figures 6.36 and 6.37 stress this.

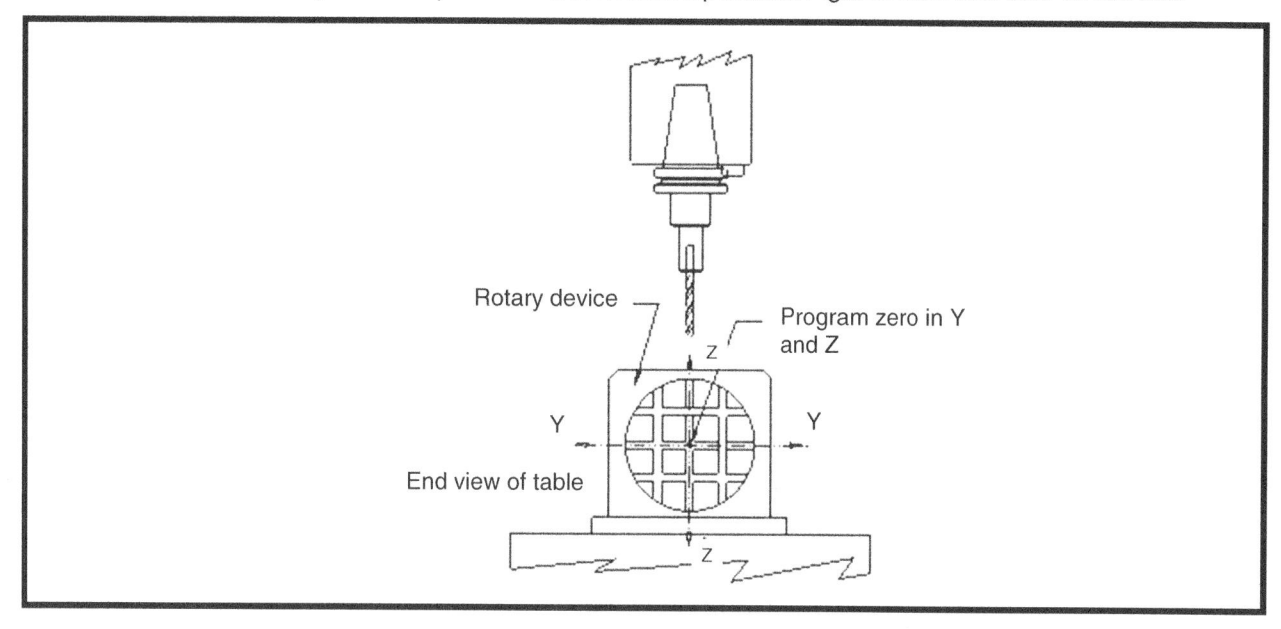

Figure 6.36 – Drawing shows center of rotation for vertical machining center

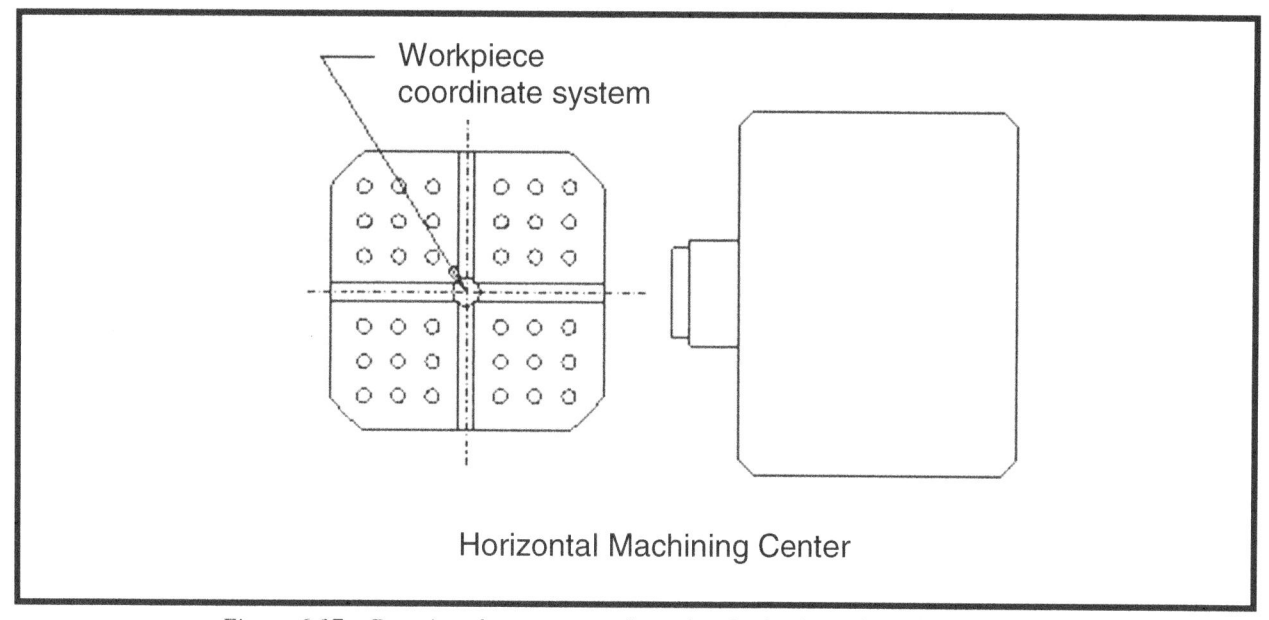

Figure 6.37 – Drawing shows center of rotation for horizontal machining center

Calculating Coordinates When Center of Rotation Is the Workpiece Coordinate System Zero Point

Calculating coordinates going into the program becomes more difficult with this method. Figure 6.38 is a drawing shown using inch mode that shows how coordinates must be calculated. Note that the fixture will have to be made correctly in order for these coordinates to work.

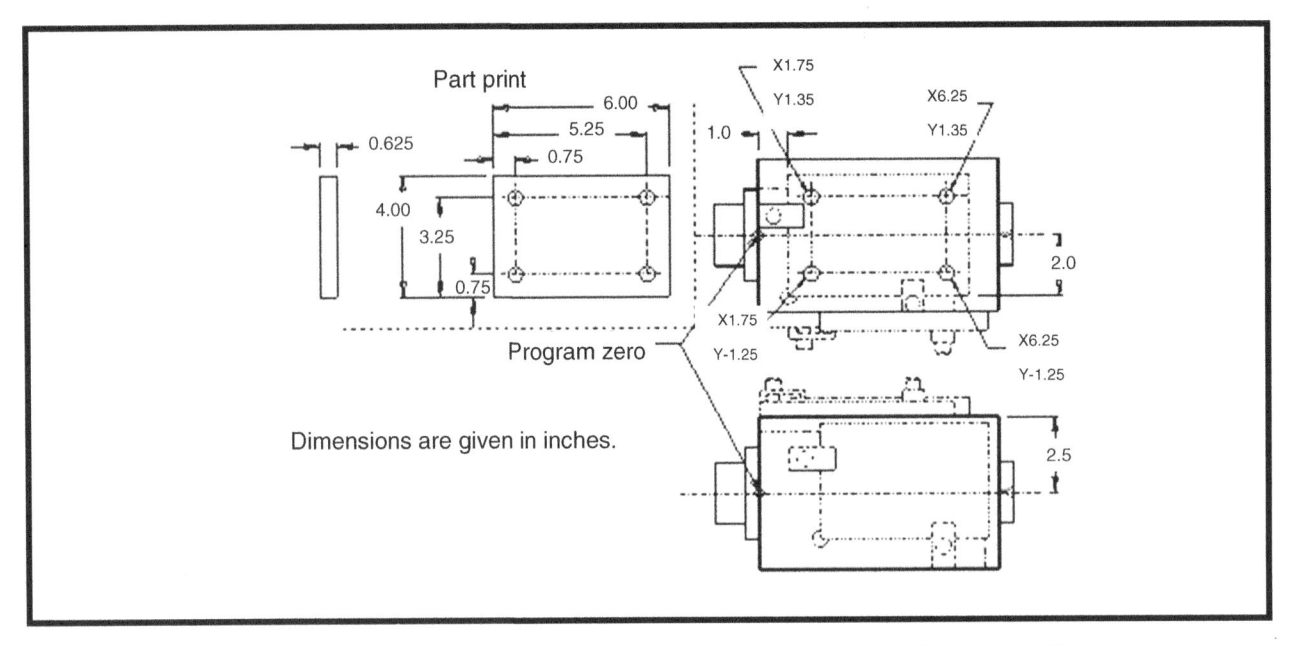

Figure 6.38 – How coordinates are calculates when center of index is used as workpiece coordinate system zero

In Figure 6.38, you can see that when you are working from a central workpiece coordinate system zero point, programmed coordinates don't make much sense. Every coordinate considers the fixture being used to hold the workpiece. If working from a central workpiece coordinate system zero, you must take into consideration the distances from the central workpiece coordinate system zero to the location surfaces of the workpiece. These distances must be added (or subtracted) to print dimensions in order to come up with the coordinates used in your program.

Example Program Using a Rotary Device

It is difficult to come up with a good example program showing the use of a rotary device that is comprehensive enough to show what can be done without becoming so complicated that you will not be able to follow. What we show is a relatively simple example program showing a rotary axis being used as an indexer.

This program uses a rotary axis designated as the A axis (for a vertical machining center when the rotary table is mounted parallel to the X-axis).

Our program will use four work offsets to assign four different workpiece coordinate system zero points. As you know, using multiple workpiece coordinate system zeros is the easier way for the programmer. But the setup person must assign each workpiece coordinate system zero. We'll say our setup is qualified, and that G10 commands will be added in the program once the workpiece coordinate system zero assignment values are determined (during the first time the setup is made).

Two separate workpieces will be machined on one fixture. Figures 6.39 and 6.40 show the workpiece drawings. We identify them as *large part* and *small part*.

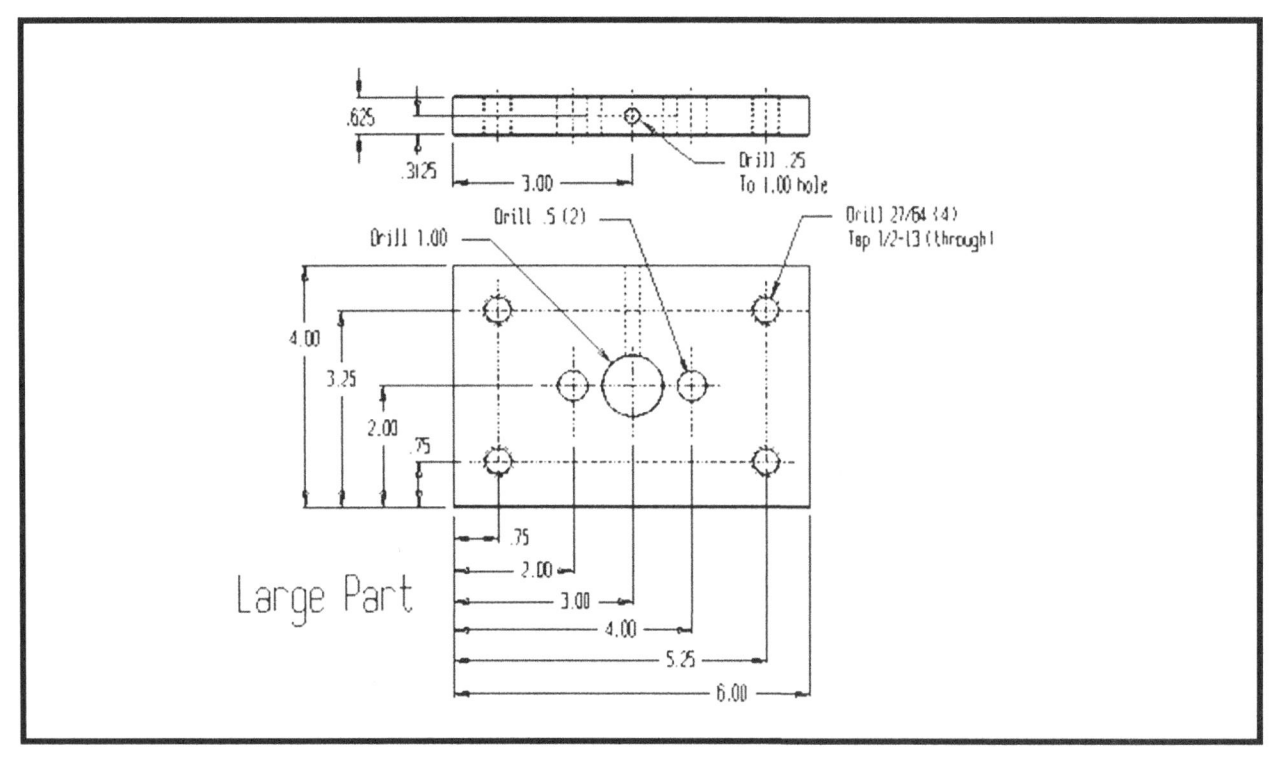

Figure 6.39 – First drawing for example program: Large Part (shown for use in inch mode)

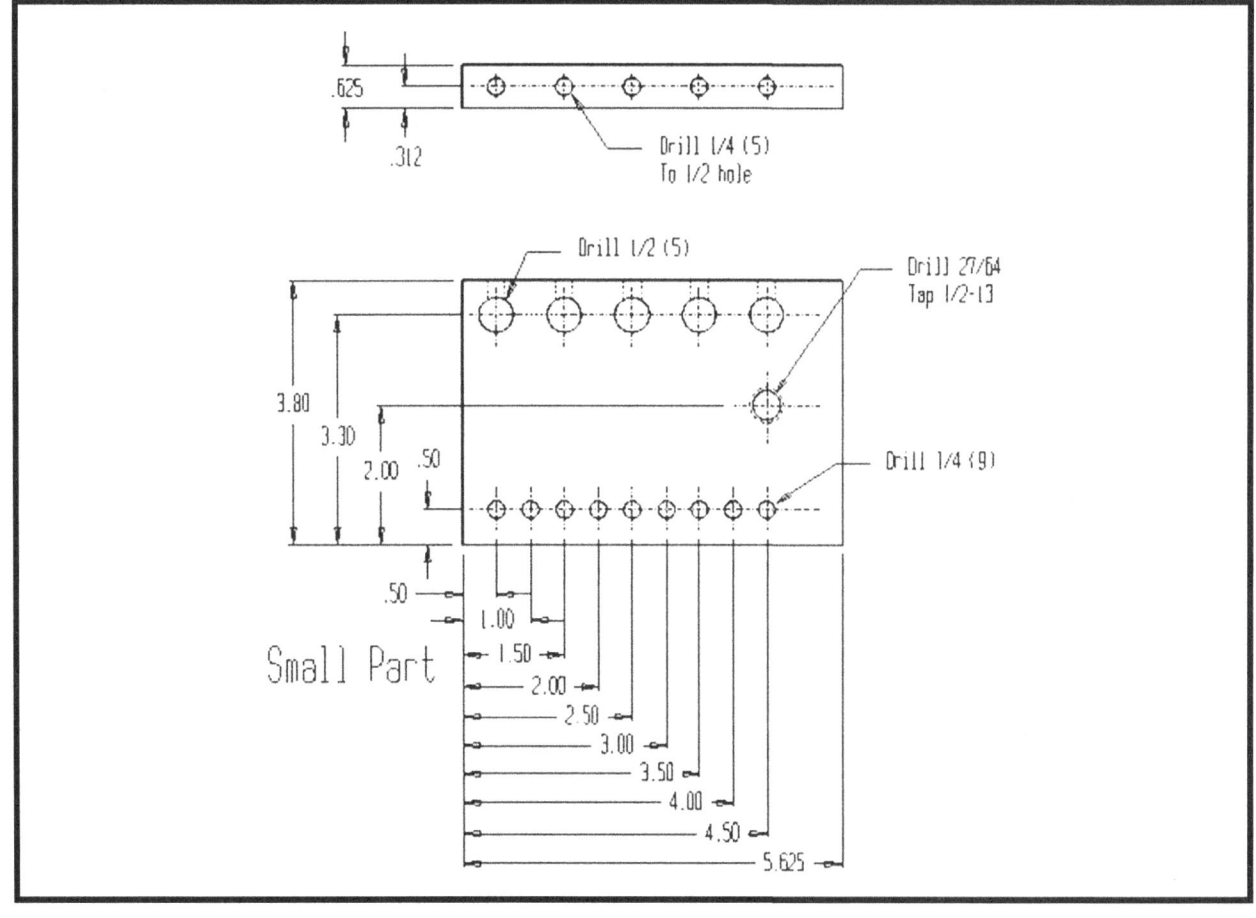

Figure 6.40 – Second drawing for example program: Small Part (shown for use in inch mode)

Key Concept 6: Features That Help Simplify Programming

Admittedly, these workpieces are very simple. We intentionally keep them simple to make it easy for you to follow what is happening in the program.

Figure 6.41 is the fixture that will hold the parts. Though it is made for a vertical machining center, rest assured that the same techniques are used on a horizontal machining center. Only the program formatting information will change.

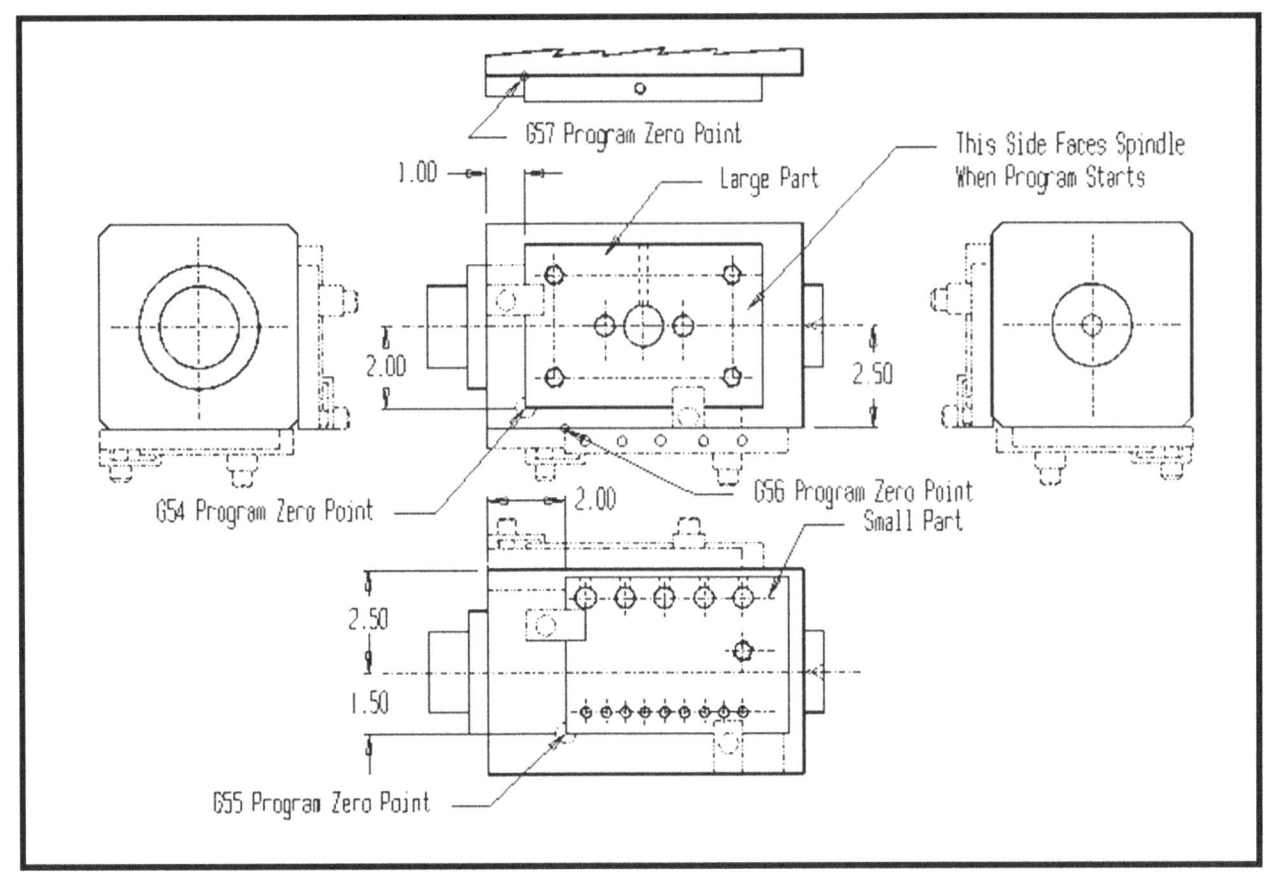

Figure 6.41 – Fixture for example program (for use in inch mode)

We have marked up the print to show the locations of all workpiece coordinate system zero points. Notice that even this simple application requires *four* workpiece coordinate system zeros, one for each surface of the part being machined. Of course, this requires the setup person to measure twelve workpiece coordinate system zero assignment values the first time the job is run. We have also specified the work offsets being used (G54 through G57).

Designations for Each Surface:
Zero degree side (A0): Large part plan view (G54) and small part side view (G56) - middle view of drawing above.

Ninety degree side (A90.0): Small part plan view (G55) - bottom view of drawing above.

Two-hundred-seventy degree side (A270.0): Large part side view (G57) - top view of drawing above.

The fixture is mounted to the rotary axis so that the zero degree side (A0) corresponds to the zero return position (a value of zero will be placed in the workpiece coordinate system offset A register). When the A axis is at its zero return position, the plan view of the large part will be facing the spindle (again, A0).

M Codes for Example Program:
M23: Force clockwise rotary axis rotation in absolute mode.

M24: Force counter-clockwise rotation in absolute mode.

M51: Rotary axis clamp (unclamp is automatic with next rotary axis motion).

Lesson 6.4: Programming Rotary Devices

Here is the process to be used for this program. Again, we are using the inch mode for this example.

Procedure for rotary axis example program				
Operation (A0 side facing spindle)	Tool	Station	Feed	Speed
Center drill holes in plan view of large part	#3 center drill	1	3.0 ipm	1,200 rpm
Center drill (5) holes in side view of small part				
Index 90 degrees ccw (A270.0)				
Center drill (1) hole in side of large part				
Index 180 degrees (A90.0)				
Center drill all holes in plan view of small part				
Drill (5) 1/2" holes in plan view of small part	1/2" drill	2	4.5 ipm	600 rpm
Index 90 degrees ccw (A0)				
Drill (2) 1/2" holes in plan view of large part				
Drill (1) 1" hole in plan view of large part	1" drill	3	5.5 ipm	300 rpm
Drill (4) 27/64 holes in plan view of large part	27/64 drill	4	4.5 ipm	600 rpm
Index 90 degrees ccw (A270.0)				
Drill (1) 27/64 hole in plan view of small part				
Drill (9) 1/4" holes in plan view of small part	1/4" drill	5	2.0 ipm	1,000 rpm
Index 90 degrees ccw (A0)				
Drill (5) 1/4" holes in side view of large part				
Tap (4) 1/2-13 holes in plan view of large part	1/2-13 tap	6	17.6 ipm	230 rpm
Index 90 degrees cw (A270.0)				
Tap (1) 1/2-13 hole in plan view of small part				
Index 90 degrees ccw (A0)				

Now you should be able to follow along with the program. Note that this program is for a machine that automatically clamps the rotary table after each rotation (no M-codes for clamping are required):

```
%
O0021
N001 G17 G20 G94 (Select XY plane, inch mode, select per-minute
feedrate mode)
N002 G40 G50 G64 (Cancel cutter comp., scaling, and select normal
cutting mode)
N003 G67 G69 G80 (Cancel custom macro, rotation, and canned cycles)
N005 G91 G28 A-15.0 (Ensure that plan view of large part is facing
spindle)

N010 T01 M06
N015 G54 G90 S1200 M03 T02 (Center drill)
N020 G00 X0.75 Y0.75 A0
N023 M51 (Clamp rotary axis)
N025 G43 H01 Z2.0
N030 M08
N035 G81 R0.1 Z-.25 F3.0 G99
```

Key Concept 6: Features That Help Simplify Programming

```
N040 X2.0 Y2.0
N045 X3.0
N050 X4.0 G98 (Clamp is close!)
N055 5.25 Y0.75 G99
N060 Y3.25
N065 X0.75 G98 (Stay above clamps for move to side view holes)
N070 G56 X0.5 Y-0.313 R0.1 Z-0.25 G99 (Note that the G56 is allowed in the canned cycle command)
N075 X1.5
N080 X2.5
N085 X3.5
N090 X4.5 G98
N095 G80 (Cancel cycle for index)
N100 G00 Z4.0 (Get some extra clearance for index)
N105 A270.0 M24 (Index 90 degrees CCW to side view of large part)
N008 M51 (Clamp rotary axis)
N110 G90 G57 G81 X3.0 Y-0.3125 R0.1 Z-0.25 G98
N115 G80
N120 G00 Z4.0
N125 A90.0 M23 (Index 180 degrees to plan view of small part)
N128 M51 (Clamp rotary axis)
N130 G90 G55 G81 X0.5 Y0.5 R0.1 Z-0.25 G99
N135 X1.0
N140 X1.5
N145 X2.0
N150 X2.5
N155 X3.0
N160 X3.5
N165 X4.0
N170 X4.5
N175 Y2.0
N180 Y3.3
N185 X3.5
N190 X2.5
N195 X1.5
N200 X.5
N205 G80 M09
N210 G91 G28 Z0 M19
N215 M01

N220 T02 M06
N225 G55 G90 S600 M03 T03 (1/2 drill)
(Still on plan view of small part)
N230 G00 X0.5 Y3.3 A90.0
N233 M51 (Clamp rotary axis)
N235 G43 H01 Z0.1
```

```
N240 M08
N245 G81 R0.1 Z-0.85 F4.5
N250 X1.5
N255 X2.5
N260 X3.5
N265 X4.0
N270 G80
N275 G00 Z4.0
N280 A0 (Index 90 degrees CCW to plan view of large part)
N283 M51 (Clamp rotary axis)
N285 G90 G54 G81 X2.0 Y2.0 R0.1 Z-0.85 G99
N290 X4.0
N295 G80 M09
N300 G91 G28 Z0 M19
N305 M01

N310 T03 M06
N315 G54 G90 S300 M03 T04 (1" drill)
N320 G00 X3.0 Y2.0 A0
N323 M51 (Clamp rotary axis)
N325 G43 H03 Z0.1
N330 M08
N335 G73 R0.1 Z-0.95 Q0.1 F5.5
N340 G80 M09
N345 G91 G28 Z0 M19
N350 M01

N355 T04 M06
N360 G54 G90 S600 M03 T05 (27/64 drill)
(Still on plan view of large part)
N365 G00 X0.75 Y0.75 A0
N368 M51 (Clamp rotary axis)
N370 G43 H04 Z2.0
N375 M08
N380 G73 R0.1 Z-0.85 F4.5 G98 (Note clamp!)
N385 X5.25 G99
N390 Y3.25
N395 X0.75
N400 G80
N405 G00 Z4.0
N410 A90.0 (Index 90 degrees CW to plan view of small part)
N413 M51 (Clamp rotary axis)
N415 G90 G55 G73 X4.5 Y2.0 R0.1 Z-0.85 G99
N420 G80 M09
N425 G91 G28 Z0 M19
N430 M01
```

```
N435 T05 M06
N440 G55 G90 S1000 M03 T06 (1/4 drill)
(Still on plan view of small part)
N445 G00 X0.5 Y0.5 A90.0
N448 M51 (Clamp rotary axis)
N450 G43 H05 Z0.1
N455 M08
N460 G73 R0.1 Z-0.8 Q0.1 F2.0
N465 X1.0
N470 X1.5
N475 X2.0
N480 X2.5
N485 X3.0
N490 X3.5
N495 X4.0
N500 X4.5
N505 G80
N510 G00 Z4.0
N515 A0 (Index 90 degrees CCW to side view of small part)
N518 M51 (Clamp rotary axis)
N520 G90 G56 G81 X0.5 Y-0.313 R0.1 Z-0.5 G99
N525 X1.5
N530 X2.5
N535 X3.5
N540 X4.5
N545 G80
N550 G00 Z4.0
N555 A270.0 (Index 90 degrees CCW to side view of large part)
N558 M51 (Clamp rotary axis)
N560 G90 G57 G83 X3.0 Y-0.3125 R0.1 Z-1.75 Q0.5
N565 G80 M08
N570 G91 G28 Z0 M19
N580 M01

N585 T06 M06
N590 G54 G90 S230 M03 T01 (1/2-13 tap)
N595 G00 X0.75 Y0.75 A0 (Index 90 degrees CW to plan view of large part)
N598 M51 (Clamp rotary axis)
N600 G43 H06 Z0.1
N605 M08
N610 G84 R0.25 Z-0.85 F17.6
N615 X5.25
N620 Y3.25
N625 X0.75
N630 G80
```

```
N635 G00 Z4.0
N640 A90.0 (Index 90 degrees CW to plan view of small part)
N643 M51 (Clamp rotary axis)
N645 G90 G55 G84 X4.5 Y2.0 R0.1 Z-0.85 G99
N650 G80 M09
N655 G91 G28 Z0 M19
N660 G90 A0 (Index 90 degrees CCW back to starting point)
N665 M01
N670 G28 X0 Y0
N675 M30
%
```

As you can see, even *simple* programs for rotary devices tend to get complicated. Note how important the sequence of machining operations is. Without this set of step-by-step instructions, even experienced programmers will become confused as they write programs.

Key Points for Lesson 6.4:

- Rotary devices can be used on vertical or horizontal machining centers to expose several surfaces of the workpiece to the spindle for machining during the CNC cycle.
- With an indexer, machining must occur after rotation. With a rotary axis, machining can occur during rotation as well.
- A rotary axis can be used as an indexer, but an indexer cannot be used as a rotary axis.
- A rotary axis shares many things in common with a linear axis (zero return position, polarity, workpiece coordinate system zero assignment, absolute versus incremental mode, and rapid versus straight line motion).
- When commanded in a G01 command, feedrate for the rotary axis must be specified in degrees per minute.

Key Concept 7: Know your Machine from an Operator's Point of View

Key Concept Number Seven begins the setup and operation portion of this text. It parallels Key Concept Number One - but in Key Concept Number Seven, we will look at a machining center from a setup person's or operator's perspective.

Key Concept 7 contains two lessons:

    7.1: Tasks related to setup and running production

    7.2: Buttons and switches on the control panels

We now begin discussions related to setup and operation. Throughout these discussions, we will assume that you have read and understand certain presentations from the programming-related lessons (in lessons 1.1 through 6.4). There are *many* setup- and operation-points made during the programming-related lessons. Without an understanding of these important points, much of what is presented from this point will make little sense.

Are You Only Interested In Setup and Operation?

Our goal with this text is to provide you with the knowledge you need to master *all three tasks* related to CNC machining center usage – programming, setup, and operation. And if you are to become proficient with all three tasks, you must read this text from beginning to end. But as we point out in lesson 1.2, companies vary with regard to what they expect of their CNC people.

Some companies expect one person to do everything related to machining center usage, including program, setup, and run production. This is common in workpiece producing companies (job shops) and tooling producing companies. Other companies divide the tasks and assign them to several people (one person programs, another makes setups, and yet another runs production). This is common in product producing companies.

If you work for a product producing company, you may not be expected to write programs. Your responsibilities may be limited to making setups and/or running out the job once a setup has been made. While we feel you can still benefit from reading *all* of the programming related presentations (again, lessons 1.1 through 6.4), your immediate interests may be solely related to setup and/or operation.

If you are only interested in setup and operation, you should, *at the very least*, read the programming-related presentations listed below. This will allow you to understand material presented from this point. And even if you *have read* all of the programming-related lessons, topics listed below will help you quickly review important material before continuing in this text. *We will not be repeating these setup- and operation-related discussions* during Key Concepts seven through ten:

From Lesson 1.1: Machine configurations

- Read from the beginning to the heading "Programmable functions of machining centers".

From Lesson 1.2: CNC job work flow

- Read the entire lesson.

From Lesson 1.4: Understanding the workpiece coordinate system

- Read from the beginning to the heading "Wisely choosing the workpiece coordinate system zero point location".

From Lesson 1.5: Determining workpiece coordinate system values

- Read the entire lesson.

Key Concept 7: Know Your Machine From An Operator's Viewpoint

From Lesson 1.6: Setting workpiece coordinate system offset values
- Read from the beginning to the heading "What if my machining center doesn't have workpiece coordinate system setting offsets?

From Lesson 4.1: Introduction to compensation
- Read the entire lesson.

From Lesson 4.2: Tool length compensation
- Read from the beginning to the topic "Programming tool length compensation".
- Read from the heading "The setup person's responsibilities with tool length compensation" to the heading "Typical mistakes with tool length compensation".
- Read from the heading "Trial machining with tool length compensation" through the end of the lesson.

From Lesson 4.3: Cutter radius compensation
- Read from the beginning to the heading "Steps to programming cutter radius compensation".
- Read from the heading "The setup persons responsibilities with cutter radius compensation" to the heading "Rough and finish milling with the same set of tool path coordinates".
- Read from the heading "Trial machining with cutter radius compensation" through the end of the lesson.

From Lesson 4.4: Workpiece coordinate system setting offsets
- Read from the beginning to the heading "Programming multiple workpiece coordinate system zero points".
- Read from the heading "Shifting the point of reference for workpiece coordinate system setting offset entries" to the topic "Programming workpiece coordinate system setting offset entries".
- Read from the heading "Some other applications for the common workpiece coordinate system setting offset" through the end of the lesson.

From Lesson 5.1: Introduction to program structure
- Read from the beginning to the topic "Machine variations that affect program structure"

You may be wondering why so many setup- and operation-related topics are presented during *programming-related* presentations. Again, many companies expect one person to program, setup, and run production – these people, must master all three tasks related to machining center use.

Even in companies that do separate these tasks and have different people perform them, a programmer must know enough about setup and running production to direct setup people and operators. Indeed, it is the programmer in these companies that creates the setup and production run documentation. *A programmer must know at least as much about setup and operation as setup people and operators.* It is a natural progression to discuss certain setup- and operation-related topics *during programming* when they apply to programming-related features and functions.

If you have read all material to this point, you may be surprised at how much you already know about setting up and operating a CNC machining center.

The Need for Hands-On Experience

Hands-on experience is extremely important to fully mastering CNC machining center setup and operation. But unfortunately, no text can provide hands-on experience. If you are using this text as part of a course you are taking at a technical school, hopefully there are machines available for practicing what you learn in class. *We cannot overstate the importance of hands-on experience* – experience we cannot provide in this text.

That said, we can provide *all of the principles* needed to set up and run CNC machining centers. As a comparison, think of what it takes to learn to fly an airplane. A future pilot must, of course, spend a great deal of time in the cockpit at the flight controls in order to master skills needed to become a pilot. But *before* a person begins hands-on flight training, they *must* understand the basics of aerodynamics and flight. Without this understanding, controls in the cockpit will have no meaning. A flight student must attend *ground school*. In ground school they will learn concepts that help understand the controls in the cockpit.

In similar fashion, a CNC setup person and/or operator *must* spend time at the CNC machine to fully master tasks related to setting and running production on a CNC machining center. But before a person can spend any meaningful time at the machine, they *must* understand concepts related to setting up and running the CNC machine. Without an understanding of the material presented from this point in the text, machine functions will have no meaning.

Think of setup and operation discussions in this text as a kind of *ground school for learning to setup and run CNC machining centers*. While you will not gain the important hands-on experience you'll need to fully master running a CNC machining center from this text alone, you will learn the *concepts* required to begin practicing. An understanding of these concepts is mandatory *before* you begin working at the machine.

Key Concept Number Seven

This Key Concept parallels Key Concept 1. But now we'll now be looking at a CNC machining center from a setup person's or operator's viewpoint. Much of what is presented in Key Concept Number One is important to setup people and operators. Again, be sure to read from the beginning of lesson 1.1 to the heading *Programmable functions of machining centers* before continuing.

More About Axis Polarity

In Key Concept Number One, we show the basic configurations of CNC machining centers – including their components and directions of motion (axes). We bring up one source of confusion that has to do with axis polarity. A programmer has the luxury of viewing polarity in a rather simplistic manner. They assign a workpiece coordinate system zero point (origin) for their program. If a position is to the right of the workpiece coordinate system zero, it requires an X plus coordinate. If it is to the left of the workpiece coordinate system zero, it requires an X minus coordinate. The same goes for the Y and Z axes. Because all coordinates are specified in the *workpiece coordinate system*, a programmer never has to be concerned with *which way the machine components must move* (table and headstock) in order to position a cutting tool.

On the other hand, a setup person or operator *must know which way is plus and which way is minus for each moving component of the machine.* When you want to move the table of a C-frame style vertical machining center to the left, you *must* know that it is the X plus direction. Figures 7.1 through 7.4 show axis polarity for each machine type shown in lesson 1.1 from a setup person's or operator's viewpoint. And again, each polarity arrow in Figures 7.1 through 7-4 shows plus versus minus for the *moving component* of the axis. Study and memorize axis polarity for the machine/s you will be working with.

Directions of motion (axes) & polarity for a C-frame style vertical machining center

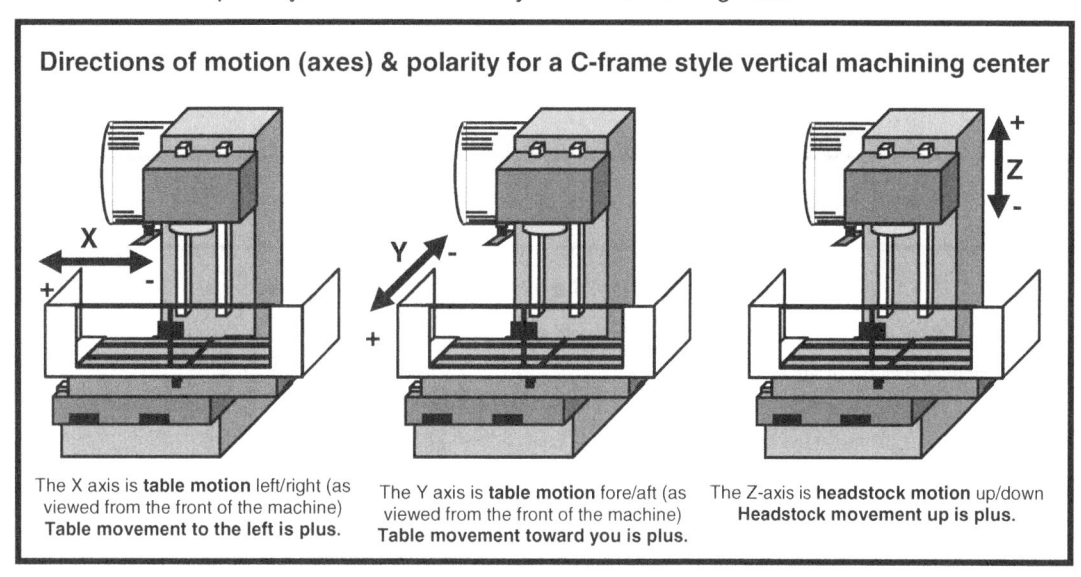

The X axis is **table motion** left/right (as viewed from the front of the machine) **Table movement to the left is plus.**

The Y axis is **table motion** fore/aft (as viewed from the front of the machine) **Table movement toward you is plus.**

The Z-axis is **headstock motion** up/down **Headstock movement up is plus.**

Figure 7.1 – Polarity for each axis of a C-frame style vertical machining center

Key Concept 7: Know Your Machine From An Operator's Viewpoint

Polarity for axes of a knee-style CNC milling machine

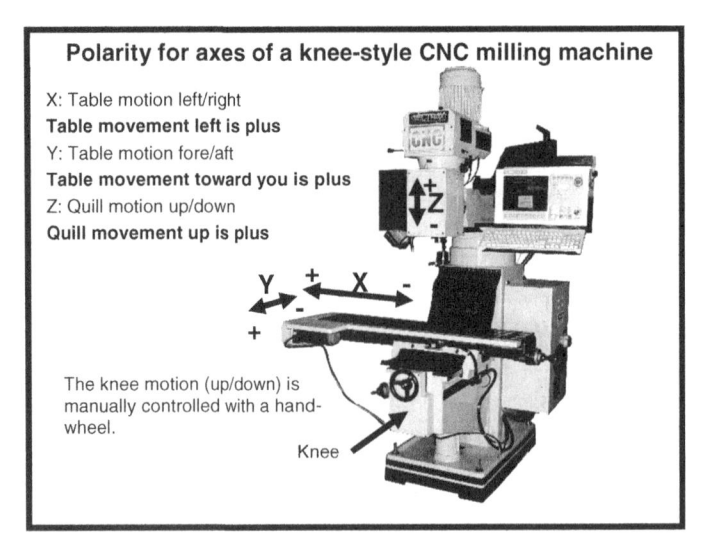

X: Table motion left/right
Table movement left is plus
Y: Table motion fore/aft
Table movement toward you is plus
Z: Quill motion up/down
Quill movement up is plus

The knee motion (up/down) is manually controlled with a handwheel.

Figure 7.2 – Polarity for each axis of a knee style milling machine

Polarity for a bridge-style vertical machining center

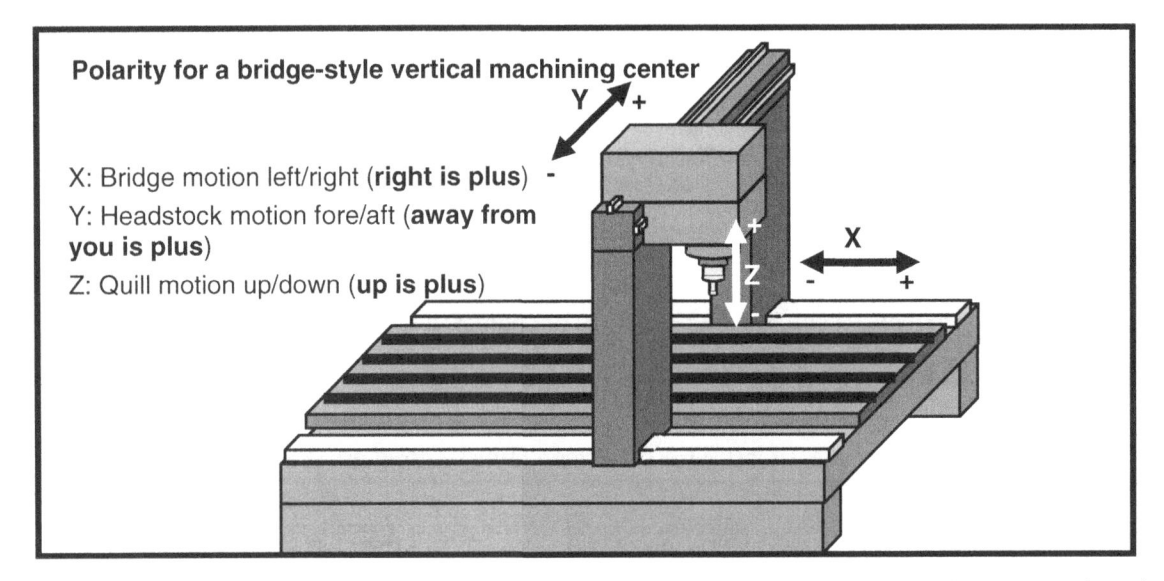

X: Bridge motion left/right (**right is plus**)
Y: Headstock motion fore/aft (**away from you is plus**)
Z: Quill motion up/down (**up is plus**)

Figure 7.3 – Polarity for each axis of a bridge style machining center (tool actually moves with each axis)

Polarity for each axis of a horizontal machining center

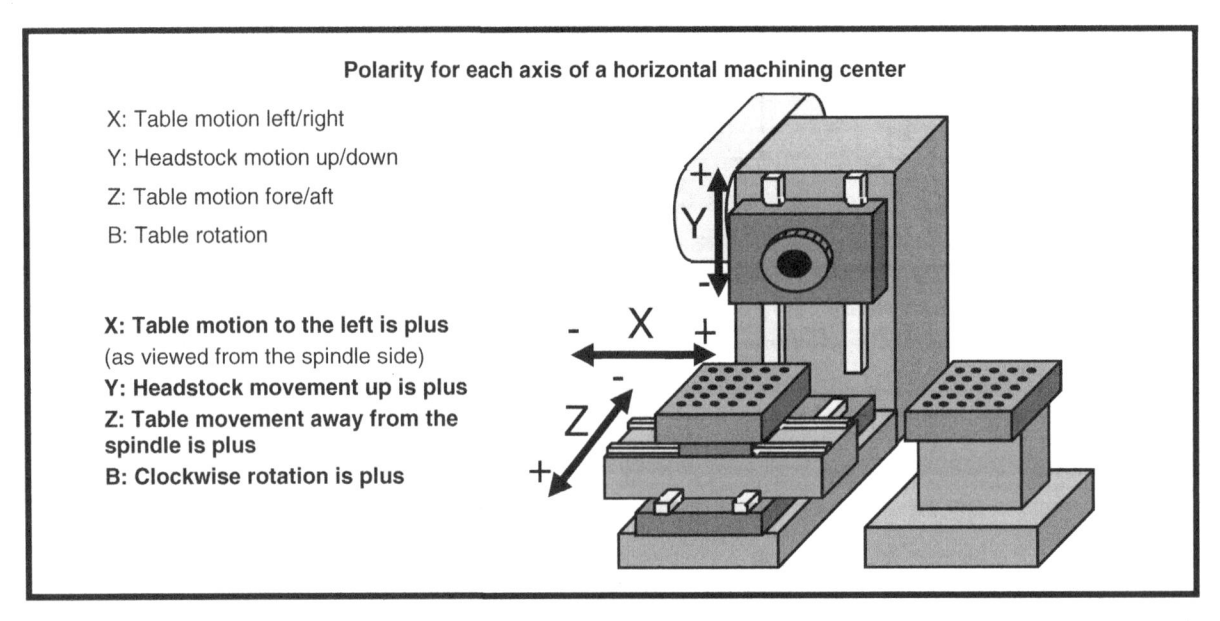

X: Table motion left/right
Y: Headstock motion up/down
Z: Table motion fore/aft
B: Table rotation

X: Table motion to the left is plus
(as viewed from the spindle side)
Y: Headstock movement up is plus
Z: Table movement away from the spindle is plus
B: Clockwise rotation is plus

Figure 7.4 – Polarity for each axis of a horizontal machining center

Procedures You Must Know

Setting up and operating a CNC machining center requires that you master some basic procedures. While we show step-by-step procedures in lesson 9.1, we want to acquaint you with some of the most rudimentary machine-usage procedures and describe when they are required.

Machine Power-Up

You must, of course, be able to turn on your machining center/s. This procedure usually involves turning on a main power breaker (a switch that is usually located on the back of the machine), then pressing the control *power on* button, and finally pressing a button to turn on the machine's hydraulic system (if it has one).

Sending the Machine To Its Zero Return (reference) Position

Some (older) machining centers require that you perform this procedure when the machine is first powered up. You may also have to send the Z-axis to its zero return position (the tool change position) for vertical machining centers prior to running a program.

This procedure varies from one machine tool builder to another. With some machines, it is as simple as selecting the zero return mode and pressing one button which causes each axis (first Z, then X, then Y) to move to its zero return position. With other (indeed most) machines, you must send each axis to it zero return position independently. After selecting the zero return mode, you select the axis you want to zero return first (usually Z for vertical machining centers). Next you press the plus button and hold it. The selected axis will rapid until it comes to within an inch or so of its zero return position. Then it will slow down, eventually reaching the zero return position. An axis origin light may come for the axis as it reaches the zero return position. This must be repeated for the other axes.

Manually Moving Each Axis

Manually moving an axis is often required. Some examples of when you must manually move an axis include measuring workpiece coordinate system zero assignment values, measuring tool length compensation values on the machine, and manually positioning a cutting tool into a position so that it can be inspected or replaced. Most machine tool builders provide you with at least two ways to manually move an axis (most machines require you to manually move only one axis at a time).

One way to move an axis is to use the *jog* function of the machine. This allows rather crude control of motion, and will be used when there is no danger of an axis component colliding into an obstruction. To jog an axis, first you select the jog mode. Next you select the axis you want to move (X, Y, or Z). You then select the motion rate with a multi-position switch (commonly called jog feedrate). And finally, you activate the axis motion in the desired direction (plus or minus). With most machines there are two buttons used for this purpose, one labeled plus – the other labeled minus. As long as you hold the button in, the axis will move. Also, you can manipulate the jog feedrate switch to speed up or slow down the motion.

Another way to move an axis is to use the *hand-wheel*. This provides a much more precise way to move an axis. It is commonly required when there is danger of an axis component (like a cutting tool) colliding with an obstruction. With the hand-wheel, you can control how quickly the machine will move when you turn the hand-wheel. In the times-one (X1) position, each increment of the hand-wheel will be 0.001-mm (or 0.0001-in). In the times-ten (X10) position, each increment will be 0.01-mm (or 0.001-in). In the times one-hundred position (X100), each increment will be 0.1-mm (or 0.010-in. To use the hand-wheel, first you select the hand-wheel mode, then you select the axis you want to move (X, Y, or Z). Finally, you select the rate of motion (X1, X10, or X100). When you turn the hand-wheel in the desired direction (plus or minus), the axis will move.

An Example of Manual Axis Movement

One common time you must manually move an axis is when measuring a tool's length. From lesson 4.2, here is the general procedure to measure tool length compensation values (in lesson 4.2 we include helpful illustrations).

1) Make the work holding setup – or place some kind of flat block on the machine table
It is important to have a nice, flat surface on which to work. The top of a vise works nicely.

Key Concept 7: Know Your Machine From An Operator's Viewpoint

2) Place a gauge block on the flat surface.

3) Without a tool in the spindle, make the spindle nose to touch the block.
Use the jog mode and hand-wheel to carefully touch the spindle nose to the block. At this point, reset (set to zero) the Z-axis relative position display.

4) Retract the Z-axis and load a cutting tool to measure. Bring the tool tip to the same block.
Use the joystick and hand-wheel, and cautiously move the spindle nose away from the block. Load a cutting tool to be measured. Now, carefully move the *tool tip* to the same block that was just touching the spindle nose. The Z-axis display will follow along. When the tool tip is touching the block, the Z-axis display will be showing you the tool's length. This is the value that must be entered into the tool length compensation offset register for this tool. Write it down.

5) Repeat step three for all cutting tools that must be measured.

In step three, the jog mode will be used to positioning the spindle over the block in X and Y. It will also be used to roughly position the spindle nose close to the block (within a few inches). The setup person will begin with the jog feedrate set pretty high to cause quick motion. But as the spindle nose gets closer, they will slow the jog feedrate. The hand-wheel will be used to actually make the spindle nose touch the block. First the times one-hundred motion rate will be selected to quickly bring the spindle nose close to the block. Then times-ten will be used to get closer. And final touching will be done in the times-one mode.

In step four, the hand-wheel will be used to cautiously move the spindle nose away from the block. Then the jog function will be used to quickly retract the tool. If the automatic tool changer is to be used to load each tool to measure, the setup person can use the zero return mode to send the Z-axis to the tool change position. After loading the tool, the jog function will be used to quickly bring the tool tip close to the block. Then the hand-wheel will be used to cautiously touch the tool tip to the block.

Manual axis motion is often required. A setup person or operator must master the use of the jog function and the hand-wheel.

Manually Starting the Spindle

This procedure will be required when using a conventional edge-finder to measure workpiece coordinate system zero assignment values as shown in lesson 1.5 and if you will be performing any manual machining operations.

Machine tool builders vary when it comes to how much manual control they provide for the spindle. With some machines, you can simply select the manual mode, set a multi-position switch to select the desired spindle speed, select the desired direction (with a two-position switch), and press a button to activate the spindle. Another button can be pressed to turn off the spindle.

If your machine does not provide manual control for the spindle, you must use a function called *manual data input* (MDI) and enter and execute a program-like command to start the spindle (like S500 M03). Another command must be entered and executed to stop the spindle (M05). We'll discuss the use of manual data input in lesson 8.1.

Manually Making Tool Changes

Machine tool builders vary when it comes to how much manual control they provide for the automatic tool changer. Frankly speaking, few will allow you to make tool changes in a completely manual fashion. Almost all will require that you use manual data input (MDI) and enter and execute a program-like command to change tools (like T04 M06). The machine must be at its tool change position before a tool change can be commanded (usually the Z-axis zero return position). We'll discuss the use of manual data input in lesson 8.1.

Manipulating the Display Screen

The display screen will show you a great deal of important information. For setup people and operators, the four most important display screen mode are *position*, *program*, *offset/setting*, and *program check*.

The Position Display Screen

As you would expect, in position mode, the display screen mode shows information about machine position. Four position pages are involved. The *absolute* position display page shows the machine's current position relative to workpiece coordinate system zero. The *relative* position display page is used to take measurements and can be set or reset to specify a point of origin for the measurement. The *machine position display page* shows the machine's current position relative to the zero return position. And a fourth page shows a combination of the previous three – along with a special *distance-to-go* position display that lets you see how much further the machine will move in the current command.

More about the relative position display page
Again, *the relative position display page is the page you will use to take measurements*. In step three of the tool length measuring procedure shown above, you must reset (set to zero) the Z-axis relative position display. This sets a point of reference for tool length measurements that follow. The relative position display is the *only* position page that can be set or reset in this manner.

More about distance-to-go
The multi-position display page includes a *distance-to-go* display. This is a very important function – one that is also displayed on the *program check* display screen. When you are verifying a program, it is often helpful to know just how much further the machine will move in the current command. Say a tool is making its first approach movement in the Z-axis. You know the tool is *supposed* to stop 2.0 millimeters (or 0.1 inches) above the work surface. But you're worried. With *dry run* and *single block* turned on (two other program verification functions), you will have total control of the machine's motion rates (including rapid) with a multi-position switch (commonly the feedrate override switch). As the tool approaches you slow motion with feedrate override. When the tool is about 15.0-mm (or 0.5-in) away from the work surface, you start getting worried, so you press the *feed hold* button to stop axis motion (feed hold is yet another program verification function). When you look at the *distance-to-go* display, the value shown in the Z-axis should be under 15..0-mm (or 0.5-in). If it is not, the current command will cause the tool to crash into the work surface (and, of course, you've found a problem with the program or setup).

The Program Display Screen

In edit or memory mode, this display screen shows you the active program. If you are running the program, this will allow you to see up-coming commands. If you need to modify the program, this display provides you with a kind of text editor. In manual data input (MDI) mode, this display screen shows the page used to enter MDI commands.

The Offset/Setting Display Screen

This display screen allows you to view, enter, and modify offsets. Any kind of offset is accessible, including tool length and cutter radius compensation offsets, as well as workpiece coordinate system setting offsets.

The Program-Check Display Screen

Used for program verification, this display screen shows important information to help you find mistakes in a program. It shows a few upcoming commands of the active program, currently active CNC words (like G and M-codes, as well as S-word and F-word), and the distance-to-go display (described during the position display screen discussion).

Loading Programs

Before programs can be run, of course, they must reside in the machine. And once a job is finished, it may be necessary to save the program for future use, especially if changes have been made to the program during the program's verification. Current FANUC controls allow programs to be loaded from a variety of external devices, including memory cards, flash drives, and distributive numerical control (DNC) systems.

The specific procedures to load and save programs are shown in lesson 9.1.

Lesson 7.1–Tasks Related to Setup and Running Production

We define setup time as the total time a machine is down between production runs. We define cycle time as the total time it takes to complete a production run divided by the number of good workpieces produced. When you think about it, there are only two general tasks that occur on CNC machining centers – machines are either in setup or they are running production.

Objectives
After completing this lesson, students should be able to:
- ✓ Describe the difference between setup-related and production-running-related tasks
- ✓ Describe tasks that are related to setup
- ✓ Describe tasks related to running production

Introduction
It is important to understand the distinction between making a setup and running production. The tasks you perform during setup are *getting the machine ready* to run production. Only when the setup is completed and a workpiece has passed inspection is it possible to run production. The person making the setup is called the *setup person* – the person completing the production run is called the *CNC operator* (though in many companies, one person makes the setup *and* runs production).

Setup-Related Tasks
- Tear down previous setup an put everything away
- Gather the components needed to make the setup
- Make the work holding setup
- Assign workpiece coordinate system zero point/s
- Assemble cutting tools
- Measure and enter tool length compensation values
- Measure and enter cutter radius compensation values
- Load cutting tools into the machine's automatic tool changer
- Load the CNC program/s
- Verify the correctness of the program and setup
- Cautiously run the first workpiece – ensure that it passes inspection
- If necessary, optimize the program for better efficiency (new programs only)
- If changes to the program have been made, save corrected version of the program

Running Production-Related Tasks
Done during every cycle:
- Load a workpiece
- Activate the cycle
- Monitor the cycle to ensure that cutting tools are machining properly (first few workpieces only)
- Remove the workpiece
- Clean/de-burr the workpiece
- Perform specified measurements (if required)

Key Concept 7: Know Your Machine From An Operator's Viewpoint

- Report measurement results to statistical process control (SPC) system

Not required in every cycle:

- Make offset adjustments to maintain size for critical dimensions
- Replace worn tools
- Remove chips from work area (if required)

In this lesson, we'll be going through this list of tasks and describing each one in detail.

There are certain setup-related tasks that must sometimes be repeated during a production run. If for example, a cutting tool gets dull and must be replaced, the same tasks required to initially assemble it, measure it, and enter offsets for it must be repeated.

Many of these tasks, of course, draw upon your basic machining practice skills. Tasks related to making work holding setups and assembling cutting tools, for example, require that you understand work holding- and cutting-tool-components. And these tasks are identical to those that must be performed on conventional (not CNC) machine tools. So, as with programming, setting up and operating a CNC machining center require a firm understanding of basic machining practices. And also as with programming, if you have experience working with conventional machine tools, you have a head start for mastering CNC machining center setup and operation.

As you look through the list of tasks shown above, notice how many have been introduced during our discussions of programming. While we're going to expand our discussions from this point on, you have already been exposed to many setup- and operation-related concepts.

This lesson is truly at the heart of the setup and operation presentations in this text. It presents the most important concepts you must understand in order to be a successful setup person or operator.

A CNC Job From Start to Finish

Figures 7.5A and 7.5B show the jobs we're going to use to discuss every task related to setup and maintaining production. One requires metric mode (7.5A) and the other requires metric mode (7.5B). They are very similar, so study the one that most applies to you. We will these jobs as if we are actually running them, and constantly referencing measurement system specifications in the drawing and process.

Metric-mode job to be run on a CNC machining center

Process:

1) Mill both ends	25.0-mm end mill	375 rpm	150.0 mmpm
2) Mill 16.0-mm slot	12.0-mm end mill	600 rpm	115.0 mmpm
3) Mill 25.0-mm slot	16.0-mm end mill	540 rpm	125.0 mmpm
4) Center drill (3) holes	#4 center drill	1200 rpm	100.0 mmpm
5) Drill (3) 3/8 holes	10.0-mm drill	800 rpm	125.0 mmpm

Tolerances: 1 & 2-place decimals +/-0.13

Notice three tight tolerances:
- Width of 16.0-mm wide slot
- Overall length of workpiece
- Depth of 16.0-mm slot

Program for job:

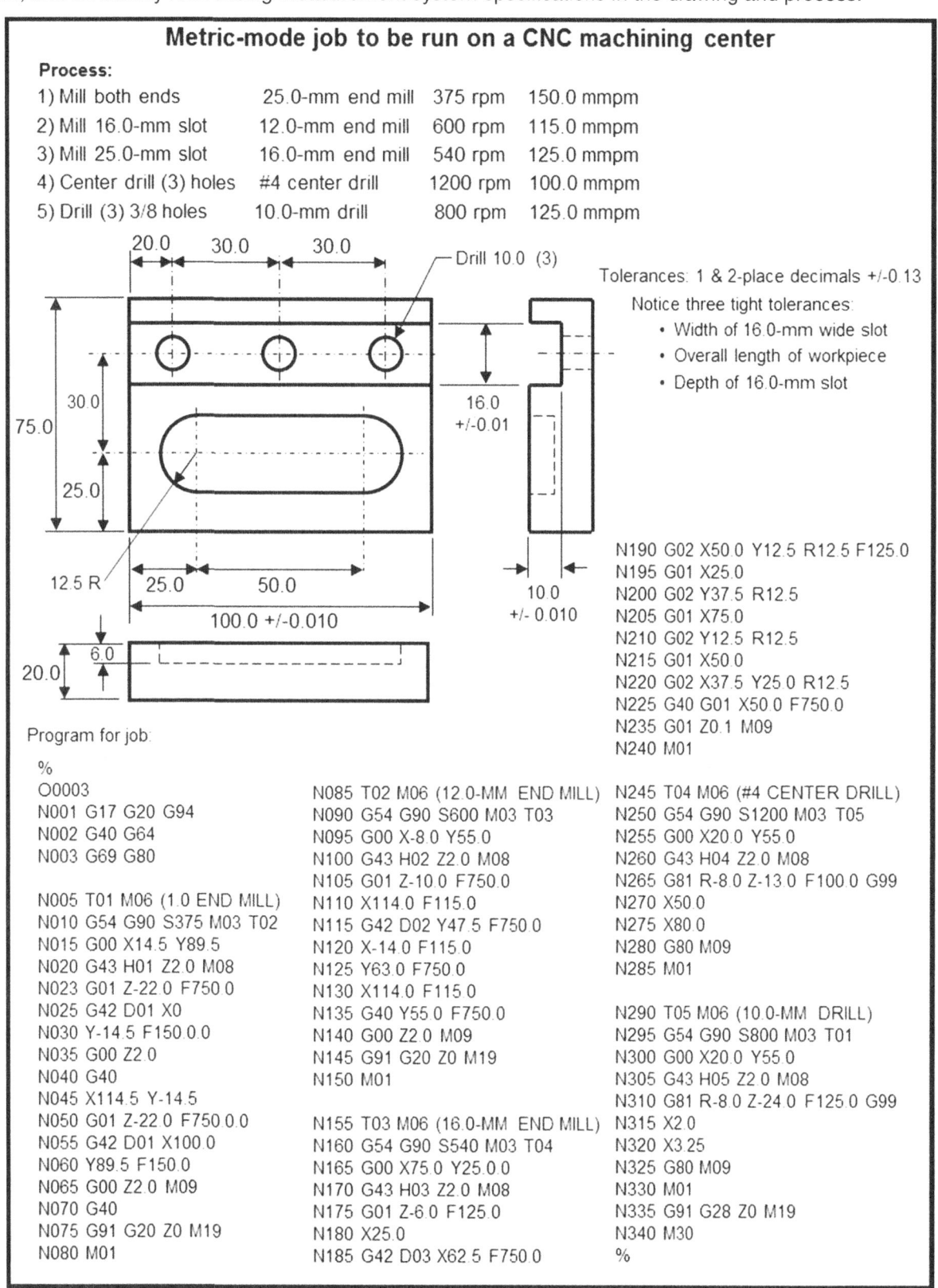

```
%
O0003
N001 G17 G20 G94
N002 G40 G64
N003 G69 G80

N005 T01 M06 (1.0 END MILL)
N010 G54 G90 S375 M03 T02
N015 G00 X14.5 Y89.5
N020 G43 H01 Z2.0 M08
N023 G01 Z-22.0 F750.0
N025 G42 D01 X0
N030 Y-14.5 F150.0.0
N035 G00 Z2.0
N040 G40
N045 X114.5 Y-14.5
N050 G01 Z-22.0 F750.0.0
N055 G42 D01 X100.0
N060 Y89.5 F150.0
N065 G00 Z2.0 M09
N070 G40
N075 G91 G20 Z0 M19
N080 M01

N085 T02 M06 (12.0-MM END MILL)
N090 G54 G90 S600 M03 T03
N095 G00 X-8.0 Y55.0
N100 G43 H02 Z2.0 M08
N105 G01 Z-10.0 F750.0
N110 X114.0 F115.0
N115 G42 D02 Y47.5 F750.0
N120 X-14.0 F115.0
N125 Y63.0 F750.0
N130 X114.0 F115.0
N135 G40 Y55.0 F750.0
N140 G00 Z2.0 M09
N145 G91 G20 Z0 M19
N150 M01

N155 T03 M06 (16.0-MM END MILL)
N160 G54 G90 S540 M03 T04
N165 G00 X75.0 Y25.0.0
N170 G43 H03 Z2.0 M08
N175 G01 Z-6.0 F125.0
N180 X25.0
N185 G42 D03 X62.5 F750.0
N190 G02 X50.0 Y12.5 R12.5 F125.0
N195 G01 X25.0
N200 G02 Y37.5 R12.5
N205 G01 X75.0
N210 G02 Y12.5 R12.5
N215 G01 X50.0
N220 G02 X37.5 Y25.0 R12.5
N225 G40 G01 X50.0 F750.0
N235 G01 Z0.1 M09
N240 M01

N245 T04 M06 (#4 CENTER DRILL)
N250 G54 G90 S1200 M03 T05
N255 G00 X20.0 Y55.0
N260 G43 H04 Z2.0 M08
N265 G81 R-8.0 Z-13.0 F100.0 G99
N270 X50.0
N275 X80.0
N280 G80 M09
N285 M01

N290 T05 M06 (10.0-MM DRILL)
N295 G54 G90 S800 M03 T01
N300 G00 X20.0 Y55.0
N305 G43 H05 Z2.0 M08
N310 G81 R-8.0 Z-24.0 F125.0 G99
N315 X2.0
N320 X3.25
N325 G80 M09
N330 M01
N335 G91 G28 Z0 M19
N340 M30
%
```

Figure 7.5A – A job requiring metric mode that must be run on a CNC machining center

Key Concept 7: Know Your Machine From An Operator's Viewpoint

Inch-mode job to be run on a CNC machining center

Process:
1) Mill both ends — 1.0 end mill — 375 rpm — 6.0 ipm
2) Mill 0.625 slot — 1/2 end mill — 600 rpm — 4.5 ipm
3) Mill 1.0 slot — 5/8 end mill — 540 rpm — 5.0 ipm
4) Center drill (3) holes — #4 center drill — 1200 rpm — 4.0 ipm
5) Drill (3) 3/8 holes — 3/8 drill — 800 rpm — 5.0 ipm

Tolerances: 1 & 2-place decimals +/-0.005

Notice three tight tolerances:
- Width of 0.625 wide slot
- Overall length of workpiece
- Depth of 0.625 slot

Dimensions shown on drawing: 0.75, 1.25, 1.25; Drill 3/8 (3); 3.0, 1.5, 1.0; 0.625 +/-0.0004; 0.5 R; 1.0, 2.0; 4.0 +/-0.0004; 0.375 +/-0.0004; 0.75, 0.25

Program for job:

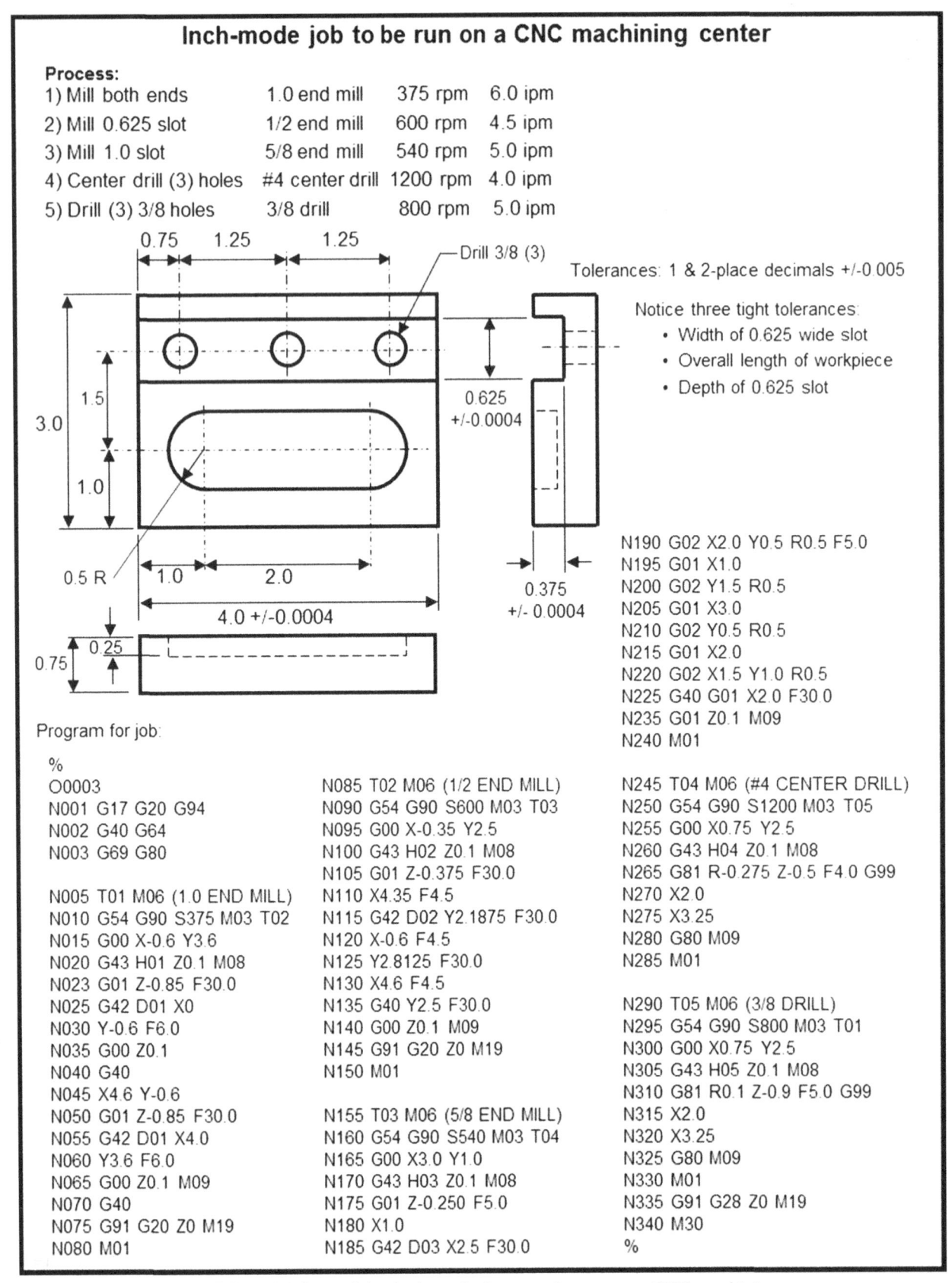

```
%
O0003
N001 G17 G20 G94
N002 G40 G64
N003 G69 G80

N005 T01 M06 (1.0 END MILL)
N010 G54 G90 S375 M03 T02
N015 G00 X-0.6 Y3.6
N020 G43 H01 Z0.1 M08
N023 G01 Z-0.85 F30.0
N025 G42 D01 X0
N030 Y-0.6 F6.0
N035 G00 Z0.1
N040 G40
N045 X4.6 Y-0.6
N050 G01 Z-0.85 F30.0
N055 G42 D01 X4.0
N060 Y3.6 F6.0
N065 G00 Z0.1 M09
N070 G40
N075 G91 G20 Z0 M19
N080 M01

N085 T02 M06 (1/2 END MILL)
N090 G54 G90 S600 M03 T03
N095 G00 X-0.35 Y2.5
N100 G43 H02 Z0.1 M08
N105 G01 Z-0.375 F30.0
N110 X4.35 F4.5
N115 G42 D02 Y2.1875 F30.0
N120 X-0.6 F4.5
N125 Y2.8125 F30.0
N130 X4.6 F4.5
N135 G40 Y2.5 F30.0
N140 G00 Z0.1 M09
N145 G91 G20 Z0 M19
N150 M01

N155 T03 M06 (5/8 END MILL)
N160 G54 G90 S540 M03 T04
N165 G00 X3.0 Y1.0
N170 G43 H03 Z0.1 M08
N175 G01 Z-0.250 F5.0
N180 X1.0
N185 G42 D03 X2.5 F30.0
N190 G02 X2.0 Y0.5 R0.5 F5.0
N195 G01 X1.0
N200 G02 Y1.5 R0.5
N205 G01 X3.0
N210 G02 Y0.5 R0.5
N215 G01 X2.0
N220 G02 X1.5 Y1.0 R0.5
N225 G40 G01 X2.0 F30.0
N235 G01 Z0.1 M09
N240 M01

N245 T04 M06 (#4 CENTER DRILL)
N250 G54 G90 S1200 M03 T05
N255 G00 X0.75 Y2.5
N260 G43 H04 Z0.1 M08
N265 G81 R-0.275 Z-0.5 F4.0 G99
N270 X2.0
N275 X3.25
N280 G80 M09
N285 M01

N290 T05 M06 (3/8 DRILL)
N295 G54 G90 S800 M03 T01
N300 G00 X0.75 Y2.5
N305 G43 H05 Z0.1 M08
N310 G81 R0.1 Z-0.9 F5.0 G99
N315 X2.0
N320 X3.25
N325 G80 M09
N330 M01
N335 G91 G28 Z0 M19
N340 M30
%
```

Figure 7.5B – A job requiring inch mode that must be run on a CNC machining center

This workpiece is the practice exercise from lesson 4.3 with three holes added. Notice that there are three tight tolerances specified on the print. The process is shown, providing you with the machining order used in the program. All coordinates in the program have been specified as mean values for each tolerance.

Notice that the program is not documented nearly as well as most we have shown in this text. Most programmers don't provide much documentation in the program (remember – documentation takes up memory space in the machine) – so this is typical of what you'll see with the jobs you'll be running. Tool naming messages are included in the program at the beginning of each tool. This program uses the format for a vertical machining center shown in lesson 5.2.

We will describe the various tasks that must be completed in order to make a setup and complete a production run. Our descriptions will assume, however, that you have some basic machining practice experience. Be sure to question an experienced person in your company if any of these presentations require more basic machining practice skills than you currently possess.

Setup Documentation

As is described in lesson 2.1, a programmer must provide setup documentation. Most programmers will use a one-page setup sheet to describe everything that must be done to make the setup. Figures 7.6A and 7.6B show the setup sheets for our example jobs, 7.6A for metric-mode discussions and 7.6B for inch-mode discussions. Most companies also supply the setup person with a copy of the print and a print-out of the CNC program.

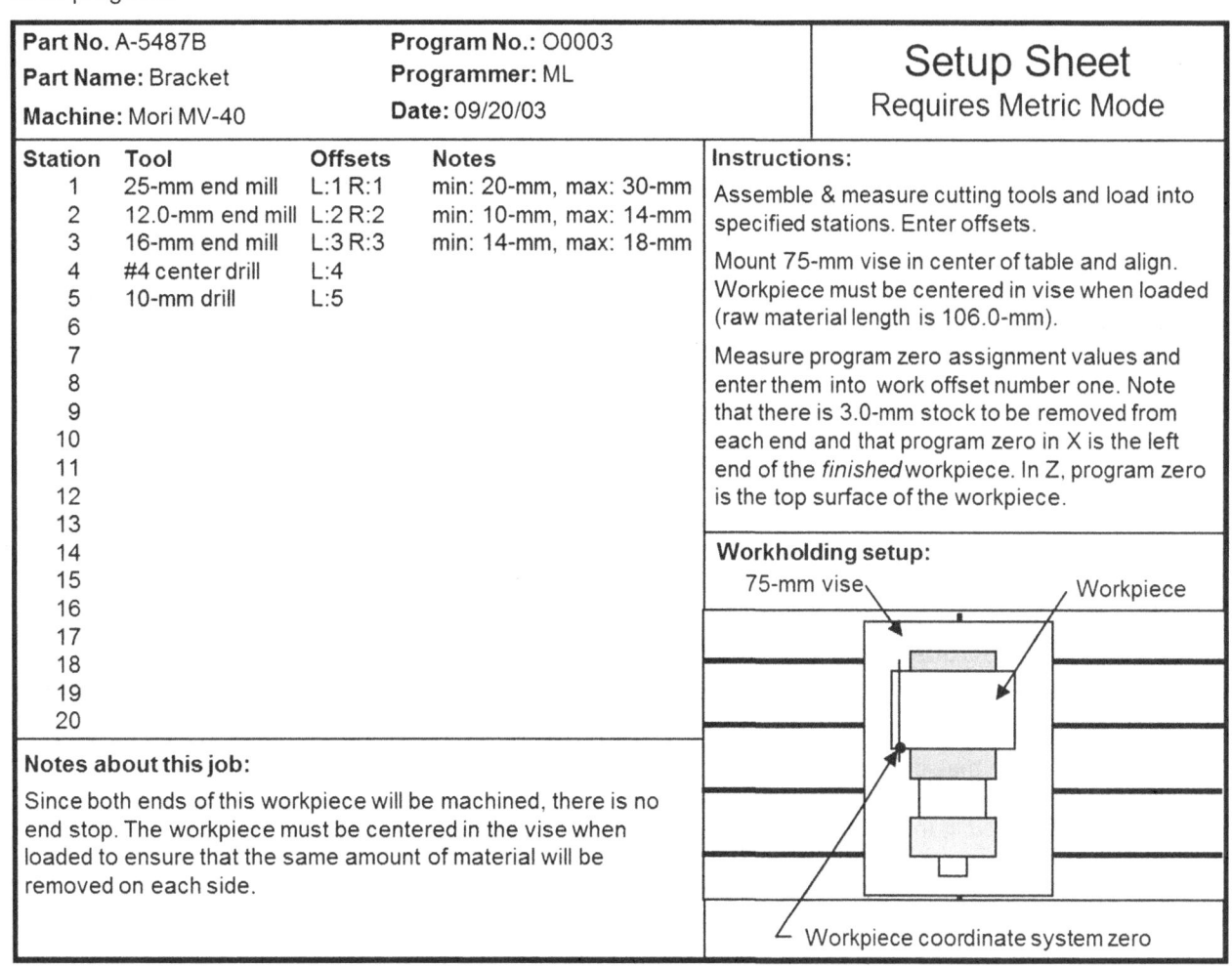

Part No. A-5487B	Program No.: O0003	**Setup Sheet**
Part Name: Bracket	Programmer: ML	Requires Metric Mode
Machine: Mori MV-40	Date: 09/20/03	

Station	Tool	Offsets	Notes
1	25-mm end mill	L:1 R:1	min: 20-mm, max: 30-mm
2	12.0-mm end mill	L:2 R:2	min: 10-mm, max: 14-mm
3	16-mm end mill	L:3 R:3	min: 14-mm, max: 18-mm
4	#4 center drill	L:4	
5	10-mm drill	L:5	
6			
7			
8			
9			
10			
11			
12			
13			
14			
15			
16			
17			
18			
19			
20			

Instructions:

Assemble & measure cutting tools and load into specified stations. Enter offsets.

Mount 75-mm vise in center of table and align. Workpiece must be centered in vise when loaded (raw material length is 106.0-mm).

Measure program zero assignment values and enter them into work offset number one. Note that there is 3.0-mm stock to be removed from each end and that program zero in X is the left end of the *finished* workpiece. In Z, program zero is the top surface of the workpiece.

Workholding setup:
75-mm vise, Workpiece, Workpiece coordinate system zero

Notes about this job:

Since both ends of this workpiece will be machined, there is no end stop. The workpiece must be centered in the vise when loaded to ensure that the same amount of material will be removed on each side.

Figure 7.6A – Setup sheet for our example job (requires metric mode)

Key Concept 7: Know Your Machine From An Operator's Viewpoint

Part No. A-5487A Part Name: Bracket Machine: Mori MV-40	Program No.: O0003 Programmer: ML Date: 09/20/03	**Setup Sheet** Requires Inch Mode

Station	Tool	Offsets	Notes
1	1.0" end mill	L:1 R:1	min: 0.75", max: 1.2"
2	1/2" end mill	L:2 R:2	min: 0.375", max: 0.55"
3	5/8" end mill	L:3 R:3	min: 0.5", max: 0.65"
4	#4 center drill	L:4	
5	3/8 drill	L:5	
6			
7			
8			
9			
10			
11			
12			
13			
14			
15			
16			
17			
18			
19			
20			

Instructions:

Assemble & measure cutting tools and load into specified stations. Enter offsets.

Mount 3" vise in center of table and align. Workpiece must be centered in vise when loaded (raw material length is 4.2").

Measure program zero assignment values and enter them into work offset number one. Note that there is 0.1" stock to be removed from each end and that program zero in X is the left end of the *finished* workpiece. In Z, program zero is the top surface of the workpiece.

Workholding setup:

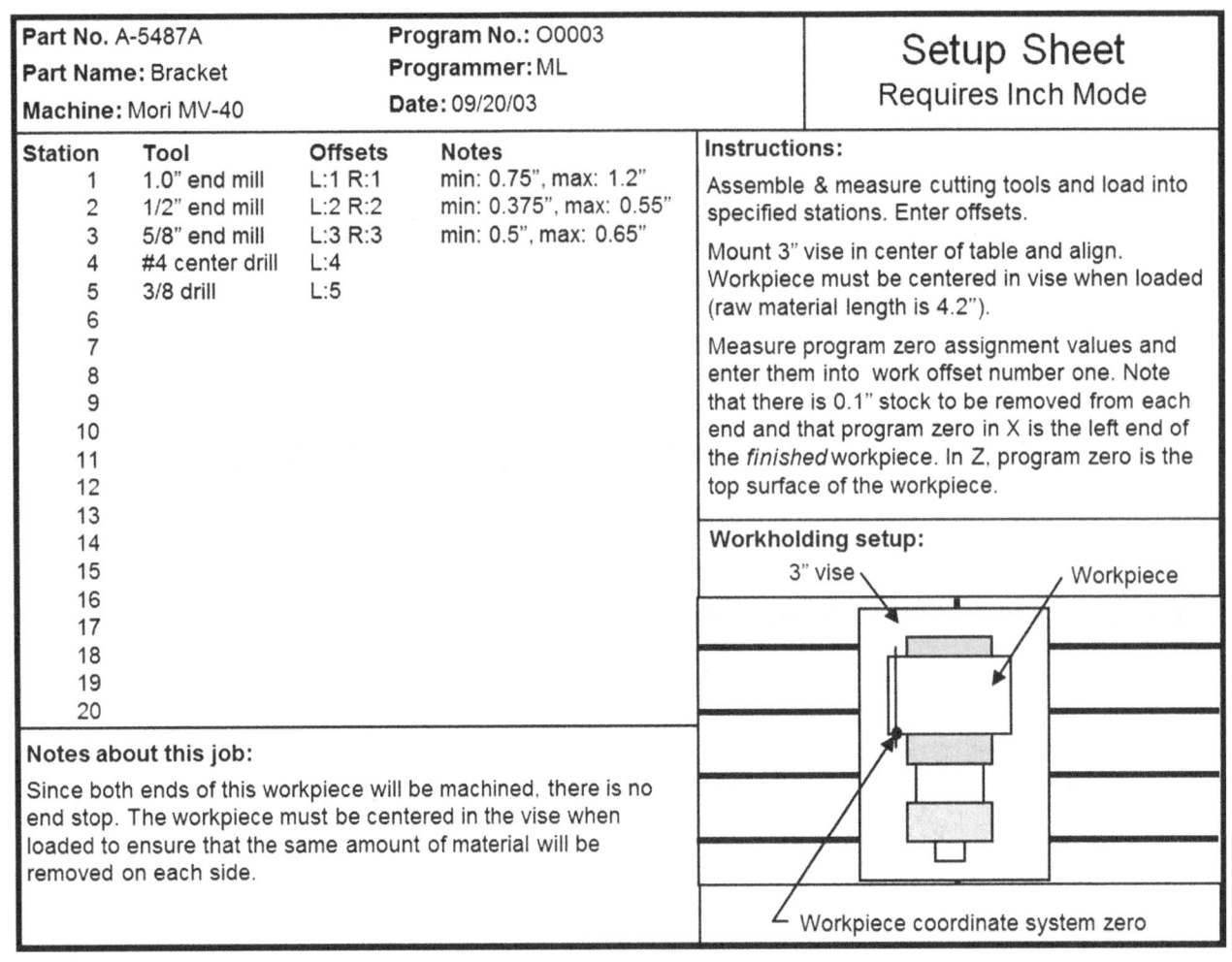

3" vise / Workpiece / Workpiece coordinate system zero

Notes about this job:

Since both ends of this workpiece will be machined, there is no end stop. The workpiece must be centered in the vise when loaded to ensure that the same amount of material will be removed on each side.

Figure 7.6 – Setup sheet for our example job (requires inch mode)

As is typical of setup documentation provided by most programmers, our setup sheets do not explain every detail of what must be done to get the job up and running. They assume quite a bit of the setup person who will be making the setup. Many setup-related tasks are quite redundant from one setup to the next – and most programmers will not be very specific about the most redundant tasks. We will point out common shortcomings of this kind of setup documentation as we continue.

Tear Down the Previous Setup and Put Everything Away

We now begin describing the specific tasks you must complete in order to make the setup for the example job. We will do so in the approximate order that setups are normally made.

Before you can begin making a new setup, you must remove components from the previous setup. Tasks include removing cutting tools that won't be used in the next job, removing the work holding device (assuming it's not used in the next job), and cleaning chips and debris from the machine table. The old CNC program should also be removed from the machine's memory (after saving a corrected version – if required). Once removed, tooling components should be put back where they belong – so you can easily find them in the future.

Gather the Components Needed To Make the Setup

The more organized you can be when making a setup, the easier the setup will be to make. Part of being organized is gathering the components you will need to make the setup. Gathering everything at one time will keep you from having to walk to and from the tool crib every time you need something.

Companies vary with regard to how well they list the components needed for a given setup. Our setup sheet isn't very good in this regard. For example, we're assuming that the setup person can figure out which components are needed to assemble the cutting tools for the job. Indeed – we're assuming they can determine which *cutting tools* will be used. This kind of setup sheet is very common in workpiece- and tooling-producing companies – when the setup person is expected possess a high degree of skill.

Some companies do much better, providing a complete list of every needed component, including work holding components, cutting tool components, hardware (fasteners like tee-nuts and bolts), gauging tools and even hand tools. This makes the gathering of needed components much easier. With this kind of list, someone other than the (highly skilled) setup person will be able to gather components. This will free up the setup person to be doing other things while components are being gathered. This kind of component list is common in product-producing companies.

Make the Work Holding Setup

Work holding devices vary from setup to setup. They commonly include fixtures, vises, and indexing heads. Our example job requires a table vise. Making a work holding setup for a CNC machining center requires the same skills needed to make work holding setups for conventional milling machines. If you've made work holding setups for manual mills, you already know what it takes to do so for CNC machining centers.

For our example job, the work holding setup requires mounting a common table vise on the machine table, like the one shown in Figure 7.7.

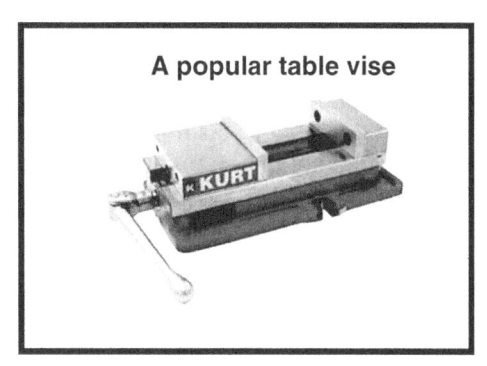

Figure 7.7 – Table vise used to hold our example workpiece

We're using a 75-mm (or 3.0-in) table vise. This means the length of the jaws will be 75.0 millimeters (or 3.0 inches). Our 106.0-mm (or 4.2-in) workpiece will overhang these jaws by 15.5-mm (or 0.6-in) on each end. Though the vise in Figure 7.7 doesn't show it, the jaws in most table vises have a small step machined to grip on a workpiece. Without this step, the workpiece must be placed on *parallels*.

While our example job doesn't need one, many vise setups require an end stop like the one shown in Figure 7.8. This end stop is held to the side of the vise with a magnet. Also, notice the machined steps in the jaws for workpiece gripping.

Figure 7.8 – Easily adjusted end stop for a table vise

After mounting the table vise on the table, the setup person will make sure it is *square with the table*. They will mount a dial indicator in the spindle and run the stylus of the dial indicator along the fixed jaw (by using the hand-wheel in the X-axis). If necessary, they will move the vise so that the jaws are perfectly parallel with the X-axis. Ask an experienced person to show the squaring of a table vise if you've never seen it done.

While many vise setups require an end stop to locate one end of the workpiece, our workpiece requires the machining of *both ends*. For this reason, the workpiece will simply be centered in the vise. While this may be possible by estimating, the setup person will probably use a scale (like a ruler) to be more accurate. For our

106.0-mm (or 4.2-in) long piece of raw material, 15.5-mm (or 0.6-in) of the workpiece must be protruding from each side of the three inch long vise jaws.

Remember that work holding setups vary from job to job. The setup documentation must be sufficient to describe how the work holding setup must be made – and directed at the lowest skill level of setup person in the company.

Assign the Workpiece Coordinate System Zero Point

In lessons 1.4, 1.5, and 1.6, we describe workpiece coordinate system zero and how it is assigned. Remember that for qualified setups, workpiece coordinate system zero assignment may be done in the program, meaning the setup person may not have to assign workpiece coordinate system zero.

With our example job, there is only one workpiece coordinate system zero point. It is located at the lower-left corner of the finished workpiece in X and Y and at the top surface of the workpiece in Z. Our setup instructions specify that workpiece coordinate system zero assignment values must be placed in workpiece coordinate system setting offset number one.

Measure Workpiece Coordinate System Zero Assignment Values

This vise setup is not qualified. That is, the setup instructions do not specify which table slots should be used in which to mount the vise. Indeed, this vise may not even have location keys that locate in table slots. (It surely does not if it must be squared with the table.) This means that workpiece coordinate system zero assignment values must be *measured* by the setup person every time this job is run.

The general procedures to measure workpiece coordinate system zero assignment values and enter them into workpiece coordinate system setting offsets is shown in lessons 1.5 and 1.6. We will review them here, showing specific techniques for our example job.

For the X and Y-Axis Workpiece Coordinate System Zero Assignment Values, the setup person will use an edge-finder (like the one shown in Figure 7.9). Also, the work holding setup must be made and a workpiece must be loaded prior to performing this procedure.

Figure 7.9 – An edge-finder (also called a wiggler)

1) The edge-finder is placed in the spindle with the 5.0-mm (or 0.2-in) diameter end down (pointed end is in the tool holder).

2) The spindle will be started at about 500 rpm and the setup person will push the edge-finder out of alignment with its shank (so the small diameter is not concentric with the shank).

3) Using the jog and hand-wheel functions of the machine, the setup person will bring the edge-finder's 5.0-mm (or 0.2-in) diameter perfectly flush with the left side of the workpiece (the raw material edge in our case).

4) Without moving the X-axis, the setup person will retract the edge-finder in the Z-axis until it is well above the workpiece.

5) The setup person will reset (set to zero) the X-axis relative position display register.

6) The edge-finder's 5.0-mm (or 0.2 in) diameter currently is flush with the left edge of the raw material. The workpiece coordinate system zero point is now precisely 5.5-mm (or 0.200-in) away in the X plus direction from the spindle center. This is calculated by adding the 2.5-mm (or 0.1-in) edge-finder radius plus the 3.0-mm (or 0.1-in) raw material that is still on the workpiece). So, using the hand-wheel and by monitoring the X-axis relative position display, the setup person will move the edge-finder in the X plus direction by 5.5-mm (or 0.2-in). (Actually the machine table will move to the left on most machines.) At this point, the center of the spindle is perfectly aligned with the X-axis workpiece coordinate system zero.

7) The setup person will reset (set to zero) the X-axis position display again.

8) The setup person will again push the edge-finder out of alignment with its shank.

9) Using the jog and hand-wheel functions of the machine, the setup person will bring the edge-finder's 5.0-mm (or 0.2-in) diameter perfectly flush with the lower side (Y minus side) of the workpiece.

10) Without moving the Y-axis, the setup person will retract the edge-finder in the Z-axis until it is well above the workpiece.

11) The setup person will reset (set to zero) the Y-axis relative position display register.

12) The edge-finder's 5.0-mm (or 0.2-in) diameter is currently is flush with the Y minus edge of the workpiece. The workpiece coordinate system zero point is now precisely 2.5-mm (or 0.100-in) away in the Y plus direction from the spindle center. So, using the hand-wheel and by monitoring the Y-axis relative position display, the setup person will move the edge-finder in the Y plus direction by 2.5-mm (or 0.100-in). (Actually, the machine table will move forward on most machines.) At this point, the center of the spindle is perfectly aligned with the Y-axis workpiece coordinate system zero.

13) The setup person will reset (set to zero) the Y-axis position display again.

14) The setup person will send the machine to its zero return position in the X and Y axes. At this point, the X and Y-axis relative position display registers will be showing the workpiece coordinate system zero assignment values in X and Y.

For the Z-Axis Workpiece Coordinate System Zero Assignment Value (distance between spindle nose and workpiece coordinate system zero in Z), there must be no tool in the spindle at the start of this procedure. We will also place a gauge block of known height on top of the workpiece. For or example, the gauge block will be precisely 50.0-mm (or 2.0-in) high.

1) Using the jog and hand-wheel functions, the setup person will bring the spindle nose into contact with the 50.0mm (or 2.0-in) gauge block.

2) The setup person will reset (set to zero) the Z-axis relative position display register.

3) The setup person will send the Z-axis to its zero return position. At this point, the setup person will add the gauge block height (50.0-mm or 2.0-inches) to the value currently shown on the Z-axis relative position display. This will be the Z-axis workpiece coordinate system zero assignment value.

Enter Workpiece Coordinate System Zero Assignment Values Into Workpiece Coordinate System Offset Number One

Since the workpiece coordinate system zero assignment values have been measured during setup, they must be manually entered into the workpiece coordinate system setting offset (workpiece coordinate system setting offset number one in our case). To do so, the setup person will press the offset soft key until the workpiece coordinate system setting offset page appears. They'll scroll the cursor down to the X register for workpiece coordinate system setting offset number one, enter the X workpiece coordinate system zero assignment value, and press the input key. They'll do the same for the Y and Z registers. (Remember that these values are the distances *from* the zero return position *to* the workpiece coordinate system zero– they are almost always *negative values*.)

Assemble the Cutting Tools Needed for the Job

In some companies, the CNC setup person is not responsible for assembling cutting tools. Some companies have a person in the tool crib (commonly called the *tool setter*) assemble cutting tools. Though this may be the case, all CNC setup people will have to assemble cutting tools from time to time – so it is important that you have the ability to perform this task.

Our setup documentation is a little sketchy when it comes to cutting tools. You haven't been told what components to use – so you're on your own to figure it out. While experienced setup people can easily do so, beginners may have some problems. Fortunately, most companies are a little more helpful when it comes to listing components needed to assemble cutting tools. They'll use some kind of identification system (usually of the company's own design) to specify cutting tool components. Or they may use the catalog numbers from their cutting tool manufacturers. The best kind of cutting tool documentation shows every component that makes up each assembled cutting tool – as is shown in Figure 7.10.

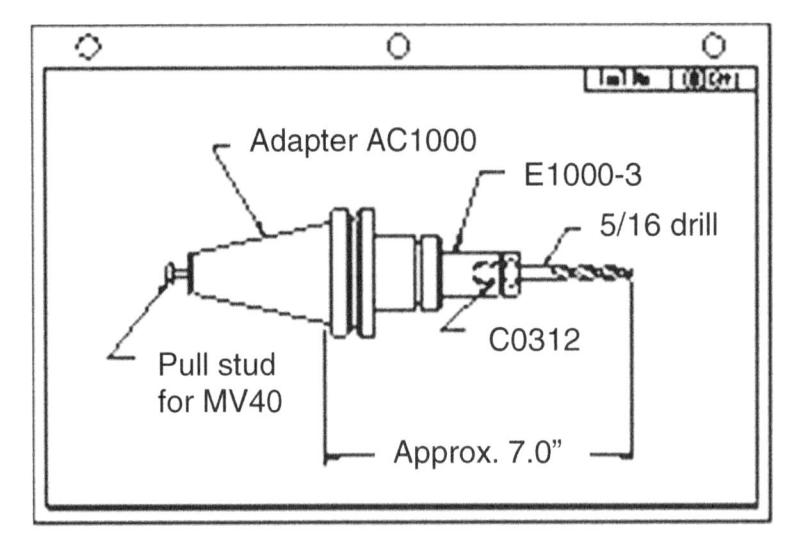

Key Concept 7: Know Your Machine From An Operator's Viewpoint

Figure 7.10 – Great cutting tool documentation

Another good system incorporates a special one or two letter abbreviation for each type of component along with four numbers to specify its size. While this system is not as easy to use as the documentation shown in Figure 7.10, it is still pretty easy to understand. Consider, for example, the following tool component identification numbers. It shows cutting tools required for inch mode programs, but a similar set of identification numbers could be developed for metric mode programs.

- DT0750 (a 3/4" twist drill)
- ME0375 (a 3/8" end mill)
- RM1000 (a 1.0" reamer)
- AC1000 (a 1.0" collet adapter)
- CT0500 (a 1/2" collet)
- AE0750 (a 3/4" end mill holder)
- AF30000 (a 3.0" face mill holder)
- EC0500 (a 1/2" collet extension)
- ET0250 (a 1/4" tap extension)

With a little ingenuity, you can come up with a naming convention for just about any cutting tool component. Instead of using one drawing per cutting tool (as is shown in Figure 7.10), a more concise tool list can be used. Using this kind of system, here is the tool list for our example setup.

Tool 1: ME1000, AE1000
Tool 2: ME0500, AE0500
Tool 3: ME0625, AE0625
Tool 4: DC0004, AC1000, EC0375, CT0375
Tool 5: DT0375, AC1000, EC0375, CT0375

While this is not nearly as explicit as the drawing in Figure 7.10, it adequately lists all cutting tool components needed for our example job.

A special bench-mounted assembly figure is used to assemble cutting tools. The tapered tool holder shank is placed into the fixture and is located from the keyways on the outside of its flange. This provides rigid support for the tool holder shank as cutting tool components are assembled and tightened. Figure 7.11 shows a bench-mounted assembly fixture.

Figure 7.11 – A bench mounted cutting tool assembly fixture

Most of the hand tools used to assemble cutting tools should be pretty familiar and include Allan wrenches, box-end wrenches, and spanner wrenches. And assembling most cutting tools is pretty straight forward. But if you are in doubt about how a particular cutting tool goes together, or how much pressure is required to adequately tighten it, you must ask an experienced person in your company.

Measure Tool Length- and Cutter-Radius-Compensation Values

As with assembling cutting tools, this task may not be the responsibility of the setup person. But even if your company employs a tool setter to measure cutting tools, all setup people must still possess the ability to do so.

All cutting tools require a tool length compensation measurement. As is shown in lesson 4.2, if you use our recommended method (offset is tool length), you can measure tool lengths either off line prior to setup or on the machine during setup. If you use the second method shown in lesson 4.2 (offset is the distance from tool tip to workpiece coordinate system zero), you must measure tool length compensation values on the machine during setup.

We show the procedures to measure tool length compensation values in lesson 4.2, so we won't repeat them here. Our example job requires that you determine five tool length compensation values. If you do the measurements off line, write down each value as you measure it. If you measure on the machine, you can transfer the Z-axis relative display screen value right into the tool length compensation offset (as is also shown in lesson 4.2). This will eliminate the possibility for entry mistakes.

For milling cutters that perform contour milling operations (milling on the periphery of the cutter) and use cutter radius compensation, you must also measure the milling cutter's radius. To do so, simply use a micrometer to measure the milling cutter's diameter and divide the diameter by two to determine its radius. For FANUC controls, it is the cutter's *radius* that is needed for cutter radius compensation offsets.

The Static Nature of Cutting Tool Measurements

Remember that when a cutting tool begins machining, it will be under the influence of *tool pressure*. Tool pressure can influence the quality your initial measurements. Another tooling-related problem that can affect the quality of your initial cutting tool measurements is *runout*. When an end mill is placed in an end mill holder, it may not be perfectly concentric with the spindle. If the milling cutter is not concentric with the spindle it will remove more material from the workpiece than it should. The smaller the tolerances you must hold on the workpiece, the more important it is that you consider using a technique called *trial machining* when you machine the first workpiece. Trial machining is presented in lessons 4.1, 4.2, and 4.3 – and we will discuss it in greater detail later in this lesson.

Enter Tool Length and Cutter Radius Compensation Values Into Offsets

If tool length compensation values are measured on the machine, you can transfer them from the Z axis position display into each offset register during the measuring procedure. There will be no need to enter them as is discussed here.

But if you measure tool length compensation offset values off line, they must be entered into offsets during setup. In similar fashion, cutter radius compensation values must be entered into offsets for milling cutters.

For our example job, the setup sheet explicitly specifies which offsets are related. Frankly speaking, most setup sheets will not. In most companies, there are unwritten rules related to how tool length- and cutter-radius-compensation offset numbers are determined. For tool length compensation, most programmers will make the tool length compensation offset number correspond to the tool *station number* (in the H GEOM offset field).

For cutter radius compensation, offset numbering depends upon how the offset display screen displays offsets. With most machines, each offset will have two values – one for the tool's length and another for its radius (the H GEOM and D GEOM fields). With these machines, the programmer will use the same offset number for both tool length- and cutter-radius-compensation. This makes it very easy to determine which offsets are related to a given cutting tool.

Unfortunately, the offset display screen for older machines has but *one* value per offset. With this kind of offset display, the offset number corresponding to the tool station number is already being used for tool length compensation – so the programmer must choose another. Most programmers will add a constant number (like thirty) to the tool station number to determine the offset number used with cutter radius compensation.

If your machine has but one value per offset, you *must* know which offsets are used with cutter radius compensation when milling cutters are used in a job. If it is not explicitly specified in the setup documentation, you must ask the programmer.

For our example job, we will enter tool length compensation values into H GEOM offsets numbered one through five. Since we're using the recommended method for tool length compensation (offset is tool length), each will be a positive value – probably between about 125.0-mm and 250.0-mm (or 5.0-in and 10.0-in) in length. And we will enter cutter radius compensation values of 12.5-mm (or 0.5-in) in D GEOM offset one, 6.0-mm (or 0.25-in) in D GEOM offset two, and 8.0-mm (or 0.3125-in) in D GEOM offset three.

Load Cutting Tools Into the Machine's Automatic Tool Changer Magazine

If you measure cutting tools on the machine, you should perform this task *before* measuring and entering tool length compensation values. This will allow you to use the machine's automatic tool changing system to load each tool that must be measured. We haven't shown the procedure to manually use the automatic tool changer. It involves using the *manual data input* (MDI) mode – and will be shown in lesson 9.1.

Machine tool builders vary when it comes to how cutting tools must be loaded into the automatic tool changer magazine. Most provide a special and convenient *tool loading station*. They also allow the magazine to be easily rotated (with two push-buttons, one for clockwise and another for counter-clockwise) right from the magazine operation panel, so you don't have to walk back to the main control panel to rotate the magazine.

Even once you've rotated the magazine to the desired loading station, machine tool builders vary when it comes to how tools are actually put into the magazine. With smaller tools (CAT-40 and smaller), most machine tool builders use a *ball-detent* system that holds each tool by its pull stud. These tools are *snapped* into the magazine station's tool pot by simply pressing them in. A special tool may be required to pry the tool out of the tool pot. For larger tools, most machine tool builders use a more positive method of clamping tools in the magazine – and at the loading station, a lever is used to release the clamping mechanism so that tools can be removed and replaced.

Even tool numbering for each station of the automatic tool changer magazine will vary from one machine tool builder to another. Some builders use hard-and-fixed tool station numbers while others allow each station to be numbered through the operation panel. You must ask an experienced person, or reference your machine tool builder's operation manual to determine how your automatic tool changer functions in this regard.

For our example job, we place the 25.0-mm (or 1.0-in) end mill in station one, the 12.0-mm (or 1/2-in) end mill in station two, the 16.0-mm (or 5/8-in) end mill in station three, the #4 center drill in station four, and the 10.0-mm (or 3/8-in) drill in station five.

Load the CNC Program

For our example job, program number O0003 must be loaded. Our setup documentation is assuming that the setup person knows where to find program number O0003. With current FANUC controls, programs can be loaded from a variety of external devices, including memory cards.

Many companies still distributive numerical control (DNC) systems. Each machine will have a special program storage folder (or directory) in the DNC system – and setup people are expected to know the related folder names. Some companies will specify the storage location right in the setup documentation.

The actual task of loading a program depends upon the company's DNC system. Most require that the machine be made ready first, by placing the mode switch to *edit* and then pressing the *read* button. At this point, the machine waits for a program to be sent from the DNC system. Then the setup person goes to the DNC system and commands it to send the program. Some DNC systems do allow the entire process to be completed right from the CNC machine tool. You must talk to an experienced person in your company to learn the specific procedure.

The Physical Tasks Related To Setup are Now Completed

At this point, you are (finally) ready to run the program. However, there could still be problems. How many problems you'll have and how severe they are depend upon several things. Generally speaking, new programs written manually by new programmers tend to present the most problems. But even proven programs (programs that have been successfully run before) can still have problems. Additionally, the setup person can make mistakes that will cause even a perfectly written program to fail. For these reasons, *all programs must be cautiously verified*. While proven programs tend to be easier to verify than new programs, you should never skip the program verification steps.

Verify the Correctness of a New or Modified Program

This task is only necessary for new programs – or for programs that have been modified since the last time they were run.

Verifying the correctness of the program is really the responsibility of the *CNC programmer*, and is usually done prior to making a setup. Indeed, it is usually done soon after the program is written. There are many *tool path verification systems* that allow a programmer to visually check the motions a CNC program will make even before the CNC program is loaded into the machine. Most are relatively inexpensive, so *there is really no excuse not to have one* – especially when you consider the expense of using a CNC machine to verify the correctness of a program.

Figure 7.12 shows a typical *solid-model-type* tool path verification screen for our example program run on a Windows computer. The system we used sells for about $200.00, which is extremely reasonable when you consider the amount of time and problems it can save during program verification on a CNC machining center.

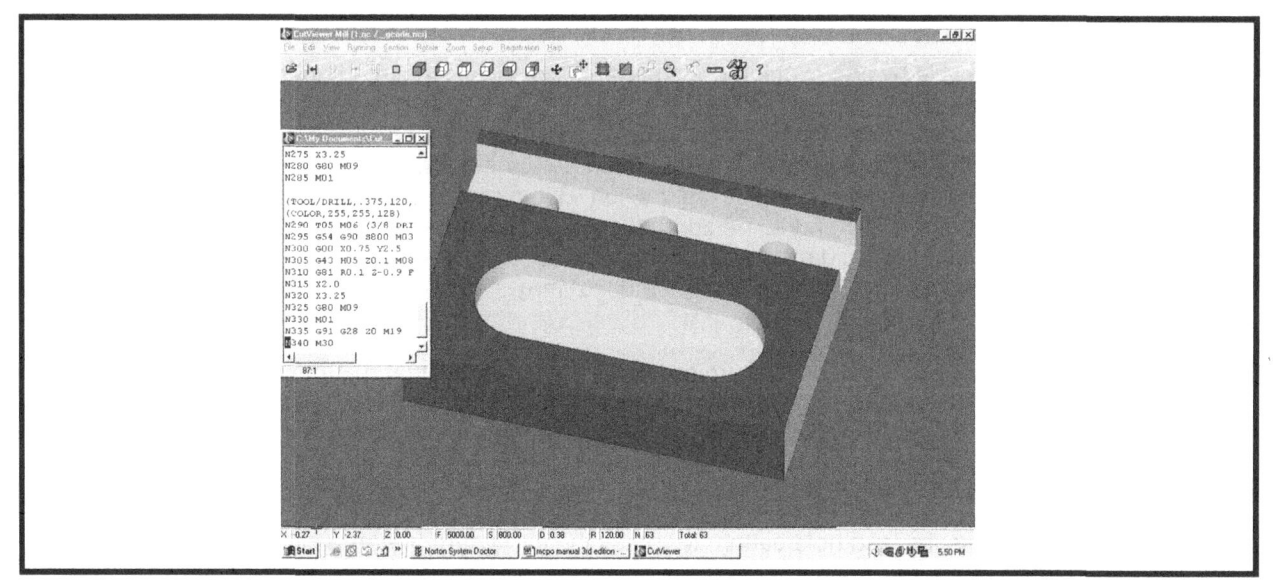

Figure 7.12 – An example of tool path verification done on a Windows computer

Once again, keep in mind that CNC-using-companies vary when it comes to what they expect of their CNC people. If you are a setup person in a product producing company, you probably won't have access to the program verification system, so you won't see the program verification being done off line. You'll have to take the programmer's word that the program's tool path is correct.

Key Concept 7: Know Your Machine From An Operator's Viewpoint

Current FANUC controls do include tool path verification capabilities that rival the best off line systems. If your machine has this capability, by all means, learn how to use it, especially if you can't see the program verification done off line. But if at all possible, this program verification step should be done *off line* (to save machine time). If mistakes are found while verifying the correctness of the program, the time required for *correcting* them can also be kept off line.

If you have neither of these tool path verification capabilities, you will not be able to see the movements that cutting tools in the program will make until you actually run the program on the machine. In this case, verifying the correctness of your programs will be extremely cumbersome, requiring that you cautiously check each movement the machine makes, using features like *dry run*, *single block*, and *feed hold*. When verifying programs in this manner, you may have to run the program several times just to begin to understand the program's movements. And you still won't be able to tell if all of the motions are correct.

Verify the Correctness of the Setup

Mistakes made during setup can cause even a perfect program to behave poorly. Consider a few of the possibilities: You could load a tool into the wrong station. You could measure or enter workpiece coordinate system zero assignment values incorrectly. You could measure or enter tool length compensation values incorrectly. You could forget to enter a cutter radius compensation offset value. You could load the wrong program.

And by the way, there are certain programming mistakes that will not show up during tool path verification. If, for example, the programmer forgets to start the spindle (or if they have it running in the wrong direction) few tool path verification systems will catch the mistake. And most tool path verification systems will ignore offset use. If the program invokes the wrong offset, most tool path verification systems won't show it.

For these reasons, *you should run the program at least once without a workpiece*. This is commonly called a *dry run*. In fact, one of the program verification functions you have to help with program verification is called *dry run*. When the dry run switch is on, you will have control of how quickly the machine moves (even during rapid movements). A multi-position switch (usually the feedrate override switch) is used for this purpose. You'll also have a *panic button* that will stop all axis motion if you get nervous (called the *feed hold* button).

When you do a dry run, again, there will be no workpiece in the setup and you'll have complete control of all motion. You'll be checking for very basic and serious mistakes – like collisions, spindle direction, tools loaded into the spindle, and general movements made by the program. If a mistake is found, you'll stop the cycle, diagnose and correct the problem, and do another dry run. You must not continue until you've seen the *entire program* run – and until you understand and agree with the movements made by the program.

A Tip That Will Save a Crash Some Day

Whenever you activate a cycle by pressing the cycle start button, always have a finger ready to press the *feed hold button*. If you don't have certain switches set properly, the machine will not behave as you expect, and you will be ready to *immediately* stop the machine when something goes wrong. For example, say you forget to turn on the dry run switch. You have feedrate override turned down to its minimum position, so you're *expecting* the machine to crawl during its first motion. But in reality, it will move at rapid (assuming rapid traverse override is set to one-hundred percent). If you have a finger ready to press the feed hold button, you'll be ready to stop the rapid motion as soon as you see that something is wrong. You won't have to look for the feed hold button – which takes time – time that may allow the machine to crash.

Dry Running Our Example Program

We remove the workpiece from the vise, and turn on dry run. We confirm that the machine is at its tool change position. We turn the feedrate override switch down to its lowest setting (usually ten percent), we place the mode switch in the auto or memory position (to run a program). We press the program check display screen key (this is the page that shows the distance-to-go function), and confirm that program number O0003 is the active program.

With a finger ready to press the feed hold button, we press the cycle start button. The first thing that happens is that tool number one (the 25.0-mm or 1.0-in mill) is placed into the spindle. Then the spindle starts – sure enough, in the forward direction. From the program check page (which also shows absolute position of the cutting tool), we can see that the machine is moving, but barely. As we crank up the feedrate override switch, sure enough, we can see the X and Y axes moving. This first XY movement is pretty safe because the machine is still at its Z-axis zero return position. So we crank the feedrate override switch up to its maximum setting and the X and Y axes move quickly into position.

When the end mill is over the vise, the Z-axis begins to move. The end mill is still quite a distance from the vise, but we're getting a little nervous. So we slow the motion a little by turning down the feedrate override switch. That's better. Not so scary.

The end mill comes into position and makes the movements needed to mill the left end of the workpiece. While it is still at the work surface in Z, we press the feed hold button. Let's confirm that the tool looks like it is at the appropriate Z position. We look into the work area and sure enough, the tool's Z position looks good. So we press the cycle start button to reactivate the cycle.

The end mill finishes milling the left end, it retracts, and then moves into position to mill the right end. It drops back down in Z and mills the right end. Everything looks good.

The first tool is now finished and it begins retracting for the tool change. We increase the feedrate override switch to maximum. As the tool change occurs, we reduce the feedrate override switch again. The 12.0-mm (or 1/2-in) end mill is placed into the spindle and the spindle starts, again, in the correct direction. It moves a little in X and Y to get to the left side of the slot and then begins moving in Z. As it gets closer, we slow the motion a little. It makes the first pass through the middle of the slot. Then it moves a little in the Y minus direction and makes the second pass. Then it moves in the Y plus direction and makes the third pass. Finally it starts retracting to the tool change position in Z.

The 16.0-mm (or 5/8-in) end mill is then placed into the spindle and the spindle starts. We cautiously monitor its movements as it mills the oval-shaped pocket. Everything looks fine.

We do the same for the center drill and 10.0-mm (or 3/8-in) drill. No problems are found.

What If You Do Find a Problem?
Our dry run went flawlessly. But here are a couple of common scenarios that signal a problem.

Say the very first tool, the 25.0-mm (or 1.0-in) end mill is making its first approach movement in the Z-axis. It's seems to be going *too far*. It's well past what seems to be the work surface and it is still moving. So you press the feed hold button to stop the tool's motion and look at the distance-to-go display. Sure enough, it says the tool is still going to move another 150.0-mm (or 6.0-in) or so in the Z negative direction. This will cause the tool to crash into the machine table. So you cancel the cycle (by pressing the *reset key*). You must find and correct the problem. What do you think it could be?

We saw the tool path display off line and it looked good. So the problem shouldn't be with the Z-axis movements in the program. Since this is the *first* tool, the problem could be with your workpiece coordinate system zero assignment in the Z-axis or with tool length compensation for this tool. (Note that if the first tool approaches properly and this problem occurs on any subsequent tool, the problem *must be* with tool length compensation.)

When we look at the tool length compensation offset for this tool (offset number one), say we find that there is a value of *zero* specified. We forgot to enter a tool length compensation value for this tool! The machine thinks we have a cutting tool that is zero millimeters (or inches) long. It was bringing the spindle nose to the work surface – which is causing the tool tip to go way past its intended position.

After correcting the problem (by entering the tool length compensation value in offset one), we must re-run the tool. Since this is the very first tool, we'll rerun the entire program. Sure enough, this time, the tool runs properly.

During tool number three (the 16.0-mm [or 5/8-in] end mill) an alarm sounds while the tool is in the pocket. We look at the program check page and find that the cursor is currently on line N185. This is the command that instates cutter radius compensation. The alarm says *over-cutting will occur during cutter radius compensation*. Now, what do you think could be wrong?

Frankly speaking, the over-cutting alarm can be generated for several reasons. But say we look at the cutter radius compensation offset value and find it to be 16.0-mm (or 0.625-in). Oops. We entered the cutter's *diameter*, not its radius. This is what caused the alarm. So we correct the offset. Now we must rerun the program, starting with tool number three. (There is no need to rerun the first two tools – doing so will waste time.) To restart, we scan to line N160 (the command after the tool change since tool number three is currently in the spindle), and restart the program. The further you are into a program when a problem is found, the more wasteful it will be to rerun the entire program once you find and correct a mistake.

Key Concept 7: Know Your Machine From An Operator's Viewpoint

Cautiously Run the First Workpiece

If the program is correct (you've seen it run on a tool path verification system or it is a program that has run before), if the process and cutting conditions are correct, and if you have made no mistakes during setup (the dry run looks good), you should be able to machine a good workpiece on your very first try.

Admittedly, there are a lot of *ifs* in the previous paragraph. And problems could still exist with *new* programs that you could not detect during the tool path verification and the dry run. If, for example, a small motion mistake is made in a tool path, you may not be able to spot it during the tool path verification. This, of course, will cause the workpiece to be machined incorrectly. This kind of mistake should not exist with proven programs – as long as the *current version* of the program is being used.

And there could also still be mistakes in your *setup* that went undetected during the dry run. If, for example, you make a small measurement mistakes during tool length compensation measurement or entry (say, a tool length compensation entry is off by 6.0-mm [or 0.25-in]), you may not be able to detect it during a dry run.

The Most Dangerous Time

You must still be very careful when running the first workpiece. During programming lessons, we recommend using a rapid approach distance of 2.0-mm (or 0.1-in). While this is a safe and acceptable approach distance, it is difficult to tell during a dry run whether the tool tip truly stops 2.0-mm (or 0.1-in) above the work surface (since there is no workpiece). If even a small mistake is made with tool length compensation, the result could be disastrous. For this reason, you must be extremely careful with each tool's first Z-axis approach movement. This is the most dangerous movement of each tool. We recommend using a special procedure during the first few commands of each tool. It assumes that the structure shown in lesson 5.2 is used to write the program. This procedure allows you to take full control of each tool's first Z-axis approach movement.

Once you confirm that the machine is at its tool change position and that the correct program is active, you can follow this approach procedure to approach with each tool. The first workpiece, of course, must be loaded at this time.

1) Turn on single block and dry run.

2) Set the feedrate override switch to its lowest setting.

3) Press the program check display screen key (this is the page that shows distance-to-go)

4) Repeatedly press cycle start until the first tool change occurs (because single block is turned on, each time you press cycle start only one command is executed).

5) Press cycle start (the spindle starts).

6) Press cycle start. The machine begins moving in the X and Y axes to the tool's first XY position. This motion is still pretty safe, since the Z-axis is still at the zero return position. Crank up the feedrate override switch until the tool comes into position in X and Y.

7) Set the feedrate override switch to its lowest setting.

8) Press cycle start. Machine now begins moving in Z. *This is the dangerous movement.* Be very careful as the tool gets close to the workpiece. When the tool comes to within about 15.0-mm (or 0.5-in) of the work surface, stop the motion by pressing the feed hold button. Look at the distance-to-go display. Confirm that the amount of motion left in the approach movement is less than the distance between the tool tip and the workpiece. As long as it is, press the cycle start button to reactivate the cycle. Let the tool come to its final approach position and stop.

9) Turn off dry run. (*Never* let cutting tools machine a workpiece under the influence of dry run. Dry run tends to slow down rapid motions, but it *speeds up* cutting motions.)

10) The tool has safely approached. What you do next will depend upon whether you're running a new or a proven program. For a new program you must carefully step through the cutting tools cutting motions. In this case, leave the single block switch on, and repeatedly press cycle start to step through the tool. For a proven program (are you *absolutely sure* that you're running the current version of the program?), turn off single block and press cycle start to let the tool machine the workpiece as it did the last time the program was run.

11) When the tool is finished, the machine will return to the tool change position. While the machine is moving to the tool change position, turn on dry run and, if you turned it off, single block. Repeat the procedure, starting at step four.

For each tool in the program, you must take control of the tool's first Z-axis approach movement with single block and dry run. Once you have seen the tool approach properly, you need not use this procedure (after running the tool for the first time). But whenever you're worried about a tool's first approach motion, this procedure will keep the tool from crashing into the workpiece.

Making Sure the First Workpiece Is a Good One

While there is a lot to think about when you run the first workpiece, it must still be your goal to make the very first workpiece being machined a good one. If there are any tiny motion-mistakes in a new program that you did not detect when you verified the tool path, it may be impossible to machine a good workpiece on your first try. The program must be correct, of course, in order for the first workpiece to be machined properly (as a *proven* program is).

We do consider the running of your first workpiece to be a *setup-related task*. If the first workpiece doesn't pass inspection, we also consider any corrections made to the program as well the running of the *second* workpiece to be a setup-related task. Indeed, *the machine is still in setup until a workpiece passes inspection* and you can run acceptable workpieces.

Frankly speaking, if the program is correct, *there is no excuse for scrapping the first workpiece you machine.* If you use the techniques we recommend, each tool will be *forced* to machine properly – and when the program is finished, each workpiece surface being machined will be within tolerance limits.

Look at step number four in the procedure shown above for cautiously approaching with each tool. This is the point at which you're going to *begin* each tool's first approach. The tool change has just occurred, and you know which tool is going to run next. At this point in the procedure, you must consider what the tool is going to do – and especially – consider the tolerance bands for dimensions that the cutting tool is going to machine. If tight tolerances must be held, you must use *trial machining* techniques.

Machining the First Workpiece In Our Example Job

Let's go through the entire example job and show how this is done.

Tool one: Tool number one is the 25.0-mm (or 1.0-in) end mill. At step four of the approach procedure shown above, this tool is placed in the spindle. We must stop to and consider what this tool will be doing. This tool will be machining both ends of the workpiece. And notice the very tight tolerance for the workpiece overall length (100.0-mm +/-0.01 [or 4.000-in +/-0.0004]). Even if we measure the diameter of the 25.0-mm (or 1.0-in) end mill perfectly and enter the cutter radius compensation value accordingly (into D GEOM offset one), this end mill may not machine the overall length within its tolerance limits. The end mill may be running out a little bit in its holder. If it is, it will machine too much stock. If this happens, we're going to scrap the first workpiece.

Since we're worried about whether our initial cutter radius compensation offset entry is accurate enough to machine the overall length of the workpiece within its tolerance limits, we must use *trial machining* techniques for this tool. This will force the end mill to leave additional stock on the two ends, ensuring that we don't machine the workpiece undersize on the first try. (Trial machining for cutter radius compensation is discussed in detail during lesson 4.3.) Before running this tool, we'll increase D GEOM offset number one by 0.25-mm (or 0.01-in). If its initial value is 12.5-mm (or 0.5-in), we will make it 12.75-mm (or 0.510-in). This will cause the end mill to leave 0.25-mm (or 0.01-in) more material on each end of the workpiece (0. 5-mm [or 0.02-in] overall).

Before continuing with the procedure, we'll turn on the *optional stop* switch. This will ensure that the machine will stop at the completion of the tool (though it will anyway if the single block switch is on).

We'll continue with the approach procedure, cautiously bringing the end mill to its approach position. Then we'll allow this tool to machine the workpiece (as is done in step ten). When the tool is finished and back at the tool change position, the optional stop function (or the single block function) will cause the machine to stop.

At this point, we will measure the overall length of the workpiece. If our initial tool measurement is perfect, the overall length should be precisely 100.5-mm (or 4.020-in) right now (remember that the end mill is leaving an additional 0.25-mm [or 0.01-in] on *each* end). But when we measure, we find the overall length to be 100.36-mm (or 4.0188-in). Sure enough, the end mill must be running out a little bit. And if we had tried to

Key Concept 7: Know Your Machine From An Operator's Viewpoint

run the workpiece with the initial offset setting, the overall length would have come out to 99.86-mm (or 3.9994-in) – and the workpiece would have been scrapped.

Assuming the target value for this dimension is 100.0-mm (or 4.000-in), we must reduce the current setting of the offset by 0.18-mm (or 0.0094-in), which is half of 0.36-mm (or 0.0188-in) - the amount the workpiece is currently oversize - and rerun the tool. This time we don't have to be so careful with the approach movement, since we've seen this tool successfully approach the workpiece. So we turn off the dry run and single block switches (leaving on the optional stop switch), set feedrate override to one-hundred percent and press cycle start. The end mill machines the overall length again. At the end of the tool, the optional stop function stops the machine. We measure the overall length again, and sure enough, it comes out to 99.998-mm (or 3.9999-in), well with the tolerance limits.

Tool two: At step four of the approach procedure for the next tool, the 12.0-mm (or 1/2-in) end mill is placed in the spindle. Again, we must stop and consider what this tool will do. This tool will machine the 16.0-mm (or 0.625-in) wide slot. And notice the tight tolerances on both the slot width and its depth (+/-0.01mm [or +/- 0.0004-in] in both cases). Again, we're worried that our initial offset settings are not accurate enough to machine the slot within its tolerance limits.

We increase the tool length offset value for this tool (H GEOM offset number two) by 0.25-mm (or 0.01-in) and increase the cutter radius compensation offset value for this tool (D GEOM offset number two) by 0.25-mm (or 0.01-in), use the safe-approach procedure, and machine the workpiece for the first time. At the optional stop (or because of single block), the machine will stop when the tool is finished.

When we measure the pocket width, we find it to be 15.56-mm (or 0.6058-in). The target value for the pocket width is 16.0-mm (or 0.625-in), so we must decrease D GEOM offset number two by 0.22-mm (or 0.0096-in). Remember, the end mill is machining on both sides of the pocket. To calculate the adjustment amount in this case, we first subtract 15.56-mm from 16.0-mm (or 0.6058-in from 0.625-in). We then divide the result by two. When we measure the pocket depth, we find it to be 9.89-mm (or 0.3653-in). The target value for pocket depth is 10.0-mm (or 0.375-in), so we must decrease H GEOM offset number two by 0.11-mm (or 0.0097-in).

Rather than running the entire program again, which would waste time and could cause the previously run tool to scuff up surface finishes (or even machine more material), we'll re-run tool number two. Since tool number two is currently in the spindle, restart block in the program is N090 (the command after the tool change). And since we have seen this tool successfully approach, there is no need to use the safe-approach procedure. We'll turn off the dry run and single block switches (leaving on the optional block switch), set feedrate override to one-hundred percent, and press the cycle start button. The 12.0-mm (or 1/2-in) end mill machines the slot again. At the optional stop, the machine stops. We measure the slot – and this time the width is 16.002-mm (or 0.6251-in) and the depth is 10.001-mm (or 0.3749-in). We're ready to go on to the next tool.

Tool three: At step four of the procedure, the next tool, the 16.0-mm (or 5/8-in) end mill, is placed in the spindle and we're ready to run it the first time. But before continuing with the approach procedure, we must consider what this tool will be doing. This tool will be machining the oval shaped pocket. Notice that there is nothing critical about this pocket. Tolerances for width and depth are +/-0.13-mm (or +/-0.005-in), and we're pretty sure that our initial tool length- and cutter-radius-compensation offset measurements are adequate to machine the pocket within tolerance limits.

Using the approach procedure, we allow the end mill to approach and machine the pocket. When it's finished, the optional stop (or single block) will cause the machine to stop. Even though we're pretty confident, we should check to see that this tool has machined properly – so we measure it. Sure enough, the pocket width is 25.02-mm (or 1.0008-in) and its depth is 5.98-mm (or 0.249-in). While are they well within tolerance limits, the width and depth dimensions are not at their target values. Most setup people will make offset adjustments to ensure that the pocket machined in the *next workpiece is perfect*. So we'll increase cutter radius compensation D GEOM offset number three by 0.01-mm (or 0.0004-in), which will make the tool remove less material) and reduce tool length compensation H GEOM offset number three by 0.02-mm (or 0.001-in), making the tool go deeper. There is no need to re-run this tool, so we proceed to the next tool using the approach procedure.

Tool four: At step four for the next tool, the #4 center drill is placed in the spindle. We stop to consider what this tool will be doing. According to the drawing, there is really nothing critical about the three holes. Tolerances are +/-0.13-mm (or +/-0.005-in). All we care about is that the center drilled holes get machined in the correct position (which is controlled by the program). So we're not going to need trial machining techniques. We'll use the approach procedure and let the tool machine the three holes. When the tool is finished, the machine stops. At this point, we check to see that the center drill has gone deep enough for the up-coming drill and that the holes are in the correct locations. We now proceed to the next tool.

Tool five: At step four for the next tool, the 10.0-mm (or 3/8-in) drill is placed in the spindle. We stop to consider what this tool will be doing. Again, there is nothing critical about the three holes. We use the safe-approach procedure and allow the drill to machine the holes. When the machine stops (this time at the end of the program), we check to see that the holes are in the correct locations and that they have broken through the bottom of the workpiece.

At this point, we're finished running the first workpiece *and it is a good one* (it will pass inspection). We're (almost) ready to begin the production run.

What if the program isn't perfect?
Again, the program must be correct in order for you to be able to make the first workpiece a good one. Say that in our example (new) program, the depth of the 16.0-mm (or 0.625-in) slot, which is supposed to be 10.0-mm (or 0.375-in), is incorrectly specified as 10.5-mm (or 0.395-in) deep. Maybe the programmer typed Z-10.5 (or Z-0.395) in the program instead of Z-10.0 (or Z-0.375). It is unlikely that we'd catch such a small mistake during the tool path verification or during the dry run. It is likely that we'd scrap the first workpiece, even with trial machining techniques. But hopefully we would be able to find and correct the mistake while we're running the first workpiece – so the *second* workpiece will be correct.

If Necessary, Optimize the Program for Better Efficiency

This task in only required with new programs and large lot sizes. If you have a large number of workpieces to machine, it may be necessary to optimize the motions in the program. As stated in lessons 5.1 and 5.2, the format we use to create programs are not as efficient as possible. As you watch your programs run, you will probably notice things that can be improved.

As you begin working with CNC machines, try not to be overly concerned about program efficiency. Concentrate on becoming familiar with running your machine and making good workpieces. As you gain confidence, by all means, apply the suggestions we offer in lesson 5.2 to make your programs run faster.

If you make optimizing changes to the program at the machine right after verifying the program, you must verify that your changes are correct. You must dry run the tools that have been changed.

If Changes Have Been Made To the Program, Save the Corrected Version of the Program

This is an important task that tends to be overlooked. While you are verifying the program – and whether it is a new program or a proven one, if you make any changes to the program, you *must* remember to save a corrected version of the program. Again, current FANUC controls allows saving of programs to a variety of devices, like a memory card. If using a distributive numerical control (DNC) system, the program must be saved back to your DNC system.

If you forget to save your corrected program, you'll have to repeat the program verification steps the next time the job is run. Worse, you may remember running the job before and you may *think* that the program being loaded is a proven program. This, of course, will lead to serious problems.

The procedure for saving programs is similar to the procedure to load programs – and again – it will vary based upon the DNC system you use. With most systems, you first go to the DNC system and get it ready to receive the program. Then you go to the CNC machine and make the command to send the program to the DNC system. With some DNC systems, the entire process can be completed from the CNC machine.

Production Run Documentation

With the setup completed and the first workpiece passing inspection, you're ready to begin the production run. We're assuming at this point, that you have more than one workpiece to machine. As with just about every facet of manufacturing, companies vary with regard to how many workpieces they commonly run per job. This can even vary within one company.

Generally speaking, CNC machining centers are most often applied to small to medium sized lots – from one to about one-thousand workpieces. While there are companies that dedicate their CNC machining centers to running one workpiece – day in and day out – the vast majority of CNC using companies, run a variety of jobs of varying lot sizes.

The number of workpieces commonly run has a big impact on how you approach production runs. Indeed, it has a big impact on how companies utilize CNC people. If a company consistently runs small lots of say,

Key Concept 7: Know Your Machine From An Operator's Viewpoint

under ten workpieces, it won't take much time to complete each production run (if program execution time is short). Companies in this situation tend to have one person setup the job and run the job out. On the other hand, if a company consistently runs larger lots of say, five-hundred workpieces, it will take some time to complete the production run. These companies tend to have one person make the setup and another (lesser skilled person) run out the job.

If one person is doing everything, including programming, setup, and running production, there won't be much of a need for production run documentation. This person will know what is intended since they planned the entire job. On the other hand, if tasks are divided (one person programs, another sets up, and yet another runs out the job), the need for good documentation is much greater.

As you have seen, most companies provide *setup* documentation in the form of a one-page setup sheet. However, many companies expect their setup people to relate what must be done to run out the job to the CNC operator. Admittedly, many of the tasks related to running production are quite basic – and very redundant if many workpieces must be produced. But adequate production-run documentation is essential if several people will be running out the job (like first, second, and third shift operator).

A Note To Programmers:

This text is providing you with information needed to master all three skills needed to use a CNC machining center – programming, setup, and running production. However, depending upon your company, you may be responsible only for programming. In this case, you must provide setup and production run documentation for other people to make setups and run production.

While you can pretty much rest assured that the setup person will have adequate basic machining practice skills to understand even minimal documentation, you can make no such assumption about CNC operators. Many companies hire people with limited (if any) basic machining practice skills to run CNC machines. Remember that you must direct your production run documentation to the *lowest skill level* of CNC operator in your company.

Many programmers do a terrible job in this regard. Even if your operators have pretty good basic machining practice skills, remember that they're not going to be nearly as familiar with the job as you are. Many programmers assume way too much of their CNC operators. While experienced operators may eventually able to figure out what is expected of them, good production run documentation will minimize the effort required to do so.

Figures 7.13A and 7-13B shows the production run documentation for our example job in the form of a one-page form (7.13A for metric mode, 7.13B for metric mode). This form must answer any questions a CNC operator will have. We refer to this form as we describe the tasks related to completing a production run.

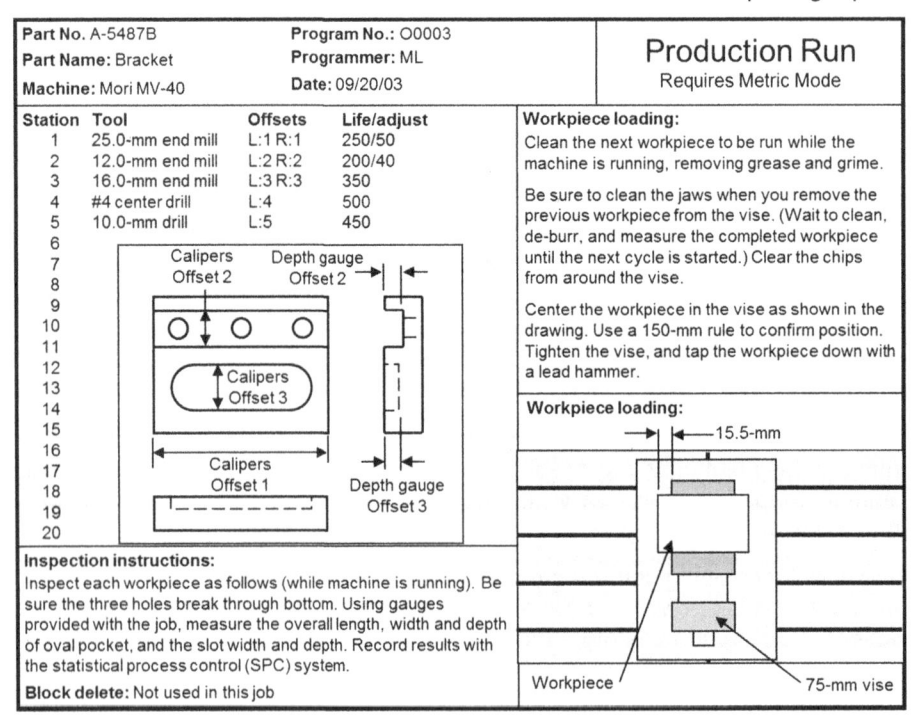

Figure 7.13A – production-run documentation for our example job (requires metric mode)

Part No. A-5487A	Program No.: O0004	**Production Run**
Part Name: Bracket	Programmer: ML	Requires Inch Mode
Machine: Mori MV-40	Date: 09/20/03	

Station	Tool	Offsets	Life/adjust
1	1.0" end mill	L:1 R:1	250/50
2	1/2" end mill	L:2 R:2	200/40
3	5/8" end mill	L:3 R:3	350
4	#4 center drill	L:4	500
5	3/8 drill	L:5	450
6-20			

Workpiece loading:
Clean the next workpiece to be run while the machine is running, removing grease and grime.

Be sure to clean the jaws when you remove the previous workpiece from the vise. (Wait to clean, de-burr, and measure the completed workpiece until the next cycle is started.) Clear the chips from around the vise.

Center the workpiece in the vise as shown in the drawing. Use a 6" rule to confirm position. Tighten the vise, and tap the workpiece down with a lead hammer.

Workpiece loading: (with diagram showing 0.6" dimension, workpiece, 3" vise)

Inspection instructions:
Inspect each workpiece as follows (while machine is running). Be sure the three holes break through bottom. Using gauges provided with the job, measure the overall length, width and depth of oval pocket, and the slot width and depth. Record results with the statistical process control (SPC) system.

Block delete: Not used in this job

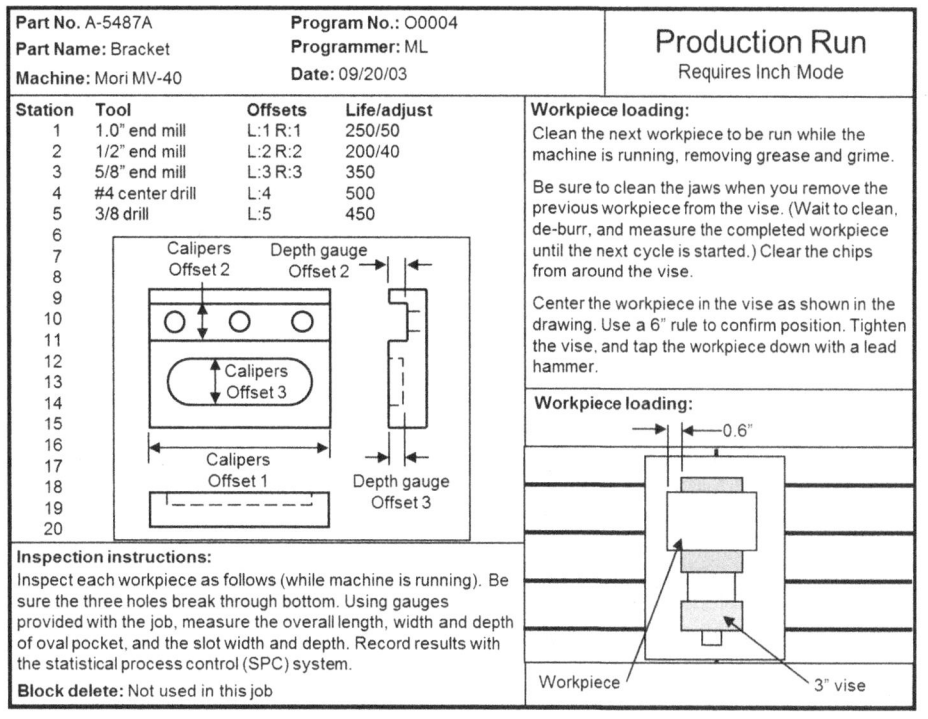

Figure 7.13B – production-run documentation for our example job (requires inch mode)

Remove the Previous Workpiece

The task of workpiece loading begins with the removal of the most recently completed workpiece from the work area. This requires the operator to understand the work holding device being used to secure the workpiece – and should be described in the production run documentation.

For our example job, the operator must remove the workpiece from the table vise, which requires the use of a vise handle. Workpiece removal usually requires that the location surfaces of the work holding device be cleaned. In most cases, the work area near the work holding device should also be cleaned to remove chips and debris.

Note that some operators get into the (bad) habit of beginning to work with the workpiece just removed before loading the next workpiece. While this workpiece must be cleaned, de-burred and measured, the operator must set it aside until the machine is in cycle for the next workpiece.

Load the Next Workpiece

While loading workpieces is usually pretty simple, the setup documentation should include workpiece loading instructions. Notice that our example production run documentation form includes explicit loading instructions requiring the use of a 150.0-mm (or 6.0-in) rule. Like workpiece removal, workpiece loading requires an understanding of the work holding device being used.

Once the workpiece is loaded, the doors (if there are any) to the work area must be closed.

Activate the Cycle

If several workpieces have just been run, this task simply involves pressing the *cycle start* button. The machine will begin running the next workpiece in the normal manner.

But if the operator has been away from the machine for any length of time (at lunch or on break), or if they are just beginning a shift, they must confirm that the machine is truly ready to run a workpiece. This involves checking (and setting) the current condition of several switches on the control panel. While this may not be a complete list, here are some of the switches that must be set.

- Dry run: off
- Single block: off
- Optional stop: off
- Machine lock (if available): off

Key Concept 7: Know Your Machine From An Operator's Viewpoint

- Block delete: As requested by programmer (probably off)
- Rapid traverse override: 100%
- Feedrate override: 100%
- Spindle override (if available): 100%
- Mode switch: Memory or Auto
- Display screen: Program mode (and correct active program is shown and cursor is at the beginning of the program)
- Machine position: At the tool change position

Again, with the machine ready to run a program, activating the cycle simply requires pressing the cycle start button. Whenever you press the cycle start button, *always* have a finger ready to press the feed hold button. If the machine behaves in an unexpected manner, you'll be ready to stop it.

Monitor the Cycle

This may not be necessary for proven programs that you have seen run many times before. But if you're new to a program, it is a good idea to get familiar with the cycle – especially if you did not program the job or make the setup. This will help you understand the tools that are being used and the general machining order.

For new jobs, you'll also want to confirm that cutting tools are machining properly for several workpieces. Certain cutting tool related problems, especially with speeds and feeds, may not present themselves when the first workpiece is being machined during setup. You'll need to stay alert while running the first few workpieces.

Clean and De-Burr the Workpiece

Once you have loaded the next workpiece and activated the cycle, you can begin working on the workpiece that has just been removed. It will probably be covered with coolant and debris. And it will probably have some razor sharp edges, so be careful handling it.

Most companies expect their CNC operators to clean and de-burr the workpieces they produce. Cleaning usually involve wiping the workpiece with a rag or shop towel. If you use any kind of air-blowing system to blow off the workpiece, be *extremely careful* – chips can fly anywhere. While you must wear eye-protection (all shops require it), nothing is protecting your ears, nose, mouth, etc.

While most companies strive to remove sharp edges in the machining cycle, there will almost always be some sharp edges on workpiece you remove from the machine. Again, be careful not to cut yourself as you handle newly removed workpieces.

A variety of hand tools is available to de-burr sharp edges, like files and hole de-burring tools. If any are unfamiliar to you, ask an experienced person to demonstrate them.

For our example workpiece, just about all machined surfaces will have sharp edges since we didn't use any de-burring tools in the program. These surfaces include all surfaces on each end of the workpiece, the top surfaces of the slot and pocket, and the top and bottom surfaces of the three holes.

Perform Specified Measurements

Most companies expect their CNC operators to inspect the workpieces they machine. They also expect the results of these measurements to be recorded in some manner. But companies vary when it comes to specific methods used to measure and record.

Many companies require one-hundred percent inspection on critical dimensions. And they use some kind of statistical process control (SPC) system to record the measurements. An SPC system usually incorporates a computer screen and keyboard at the measuring station for data entry.

Taking measurements, of course, requires basic machining practice skills. The operator must know how to use the various gauging tools needed to take the measurements. Production run documentation must specify the gauging tools to be used (as our example documentation does). If you are unfamiliar with the required gauges, you must ask to be shown how to use them. And you must practice with them to ensure that you can take accurate measurements.

For our example job, notice that production run documentation specifies that the overall length of the workpiece must be measured. So must the width and depth of the 16.0-mm (or 0.625-in) slot must be measured on every workpiece. The operator is also asked to check the width and depth of the 25.0-mm (or

1.0-in) pocket. And they're asked to visually check that the holes break through. They must record measured dimensions with the company's SPC system.

Make Offset Adjustments To Maintain Size for Critical Dimensions (sizing)

The tasks shown so far must be performed in every cycle. The tasks shown from this point are only performed if and when they are required.

The tighter the tolerances a cutting tool must machine, the more likely it will be that it will not machine surfaces within the tolerance band for its entire life (at which time it must be replaced). With small lots, this probably won't present a problem. Every cutting tool will machine properly until the production run is completed.

But with larger lots, the wear a cutting tool experiences may cause the surfaces it machines to change by a small amount. And with tight tolerances, this may place the workpiece in jeopardy. In this case, the CNC operator must make an offset adjustment (called a sizing adjustment) to keep the cutting tool machining properly.

With our example job, there are three very tight tolerances to hold: the overall workpiece length, the 15.0-mm (or 0.625-in) slot width, and the 16.0-mm (or 0.625-in) slot depth. Say we have a large lot of two-thousand workpieces to produce. After the setup is made and the first workpiece passes inspection, each of these critical dimensions will be machined at its target value (its mean value in our example).

As the operator continues to run workpieces, they notice that the overall workpiece length is *growing*. It started out at 99.998-mm (or 3.9999-in), which is slightly smaller than its target value. But after twenty workpieces, they find it to be 100.001-mm (or 4.0001-in). After forty workpieces, it is 100.004-mm (or 4.0002-in). And after fifty workpieces, it is 100.008-mm (or 4.0003-in). This growth in overall length, of course, is being caused by wear of the 25.0-mm (or 1.0-in) end mill. If this trend continues, the overall length of a future workpiece will be out of tolerance (scrap).

An adjustment must be made to the 25.0-mm (or 1.0-in) end mill's cutter radius compensation value (D GEOM offset number one). If the overall length of the workpiece is currently 100.006-mm (or 4.0002-in), the current offset value must be reduced by 0.003-mm (or 0.0001-in) to bring the dimension to its target. This will cause the end mill to machine 0.003-mm (or 0.0001-in) more stock from each end of the next workpiece, making the overall length of the next workpiece 100.0-mm (or 4.0-in) – back to its target value.

The same situation will exist with the width (and possibly the depth) of the 16.0-mm (or 0.625-in) wide slot. As the operator continues to run workpieces, they must be alert for sizing adjustments required because of tool wear.

If you expect the operator to make sizing adjustments during a production run, they must be told *which* offsets are related to each tool. In our production run documentation, we list the offsets related to each tool in the tool list. But we further clarify with a nice drawing which shows the offsets related to each workpiece surface. This allows the operator to make sizing adjustments without having to know which tool machines each surface. Unfortunately, most production run documentation doesn't provide this level of clarity. Most operators are on their own to figure out which tool machines each critical surface and which offsets are related to the tool.

Since the 25.0-mm (or 1.0-in) wide pocket and the holes have no tight tolerances, the tools that machine them should last for their entire lives without requiring sizing adjustments (with drills, by the way, there is nothing that can be done by offsets that will change the diameter they machine).

If possible, production run documentation should help the CNC operator know *when* sizing adjustments are needed. This, of course, requires experience running the job. During the first production run for a new program, the operator will learn if sizing adjustments are necessary. If they are, this information should be added to the production run documentation. Notice in the tool list for this job, we include the approximate number of workpieces that can be machined before a sizing adjustment is necessary (in the Live/adjust column). This lets the operator know if sizing adjustments will be required before a tool must be replaced.

Replace Worn Tools

As with sizing adjustments, this task will only be required with larger lot sizes – when cutting tools will wear out before the job is completed.

When a cutting tool must be replaced, the same tasks required during the initial setup must be repeated. These tasks include assembly, measurement of tool length compensation values, measurement of cutter radius compensation values for milling cutters, offset entry, and placement into the machine's automatic tool changer magazine. All of these tasks are presented during our discussion of setup and will not be repeated here.

Keep in mind that if a cutting tool requires trial machining during the initial setup, it will require trial machining during replacement. Tools one and two in our example job (the 25.0-mm [or 1.0-in] end mill and the 12.0-mm [or 1/2-in] end mill) both require trial machining during setup. If they are replaced during the production run, they will require trial machining again to ensure that the surfaces they machine come out to size.

All of this means, of course, that if a CNC operator is responsible for replacing tools during a production run, they must possess many of the same skills possessed by the setup person. For this reason, some CNC using companies do not expect their CNC operators to replace worn tools. Instead, they have the setup person do so.

If possible, production run documentation should specify the expected life for each tool. Again, this requires some experience with the job (or at least, experience with the related cutting tools and the material being machined). In our example documentation, we specify an expected tool life for each tool – so the operator will know if the job can be completed before any tools wear out.

Clean the Machine

Most companies expect their CNC operators to keep their machines clean. Every so often (commonly at the end of each shift), the operator will remove all chips from the work area and clean the machine table. Chips machined during the shift will be dumped (from the chip disposal drum).

Preventive Maintenance

Some CNC using companies expect their CNC operators to perform basic preventive maintenance – like maintaining coolant levels, way lube levels, hydraulic oil levels, and filters. If this is required, instructions must be provided that describe the required procedures and their frequency.

Key Points for Lesson 7.1:

- There are only two general tasks that occur on CNC machining centers – machines are either in setup or they are running production.
- You must understand the tasks needed to make setups – we've provided a complete explanation of each task in the approximate order that setups are made.
- You must know how to tell when trial machining is required and know how to perform trial machining.
- You must know how to perform each task that is required to complete a production run – we've provided a complete explanation of each task.

Metric-mode exercise: Practice running the first workpiece

Instructions: Study the following drawing, paying particular attention to the dimensions and tolerances. Note that this drawing is not fully dimensioned. Only the dimensions and tolerances that are related to this exercise are provided. Next, read the description of the process and offset settings. Finally, answer the questions that follow.

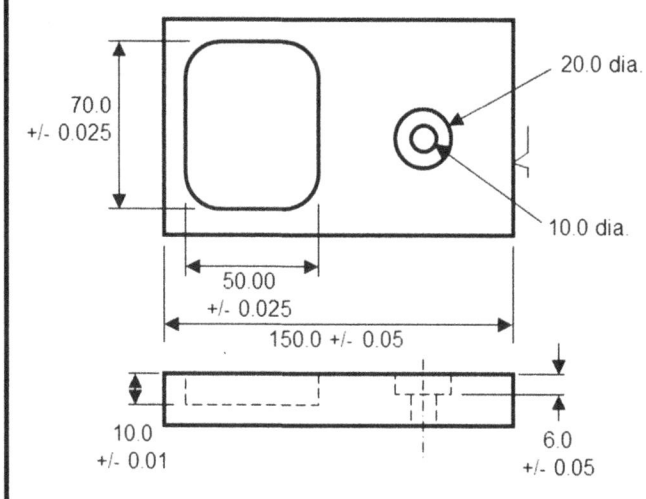

Process:

Description	Tool	Station
Mill right side	25.0 end mill	1
Mill 50.0 X 70.0 pocket	30.0 end mill	2
Drill 10.0 hole	10.0 drill	3
Counter-bore 20.0 hole	20.0 end mill	4

After measuring cutting tools, you enter the offsets as follows:

Offset # Value:
H GEOM 1: 152.435 (length of tool 1)
H GEOM 2: 126.443 (length of tool 2)
H GEOM 3: 177.332 (length of tool 3)
H GEOM 4: 152.948 (length of tool 4)
D GEOM 1: 12.5 (radius of tool 1)
D GEOM 2: 15.0 (radius of tool 2)

Questions:

1) You're ready to run tool number one (the 25.0-mm end mill that mills the right side) for the first time. You notice that there is a pretty close tolerance (+/- 0.05-mm) on the 150.0-mm dimension that this milling cutter machines. You're worried that this tool may machine too much stock on its first try, so you decide to trial machine. You increase the value of offset D GEOM number 1 by 0.25-mm (making it 13.0) and allow this milling cutter to machine the right end. When the tool is finished, you stop the machine (with the optional stop switch) and measure the 150.0-mm dimension. You find that it comes out to 150.21-mm. What must you do to D GEOM of offset 1 to make this milling cutter machine properly when you re-run the tool? Be specific.

2) After re-running the 25.0 end mill, the 150.0-mm dimension comes out right on size (to 150.0-mm). You're ready to run tool two, the 30.0-mm end mill that machines the pocket. Again you notice the tight tolerance (+/- 0.025-mm) and decide to trial machine. You increase the value of D GEOM offset number 2 by 0.25-in (making it 15.25). You also notice the even tighter tolerance on the pocket depth (+/- 0.01-mm) and decide to trial machine this surface as well. So you increase the value of H GEOM offset number two by 0.25 (making it 126.693). After this tool machines you stop the machine. You measure the pocket size in XY and it comes out to 49.8-mm along X and 69.8-mm along Y. You measure the pocket depth and find it to be 9.82. What must you do to D GEOM offset 2? What must you do to H GEOM offset 2? Be specific.

3) After re-running the 30.0-mm end mill, you're ready to run tool number three (the 10.0-mm drill). Does this tool require trial machining? (yes or no) Why?

4) You move on to tool number four (the 20.0-mm end mill that plunges the 20.0-mm counter-bore). This tool machines in a plunging fashion, just like a drill. Again, the depth of this pocket (6.0-mm) has a tight tolerance (+/- 0.05-mm), so you decide to trial machine. You increase the value of offset H GEOM number four by 0.25-mm (making it 153.198) and run the tool. After machining, you stop the machine. When you measure the depth of the counter-bored hole, you find it to be 5.81-mm. What must you do to offset H GEOM number four?

Answers: 1) Reduce D GEOM offset number one by 0.21-mm. 2) Reduce D GEOM offset two by 0.2-mm and reduce H GEOM offset number two by 0.18-mm. 3) No, this tool does not require trial machining because none of the dimensions it machines have tight tolerances. 4) Reduce offset number four by 0.19-mm.

Key Concept 7: Know Your Machine From An Operator's Viewpoint

Inch-mode exercise: Practice running the first workpiece

Instructions: Study the following drawing, paying particular attention to the dimensions and tolerances. Note that this drawing is not fully dimensioned. Only the dimensions and tolerances that are related to this exercise are provided. Next, read the description of the process and offset settings. Finally, answer the questions that follow.

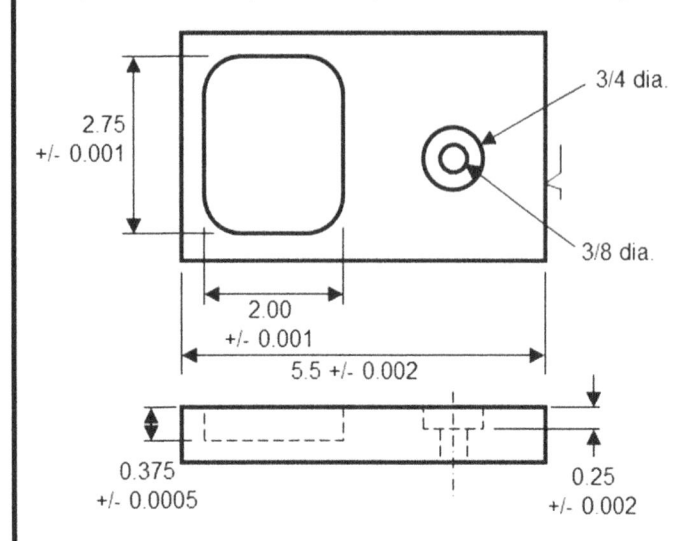

Process:

Description	Tool	Station
Mill right side	1.0 end mill	1
Mill 2 x 2.75 pocket	1.25 end mill	2
Drill 3/8 hole	3/8 drill	3
Counter-bore 3/4 hole	3/4 end mill	4

After measuring cutting tools, you enter the offsets as follows:

Offset # Value:
H GEOM 1: 6.4435 (length of tool 1)
H GEOM 2: 5.5443 (length of tool 2)
H GEOM 3: 7.3342 (length of tool 3)
H GEOM 4: 6.9448 (length of tool 4)
D GEOM 1: 0.5000 (radius of tool 1)
D GEOM 2: 0.6250 (radius of tool 2)

Questions:

1) You're ready to run tool number one (the 1.0-in end mill that mills the right side) for the first time. You notice that there is a pretty close tolerance (+/- 0.002-in) on the 5.5-in dimension that this milling cutter machines. You're worried that this tool may machine too much stock on its first try, so you decide to trial machine. You increase the value of offset D GEOM number 1 by 0.01-in (making it 0.510) and allow this milling cutter to machine the right end. When the tool is finished, you stop the machine (with the optional stop switch) and measure the 5.5-in dimension. You find that it comes out to 5.509-in. What must you do to D GEOM of offset 1 to make this milling cutter machine properly when you re-run the tool? Be specific.

2) After re-running the 1.0-in end mill, the 5.5-in dimension comes out right on size (to 5.5-in). You're ready to run tool two, the 1.25-in end mill that machines the pocket. Again you notice the tight tolerance (+/- 0.001) and decide to trial machine. You increase the value of D GEOM offset number 2 by 0.010-in (making it 0.635). You also notice the even tighter tolerance on the pocket depth (+/- 0.0005-in) and decide to trial machine this surface as well. So you increase the value of H GEOM offset number two by 0.010-in (making it 5.5543). After this tool machines you stop the machine. You measure the pocket size in XY and it comes out to 1.982-in along X and 2.732-in along Y. You measure the pocket depth and find it to be 0.366-in. What must you do to D GEOM offset 2? What must you do to H GEOM offset 2? Be specific.

3) After re-running the 1.25-in end mill, you're ready to run tool number three (the 3/8-in drill). Does this tool require trial machining? (yes or no) Why?

4) You move on to tool number four (the 3/4 end mill that plunges the 3/4-in counter-bore). This tool machines in a plunging fashion, just like a drill. Again, the depth of this pocket (0.25-in) has a tight tolerance (+/- 0.002-in), so you decide to trial machine. You increase the value of offset H GEOM number four by 0.01-in (making it 6.9548) and run the tool. After machining, you stop the machine. When you measure the depth of the counter-bored hole, you find it to be 0.238-in. What must you do to offset H GEOM number four?

Answers: 1) Reduce D GEOM offset number one by 0.009-in. 2) Reduce D GEOM offset two by 0.009-in and reduce offset number two by 0.009-in. 3) No, this tool does not require trial machining because none of the dimensions it machines have tight tolerances. 4) Reduce offset number four by 0.012-in.

Lesson 7.2–Operation Panels

While there are many buttons and switches on a CNC machining center, you must try to learn the reason why each one exists. If you don't, you may be overlooking a helpful – if not necessary – machine function.

Objectives
After completing this lesson, students should be able to:
- ✓ Describe the two most important operator's panels
- ✓ Describe the functions on the control panel
- ✓ Describe the functions on the machine panel

Introduction
You now know the tasks required to setup and run a CNC machining center. During our discussions of these tasks, we have mentioned several specific buttons and switches. So you also know the function of some of the most important buttons and switches on the machine. But there are still some buttons and switches that you have not yet been exposed to. In this lesson, we're going to introduce *all* of the buttons and switches that are found on a typical CNC machining center.

The Two Most Important Operation Panels
Most machining centers have at least two distinct operation panels. We'll be calling them the ***control panel*** (designed by FANUC) and the ***machine panel*** (usually designed by the machine tool builder).

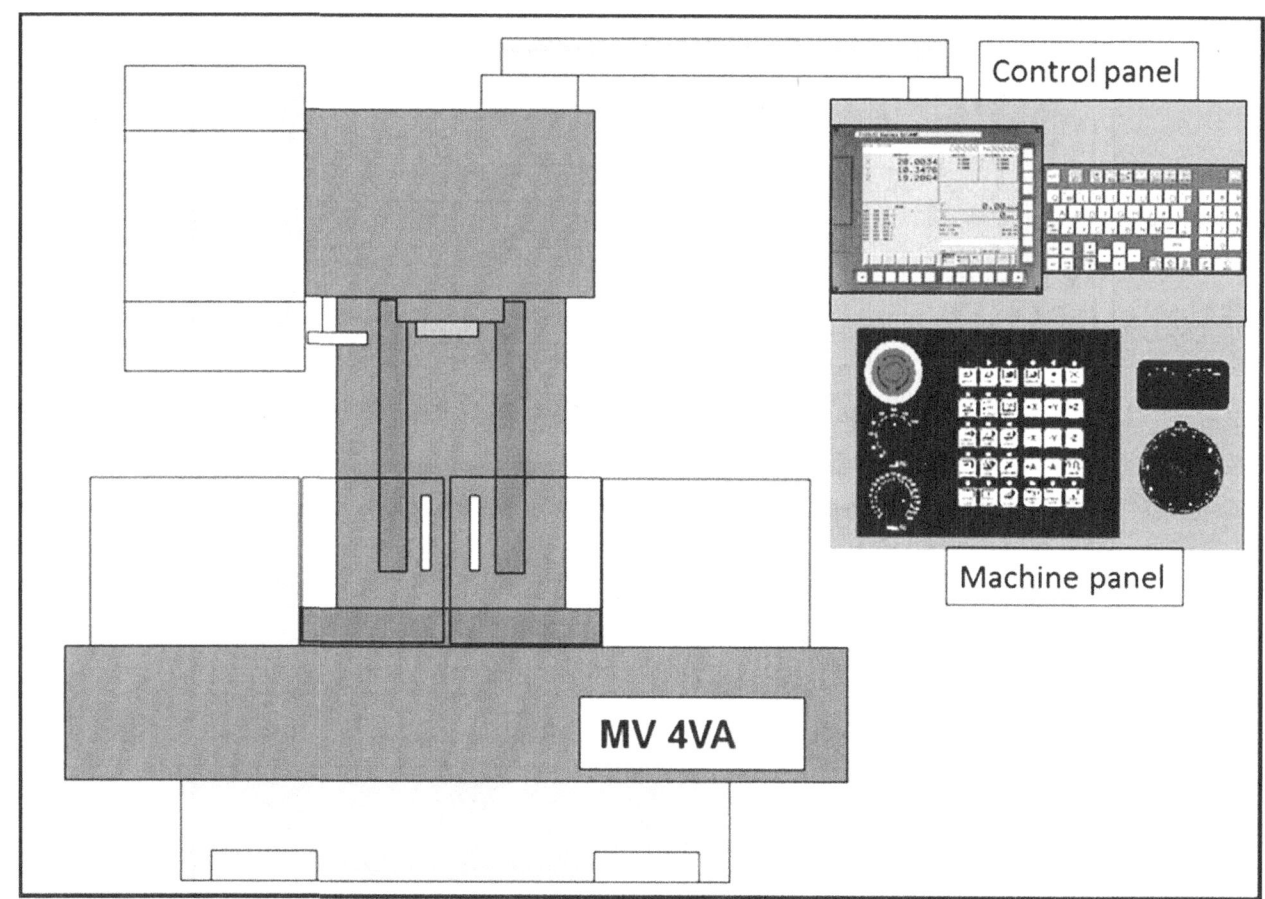

Key Concept 7: Know Your Machine From An Operator's Viewpoint

The control panel will be remarkably similar from one FANUC control mode to another. But since the machine panel is usually designed by the machine tool builder, machining centers that have been manufactured by different machine tool builders will have substantially different machine panels. To compound this problem, machine tool builders can't seem to agree on the functions that should be provided on a machine panel.

While we can be specific about the function of buttons and switches on the control panel, we remain a little vague about machine panel buttons and switches. Also, there may be buttons and switches on your machining center's machine panel that we don't explain in this text. If you find one, be sure to reference the machine tool builder's operation manual to determine the function for the button or switch.

And by the way, *a proficient setup person or operator knows the function of all buttons and switches* on their machine/s. While some may be seldom or never used, you must not consider yourself fully capable of running a machine until you know the function of all buttons and switches. Again, if we don't cover a given button or switch in this text, you must read the machine tool builder's operation manual – or ask an experienced person. Don't give up until you know the function of every button and switch on your machining center.

The Control Panel Buttons and Switches

Figure 7.14 shows the control panel for one popular FANUC control model. For the most part, the keyboard is used in conjunction with the display screen.

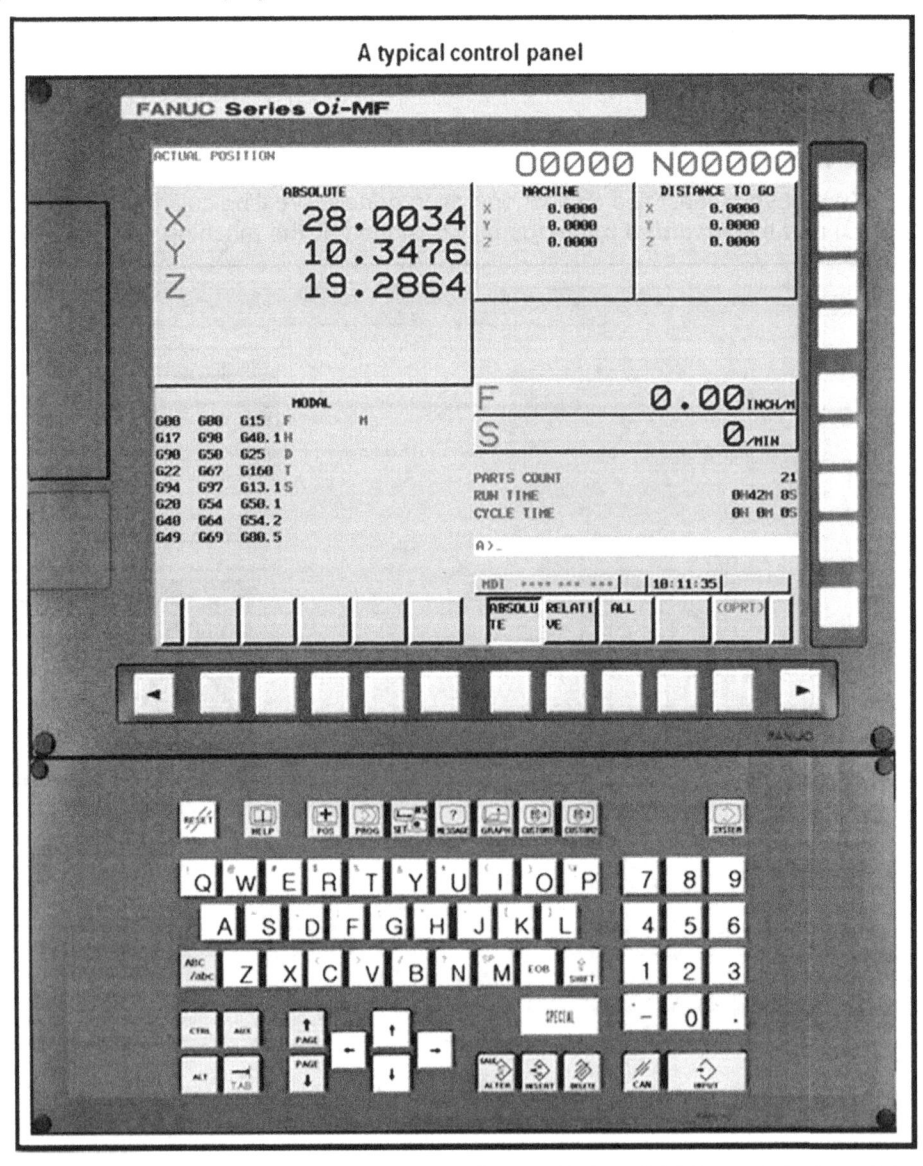

Figure 7.14 – The control panel for a popular FANUC model (shown in inch mode)

Lesson 7.2: Operation Panels

This particular keyboard resembles the (QWERTY) keyboard used with personal computers, though some FANUC control panel keyboards do not. With this keyboard, all letter addresses are readily available. Lesser used characters require that you first press the SHIFT key for access.

Display Screen Mode Keys

With current FANUC controls, display screen mode keys are located on the MDI panel keyboard.

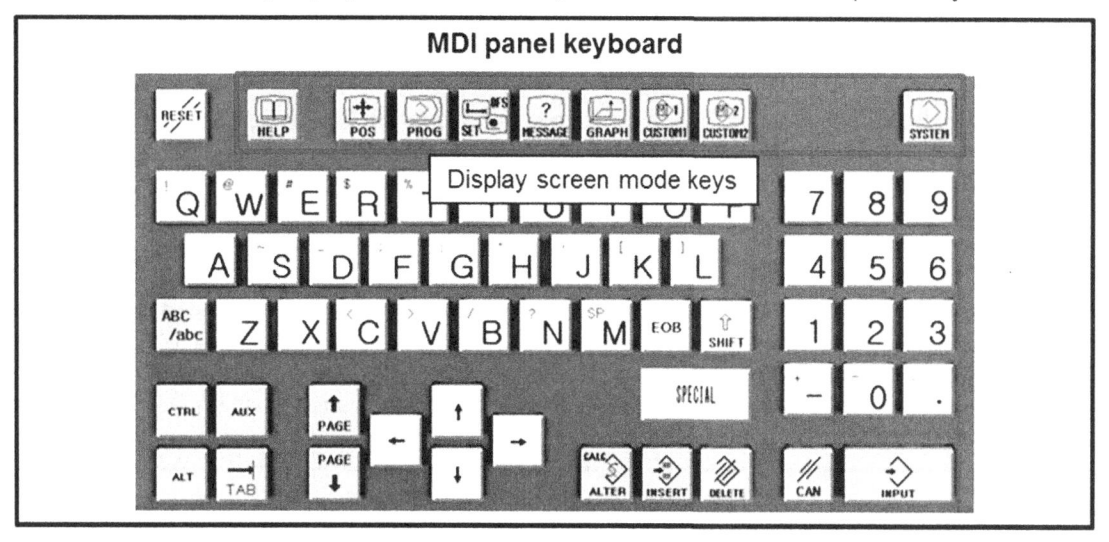

MDI panel keyboard — Display screen mode keys

The five most important display screen mode keys for setup people and operators include POS (for position), PROG (for program), OFF/SET (for offset and setting), HELP, and GRAPH. You will be pressing these keys on a regular basis.

The display screen mode key labeled SYSTEM is used, among other things, for displaying and modifying parameters, and is not commonly needed by setup people and operators. The MESSAGE key can be used to display alarms and messages, but when the control has an alarm or message to display, it does so. There is no need to press this key.

Notice the row of buttons under the display screen in Figure 7.15.

Display screen with soft keys

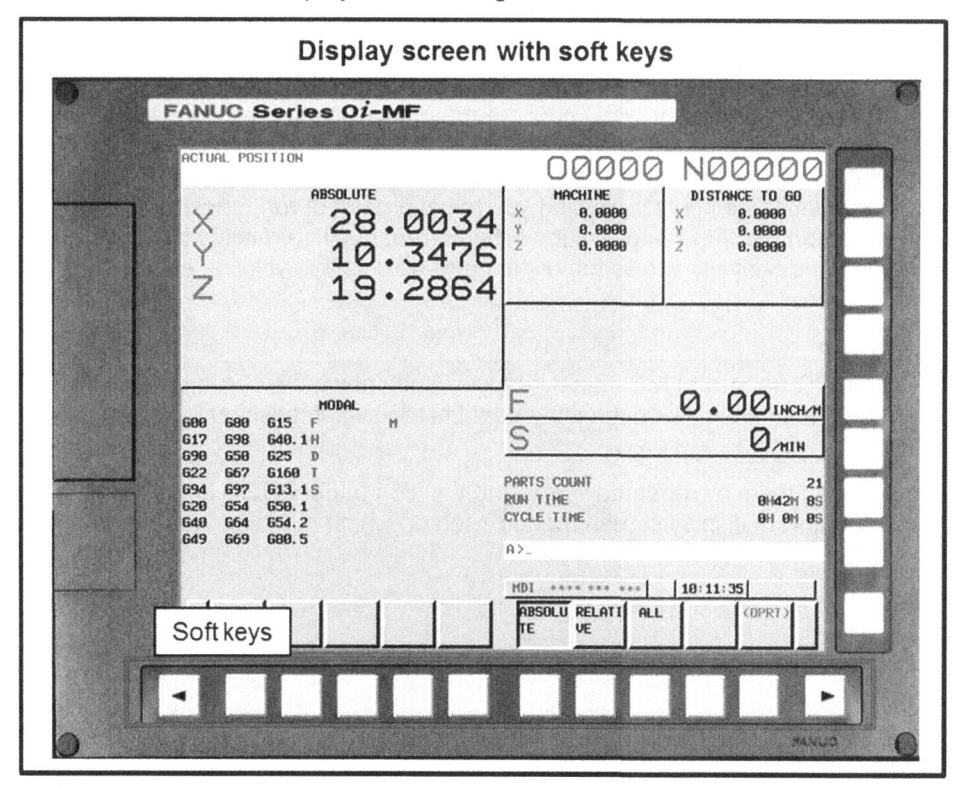

Figure 7.15 – The display screen of a CNC control – notice the soft keys at the bottom of the screen

FANUC CERT – Machining Center   ©2017 CNC Concepts, Inc.   Page 373

Key Concept 7: Know Your Machine From An Operator's Viewpoint

These buttons are called *soft keys*. Their functions will change based on the current display screen mode and the function being performed. Right now the display screen is in the POS (position) mode. The current position mode selection is ABS (absolute). Other current soft keys currently allow you to select the REL (relative) position page and the ALL position page.

The OPRT (operations) soft key, if pressed, will cause the soft keys to change, displaying a set of functions that can be performed with the current display page. In many cases, you won't have to press the OPRT key, since as soon as you begin typing for the function you wish to perform the choices of the operations soft key will be automatically displayed.

The left-most soft key (the one with the left arrow) brings you back to the most basic choices in the current display screen mode. If this soft key is active, a left arrow will appear on the display screen above this soft key.

The right-most soft key (the one with the right arrow), if active, will show more choices in the current display screen mode. A plus sign (+) will appear above this soft key when it is active.

Position Display Pages

When you press the POS (position) MDI panel key, you will be shown the current position display page. With most controls, there are three position display pages, ABS (absolute), REL (relative), and ALL (a combination of all position pages along with the *distance-to go* display). To toggle from one position display page to another, repeatedly press the POS key. Or you can select the desired position page by pressing the related soft key. The next illustration shows the position display pages.

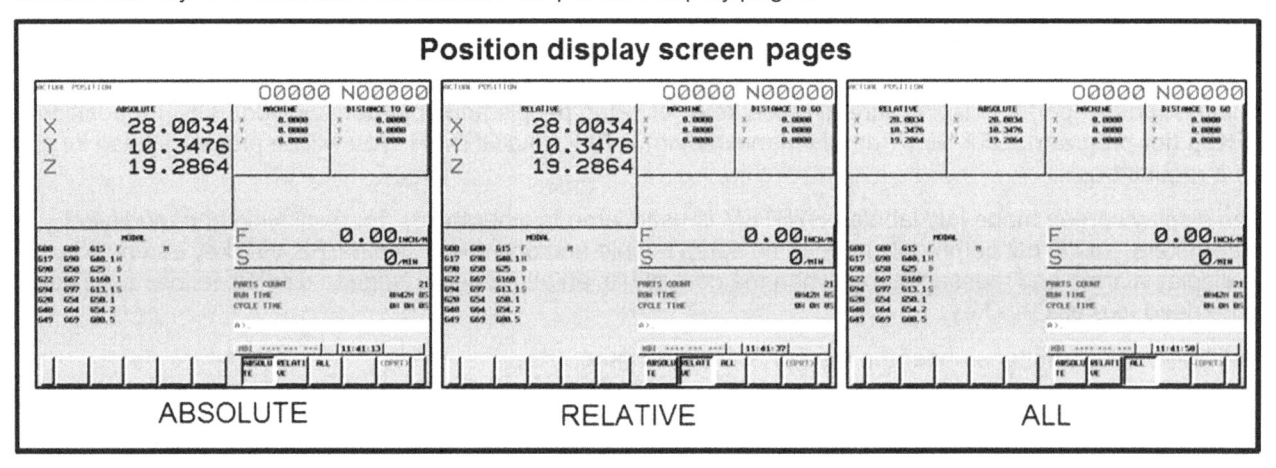

Position display screen pages

ABSOLUTE | RELATIVE | ALL

If you've read the entire text to this point, you know the meaning of each position display screen page. The *absolute* position display page shows the current position relative to workpiece coordinate system zero. The *relative* position display page is used to take measurements on the machine (like workpiece coordinate system zero assignment measurements), allowing you to set a point of reference for the measurement. The *machine* position (shown on the "ALL" page) shows the current position relative to the zero return position. And the *all* position page additionally shows the distance-to-go display, which is extremely important when you are verifying programs.

Program Display Pages

When you press the PROG (program) MDI panel key, you'll be shown the current program display screen page. This page varies based upon which machine panel mode switch position is selected (AUTO, MDI, or EDIT).

And again, toggle between them by repeatedly pressing the MDI panel PROG key. Figure 7.17 shows the pages available in the program display screen mode based on which machine panel mode switch position is selected.

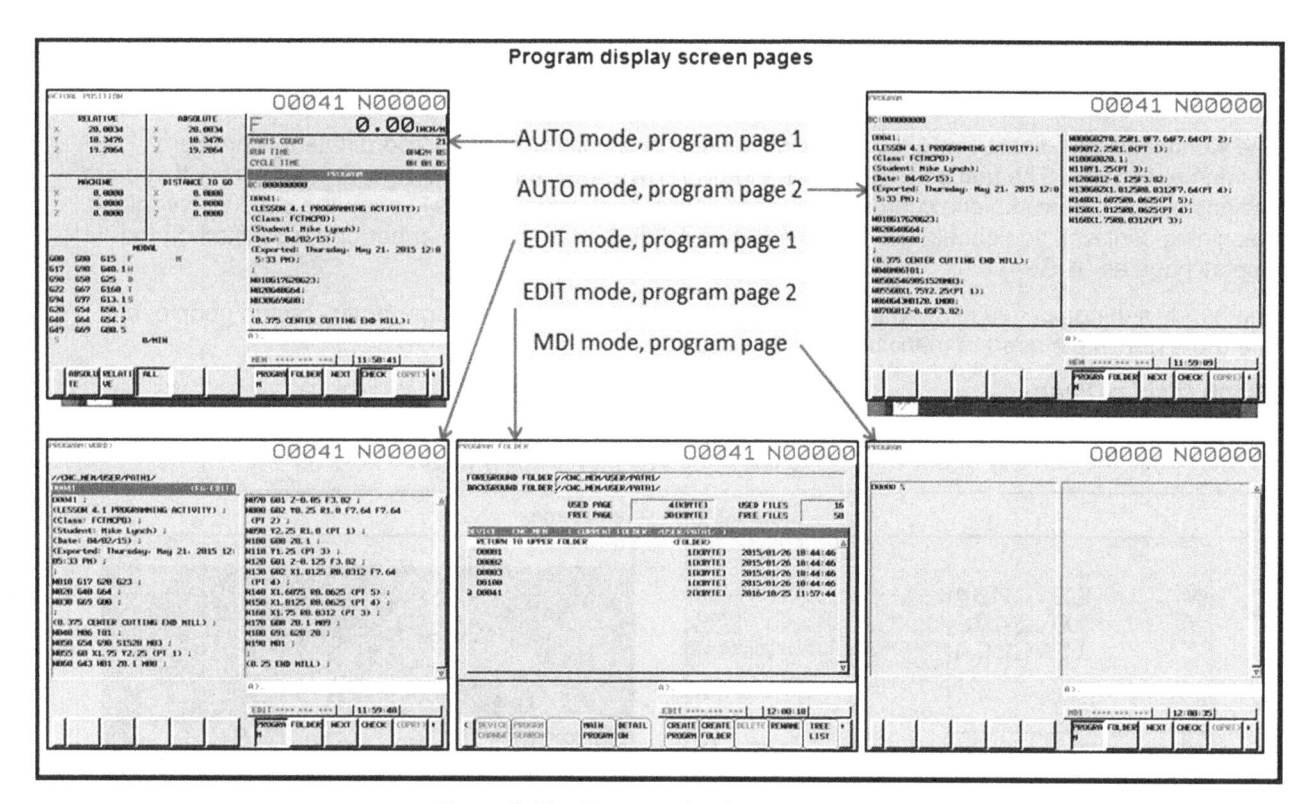

Figure 7.17 – Program display screen pages

The upper-left screen shows the CHECK page, which is handy when verifying programs. The upper-right and lower-left pages look similar, but the upper-right screen shows the program in the AUTO mode while the lower-left screen shows the program in the EDIT mode. The lower middle screen shows a directory of programs, helpful for finding programs. The lower right screen shows the MDI (manual data input) display screen page.

Offset Display Pages

When you press the OFF/SET (offset and setting) MDI panel key, you'll see the current offset display screen pages, as are shown in Figure 7.18.

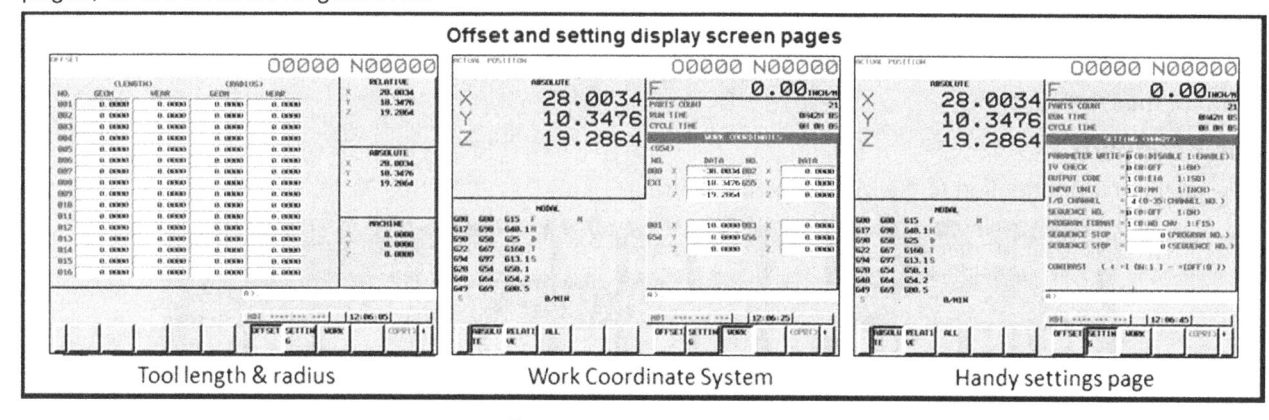

Figure 7.18 – Offset display screen pages

Once the OFF/SET key is pressed, If you press the soft key under offset, you will be able to choose between the tool offset and work offset pages. If you press the OFF/SET key again, the handy settings page will be shown. The tool length and radius offset page shows four registers per offset number. The GEOM(H) register is used to specify the cutting tool's length (height) and the WEAR(H) register is used to specify sizing adjustments (possibly required by tool wear).

Key Concept 7: Know Your Machine From An Operator's Viewpoint

The GEOM(D) and WEAR(D) are used for similar purposes for milling cutters to specify the milling cutter's diameter and wear amounts. GEOM(D) and WEAR(D) are only required when cutter radius compensation is being used or when you are using the Manual Guide i program verification function.

The workpiece coordinate system offset page is used to specify workpiece coordinate system zero assignment values. The first page (shown in Figure 7.18) shows the offsets zero through three. If you are assigning only one workpiece coordinate system zero for you program (as is often the case with vertical machining centers), you should use workpiece coordinate system offset number one (specified on the display page as 01 G54).

The handy settings pages allow you to specify certain parameters that programmers, setup people, and operators may have need to manipulate pretty regularly.

Graph Display Pages

When you press the GRAPH MDI panel key, you will see the tool path display screen page, as Figure 7.14 shows. Press the soft key under GRAPH to show the tool path display page.

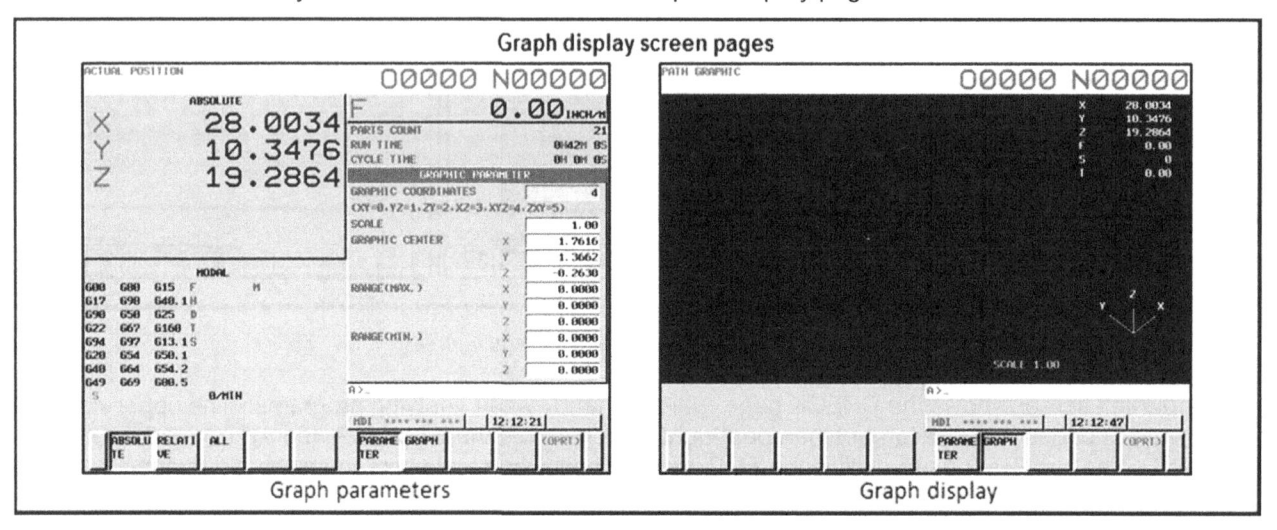

Graph display screen pages

Graph parameters | Graph display

The graph parameters page (left screen) lets you specify the position and size for the tool path display. The graph page (right screen) lets you view the tool paths for cutting motions during a program's execution.

Other Display Screen Modes

The HELP display screen pages provide documentation for many control functions. The SYSTEM display screen pages are used to specify parameter settings. The MESSAGE display screen pages can be used to display alarms and messages, but again, this function is automatically selected when the control has an alarm or message to show.

The Keyboard

Now let's turn our attention to the MDI panel keyboard.

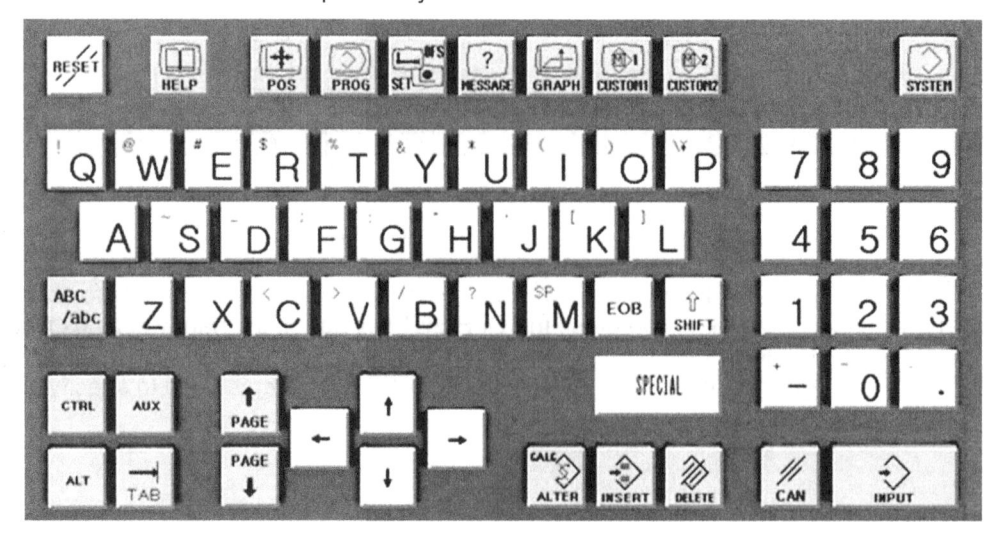

Lesson 7.2: Operation Panels

Letter Keys

This part of the keyboard allows character entry. With this keyboard, lesser used characters (shown smaller on each key) require that you first press the SHIFT key for entry. Some FANUC models provide a smaller keyboard, requiring the SHIFT key more often.

The Slash Key (/)

This key allows the entry of the slash code into programs. The slash code is the block delete character. Block delete is described in lesson 6.7.

Number Keys

These keys allow numeric data entry, as is required when entering and modifying offsets and programs.

Decimal Point Key

This key allows numeric entry with a decimal point. Setting offsets and entering CNC programs are examples of when it will be needed.

CAN Key

Standing for cancel, this key is your backspace key. If you make a mistake while entering data, use this key to backspace one character at a time.

EOB Key

Standing for end of block, this key allows you to type the end of command delimiter in a program. It is displayed as a semi colon (;) at the end of each CNC command in a program.

The INPUT Key

This key is pressed to enter data. Examples of when this key is pressed include entering wear offsets, geometry offsets, and the work shift value. This key is *not* used for entering words and commands into a program. Instead the *insert* key is used for inserting words and commands.

Note that an INPUT soft key will appear on certain display screen pages. It will have the same effect as the MDI panel INPUT key.

Cursor Control Keys

The display screen often shows a *cursor* that designates the current entry position. Examples of its use include when working on the active program or when entering offset data.

The up and down PAGE arrows will move the cursor from page to page. The arrows left and right will move the cursor left and right one field/word per press. And the arrows up and down will move the cursor up and down one field/command per press.

Program Editing Keys

Programs can be modified right at the machine using these keys. With some controls, they are soft keys that only appear when the display screen is in the program mode (and when the mode switch is set to edit). See lesson 9.1 for step-by-step procedures for entering and modifying programs.

INSERT key
Not to be confused with *input*, this program editing key allows new words and commands to be entered into a program. Most controls will insert the entered word or command *after* the cursor's position.

ALTER key
This program editing key allows words in a program to be altered. After positioning the cursor to the *incorrect* word in the CNC program, enter the correct word, and press the alter key.

DELETE key
This program editing key allows program words and commands to be deleted. You can delete a word, a command, a series of commands, or an entire program.

RESET Key

This very important key has three functions. First, in the EDIT mode, the reset key will send the cursor to the beginning of the program.

Second, in the AUTO or MEM mode while executing programs, the reset key will clear the *look-ahead buffer* and stop execution of the program. This cancels the cycle's execution and is required when there is

Key Concept 7: Know Your Machine From An Operator's Viewpoint

something wrong in the program and you wish to stop executing the program. This is common when verifying a new program. *Be careful with the RESET key*. Again, the reset key will clear the look-ahead buffer. If the program is reactivated (by pressing cycle start) immediately after the reset key is pressed, the control will skip the commands that were in the look-ahead buffer. This, of course, can lead to serious problems.

Third, when the control is in alarm state, the reset key will *clear the alarm* as long as the problem causing the alarm has been corrected.

The Machine Panel

As stated, the machine panel is designed and built by the machine tool builder. And machine tool builders vary dramatically with regard to the functions they provide on the machine panel. They vary most with regard to how much manual control they provide for machine functions (like spindle activation and automatic tool changer control). Figure 7.20 shows a typical machine panel for a vertical machining center. As you look at this operator's panel, notice how many of its functions have been introduced in previous lessons.

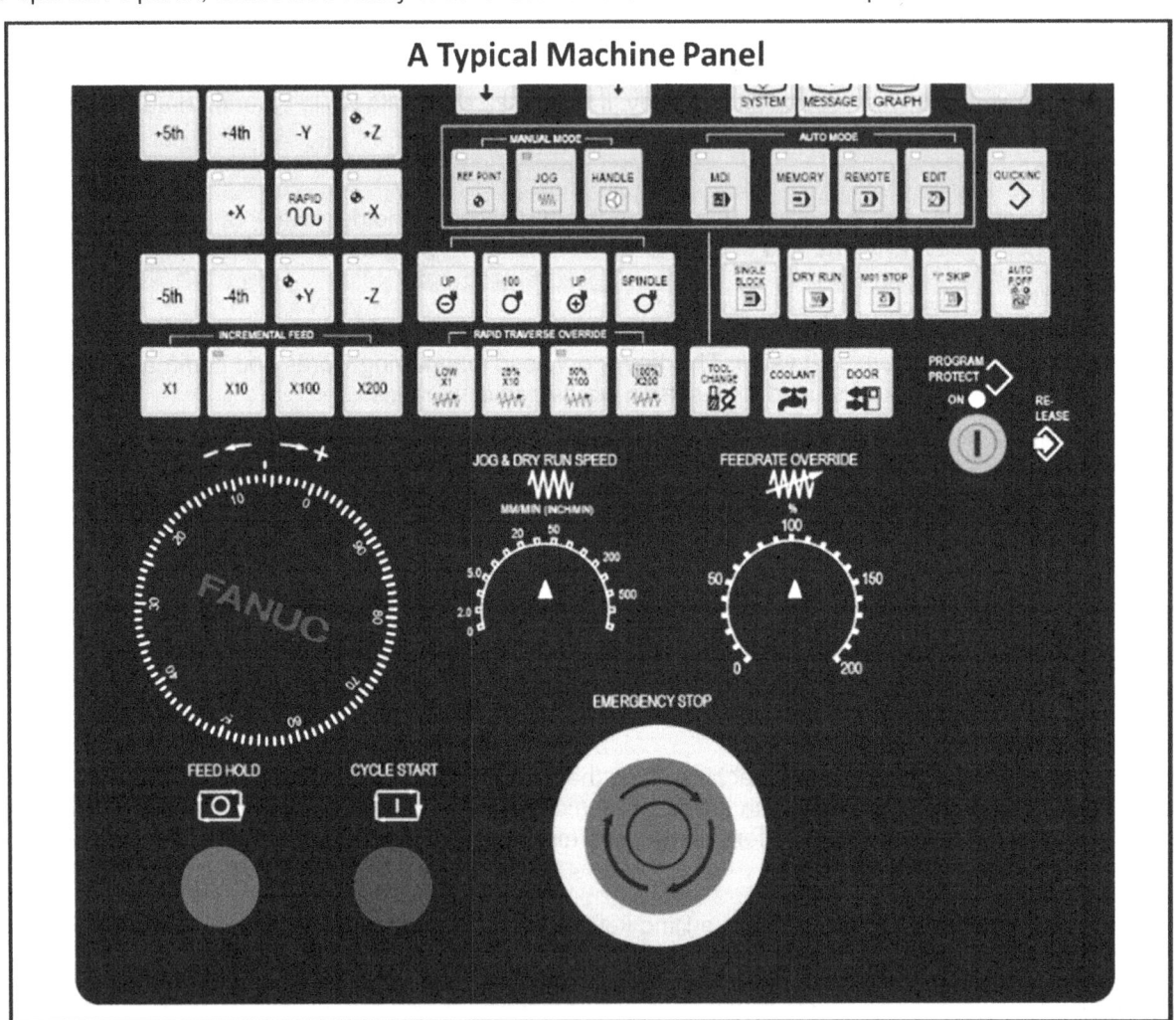

Figure 7.20 – A typical machine panel

Power Buttons

While machine tool builders vary, most will place power on and off buttons on the machine panel. POWER ON will be used when powering up the machine and POWER OFF will be used when turning it off.

MODE switch

The MODE switch is the heart of a CNC machining center. It will be the very first switch you set when performing any function on the machine. The mode switch must be positioned properly in order for the desired function to be performed. If it is not, the machine will not respond to your action. Again, we discuss the mode switch in great detail during lesson 8.1.

CYCLE START Button

This button has two functions. First, it is used to activate the active program when the machine is in the memory or auto mode. Second, the cycle start button is used to activate manual data input (MDI) commands. We recommend that you get in the habit of having a finger ready to press the FEED HOLD button (which is usually in close proximity to the cycle start button) whenever you press the CYCLE START button.

FEED HOLD Button

While a program (or MDI command) is being executed, this button allows you to halt axis motion. The CYCLE START button can be used to reactivate the cycle. Note that all other functions of the machine (coolant, spindle, etc.) will continue to operate even when the machine is in FEED HOLD state.

You should think of FEED HOLD as your *first panic button*. You should *always* have a finger on this button whenever you press the cycle start button. You should also keep a finger ready to press the FEED HOLD button for the entire time that you are verifying a program.

Some people feel the EMERGENCY STOP button should be considered the panic button. However, the emergency stop button may actually turn the power off to the machine tool, which sometimes causes more problems than it solves. For example, when the emergency stop button is pressed and if the machine power is turned off, the axes of the machine will drift until a mechanical break can lock them in position. Axes bearing a great deal of weight (like the Z-axis of most vertical machining centers) are most prone to drift. The amount of drift is usually quite small (less than 0.25-mm [or 0.010-in] in most cases). But if a cutting tool is actually machining a workpiece when the emergency stop button is pressed, the Z-axis drift could cause damage to the tool and workpiece.

FEEDRATE OVERRIDE Switch

This multi-position switch usually has three functions. In automatic mode (AUTO or MEM), this switch allows you to change the programmed feedrate during cutting motions. The FEEDRATE OVERRIDE switch usually allows the programmed feedrate to be changed in 10% increments and will usually range from 0% through 200%. This means the operator can stop cutting motions (at the 0% setting) and double the programmed federate (at the 200% setting).

If the program is written correctly, the entire cycle should run at 100% feedrate, meaning once a program is verified, you should be able to set this switch at 100% and leave it alone. But when you are verifying a program, running the very first workpiece, this switch can be very helpful to confirm that the programmed feedrate for each tool is correct.

Most machine tool builders use the FEEDRATE OVERRIDE switch for a second purpose. When the DRY RUN switch is turned on, the FEEDRATE OVERRIDE switch is used to take control of the motion rate for all movements (including rapid motions). This function is necessary during program verification, when you're running the program for the first time without a workpiece to check for setup-related mistakes. Some machine tool builders use another switch called JOG FEEDRATE for this purpose.

With some machines, the FEEDRATE OVERRIDE switch will override rapid motions when the machine is in automatic mode. This eliminates the need for the DRY RUN function.

Finally, many machine tool builders use the FEEDRATE OVERRIDE switch to control axis motion rate when manually jogging an axis, though some use a special switch (called JOG FEEDRATE) for this purpose. The machine panel shown above has a special switch (JOG FEEDRATE) for this purpose.

RAPID OVERRIDE Switch

As the name implies, this function is used to slow the machine's rapid rate. The rapid rate on current model machining centers is very fast (some machines rapid at well over 38 meters per minute [or 1,500 inches per minute], and this switch brings some welcomed control over the very fast rapid rate.

With most machines, the RAPID OVERRIDE switch is a multi-position switch (as is the case with the one on our example machine panel). But we have seen some that are simple on/off switches. When this switch is on, rapid rate is slowed to about twenty-five percent of its normal rate. When it is off, rapid motions will at 100%.

With many machines, the RAPID OVERRIDE switch has four settings, F0, 10%, 25%, and 100%. At the F0 setting, the machine will move at about 1% of its rapid rate when rapid movements are commanded.

Generally speaking, the RAPID OVERRIDE switch is used for program verification. It's nice to leave it at 10% or 25% while you're verifying programs. You can rest assured that the machine will never achieve its true rapid rate. This will give you more time to react should something go wrong.

EMERGENCY STOP Button

This button will commonly turns power off to the machine tool. Usually, power remains turned on to the control. See the description of the feed hold button for more information about how emergency stop is used.

Conditional Switches

There are several on/off switches on the machine panel that control how the machine behaves during automatic and manual operation. They could be toggle switches, locking push-buttons, or lighted buttons (a light within the button comes on when the function is on).

Frankly speaking, these functions are more related to the control panel than they are the machine panel, so they remain the same from one machining center to another, regardless of the machine tool builder. These switches are very important. If one is improperly set, the machine will not perform as expected. You must get in the habit of checking each of these switches before executing a CNC program.

DRY RUN On/Off Switch

As described in lesson 7.1, this conditional switch is used during program verification. When this switch is on, it gives you full control of the motion rate for all movements the machine makes (including cutting movements and rapid movements).

When the DRY RUN switch is turned on, another switch (usually feedrate override) will act as a rheostat – and will allow you control of all motion rates. DRY RUN will slow rapid motions, but it will tend to speed up cutting motions. This means you should *never allow a cutting tool to machine a workpiece when dry run is turned on* (you'll have no idea as to what the actual feedrate is).

SINGLE BLOCK On/Off Switch

As described in lesson 7.1, this conditional switch will force the machine to stop after it executes each command. To reactivate the cycle (execute the next command), you must push the cycle start button.

This switch is helpful during program verification. With a new program, you must cautiously check each motion the machine makes, especially movements that have a cutting tool approach the workpiece. With single block turned on, you can rest assured that the machine will stop at the end of each command, giving you a chance to check the motion just made and to check the program to see what is going to happen next. It is commonly used with the distance-to-go display, which shows how much further the machine is going to move in the current command.

BLOCK DELETE On/Off Switch (also called optional block skip)

The applications for block delete are described in lesson 6.3. The block delete switch works in conjunction with slash codes (/) in the program. If the control sees a slash code in the program, it will look to the position of the BLOCK DELETE switch. If the switch is on, the control will ignore the words to the right of the slash code. If the BLOCK DELETE switch is off, the control will execute the words to the right of the slash code. This function is used to give the operator a choice between one of two possibilities. (See lesson 6.3 to learn more about block delete applications.)

OPTIONAL STOP On/Off Switch

This conditional switch works in conjunction with an M01 word in the program. When the control sees an M01, it looks to the position of the OPTIONAL STOP switch. If the switch is on, the control will stop the execution of the program and turn off certain machine functions (like spindle and coolant). You must press the cycle start button to reactivate the program. If the OPTIONAL STOP switch is off, the control will ignore the M01 and continue executing the program.

In the programming format shown in lesson 5.2, we recommend placing an M01 in the program at the end of each tool. This gives you the ability to stop the machine at the end of each tool by simply turning on the OPTIONAL STOP switch. Stopping at the end of each tool is important during program verification (as described in lesson 7.1) to let you check what one tool has done before going on to the next.

Buttons and Switches for Manual Functions

The machine panel for CNC machining centers will include some buttons and switches related to manual control of the machine's functions. These buttons and switches vary dramatically from one builder to the next.

Axis Jogging Controls

On our example machine panel (Figure 7.20), notice the multi-position switch labeled JOG FEEDRATE, the two push-buttons (PLUS and MINUS) and the AXIS SELECTOR (X, Y, and Z). These functions are active when the machine is in a manual mode (jog or zero return on the mode switch). To jog an axis, after selection a manual mode, you will first select the axis to be jogged (with the X, Y, Z switch). Then you will select the motion rate (with the JOG FEEDRATE switch). Finally, you will press the desired direction button (PLUS or MINUS). As you press the button, the selected axis will begin moving in the selected direction at the selected feedrate. This, of course, requires that you know which axis you want to move and which way is plus and which way is minus (as is described in lesson 1.1 and at the beginning of the discussions for Key Concept Number 7).

Jogging an axis is often required. Applications include measuring workpiece coordinate system zero assignment values and tool length compensation values on the machine, sending the machine to its zero return position, and moving an axis into a convenient position so the cutting tool in the spindle can be inspected or replaced.

Positioning is not very precise with jog functions. For this reason, most machine tool builders also provide a *hand-wheel* on the control panel.

Hand-Wheel Controls

On the example machine panel, notice the hand-wheel in the lower-left corner of the panel. The hand-wheel is active when the mode switch is set to hand-wheel (with many machines, a light next to the hand-wheel will come on when the hand-wheel is active). Also notice the axis selector switch (X, Y, Z) and the rate selector switch (X1, X10, and X100).

To use the hand-wheel, after you select the hand-wheel mode, you first place the axis selector to the desired position (X, Y, or Z). Then you select the motion rate. In the X1 (times one) position, each increment of the hand-wheel is 0.001-mm (or 0.0001-in), and motion is very slow – barely detectable except with the position displays. In the X10 (times ten) position, each increment is 01-mm (or 0.001-in), and motion is faster – easily detectable. In the X100 (times one hundred) position, each increment is 0.1-mm (or 0.010-in), and quite fast.

Using the hand-wheel is often required after using the jog function to manually move an axis. Again, the jog function does not allow precise positioning control. So you will commonly use jog to bring an axis close to its intended position, and then use the hand-wheel for the rest of the motion. Applications include measuring workpiece coordinate system zero assignment values and tool length compensation values on the machine.

Spindle Control

Manual spindle control is not often required. About the only time we've needed to manually activate the spindle is when using an edge-finder to measure workpiece coordinate system zero assignment values.

Some will provide a speed selector rheostat and buttons to start and stop the spindle. But they may not allow you to select the spindle direction.

Other machining centers do not allow manual control of the spindle (as our example machine panel shows). For any machine function that you must activate manually – but for which there are no manual controls – you must use the manual data input mode (described in lesson 8.1).

Automatic Tool Changer Control

Notice that there are no manual controls for the automatic tool changer on our example machine panel. Very few machine tool builders provide manual controls that allow you to make tool changes. Yet you will probably have to make manual tool changes on a regular basis. If, for example, you measure tool length compensation values on the machine during setup, it is usually best to load the tools into the automatic tool changer magazine and then use the automatic tool changer system to load each tool into the spindle for measuring.

For any machine function that you must activate manually – but for which there are no manual controls – you must use the manual data input mode (described in lesson 8.1).

Indicator Lights and Meters

Most CNC machining centers have a series of lights and meters that allow the operator to monitor the condition of important machine functions.

Key Concept 7: Know Your Machine From An Operator's Viewpoint

Spindle RPM and Horsepower Meters

Most machining centers have two meters that show you key information about the spindle. The first meter is the rpm meter, which shows you how fast the spindle is currently rotating. Some machines show spindle speed on the display screen. The second spindle-related meter is a *load meter*. If the machine tool has this meter, you can monitor the amount of stress on the spindle drive system. Usually this meter shows a percentage-of-load, ranging from 0% through 150%. This means you can easily tell to what extent a machining operation is taxing the spindle of the machine.

Axis Drive-Motor Horsepower Meter

This meter, if equipped, allows the operator to see how much horsepower is being drawn by any of the machine axes drive motors. Usually there is only one meter and you must select which axis (X, Y, or Z) you wish to monitor by a three-position switch. Like the spindle horsepower meter, this meter allows you to see how much stress is on a drive motor during a machining operation.

Cycle Indicator Lights

Most CNC machines have two indicator lights to show whether the machine is in cycle. One is above (or close to) the cycle start button and will stay on as long as the machine is in cycle. The other is above (or close to) the feed hold button and comes on when the machine is in feed hold state (when you press the feed hold button).

Zero Return Position Indicator Lights

Some machining centers have indicator lights that come on when an axis is at its zero return position. These lights are commonly labeled *axis origin lights*. If a program is planned to start from the zero return position, these lights can be helpful to an operator. They will tell you whether it is safe to activate the cycle.

Optional Stop Indicator Light

Some machining centers have an indicator light that is close to the optional stop switch. If the machine has been halted by an optional stop in the program (M01), this indicator light comes on to tell you why the machine has halted operation.

Other Buttons and Switches on the Machine Panel

Remember that the machine panel on your particular machine may have more buttons and switches than we have described – especially if the machine is equipped with special accessories (like a pallet changer or automatic loading system). When you come across a button or switch that you don't recognize, be sure to reference the machine tool builder's operation manual to find out what it does.

Other Operation Panels on Your Machining Center

Many machine tool builders include additional operation panels. Because the automatic tool changer is mounted quite a distance from the main operation panels, many machine tool builders place a special automatic tool changer control panel right next to the automatic tool changer. Buttons and switches on this panel will commonly allow magazine rotation, tool station numbering, and if required, clamp & unclamp of the tool in the loading station.

If your machine has a pallet changer, it will likely have a special operation panel mounted near the pallet changer to allow manual pallet changing, and to allow you to press a button (commonly called the *standby button*) to inform the machine that you have finished loading the workpiece and that it is safe to make a pallet change.

Key points for lesson 7.2:

- You should know the function all of the buttons and switches on the machines you run.
- The control panel is made by the control manufacturer (like FANUC) and the machine panel is made by the machine tool builder.
- Machine panels vary, even among similar machines made by different builders.
- The display screen has a series of soft keys whose meanings change based upon the selected display screen mode and the function being performed.
- The mode switch is at the heart of any CNC machine.

Key Concept 8: Know the Basic Modes of Operation

From a setup person's or operator's perspective, every button and switch on the machine can be divided into one of four categories. It is related to manual mode, manual data input mode (MDI), edit mode, or program operation mode.

Key Concept 8 contains one lesson:

    8.1: Know the Basic Modes of Operation

We mentioned the mode switch in lesson 7.2. You know it is the most important switch on a CNC turning center – and it is the first switch you should set when you perform any operation on the machine.

We compare the mode switch of a CNC machine to the remote device for a home entertainment system.

You know that if you have selected the CABLE box, only buttons for that device will be active. Buttons for other devices, like the DVD player, will be inactive. If you press any of them, nothing will happen.

In similar fashion, when you set the machine panel mode switch to a given mode, only buttons and switches that are related to that mode will be active. Buttons and switches related to other modes will be inactive. Generally speaking, if you have the mode switch in the wrong mode when you press a button or activate a switch, nothing will happen.

Lesson 8.1–Operation Modes

The most common operation mistake is having the mode switch in the wrong position. Fortunately, this mistake will not cause serious problems. The machine will not respond to your action.

Objectives

After completing this lesson, students should be able to:

- ✓ Describe the functions and mode switch positions of the manual mode
- ✓ Describe the function of the manual data input (MDI) mode
- ✓ Describe the function of the edit mode
- ✓ Describe the functions and mode switch positions of the program operation mode

Introduction

You know the mode switch of a machining center is a multi-position switch. The actual positions of a typical mode switch include AUTO (or MEM), EDIT, MDI, REMOTE, REF (or Zero Return), JOG, INC, and HANDLE.

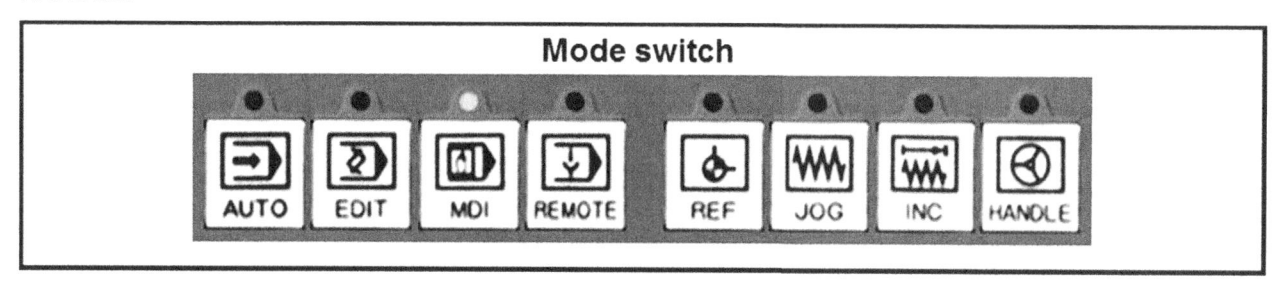

Mode switch

[Mode switch panel showing buttons: AUTO, EDIT, MDI, REMOTE, REF, JOG, INC, HANDLE]

This mode switch is a series of lighted buttons. A light will come on to indicate which mode is selected. Abbreviations include AUTO for automatic (this position is sometimes labeled MEM for memory), MDI for manual data input, and REF for reference (this position is sometimes labeled ZR for Zero Return), INC for incremental jog, and HANDLE for hand-wheel.

The Manual Mode Switch Positions

In the manual mode, a CNC machining center behaves much like a conventional milling machine. The manual mode positions of the mode switch include *JOG* (sometimes called *continuous jog*) mode, *INC* mode, *HAND-WHEEL* mode, and *REF* mode (often labeled Zero Return).

You know from previous lessons that you will often need to perform manual functions. Examples include measuring the workpiece coordinate system zero assignment values, measuring tool length compensation values, and when replacing dull tools.

With the manual mode, you will press a button, turn a hand-wheel, or activate a switch that will cause an *immediate response* from the machine. An axis will move, the spindle will start, the coolant will come on, or some other machine function will respond to your action.

As you know from lesson 7.2, the **JOG** mode switch position will allow you to manually move a selected axis. After selecting the jog mode, you select the axis to move (X, Y, or Z), select the motion rate (with the JOG FEEDRATE switch) and press a button that corresponds to the direction you want the axis to move (plus or minus). The machine will immediately respond by moving the selected axis at a selected motion rate in the selected direction.

In similar fashion, the **HANDLE** mode switch position will allow you to move an axis with the hand-wheel. Again, after placing the mode switch to the HANDLE position, you select the axis to move (X, Y, or Z), select the rate of motion (with X1, X10, or X100), and turn the hand-wheel. The machine will respond by moving the selected axis.

Key Concept 8: Know The Operation Modes

In the **REF** mode switch position, you can manually send the machine to its zero return position. Machining centers vary when it comes to the actual procedure needed to send an axis to its zero return position. With most machines, after placing the mode switch to the REF position, you select the axis to be zero returned (X, Y, or Z) and then press the plus button. Hold it until the axis reaches its zero return position (and possibly an axis origin light for the axis comes on).

The **INC** mode switch position allows you to manually move an axis by a designated amount with every press of a push-button. You first select the axis to move with the X, Y, or Z selector and the motion distance with X1 (0.001-mm [or 0.0001-in]), X10 (0.01-mm [or 0.001-in]), or X100 (0.1-mm [or 0.010-in]). Every time you press the plus or minus button, the axis will move the chosen distance. Frankly speaking, the handwheel provides much more precise control of manual movement, so we seldom recommend the use of the INC mode.

In any of the manual mode switch positions, certain other buttons and switches may be active. Some machining centers allow manual control of the spindle and coolant. Some allow manual control of the automatic tool changer (ATC) function. The related buttons and switches will be aptly named and placed on the machine panel.

But as you know from lesson 7.2, there are usually some machine functions that you do need to control manually that the machine tool builder has not provided buttons or switches to control. The automatic tool changer function is a classic example. Many machining centers do not allow you to manually change tools.

For functions that you need to activate manually but for which you have no manual buttons and switches, you must use the MDI mode switch position.

The Manual Data Input (MDI) Mode Switch Position

Manual data input (MDI) mode switch position is used for two reasons. First, it is used to perform manual functions that cannot be done by any other means. Again, machine tool builders vary with regard to what can be done in a completely manual manner. For those machine functions of which you have no manual control, you must use the MDI mode.

Second, MDI mode can be used to perform certain manual function of which you do have manual control, but can do faster or easier in the MDI mode. A manual zero return on most machines, for example, takes much more time and effort to complete than a zero return done in the MDI mode.

Commanding an MDI Zero Return

The MDI mode requires that you know the CNC command to activate the manual function you want to perform. This is the same command used in a CNC program to activate the function. If, for example, you want to use the MDI mode to command a zero return in all three axes, you must know the program command needed to do so, which happens to be

```
G91 G28 X0 Y0 Z0;
```

When this command is entered and executed in the MDI mode, the machine will rapid all three axes (simultaneously) to the zero return position.

Notice the semi-colon (;) at the end of the G28 command example. This is the character that most (FANUC) controls use to represent an *end-of-block*. There is a key on the keyboard labeled EOB, which stands for end-of block. When you enter an MDI command for most controls, it must end with the end-of-block character. With these control models, when you are finished entering an MDI command, you must remember to press the EOB key prior to inserting the command into the MDI buffer.

If you want only the Z-axis to be sent to the zero return position (which is the tool change position on most vertical machining centers), the command will be

```
G91 G28 Z0;
```

The Procedure to Give an MDI Command

The step-by-step procedure to use the MDI mode with current model FANUC controls is as follows:

1) Place the mode switch to [MDI].

2) Press the MDI panel [PROG] key until the MDI display screen page is shown.

3) Using the keyboard, enter the MDI command you wish to execute (for the zero return, this is `G91 G28 X0 Y0 Z0`). There is no need for spaces between each word.

4) Press the MDI panel [EOB] key (this places a semi-colon at the end of the command).

5) Press the MDI panel [INSERT] key. The command moves up into the active MDI buffer.

6) Press the [CYCLE START] button on the machine panel. Again, always have a finger ready to press the [FEED HOLD] button when you execute an MDI command. As soon as you press the [CYCLE START] button, the machine will perform the commanded action. In our example, the machine will rapid to the zero return position in all commanded axes.

Commanding a Tool Change

We've mentioned several times to this point that most machining centers provide no manual control of the automatic tool changer. If you want to make a manual tool change, you must use the MDI mode to do so. Just as in a CNC program, most machines must be at the tool change position. And again, you must know the CNC words related to your automatic tool changer. For most machining centers, a *T word* places the tool in the ready station of the magazine and an M06 word actually makes the tool change. As long as the machine is at the tool change position, the command

`T04 M06;`

will place tool number four into the spindle. This command can be entered in step number three of the procedure just shown. But again, the machine must be at its tool change position (usually the Z-axis zero return position) for this command to work.

Commanding Spindle Activation with MDI

While some machining centers provide adequate manual control of the spindle, others do not. If yours does not, you can use the MDI mode to activate the spindle. As you know from the programming lessons, an *S-word* is used to specify the desired speed in rpm. An M03 will start the spindle in the forward direction (needed for right-hand tools) and an M04 will start the spindle in the reverse direction (needed for left-hand tools). When you want to stop the spindle, an M05 must be commanded. The command

`S700 M03;`

will start the spindle at 700 rpm in the forward direction. The command

`M05;`

will stop the spindle. Again, enter these commands in step three of the procedure shown above.

Other Times When MDI is Used

Again, any time you need to perform a manual function, you can use MDI to do so. Any machine function can be activated in MDI mode – as can any G-code. Truly, if a command works in a program, it will work in the MDI mode. Some examples include

- Coolant (M08 and M09)
- Pallet change (you must know the related command word/s, commonly M60)
- Door open and door close (if the machine has automatic doors)
- Chip conveyor on and off (if the chip conveyor is programmable)
- Switching between inch and metric modes (G20 and G21)

Can You Make Motion Commands with MDI?

The MDI mode can even be used to machine a workpiece. Since just about all CNC commands that run in a program will work in the MDI mode (including G00, G01, G02, and G03), you can command machining commands in the same way they are commanded in a CNC program. You can even enter several commands at a time – just remember to enter the EOB key at the end of each command.

You must, however, be extremely careful when using MDI to command axis motion. While some operators get very good at making motion commands with MDI, your command/s will be executed just as you enter them. If you make a mistake while entering a CNC command in the MDI mode, the results could be

Key Concept 8: Know The Operation Modes

disastrous. You will have no chance to verify your MDI commands as you can with a CNC program. And MDI commands cannot be saved. As soon as you press the cycle start button to execute MDI commands, they will be lost.

The Edit Mode Switch Position

With the *edit* mode switch position, you can enter new programs (which can be time-consuming) or modify programs that are currently in the machine's memory. Aptly named, editing functions include *insert*, *alter*, and *delete*. *Insert* allows you to enter new words and commands into a program. *Alter* allows you to modify words in the program. And *delete* allows you to delete a word, a command, a series of commands, or an entire program.

Some machines have a feature called *memory protect* that can be used to keep the operator from making changes in a program. A special *key* (like the key to the door of your home) is used to turn this function on and off. If the memory protect function is turned on and the key is removed from the machine, you will not be able to modify programs.

In order to modify programs, you must understand programming words and commands. If you have read the programming lessons in this text, you should have a good understanding of programming. Without this understanding, an operator cannot make safe and correct changes to a CNC program (which is the reason many companies use the memory protect function – they have CNC operators that are unfamiliar with programming).

The step-by-step procedures to use program editing functions are shown in lesson 9.1. But we want to give a few examples here.

All program editing procedures begin with:

   1) Place the machine panel mode switch to EDIT.

   2) Press the MDI panel PROG key (until the active program is shown). Be sure the active program is the one you want to edit.

   3) Turn off the memory protect function (if the machine has this function). This requires a key.

You will notice a cursor somewhere on the page (a highlighted or back-lit word in the program). Before performing any editing function you must to move the cursor to the desired location. One way to do so is with the cursor control keys (described during the control panel discussions in lesson 7.2). Cursor control keys consist of a series of arrow keys that allow you to move the cursor – one word, one command, or one page at a time and in either direction.

Another way to move the cursor is to use the *SRH down and SRH up* functions. You can type the CNC word to which you want the cursor positioned and then press the soft key under SRH down or SRH up (depending upon the cursor's current position). The machine will bring the cursor to the next occurrence of the word you typed. Again, procedures for editing programs, including cursor movement, are shown in lesson 9.1.

Once the cursor is at the desired position, you will be able to INSERT a new word or command after the current cursor position, ALTER the word on which the cursor is placed, or DELETE words or commands starting from the cursor's current position.

Say, for example, the cursor is currently at the beginning of the program (on the O-word). You want to change the first feedrate word in the program to F5.0. If you can see the F-word on the current page of the display screen, the easiest way to position the cursor may be to use the arrow keys. Or you could type F and press the SRH down soft key. The control will scan to the next occurrence of an F-word – which is the F-word you want to change. With the cursor on the F-word to be changed, type F5.0 and press the ALTER key. The feedrate word will be changed to F5.0.

To Make a Program in Memory the Active Program (to call up a program)

We have been referring to the *active program*. FANUC controls can hold several programs, but only one of them will be the active program. A program must be the active program in order for you to modify it or run it. To make a program the active program, first follow the three steps above. Then:

   4) Type the letter address O (letter O, *not* number zero) and the program number for the program in memory that you want to make active.

5) Press the down arrow key or the soft key under SRH down. As long as the program you've typed is in the machine's memory, it will be shown on the display screen. It is now the active program. If the program number you've typed is not in the machine's memory, an alarm will be sounded.

To Enter a New Program

Entering programs using the display screen and keyboard is time-consuming. Hopefully your company uses memory cards or has a distributive numerical control (DNC) system to eliminate this time consuming task.

First, follow the three steps given above. Then:

4) Press the letter address O key (letter O, *not* the number zero) and type program number to be entered.
5) Press the MDI panel INSERT key.
6) Press the EOB key and press the MDI panel INSERT key.
7) Type the first command of your program followed by the EOB key and then press the MDI panel INSERT key.
8) Enter the rest of the commands in the program, ending each by pressing the EOB key and then the MDI panel INSERT key.

What if I make a mistake when typing?
Use the MDI panel CAN (standing for cancel) key. When you press it once for each backspace, the cursor on the command entry line will back up one space, deleting the last character you typed.

Remember that a CNC machine makes a very expensive typewriter. Many machines cannot be running a workpiece at the same time a program is entered (unless you use the BACKROUND EDIT function). Even with background edit, it is somewhat cumbersome to enter programs while the machine is running production. Also, the control panel for most machines is not mounted in a comfortable position (most are mounted in a vertical attitude). Your arm will soon get tired when typing in this position.

The Program Operation Mode

This mode of operation involves actually *running programs*. Current FANUC controls may have two mode switch positions for this purpose. MEM (or AUTO) will cause the machine to run a program from internal memory. REM (or REMOTE) will cause the machine to run a program from an external device, like a memory card.

When in a program operation mode, the CYCLE START button is used to activate the program and the FEED HOLD button can be used to stop axis motion at any time during the cycle. (Again, keep a finger ready to press the feed hold button whenever you press the cycle start button.) When the CYCLE START button is pressed, the control will begin executing the program from the cursor's current position in the program, so be sure you have it in the right place.

Several conditional switches (discussed in lessons 7.1 and 7.2) determine how the machine will behave in the program operation mode. The DRY RUN switch gives the operator control of the motion rate. SINGLE BLOCK forces the control to execute only one command at a time. OPTIONAL STOP (when on) will cause the control to stop the program's execution when an M01 word is executed in the program. BLOCK DELETE (when on) will cause the control to skip words to the right of the slash code (/). (Again, see lesson 7.2 for more information about these conditional switches.)

To Run the Active Program from the Beginning

This procedure will only make sense if you understand the presentations made in lessons 7.1 and 10.1. You must thoroughly understand these presentations in order to safely run a CNC program. If you don't, it is likely that you will incorrectly set one of the switches listed in this procedure – and the results will be disastrous.

1) Place the mode switch to the program operation mode (AUTO)
2) Press the MDI panel PROG key until the active program is displayed. Be sure the active program is the one you want to run and that the cursor is on the program number.
3) Check the position of all conditional switches (DRY RUN, SINGLE BLOCK, OPTIONAL STOP, etc.).
4) Place the FEEDRATE OVERRIDE switch to 100% (assuming the program is proven).
5) Place the RAPID OVERRIDE switch to the desired position (100% if running production).

Key Concept 8: Know The Operation Modes

6) Be sure the machine is in an appropriate position for the program to begin (the tool change position).
7) With a finger ready to press the FEED HOLD button, press the CYCLE START button. The program will run, behaving in accordance with how the conditional switches are set.

Key Points for Lesson 8.1:

- Though there are more than four positions on the mode switch, there are really only four basic modes of operation – manual, manual data input, edit mode, and program operation. All buttons and switches can be placed into one of these three categories.
- Manual mode is used to perform an immediate action.
- Manual data input mode is used to use the display screen and keyboard to enter and modify commands.
- Edit mode is used to modify programs.
- Program operation mode is used to run programs.

Key Concept 9: The Importance of Procedures

Running a CNC machining center requires little more than following a series of step-by-step procedures. The trick lies in knowing when a given procedure is required. From the material presented to this point, you should now know when each procedure is needed.

Key Concept 9 contains one lesson:

9.1: The Key Operation Procedures

From what has been presented to this point, you should be pretty comfortable with what you must do to make setups and run production – at least in theory. But when you step up to a CNC machining center for the first time, you'll probably be quite intimidated. Things you thought you knew won't seem so clear.

This will happen because you are still lacking hands-on experience – experience that we cannot provide in this text. Though we can't provide hands-on training, we can provide you with a description of the *procedures* that you are going to need in order to run your machining center.

Procedures will help you get familiar with your machining center. For example, if you need to power-up the machine at the beginning of your shift and make it ready to run production, what will you do?

Without a machine start-up procedure – or at least someone to demonstrate how this procedure is performed – you'll be lost. And trying to operate a CNC machining center without being sure of what you are doing can be very dangerous.

Even with a person available to help, there can be problems. This person may not be available every time you need them. And they will soon tire of repeating demonstrations that they feel you should have memorized. While you will *eventually* memorize often-used procedures, it may take you longer than it took the person helping you. And this can be frustrating – for both of you.

When you are shown a procedure for the first time, **write it down**. This will keep you from having to keep asking someone for help every time you need to perform the procedure. In lesson 9.1, we provide an example of an operation handbook that shows most of the procedures you will need in order to run a CNC machining center. Though you will surely find it quite helpful, it is just intended to get you started. You will surely come across machine functions for which we have not provided procedures. You'll have to develop the procedure for yourself, and for others that come after you. We provide a blank procedure form to help get you started.

Lesson 9.1–The Key Operation Procedures

Step-by-step procedures can keep you from having to memorize every function that you must perform on your CNC machining center. You will soon memorize procedures for tasks that you perform on a regular basis – but written procedures will still help you perform lesser used tasks.

Objectives

After completing this lesson, students should be able to:

- ✓ Describe the five categories of operation procedures
- ✓ Follow procedures in each of the categories
- ✓ Develop procedures for tasks you must perform that have not been previously documented
- ✓ Develop an operation handbook for the machine/s you must run

Introduction

We divide the procedures needed for CNC machining center usage into five categories:

- Manual procedures
- Setup procedures
- Manual data input (MDI) procedures
- Program editing procedures
- Program operation procedures

Develop Your Own Operation Handbooks

On the pages that follow, we provide an example operation handbook to help you with the most commonly used procedures. This will help with FANUC's current control models. The machine used for our example is the FANUC Certified Education CNC Training Mill Certification Cart, made by Levil (assuming it is equipped with a FANUC 0*i*F CNC.

We recommend that you develop operation handbooks for the machine/s you will be expected to run. Use our example as a template.

Here is a list of procedures in each category:

Manual Procedures:
- To start the machine
- To do a zero return
- To index the turret
- To start the spindle
- To jog axes (using continuous jog)
- To jog axes (using incremental jog)
- To use the hand-wheel
- To set axis displays
- To enter tool offsets
- To enter workpiece coordinate system offsets

MDI Procedures
- To execute an MDI command

Key Concept 9: The Importance Of Procedures

Program Manipulation Procedures
- To get ready to edit programs
- To show a directory of programs
- To call up a program from within the CNC memory (make it the active program)
- To load programs
- To delete programs
- To search within a program
- To alter, insert, & delete
- To save programs
- To use background edit

Setup Procedures
- To mount work holding devices on the table
- To measure workpiece coordinate system zero assignment values in the X and Y axes
- To measure workpiece coordinate system zero assignment values in the Z-axis
- To measure and enter tool length compensation values

Program Running Procedures
- To run the program (in normal production - no verification techniques)
- To rerun tools
- To do a free flowing dry run
- To do an normal air cutting run
- To run the first workpiece
- To cancel a cycle
- To clear an alarm

The goal with any step-by-step procedure is to keep you from having to memorize. But you must, of course, understand *when* and *why* to perform each procedure. If you have read this text, you should be pretty comfortable with the when and why.

For example, say you're beginning a new setup and you've just mounted a work holding device, like a vise, on the table. Now you must measure workpiece coordinate system zero assignment values. According to the general procedure in lesson 1.5, if you don't have a spindle probe, you must use an edge-finder and measure the distance between the zero return position and the X and Y workpiece coordinate system zero surfaces on the workpiece. This means you must start the spindle and manually move the machines axes.

As you're standing in front of the machine, scratching your head, you remember that we've provided procedures to manually start the spindle and jogging the axes. There is also a procedure to use the hand-wheel. With these rudimentary step-by-step procedures, even a newcomer can perform the needed tasks. But again, you must know when and why to perform them.

Blank Procedure Form

If you come across a task for which we have not supplied a step-by-step procedure, get someone to help you. As they demonstrate what it takes to perform the task, write down what they show you in step-by-step fashion. You'll have a procedure for the *next time* you must perform the task – and you won't have to ask for help. Figure 9.1 shows a form you can use to create your own procedures.

Procedure name: _____ Machine: _____
Needed when: _____
Procedure:
1) _____ 9) _____
2) _____ 10) _____
3) _____ 11) _____
4) _____ 12) _____
5) _____ 13) _____
6) _____ 14) _____
7) _____ 15) _____
8) _____ 16) _____
Notes: _____

Figure 9.1 – Procedure form

The next few pages show an example operation handbook, and should provide a good template for developing operation handbooks the machines you will be expected to run.

Key Concept 9: The Importance Of Procedures

Sample Operation Handbook: Levil Certification Cart (mill) with 0*i*F

Manual Procedures

To Power-Up the Machine

Needed: Whenever the machine is used

1. Ensure that the power cord is connected to power outlet
2. Turn on the main breaker (labeled MAIN SWITCH)
3. Wait for the display screen to show position page

To Do a Reference Return (send each axis to its reference position)

Needed: To move headstock to a clearance position - with other machines, the Z-axis zero return position may be the tool change position, to measure workpiece coordinate system zero assignment values

Note: Pushing plus *or minus* axis buttons causes plus axis motion

1. Place the mode switch to REF
2. Set motion rate switch, if desired
3. Hold Z+ pushbutton until Z-axis stops
4. Hold X+ pushbutton until X-axis stops
5. Hole Y+ pushbutton until Y-axis stops

To Start the Spindle

Needed: To measure workpiece coordinate system zero assignment values in XY with a "wiggler" type edge-finder

1. Place the mode switch to JOG or HAND-WHEEL
2. Adjust spindle override switch to select/change speed
3. Press the SPINDLE FWD button
4. To stop the spindle, press the SPINDLE STOP button

To jog axes (using continuous jog)

Needed: Often, measuring workpiece coordinate system zero assignment values, measuring tool lengths

1. Place the mode switch to JOG
2. Place motion rate switch in the desired position
3. Press and hold the desired axis pushbutton (X+, X-, Y+, Y-, Z+, or Z-)
4. Release the pushbutton to stop the axis motion

To Jog Axes (using incremental jog)

Needed: Not preferred, the hand-wheel is usually better

Note: Each time you press an axis button (X+, X-, Y+, Y-, Z+, or Z-), the axis will move a selected distance

1. Place the mode switch to INC (for incremental jog)
2. Repeatedly press the MDI keyboard X1, X10, X100 button to select the desired motion distance (slow blinking is 0.001mm or 0.0001in, medium blinking is 0.01mm or 0.001in, fast blinking is 0.1mm or 0.01in)
3. Press and release the desired axis pushbutton (X+, X-, Y+, Y-, Z+, or Z-) and the axis will move the selected amount

To Use the Hand-Wheel

Needed: Often, measuring workpiece coordinate system zero assignment values, measuring tool lengths

Note: Clockwise is plus axis motion, counter clockwise is minus

1. Place the mode switch to HANDLE
2. Place axis selector switch near the hand-wheel in the desired position (X, Y, or Z)
3. Place rate switch near the hand-wheel in the desired position (X1, X10, X100)
4. Turn the hand-wheel in the desired direction

To Set Axis Displays

Needed: When taking measurements for workpiece coordinate system zero assignment and tool length measurements

Note: Machine panel mode switch can be in any position

Origin

1. Using JOG and/or HAND-WHEEL procedures, move the axes to the point of reference for the measurement
2. Press POSITION display key until the RELATIVE page is shown
3. Type the desired axis address (X, Y, or Z)
4. Press the soft key under ORIGIN
    a. If necessary, press the soft key under EXEC (for execute) to confirm

Preset

1. Using JOG and/or HAND-WHEEL procedures, move the axes to the point of reference for the measurement
2. Press POS display key until the RELATIVE page is shown
3. Type the desired axis address (X, Y, or Z) and the value to preset
4. Press the soft key under PRESET

To Enter Tool Offsets (GEOM(H), WEAR(H), GEOM(D), WEAR(D))

+INPUT *(Modifies current offset value)*
Needed: When measuring cutting tool lengths and diameters, when making sizing adjustments during setup and production run

Note: Machine panel mode switch can be in any position

1. Press OFS / SET display key, and if the tool offset page is not displayed, press OFFSET soft key
2. Using the cursor page and arrow keys, position the cursor to the desired register
3. Type the deviation amount and press the soft key under +INPUT
4. If asked for confirmation, press the soft key under execute

INPUT *(Overwrites current offset value)*
Needed: When trial machining or when you know the resulting offset value

Note: Machine panel mode switch can be in any position

1. Press OFS / SET display key, and if the tool offset page is not displayed, press OFFSET soft key
2. Using the cursor page and arrow keys, position the cursor to the desired register
3. Type the offset value and press the soft key under INPUT

INP. C. *(Input coordinate: transfers relative axis display coordinate to offset register)*
Needed: When measuring tool length compensation values

Note: Machine panel mode switch can be in any position

1. Using the jog, hand-wheel, and origin/preset functions, make the relative page Z-axis display show the tool length
2. Press OFS / SET display key, and if the tool offset page is not displayed, press OFFSET soft key
3. Using the cursor page and arrow keys, position the cursor to the desired register, a GEOM (H) register
4. Type Z and press the soft key under INP. C.

To Enter Workpiece Coordinate System Offsets

INPUT *(Overwrites current offset value)*
Needed: When measuring workpiece coordinate system zero assignment values

Note: Machine panel mode switch can be in any position

1. Press OFS / SET display key, and if the work page is not displayed, press WORK soft key
2. Using the cursor page and arrow keys, position the cursor to the desired register
3. Type the offset value and press the soft key under INPUT

MDI Procedures

To Execute an MDI Command

Needed: When there are no manual buttons & switches, when action can be done faster than with manual procedures

Note: You must know the related CNC command

1. Place the mode switch to MDI
2. Press the PROG display key until MDI page is displayed
3. Type the CNC command, ending with the MDI keyboard EOB key (which inserts a semicolon)
4. Press the INSERT key
5. Be ready to press the FEED HOLD button and press the CYCLE START button

Program Manipulation Procedures

To Get Ready to Edit Programs

Needed: Often, whenever programs must be modified or run

Note: All program modifying procedures begin with:

1. Place the mode switch to EDIT
2. Press the PROG display key until the active program is shown

To Show a Directory of Programs

Needed: To find programs

1. Place the mode switch to EDIT
2. Press the PROG display key until the directory page is shown
3. To show programs stored on other devices (like the memory card), press the OPRT soft key
4. If necessary, press the right soft key until the DEVICE CHANGE soft key is shown, press it
5. Press the soft key under the desired device (like MEMORY CARD)
6. To get back to displaying programs in the CNC memory, press the DEVICE CHANGE soft key, then press the soft key under CNCMEM

To Call Up a Program from Within the CNC Memory (make it the active/main program)

Needed: To begin editing, to run workpieces

Method 1: *(from the program page, assumes you know the program number)*
1. Place the mode switch to EDIT
2. Press the PROG display key until the current active program is shown
3. Type O and the program number for the program to be made active
4. Press the soft key under O SRH, SRH down, SRH up, or down arrow key

Method 2: *(from the directory/folder page)*
1. Place the mode switch to EDIT
2. Press the PROG display key until the PROGRAM FOLDER page is shown
    a. Confirm that the DEVICE is set to CNC_MEM
3. Using the cursor page and arrow buttons, position the cursor to the program you want to make the active/main program
4. Press the soft key under (OPRT) (for operation)
5. Press the soft key under MAIN PROGRAM
    a. To see the program, press the PROG display key

To Load a Program

Needed: During setup

Note: You must first set the device from which programs will be loaded and know the file name of the file containing the CNC program (you can use directory procedure to find the file name)

Note: Once a program is loaded, it will automatically become the active/main program

1. Place the mode switch to EDIT, press PROG key until the PROGRAM FOLDER page is shown
2. Select the device from which you want to load a program
    a. Press the (OPRT) soft key until the DEVICE CHANGE soft key is shown, press it
    b. Press the soft key under the desired device (like MEMORY CARD)

3. Using the cursor page and arrow buttons, position the cursor to the program you want to load
4. Press the F INPUT soft key
5. Press the F GET soft key
6. Press the F SET soft key
7. Press the EXEC soft key
    a. To see the program just loaded, press the PROG display key

To Delete a Program

Needed: When you need to make room for more programs

Note: Be careful, you can delete any program in CNC memory (not just the active program)

1. Place the mode switch to EDIT, press PROGRAM key until active program is shown
2. Type O and the number for the program to delete
3. Press the soft DELETE key
4. If asked to confirm, press the soft key under EXEC

To Search Within a Program

Needed: When you need to modify a program, or to rerun a tool

Note: Assumes the EDIT mode switch position is selected and the active program is being displayed

Search with Cursor Keys
1. Press cursor page, left, right, up, and down keys to move cursor to the desired CNC word

Find the Next Word Occurrence Using Only the Letter Address
1. Type the letter address (S, F, T, etc.)
2. Press the soft key under SRH down (or SRH up)

Find the next word occurrence using the whole word
1. Type the word to be searched (N015, T0303, etc.)
2. Press the soft key under SRH down (or SRH up)

To Alter, Insert, & Delete

Needed: When corrections must be made to a program

Alter
1. Place the mode switch to EDIT, press PROG key until program is shown
2. Place the cursor on the word to be altered
3. Type the correct word
4. Press the MDI keyboard ALTER key

Insert
1. Place the mode switch to EDIT, press PROG key until program is shown
2. Place the cursor on the word before word/s to be inserted (if inserting a complete command, place the cursor on the EOB character (;) of the command prior to the one you want to insert)
3. Type the word/s to be inserted
4. Press the MDI keyboard INSERT key

Delete
Delete a word

1. Place the mode switch to EDIT, press PROG key until program is shown
2. Place the cursor on the word to be deleted
3. Press the MDI keyboard DELETE key

Delete from one word through another word

1. Place the mode switch to EDIT, press PROG key until program is shown
2. Place the cursor on the first word to be deleted
3. Type the last word to be deleted (must be the first occurrence of the word)
4. Press the MDI keyboard DELETE key

Delete a command
1. Place the mode switch to EDIT, press PROG key until program is shown
2. Place the cursor on the first word of the command to be deleted
3. Press the MDI keyboard EOB key
4. Press the MDI keyboard DELETE key

Delete a series of commands
1. Place the mode switch to EDIT, press PROG key until program is shown
2. Place the cursor on the sequence number of the first command to be deleted
3. Type the sequence number for the last command to be deleted
4. Press the MDI keyboard DELETE key

To Save Programs to an External Device

Needed: After changes have been made to programs in memory

Note: Programs will be saved to the currently selected device (like the memory card)

1. Place the mode switch to EDIT, press PROG key until program is shown
2. Press the OPRT soft key, then press right soft key until F OUTPUT is shown
3. Press the soft key under F OUTPUT
4. Type the file name for the program to be saved, press soft key under F SET
5. Press the soft key under EXEC

To Use Background Edit

Needed: To manipulate programs while the machine is in cycle

Note: Be careful, programs displayed in background edit look much the same as the active program

WARNING: DO NOT use the RESET key to rewind the program - use the soft key under REWIND

1. Mode switch can be in EDIT, AUTO, or MDI, press PROG key
2. Press the OPRT soft key, then press soft key under BG-EDIT
3. Type O and the program to manipulate
4. Press the soft key under EDT EXE (functions of edit mode are now available)
5. To exit background edit, press the OPRT soft key, then press BG END

Setup Procedures

These procedures are more complex, requiring that you understand many of the rudimentary procedures shown to this point.

To Mount the Work Holding Device on the Table

Needed: During setup

Note: Assumes the work holding device from the previous job have already been removed

1. Move the table forward so you have room to work
2. If necessary, assemble tee nuts to bolts
3. Place the vise on the table so the table tee slots line up with the tee slots in the vise
4. Slide the tee nut/bolt assemblies through the table tee slots and engage the vise tee slots
5. Tighten the bolts to secure the vise to the table

To Measure the Workpiece Coordinate System Zero Assignment Values in the X and Y Axes

Needed: During setup

Note: This procedure is better illustrated during lesson 1.5, we provide a general procedure here

Note: You must first make the work holding setup and (usually) load a piece of raw material as it will be loaded during the production run, you must also load an edge-finder or dial indicator (depending upon the workpiece to be run) into the spindle

1. Using the jog and hand-wheel functions, align the spindle center with the X and Y workpiece coordinate system zero surfaces (again, the method to do this varies based upon workpiece configuration and is better shown in lesson 1.5)
2. Using the procedure to set axis displays, set (using ORIGIN) the X and Y registers of the RELATIVE position display page to zero

3. Using the zero return procedure, send the X and Y axes to the zero return position
4. The RELATIVE position display page X and Y registers will show the magnitude of the workpiece coordinate system zero assignment values (the values entered into workpiece coordinate system offsets must be negative)

To Measure the Workpiece Coordinate System Zero Assignment Value in the Z-Axis

Needed: During setup

Note: This procedure is better illustrated during lesson 1.5, we provide a general procedure here

Note: You must first make the work holding setup and (usually) load a piece of raw material as it will be loaded during the production run

Note: The spindle must be empty when you begin this procedure

1. Place a gauge block of known length on the Z-axis workpiece coordinate system zero surface (usually the top of the workpiece)
2. Using the jog and hand-wheel functions, bring the spindle nose to the top of the gauge block
3. Using the procedure to set axis displays, set (using PRESET) the Z-axis register of the RELATIVE position display page to the length of the gauge block (this is the current distance between the Z-axis workpiece coordinate system zero surface and the spindle nose)
4. Using the zero return procedure, send the Z-axis to its zero return position
5. The RELATIVE position display page Z register will show the magnitude of the workpiece coordinate system zero assignment value (the value entered into workpiece coordinate system offsets must be negative)

To Measure and Enter Tool Length Compensation Values

Needed: During setup

Note: This procedure is better illustrated during lesson 4.2, we provide a general procedure here

Note: This procedure renders the tool length (our recommended method), which is the distance from the tool tip to the spindle nose

Note: The spindle must be empty when you begin this procedure

1. Place a gauge block on a flat surface (like the top of the workpiece)
2. Using the jog and hand-wheel functions, bring the spindle nose to the top of the gauge block
3. Using the procedure to set axis displays, set (using ORIGIN) the Z-axis register of the RELATIVE position display page to zero (this sets a point of reference for tool length measurements)
4. Using the jog function, retract the Z-axis and load the first tool to measure into the spindle
5. Using the jog and hand-wheel functions, bring the tip of the cutting tool to the top of the gauge block
6. The RELATIVE position page Z register currently shows the tool length

To Transfer the RELATIVE Page Z Register Value to the Offset Register
7. Using the procedure to enter offsets, call up the tool offset page
8. Position the cursor to the offset register to be set (the GEOM(H) register for the offset)
9. Type Z and press the soft key under INP. C. (for input coordinate), the offset register will be set to the Z-axis RELATIVE display screen value.
10. Repeat steps 4-10 for other cutting tools to be measured

Program Running Procedures

To Run the Program (in normal production - no verification techniques)

Needed: Once program has been verified, after power-up to continue running production

Note: Assumes the program has been verified

1. Load a workpiece in the work holding device
2. Send machine to the tool change position (the Z-axis zero return position)
3. Place the machine panel mode switch to the EDIT position
4. Press the MDI panel PROG display key
5. Press the MDI panel RESET key
6. Turn off other conditional switches (SINGLE BLOCK, DRY RUN, etc.)

Key Concept 9: The Importance Of Procedures

7. Place the motion rate switch to 100%
8. Place the RAPID OVERRIDE switch to desired position (like 100%)
9. Place the machine panel mode switch to the AUTO position
10. While ready to press the FEED HOLD button, press the CYCLE START button

To Rerun Tools

Needed: When mistakes are found during program verification, trial machining, a tool does not machine correctly

Note: Assumes the program (or at least the tool to be rerun) has been verified

1. A partially machined workpiece is in the work holding device
2. Send machine to the tool change position (the Z-axis zero return position)
3. Place the machine panel mode switch to the EDIT position
4. Press the MDI panel PROG display key
5. Place the cursor on the first command of the tool to be rerun
6. Turn on the OPTIONAL STOP switch
7. Turn off other conditional switches (SINGLE BLOCK, DRY RUN, etc.)
8. Place the motion rate switch to 100%
9. Place the RAPID OVERRIDE switch to desired position (like 100%)
10. Place the machine panel mode switch to the AUTO position
11. If the CHECK display screen page is not shown, press the PROG key until it is
12. While ready to press the FEED HOLD button, press the CYCLE START button

To Do a Free Flowing Dry Run

Needed: During program verification

1. If necessary, remove workpiece from work holding device
2. Send machine to the tool change position (the Z-axis zero return position)
3. Place the machine panel mode switch to the EDIT position
4. Press the MDI panel PROG display key
5. Press the MDI panel RESET key (cursor must be at program beginning)
6. Turn on DRY RUN switch
7. Turn off other conditional switches (SINGLE BLOCK, OPTIONAL STOP, etc.)
8. Place the motion rate switch to its lowest position (usually feedrate override)
9. Place the machine panel mode switch to the AUTO position
10. If the CHECK display screen page is not shown, press the PROG key until it is
11. While ready to press the FEED HOLD button, press the CYCLE START button
12. Turn up the motion rate switch to attain a comfortable motion rate
13. Press the FEED HOLD button at the first sign of trouble

To Do a Normal Air Cutting Run

Needed: During program verification

1. If necessary, remove workpiece from work holding device
2. Send machine to the tool change position (the Z-axis zero return position)
3. Place the machine panel mode switch to the EDIT position
4. Press the MDI panel PROG display key
5. Press the MDI panel RESET key (cursor must be at program beginning)
6. Turn off DRY RUN switch
7. Turn off other conditional switches (SINGLE BLOCK, OPTIONAL STOP, etc.)
8. Place the motion rate switch to 100%
9. Place the RAPID OVERRIDE switch to desired position (like 100%)
10. Place the machine panel mode switch to the AUTO position
11. If the CHECK display screen page is not shown, press the PROG key until it is
12. While ready to press the FEED HOLD button, press the CYCLE START button

To Run the First Workpiece

Needed: During program verification

1. Load raw material into the work holding device
2. Send machine to the tool change position (the Z-axis zero return position)

3. Confirm that the cursor is at program beginning (EDIT, PROG, RESET)
4. Turn on the SINGLE BLOCK, DRY RUN and OPTIONAL STOP switches
5. Turn off other conditional switches
6. Place the motion rate switch to its lowest position (usually feedrate override)
7. Place the machine panel mode switch to the AUTO position
8. If the CHECK display screen page is not shown, press the PROG key until it is
9. Keep a finger ready to press the FEED HOLD button
10. Repeatedly press the CYCLE START button until the axes begin moving
11. Turn up the motion rate switch to attain a comfortable motion rate
12. As the tool gets close to the workpiece (within 25.0-mm [or 1.0-in]), press the FEED HOLD button and check the DISTANCE TO GO display
13. If remaining distance looks good, press cycle start and carefully finish the approach movement
14. Turn off the DRY RUN switch
15. Repeatedly press the CYCLE START button to step through the tool's cutting motions
16. The machine will stop at the M01 command, so check what the tool has done and make offset adjustment/s as necessary, if necessary, rerun the tool
17. To continue to next tool, turn on the DRY RUN switch and repeat steps 9-16

To Cancel a Cycle

Needed: When aborting the program's execution, possibly you found a mistake during the program's execution

1. Press the MDI panel FEED HOLD key to stop axis motion
2. Press the MDI panel RESET key (this will stop spindle and coolant)
3. Using the JOG and/or HAND-WHEEL functions carefully move the axes to a safe position (commonly the zero return position)

To Clear an Alarm

Needed: When the machine goes into alarm state, commonly when verifying a program

1. Diagnose and correct the cause of the alarm (alarm message may be sufficient, but you may have to look it up)
2. Some alarms require that you correct the problem before you can clear the alarm
3. Press the MDI panel RESET key

Key Points for Lesson 9.1:

- Running a CNC machining center is little more than following a series of procedures. The trick is knowing when and why to perform a given procedure.
- Writing down procedures will keep you from having to ask someone for help every time you need to perform the procedure.
- Write down a procedure for every task you must perform on your machining center.

Key Concept 10: Know How to Safely Verify Programs

You cannot begin a production run until the CNC program is verified and a workpiece passes inspection. Safely verifying programs will be the focus of Key Concept Number Ten.

Key Concept 10 contains one lesson:

    10.1: Program Verification

You know that program verification is part of setup. Indeed, we provide some pretty good explanations in lesson 7.1 about how programs are verified. But since program verification is the most dangerous part of running a CNC machine, we want to spend more time discussing it. Lesson 10.1 will show you how to safely verify programs, even if serious mistakes have been made while making the setup and writing the program.

Safety Priorities

When verifying new programs, remember CNC machines will follow programmed instructions *precisely as they are given*, even if there are mistakes in the program. In lesson 2.1, we present some of the most common programming mistakes. With the exception of basic syntax (program formatting) mistakes, the machine will rarely alert you when a mistake has been made. While verifying any new program, you must be ready for just about anything. If you make a mistake in the program which tells the machine to rapid a tool into the workpiece, the machine will follow your commands and do so, causing what is commonly referred to as a *crash*.

And even with a proven program, remember that you can make serious mistakes during setup. Mistakes with tool loading, tool length measurement, workpiece coordinate system zero assignment value measurement, and offset entry can be every bit as serious as programming mistakes. So you must also be very careful when running programs that you have run before.

We cannot overstress the need for using safe procedures and staying alert when working with CNC equipment. While we are not trying to scare you, we do want to instill in you a very high level of respect for your very powerful and potentially dangerous machine tool.

There are three levels of priority that you must adhere to when you work with any machine tool, including CNC machining centers.

Operator Safety

The first priority must be your safety. You must use every opportunity to ensure your safety and the safety of the people around you. The verification procedures we provide stress operator safety as the number one priority. As time goes on and you start gaining experience, your tendency will be to *short cut* these procedures in order to save some time. We urge you to avoid this tendency. When you begin relaxing your guard, you open the door to very dangerous situations.

Compare this to a person that enjoys snow skiing. The first few times a person goes skiing, they tend to be very careful, and rarely does a new skier out for the first time get seriously injured. It is only after a skier gains confidence that they become bold and careless. More experienced skiers sustain serious injuries than do beginners.

In similar fashion, most entry level setup people and operators tend to be very careful when running a CNC machine. It is only after they gain some experience that some people become bolder and less careful. This tendency is inspired by the need to work faster. Again, don't compromise safety in order to go faster. As you gain experience, you'll naturally gain the ability to perform more efficiently. You don't have to shortcut safety-related procedures to do so.

Machine Tool Safety

The second safety priority is the CNC machining center itself. Every operator must do their best to ensure that no damage to the machine can occur. Obviously, CNC machine time is very expensive. When a CNC machine goes down for any reason, the actual cost of repairing the machine is usually very small compared to the lost production time.

There is no excuse for machine downtime caused by operation mistakes. If the verification procedures we give are followed, you can ensure that the machine will not be placed in dangerous situations. While no method is completely failsafe, our recommendations will truly minimize the potential for machine damage.

Workpiece Safety

The third safety priority is making all of your workpieces within specifications. The effort that companies put forth to ensure zero scrap varies from one company to another. One company may be machining extremely expensive material, like titanium or stainless steel. The raw material for large workpieces can be expensive – especially for workpieces that require machining operations *prior to* the CNC machining center operation. For this company, *anything* they can do to minimize the potential for even one scrap workpiece will be done.

In another company, the raw material cost may be very low. Consider machining a small 20.0-mm (or 0.75-in) long, 12.0-mm (or 0.5-in) diameter piece of steel bar. The total cost of raw material may be less than 0.50 cents. In this company, the setup person's time may be more valuable than the time it takes to run one (the first) workpiece. This company may be less concerned with attaining zero scrap. Some companies in this situation even supply the setup person with extra pieces (commonly called *practice parts*). They don't expect every workpiece machined to be a good one.

While things can happen during a production run that will cause scrap workpieces, the most critical time is during the machining of the very *first* workpiece. Knowing this, and by knowing which workpiece attributes are the most critical, a CNC setup person can minimize the potential for scrap workpieces by using *trial machining techniques* (as described in lessons 4.1, 4.2, and 7.1). If you consider what each tool in the program will be doing when you come to it – and if you use trial machining techniques when appropriate – each tool will machine the workpiece properly. When you're finished with the last tool in the program, you'll have a good workpiece that will pass inspection.

There are people in our industry that feel that trial machining is wasteful. People that feel this way tend to come from companies that machine very *inexpensive* raw material. They do not care about scrapping the first few workpieces as long as they *eventually* learn enough about offset settings to get a workpiece to come out to size.

But regardless of whether your company believes in trial machining or not, *all* setup people should have the ability to machine the very first workpiece correctly for the times when it is necessary to do so. Even if workpiece material cost is very low, the day will come when you have five pieces of raw material and you must machine five good workpieces.

Lesson 10.1–Program Verification

The most dangerous time for a CNC setup person is when verifying programs. You must stay alert and be ready for just about anything. You must master the program verification procedures – they must truly become second nature.

Objectives

After completing this lesson, students should be able to:

- ✓ Cancel a CNC cycle and stop executing a program
- ✓ Rerun a tool
- ✓ Verify a job that contains mistakes

Introduction

In lesson 7.1, we show all of the tasks related to making setups and running production. We do so by using an example job to stress how each task is done – including program verification. In this lesson, we're going to do so again – but we'll only discuss tasks related to program verification. The setup and program we show in lesson 7.1 is perfect – there are no mistakes. By comparison, the setup and program we use *in this lesson* contain many mistakes. Most of the mistakes are typical of mistakes a beginner is likely to make.

While you may be able to spot some of the mistakes in our example job as soon as we show them, remember that when *you* make mistakes, you won't know it (if you did, you wouldn't make the mistake). Our objective is to show you how to catch even very serious mistakes as you verify CNC programs – and of course – to catch them before a crash can occur.

Two More Procedures

In lessons 7.1 and 9.1, we show several procedures that are related to verifying programs, including how to perform a dry run, how to cautiously run the first workpiece using a special approach procedure for each tool, how to trial machine, and how to run a verified program. These procedures are extremely important, and you must master them. But they don't show you how to *handle problems* when mistakes are found. Here are two procedures that are needed when you *do find mistakes*.

Canceling the CNC Cycle

You know that feed hold is your panic button. As you're running a program, you can press it any time you are worried. It causes all axis motion to stop. If you find that nothing is wrong, you can press the cycle start button to continue. *But what if something is wrong* – something that is so serious that you cannot allow the program to continue?

Say for example, during the running of the first workpiece, you are allowing a tool to approach the workpiece for the first time. You are using the approach procedure shown in lesson 7.1, so you have dry run and single block turned on, and you're controlling the tool's motion rate with feedrate override. As the tool gets within 25.0-mm (or 1.0-in) of the work surface, you stop the cycle. You're worried that the tool will not stop at a position 2.0-mm (or 0.1-in) above the work surface. So you look at the distance to go page. It says the Z-axis is still going to move another 32.0-mm (or 1.3-in). As you look at the tool's position relative to the work surface, you can easily tell that the tool cannot move another 32.0-mm (or 1.3-in) without contacting the workpiece (you've just saved a crash).

You cannot allow the program to continue. You must, of course, find and correct the problem. But before you go any further, you must *cancel the cycle*. Here is the procedure to do so:

    1) Press the reset key on the control panel. For most machines, this will stop the spindle and coolant (if they're on).

    2) Place the mode switch to edit (if the spindle and coolant don't stop when the reset button is pressed, they will now).

    3) Select the program display screen page. The cursor is somewhere in the middle of the program.

Key Concept 10: Know How To Safely Verify Programs

4) Press reset again (this sends the cursor back to the program number at the beginning of the program).

5) Send the machine to its tool change position (usually the zero return position). This can be done manually or by using an MDI command.

6) Find and correct the problem that caused you to have to cancel the cycle.

7) When the problem is corrected, follow the procedure to re-run the tool.

To Re-Run a Tool

This procedure assumes that you can identify the appropriate *restart command* for the tool you want to re-run. It also assumes the programmer has used the format shown in lesson 5.2. As we show in lesson 5.2, the restart command depends upon whether the tool you want to re-run is currently in the spindle. If the tool *is* in the spindle, the restart command will be the command *after* the tool change command that places the tool in the spindle. If the tool *is not* in the spindle, the restart command is the tool change command that places the tool in the spindle. This series of commands (shown in metric mode) should help to clarify which command is the restart block:

```
N135 G40 Y60.0 F750.0
N140 G00 Z2.0 M09
N145 G91 G28 Z0 M19
N150 M01

N155 T03 M06 (5/8 END MILL) <---- Restart command if tool three is
  not in the spindle
N160 G54 G90 S540 M03 T04 <----- Restart command if tool three is in
  the spindle
N165 G00 X75.0 Y25.0
N170 G43 H03 Z2.0 M08
   N175 G01 Z-6.0 F125.0
```

This procedure also varies based upon whether you have seen the tool run before. If you *have* seen the tool run properly (as is the case when you re-run a tool after trial machining), you need not be careful with its first approach movement and the balance of its cutting motions.

But if you *have not* seen the tool successfully complete its operation (as is the case in the scenario above), you must still be very careful with its first approach movement and all of its motions.

Here is the procedure to re-run a tool. The machine must be at its tool change position when this procedure is used (if you canceled the cycle with the procedure just shown, it will be):

1) Consider the condition SINGLE BLOCK (if you've seen the tool run before, turn it off, if not, turn it on).

2) Consider the condition of DRY RUN (if you've seen the tool run before, turn it off, if not, turn it on).

3) Consider the condition of FEEDRATE OVERRIDE (if you have seen the tool run before, set it to 100%. If not, set it at its lowest setting.

4) Consider the condition of OPTIONAL STOP (if you want the machine to stop when the tool is finished, as is normally the case when verifying programs, turn it on. If not, turn it off).

5) Place the machine panel mode switch to EDIT.

6) Press the PROG display screen key. The display will show the active program.

7) Press the RESET key. This places the cursor on the first word in the program (the program number).

8) Scan to the restart command. To do so, type the sequence number for the restart command and press the SRH down key.

9) Place the mode switch to AUTO (or MEM).

10) Press the CYCLE START button to activate the cycle. If you've seen the tool run before (SINGLE BLOCK is off), the tool will perform its machining operation. As long as you have the OPTIONAL

STOP switch turned on, the machine will stop when the tool is finished so you can check what the tool has done.

11) If you have never seen this tool run before (SINGLE BLOCK is on), you must press the CYCLE START button repeatedly until the machine begins to move. You'll use the FEEDRATE OVERRIDE switch to control motion rate.

12) Once the tool has successfully approached the workpiece, turn off DRY RUN. (*Never* let cutting tools machine a workpiece under the influence of dry run. Dry run tends to slow down rapid motions, but it *speeds up* cutting motions.)

13) The tool has safely approached. What you do next will depend upon whether you're running a new or a proven program. For a new program you must carefully step through the cutting tool's cutting motions. In this case, leave the SINGLE BLOCK switch on, and repeatedly press CYCLE START to step through the tool. For a proven program (are you *absolutely sure* that you're running the current version of the program?), turn off SINGLE BLOCK and press cycle start to let the tool machine the workpiece as it did the last time the program was run.

14) When the tool is finished, the machine will return to its tool change position.

Key Concept 10: Know How To Safely Verify Programs

Verifying a Job that Contains Mistakes

> Remember, there are mistakes contained in the setup and program we are about to show.

Figure 10.1 shows the print and process for our example job. It is the exercise from lesson 6.3. We have added two tight tolerances (the depth of the 12.0-mm (0.5-in) counter-bored holes and the depth of the two pockets). Figure 10.1 shows the print and process. Figures 10.2A and 10.2B show the programs. Figure 10.3 shows the setup sheet.

Program Verification Example Job

- 25.0-mm *(1.0-in)* high clamp (2)
- Drill 6.0 *(0.25)*
- Counter-bore 12.0 *(0.5)* — 4 holes
- Drill 5.5 *(15/64)*
- Ream 6.0 *(0.250)* — 2 holes
- 10.0-mm *(0.375)* R — Two slots
- 8.8 *(5/16)* drill
- M10-1.25 *(3/8-16)* tap — 2 holes

Tolerances: Unless otherwise noted, all tolerances are +/-0.125-mm *(+/-0.005-in)*

Values **not** in parentheses are for the metric mode discussion.
Values in parentheses are for the inch mode discussion.

Dimensions: 75.0 (3.0), 15.0 (0.625), 25.0 (1.0), 15.0 (0.625), 10.0 (0.375), 10.0 (0.375), 15.0 (0.625), 37.5 (1.5), 37.5 (1.5), 15.0 (0.625), 125.0 (5.0), 20.0 (0.75)

6.0 +/-0.02 *(0.25 +/-0.0008)*
10.0 +/-0.02 *(0.375 +/-0.0008)*

Metric-mode process:

Step	Tool	Speed	Feed
1) Mill (2) 10.0 radius slots	20.0 end mill	500 rpm	150.0 mmpm
2) Center drill all holes	#4 center drill	1200 rpm	100.0 mmpm
3) Drill (4) 6.0 holes	6.0 drill	1100 rpm	85.0 mmpm
4) Counter-bore (4) 12.0 holes	12.0 end mill	800 rpm	110.0 mmpm
5) Drill (2) 5.5 holes	5.5 drill	1,150 rpm	80.0 mmpm
6) Ream (2) 6.0 holes	6.0 reamer	800 rpm	110.0 mmpm
7) Drill (2) 8.8 holes	8.8 drill	700 rpm	125.0 mmpm
8) Tap (2) M10-1.25 holes	M10-1.25 tap	400 rpm	500.0 mmpm

Inch-mode process:

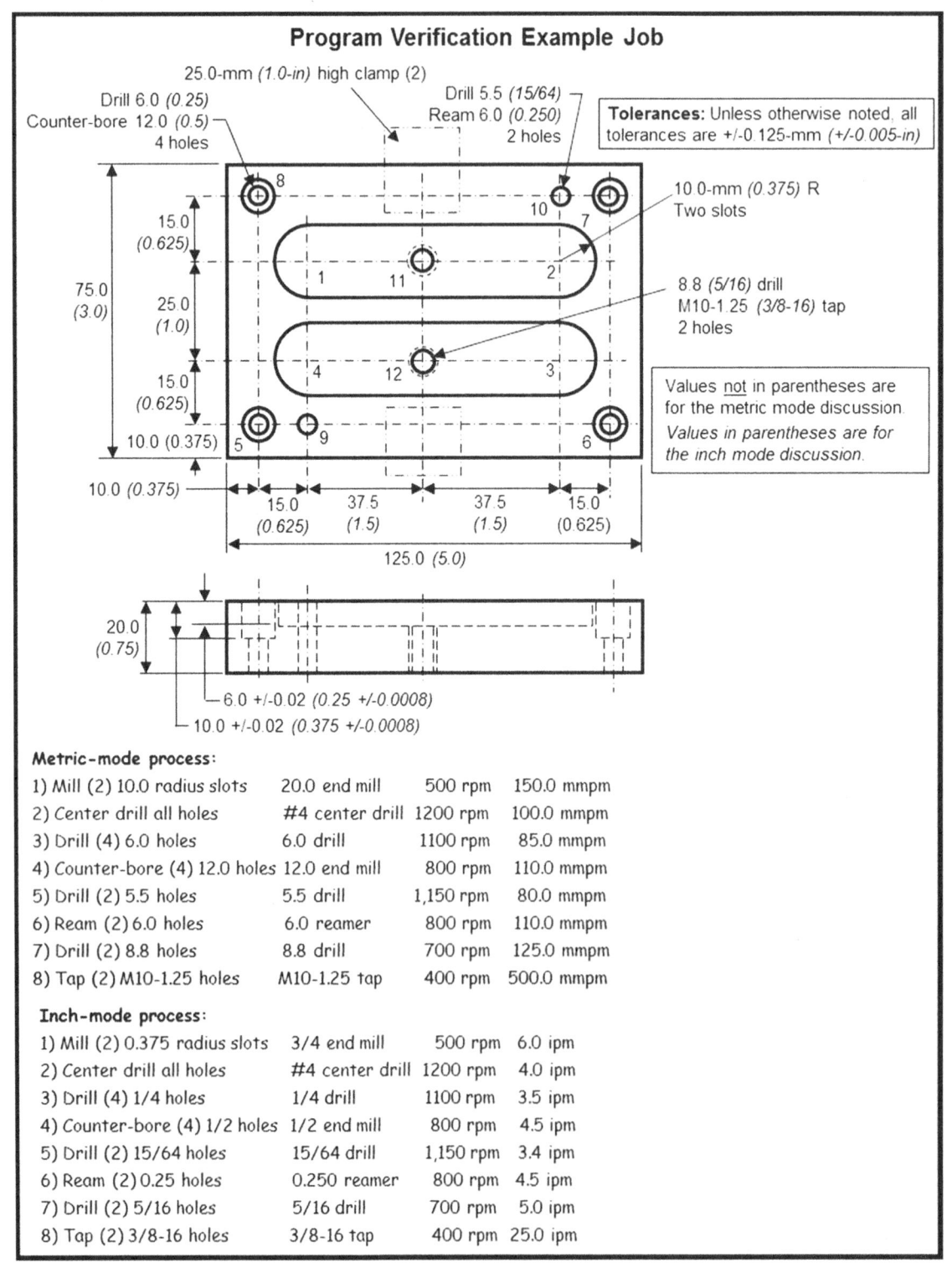

Step	Tool	Speed	Feed
1) Mill (2) 0.375 radius slots	3/4 end mill	500 rpm	6.0 ipm
2) Center drill all holes	#4 center drill	1200 rpm	4.0 ipm
3) Drill (4) 1/4 holes	1/4 drill	1100 rpm	3.5 ipm
4) Counter-bore (4) 1/2 holes	1/2 end mill	800 rpm	4.5 ipm
5) Drill (2) 15/64 holes	15/64 drill	1,150 rpm	3.4 ipm
6) Ream (2) 0.25 holes	0.250 reamer	800 rpm	4.5 ipm
7) Drill (2) 5/16 holes	5/16 drill	700 rpm	5.0 ipm
8) Tap (2) 3/8-16 holes	3/8-16 tap	400 rpm	25.0 ipm

Figure 10.1 – Print and process for example job

Metric-mode program listing for program verification example job

%
O0005 (Program verification practice)
N001 G17 G21 G23
N002 G40 G64
N003 G69 G80

(20.0-MM END MILL)
N005 TO1 M06
N010 G54 G90 S500 M03 T02
N015 G00 X25.0 Y50.0 (pt 1)
N020 G43 H01 Z2.0
N025 M08
N030 G01 Z-6.0 F150.0
N035 X100.0 (pt 2)
N040 G00 Z2.0
N045 Y25.0 (pt 3)
N050 G01 Z-6.0
N055 X25.0 (pt 4)
N050 G00 Z2.0 M09
N055 G91 G28 Z0 M19
N060 M01

(CENTER DRILL)
N065 T02 M06
N070 G54 G90 S1200 M03 T03
N075 G00 X10.0 Y10.0 (pt 5)
N080 G43 H02 Z50.0
N085 M08
N090 G81 R2.0 Z-3.0 F100.0 G99
N095 X25.0 G98 (pt 9)
N100 X115.0 G99 (pt 6)
N105 Y65.0 (pt 7)
N110 X100.0 G98 (pt 10)
N115 X10.0 G99 (pt 8)
N120 X62.5 Y50.0 R-4.0 Z-9.0 G98 (11)
N125 Y25.0
N130 G80 M09
N135 G91 G28 Z0 M19
N140 M01

(6.0-MM DRILL)
N145 T03 M06
N150 G54 G90 S1100 M03 T04
N155 G00 X10.0 Y10.0 (pt 5)
N155 G43 H01 Z50.0
N160 M08
N165 G83 R2.0 Z-22.8 Q12.0 F85.0G98
N170 X115.0 G99 (pt 6)
N175 Y65.0 G98 (pt 7)
N180 X10.0
N185 G80 M09
N190 G91 G28 Z0 M19
N195 M01

(12.0-MM END MILL)
N200 T04 M06
N005 G54 G90 S800 M03 T05
N210 G00 X10.0 Y10.0 (pt 5)
N215 G43 H04 Z50.0
N220 M08
N225 G82 R2.0 Z-6.0 P500 F110.0 G98
N230 X115.0 G99 (pt 6)
N235 Y65.0 (pt 7)
N240 X10.0 (pt 8)
N245 G80 M09
N250 G91 G28 Z0 M19
N255 M01

(5.50-MM DRILL)
N260 T05 M06
N265 G54 G90 S1150 M03 T06
N270 G00 X25.0 Y10.0 (pt 9)
N275 G43 H05 Z50.0
N280 M08
N285 G83 R2.0 Z-22.65 Q12.0 F80. G98
N290 X100.0 Y65.0 (10)
N295 G80 M09
N300 G91 G28 Z0 M19
N305 M01

(0.6.0-MM REAMER)
N310 T06 M06
N315 G54 G90 S800 M03 T07
N320 G00 X25.0 Y10.0 (pt 9)
N325 G43 H06 Z50.0
N330 M08
N335 G81 R-2.0 Z-21.0 F110.0 G98
N340 X100.0 Y65.0 (pt 10)
N345 G80 M09
N350 G91 G28 Z0 M19
N355 M01

(8.8-MM DRILL)
N360 T07 M06
N365 G54 G90 S700 M03 T08
N370 G00 X62.5 Y25.0 (pt 12)
N375 G43 H01 Z2.0
N380 M08
N385 G81 R-4.0 Z-23.64 F125.0 G98
N390 Y50.0 (pt 11)
N395 G80 M09
N400 M01

(M10-1.25 TAP)
N405 T08 M06
N410 G54 G90 S400 M03 T01
N415 G00 X2.5 Y1.0 (pt 12)
N420 G43 H08 Z50.0
N425 M08
N428 M29 S400
N430 G84 R-4.0 Z-24.0 F500.0 G98
N435 Y50.0 (pt 11)
N440 G80 M09
N445 G91 G28 Z0 M19
N450 G28 Y0
N455 M30
%

Figure 10.2A – Program for example job

Key Concept 10: Know How To Safely Verify Programs

Inch-mode program listing for program verification example job

```
%
O0005 (Program verification practice)
N001 G17 G20 G94
N002 G40 G64
N003 G69 G80
(3/4 END MILL)
N005 T01 M06
N010 G54 G90 S500 M03 T02
N015 G00 X1.0 Y2.0 (pt 1)
N020 G43 H01 Z0.1
N025 M08
N030 G01 Z-0.25 F6.0
N035 X4.0 (pt 2)
N040 G00 Z0.1
N045 Y1.0 (pt 3)
N050 G01 Z-0.25
N055 X1.0 (pt 4)
N050 G00 Z0.1 M09
N055 G91 G28 Z0 M19
N060 M01
(CENTER DRILL)
N065 T02 M06
N070 G54 G90 S1200 M03 T03
N075 G00 X0.375 Y0.375 (pt 5)
N080 G43 H02 Z2.0
N085 M08
N090 G81 R0.1 Z-0.12 F4.0 G99
N095 X1.0 G98 (pt 9)
N100 X4.625 G99 (pt 6)
N105 Y2.625 (pt 7)
N110 X4.0 G98 (pt 10)
N115 X0.375 G99 (pt 8)
N120 X2.5 Y2.0 R-0.15 Z-0.37 G98 (11)
N125 Y1.0
N130 G80 M09
N135 G91 G28 Z0 M19
N140 M01
(0.25 DRILL)
N145 T03 M06
N150 G54 G90 S1100 M03 T04
N155 G00 X0.375 Y0.375 (pt 5)
N155 G43 H01 Z2.0
N160 M08
N165 G83 R0.1 Z-0.855 Q0.5 F3.5 G98
N170 X4.625 G99 (pt 6)
N175 Y2.625 G98 (pt 7)
N180 X0.375
N185 G80 M09
N190 G91 G28 Z0 M19
N195 M01

(0.5 END MILL)
N200 T04 M06
N005 G54 G90 S800 M03 T05
N210 G00 X0.375 Y0.375 (pt 5)
N215 G43 H04 Z2.0
N220 M08
N225 G82 R0.1 Z-0.375 P500 F4.5 G98
N230 X4.625 G99 (pt 6)
N235 Y2.625 (pt 7)
N240 X0.375 (pt 8)
N245 G80 M09
N250 G91 G28 Z0 M19
N255 M01
(15/64 DRILL)
N260 T05 M06
N265 G54 G90 S1150 M03 T06
N270 G00 X1.0 Y0.375 (pt 9)
N275 G43 H05 Z2.0
N280 M08
N285 G83 R0.1 Z-0.855 Q0.5 F3.4 G98
N290 X4.0 Y2.625 (10)
N295 G80 M09
N300 G91 G28 Z0 M19
N305 M01
(0.250 REAMER)
N310 T06 M06
N315 G54 G90 S800 M03 T07
N320 G00 X1.0 Y0.375 (pt 9)
N325 G43 H06 Z2.0
N330 M08
N335 G81 R-0.1 Z-0.8 F4.5 G98
N340 X4.0 Y2.625 (pt 10)
N345 G80 M09
N350 G91 G28 Z0 M19
N355 M01
(5/16 DRILL)
N360 T07 M06
N365 G54 G90 S700 M03 T08
N370 G00 X2.5 Y1.0 (pt 12)
N375 G43 H01 Z0.1
N380 M08
N385 G81 R-0.15 Z-0.87 F5.0 G98
N390 Y2.0 (pt 11)
N395 G80 M09
N400 M01

(3/8-16 TAP)
N405 T08 M06
N410 G54 G90 S400 M03 T01
N415 G00 X2.5 Y1.0 (pt 12)
N420 G43 H08 Z0.1
N425 M08
N430 G84 R0 Z-1.0 F25.0 G98
N435 Y2.0 (pt 11)
N440 G80 M09
N445 G91 G28 Z0 M19
N450 G28 Y0
N455 M30
%
```

Figure 10.2B – Program for example job

Setup Sheet

Part No. A-6324
Part Name: End cap
Machine: Mori MV-40
Program No.: O0005
Programmer: ML
Date: 09/25/04

Station	Tool	Offsets	Notes
1	20.0 (3/4) end mill	L:1	Use center-cutting type
2	#4 center drill	L:2	
3	6.0 (0.25) drill	L:3	
4	12.0 (0.5) end mill	L:4	
5	5.5 (15/64) drill	L:5	
6	6.0 (0.250) reamer	L:6	
7	8.8 (5/16) drill	L:7	
8	M10-1.25 (3/8-16) tap	L:8	
9			
10			
11			
12			
13			
14			
15			
16			
17			
18			
19			
20			

Instructions:

Assemble & measure cutting tools and load into specified stations. Enter offsets.

Mount the fixture F-3387 in center X slot and third Y slot.

Measure program zero assignment values and enter them into workpiece coordinate system setting offset number one.

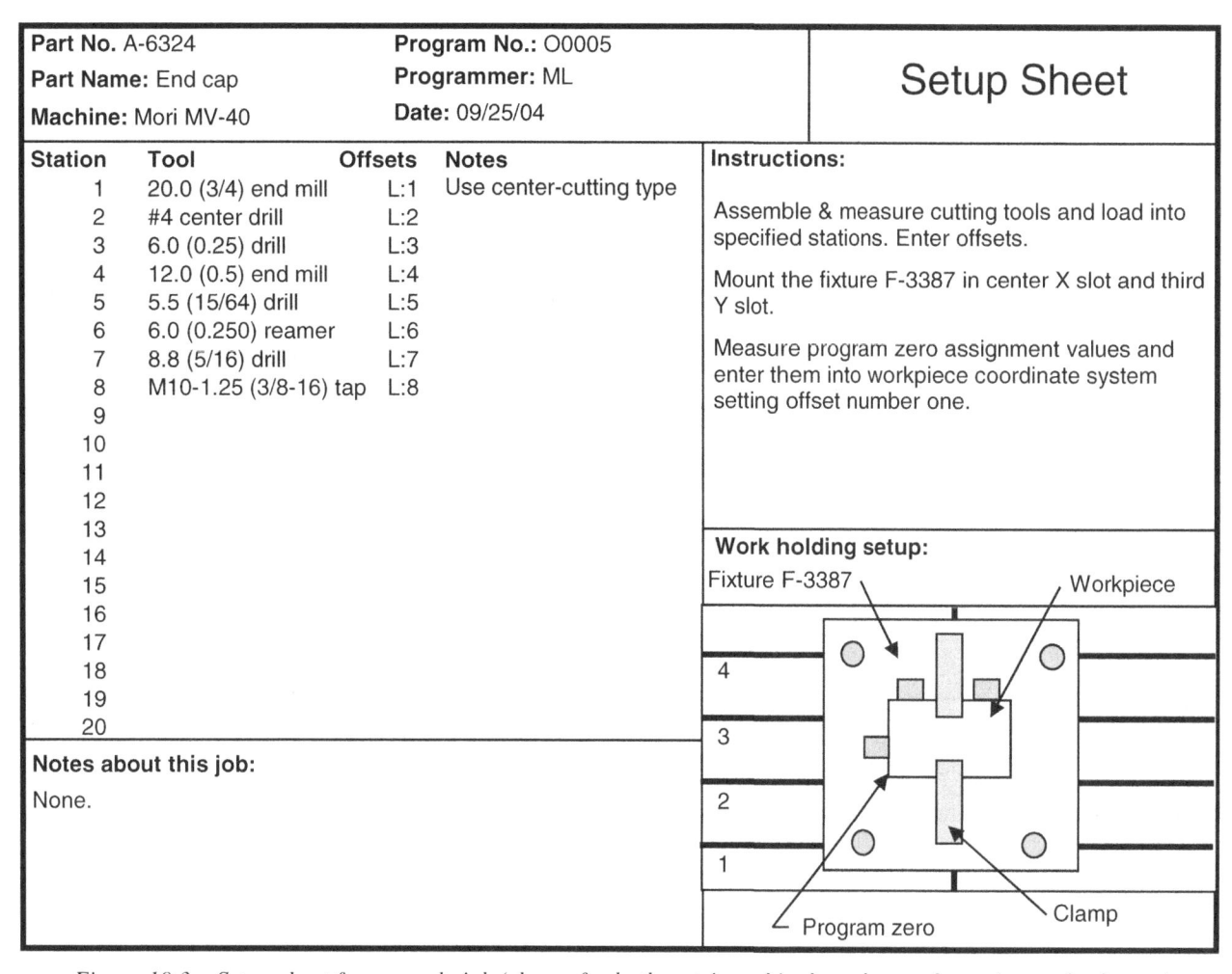

Work holding setup:
Fixture F-3387, Workpiece, Clamp, Program zero

Notes about this job:
None.

Figure 10.3 – Setup sheet for example job (shown for both metric and inch modes - only cutting tools change)

Before you can begin verifying a program, of course, the *physical tasks* related to making the setup must be completed. Here is a checklist of setup-related tasks in the approximate order that setups are made. This is the same set of tasks shown in lesson 7.1.

1. Tear down previous setup an put everything away
2. Gather the components needed to make the setup
3. Make the work holding setup
4. Measure workpiece coordinate system zero assignment values
5. Enter measured values into workpiece coordinate system setting offsets
6. Assemble cutting tools
7. Measure and enter tool length compensation values into tool offsets
8. Measure and enter cutter radius compensation values into tool offsets
9. Load cutting tools into the machine's automatic tool changer
10. Load the CNC program/s
11. Do a dry run to verify the correctness of the setup
12. Cautiously run the first workpiece – ensure that it passes inspection
    a. Use approach procedure with each tool
    b. Trial machining with tools machining tight tolerances
13. If necessary, optimize the program for better efficiency (new programs only)
14. Save corrected version of the program (if changes have been made)

Key Concept 10: Know How To Safely Verify Programs

After you enter your workpiece coordinate system setting offset and tool length compensation offsets, the related offset pages for inch-mode discussions look like this:

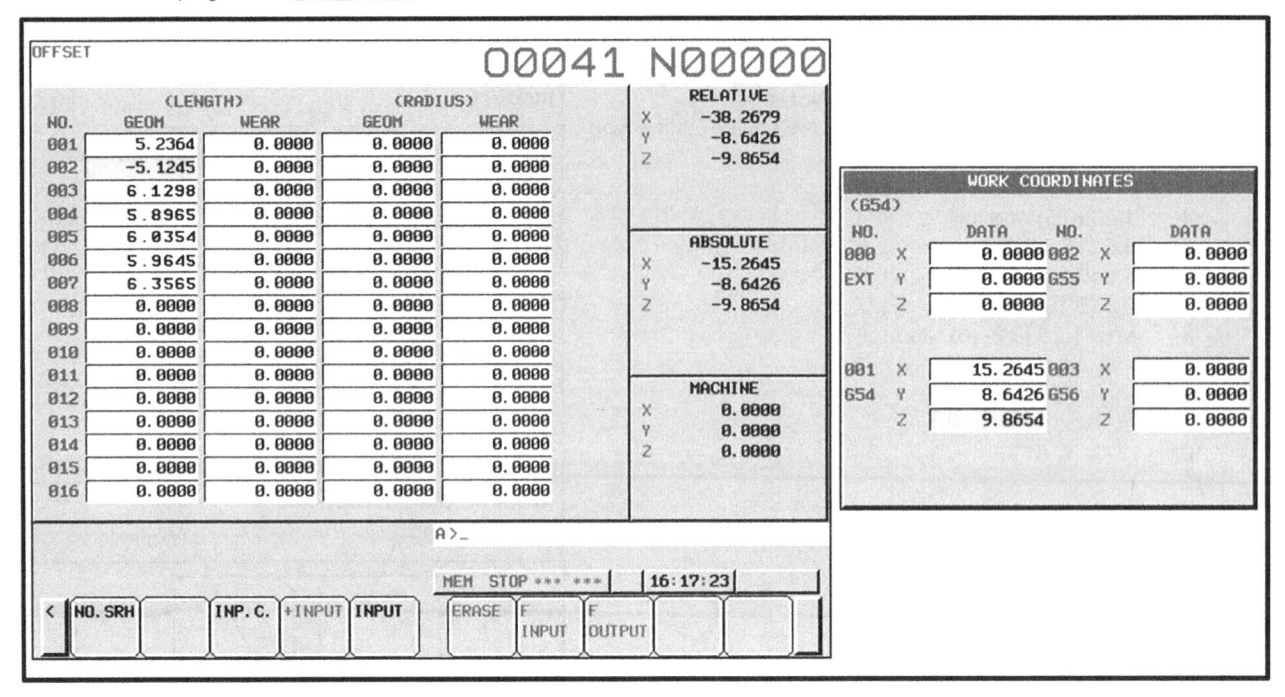

OFFSET										
	(LENGTH)		(RADIUS)		RELATIVE					
NO.	GEOM	WEAR	GEOM	WEAR	X	-38.2679				
001	5.2364	0.0000	0.0000	0.0000	Y	-8.6426				
002	-5.1245	0.0000	0.0000	0.0000	Z	-9.8654				
003	6.1298	0.0000	0.0000	0.0000				WORK COORDINATES		
004	5.8965	0.0000	0.0000	0.0000			(G54)			
005	6.0354	0.0000	0.0000	0.0000	ABSOLUTE		NO.	DATA	NO.	DATA
006	5.9645	0.0000	0.0000	0.0000	X	-15.2645	000 X	0.0000	002 X	0.0000
007	6.3565	0.0000	0.0000	0.0000	Y	-8.6426	EXT Y	0.0000	G55 Y	0.0000
008	0.0000	0.0000	0.0000	0.0000	Z	-9.8654	Z	0.0000	Z	0.0000
009	0.0000	0.0000	0.0000	0.0000						
010	0.0000	0.0000	0.0000	0.0000			001 X	15.2645	003 X	0.0000
011	0.0000	0.0000	0.0000	0.0000	MACHINE		G54 Y	8.6426	G56 Y	0.0000
012	0.0000	0.0000	0.0000	0.0000	X	0.0000	Z	9.8654	Z	0.0000
013	0.0000	0.0000	0.0000	0.0000	Y	0.0000				
014	0.0000	0.0000	0.0000	0.0000	Z	0.0000				
015	0.0000	0.0000	0.0000	0.0000						
016	0.0000	0.0000	0.0000	0.0000						

Here are the offset settings for **metric-mode** discussions:

Tool offsets (LENGTH) in the GEOM registers

1: 133.004

2: -130.149

3: 155.696

4: 149.771

5: 153.299

6: 151.498

7: 161.455

Work coordinates in 001 (G54)

X: 387.718

Y: 168.722

Z: 250.581

Loading the Program

At the tenth step in the checklist shown above, you're ready to load the program. You follow the procedure shown in lesson 9.1, and load the program for this job (O0005). But when you finish loading the program, you notice that something very strange has happened. As you look at the program shown on the display screen, it begins like this:

```
O0001 M06
N010 G54 G90 S500 M03 T02
N015 G00 X1.0 Y2.0 (pt 1)
N020 G43 H01 Z0.1
N025 M08
N030 G01 Z-0.25 F6.0
N035 X4.0 (pt 2)
N040 G00 Z0.1
N045 Y1.0 (pt 3)
```

```
N050 G01 Z-0.25
N055 X1.0 (pt 4)
   .
   .
   .
```

It continues through to the end of the program. When you look at the directory page, you notice that *there is* a program named O0005. When you call it up, here's what you see:

```
O0005 (Canned cycles practice)
N001 G17 G20 G23
N002 G40 G50 G64
N003 G67 G69 G80

(3/4 END MILL)
N005 T
```

Obviously, something is wrong. Before reading on, can you determine what it is (by looking at the program listing for the job)?

Here is a hint: The problem is in block N005.

Actually, this is a very common beginner's mistake. The person typing this program used the upper case letter O in the T word in block N005, typing TO1 (T-oh-one) instead of T01 (T-zero-one). As you know, the only letter O in a program (other than in comments) is the program number.

As this program is being loaded, the CNC sees the letter O in block N005 and thinks you want to begin loading *another* program. Since the number following the letter O in block N005 is 1, the balance of the program being loaded will be placed in a program named O0001. When the program loading process is finished, *two* programs have been loaded, O0005 and O0001. And since O0001 is the last program loaded, it is the active program when the program loading process is completed.

To correct this mistake, of course, you must change block N005 from TO1 to T01. You must then delete both programs O0001 and O0005 from memory and repeat the program loading process. When you do, the program will be loaded properly (there will be no other program-loading problems).

Note that a good off-line program verification system will find and display this problem well before you're ready to load the program into the machine – but since this is such a common mistake, we wanted to show you what will happen when you type the letter O in a program instead of a number zero.

The Dry Run to Check for Setup Mistakes

We'll say this program has been verified on an offline program verification system, and the general motions made by the program look pretty good to the person performing the off-line verification. So during the dry run, we're looking, first and foremost, for setup mistakes.

With the setup complete and the program loaded, you're ready to do the dry run. At this point, of course, you have removed the workpiece from the setup. You roughly position the two clamps as they will be when a workpiece is in position.

You follow the procedure used in lesson 7.1 for doing a dry run. You press cycle start and the first tool change places the 20.0-mm (3/4-in) end mill in the spindle. But in the first XY motion command (block N015) the machine over-travels in the plus X and Y directions. The machine is going in the wrong direction in both the X and Y axes. Before reading on, can you determine what the problem is? Hint: The problem is *not* with the program.

On the workpiece coordinate system offset display (shown above), all three workpiece coordinate system zero assignment values have been entered as *positive* values. The machine thinks workpiece coordinate system zero is on the positive side of the zero return position. As you know from lesson 1.6, workpiece coordinate system setting offsets must be specified *from* the zero return position *to* workpiece coordinate system zero. Since the zero return position is at the extreme plus end of each axis, workpiece coordinate system setting offset values will be *negative* in all axes.

Key Concept 10: Know How To Safely Verify Programs

To correct this problem, you must change all three workpiece coordinate system setting offset entries to negative values (retype them). You must also move the machine off its limits in X and Y, clear the alarm (by pressing reset), and send the machine to its tool change position. You do so and now you are ready to continue. (By the way, since this mistake is a *setup mistake*, a program verification system cannot find it off line.)

Since tool number 1 is already in the spindle, you follow the procedure to rerun the tool, using block N010 as the restart command. You are still doing a dry run, so you still have control of all motions with the feedrate override switch. And you're keeping a finger ready to press feed hold at all times.

You crank up the feedrate override switch, and the end mill moves over to the workpiece. Then it moves down in Z and begins its machining motions. As far as you can tell, everything looks good. When the tool is finished, the machine returns to the tool change position.

Tool number two (the #4 center drill) is placed in the spindle. It moves a little in X and Y and then starts moving down in Z. As it comes close to the fixture, you think it has moved past the work surface in Z (there is no workpiece in position so it may be a little hard to tell). But you're worried, so you press the feed hold button and look at the distance-to-go display. It says the tool is still going to move another 250.8-mm (or 9.8756-in) in the Z minus direction. You've just saved a crash. You follow the procedure to cancel the cycle. Now you must determine what is wrong. Before reading on, can you spot what is wrong? Hint: The problem is not in the program.

When you check the tool offset page, you notice that the value of offset number two is -130.149 (or -5.1245). This offset should be a positive value (the length of tool number two). The setup person typed it incorrectly. You must re-enter offset number two as a positive value. (Again, this mistake is a setup mistake so no off-line program verification system will show it.)

After you cancel the cycle and correct the problem, you follow the procedure to re-run tool number two. This time, it approaches properly and the rest of the motions look fine.

Tool three (the 6.0-mm [1/4-in] drill) is placed into the spindle. It moves a little in X and Y and then begins the Z approach. This tool also looks like it's going too far. You press feed hold when the tool looks like it has moved down past the work surface in Z. You look at the distance-to-go page and it says the tool is still going to move another 18.933-mm (or 0.7454-in) in the Z minus direction. While this movement may not cause the tool to crash into the fixture, something is definitely wrong. Can you tell what it is? Hint: This time the problem is in the program.

Look at block N155 in the program. It is the 6.0-mm (or 1/4-in) drill's first Z-axis approach movement and includes the tool length compensation command. But instead of using H03 (this tool's offset number), the program specifies H01 (the tool length offset for tool number *one*). If you look at the offset page, you'll notice that tool number one is shorter than tool number three, which is why this (longer) tool is moving past the work surface. To correct the problem, you must change the H01 in block N155 to H03. (While this mistake is a programming mistake, most off line program verification systems ignore offsets, so it will not be found when verifying the program off line.)

After you cancel the cycle and correct the problem, you follow the procedure to re-run tool number three. This time it runs fine.

Tool number four (the 12.0-mm [or 1/2-in]) end mill is placed in the spindle. It approaches properly and appears to machine the first three holes just fine. But when it comes out of the third hole, it doesn't appear to come up high enough to clear the clamp between the third and fourth hole. It begins moving in the X-axis to get to the fourth hole, but it is going to hit the clamp before it gets there. Again, you halt motion with the feed hold button. What's wrong this time? Hint: The problem is in the program.

Block N235 is the command that machines the third counter-bored hole. It should include a G98 word to tell the machine to retract the end mill to the initial plane (notice the G99 in block N230). To correct this problem, you must add a G98 to block N235. (With a watchful eye, the person doing the off line program verification *should* be able catch this mistake. This is an example of a motion mistake that you might not catch even when verifying programs off line.)

After you cancel the cycle and correct the problem, you follow the procedure to re-run tool number four. This time it runs fine.

The next three tools (5.5-mm [or 15/64-in] drill, 6.0-mm [or 0.250-in] reamer, and 8.8-mm [or 5/16-in] drill) run fine. At least you can't spot any mistakes with the setup or in the program.

But while the last tool, the M10-1.25 (or 3/8-16) tap, is making its approach movement in the Z-axis, you find another problem. It gets within 0.25 of the work surface and is still moving. You press the feed hold button and check the distance to go. It says the Z-axis is going to move another 178.714-mm (or 7.0376-in) in the negative direction. This is another severe problem that would have caused a crash. You must cancel the cycle and correct the problem. Before you read further, can you determine what is wrong?

When you look at the tool offset display screen page, you notice that offset number eight (the tool length offset for the tap) is set to zero. The setup person has forgotten to enter a tool length compensation value for the tap. You correct this problem (by measuring the tap's length and entering its tool length compensation value in offset number eight) – then you re-run the tap. This time, everything looks good.

During this dry run, you have found five mistakes that, if not detected, would have caused crashes. And you did so in a very safe manner. Some setup people will eventually skip the dry run step in order to try to save some time – especially after they have run several jobs in a row without mishap. But as you have seen in this example, doing so can be a terrible mistake.

Cautiously Running the First Workpiece

With the dry run complete, you load the first workpiece. You will follow the safe approach procedure for each tool. When each tool is placed into the spindle, the machine will stop (because single block is turned on) and you'll consider whether it requires trial machining. You'll also turn on the optional stop switch to ensure that the machine will stop when the tool is finished.

When tool one (the 20.0-mm [or 3/4-in] end mill) is placed into the spindle, you consider what it will be doing. It will be machining the two pockets. You notice the tight tolerance for the pocket depth and decide to trial machine. So you increase the value of offset number one by 0.25-mm (or 0.010-in) and run the tool. When it's finished, the machine stops (since optional stop is on). You measure the pocket depth and find it to be 5.772-mm (or 0.241-in). So you reduce offset number one by 0.228-mm (or 0.009-in) and follow the procedure to re-run tool number one. When it's finished, the machine stops again. This time, when you measure the pocket depth, you find it to be 6.003-mm (or 0.2501-in). On to tool number two.

With tool number two in the spindle (the #4 center drill), you stop and consider what it is doing. There are no tight tolerances being machined by this tool. Actually, there is only one more tight tolerance. It is machined by the 12.0-mm (or 1/2-in) end mill. So of the remaining tools, only the 12.0-mm (or 1/2-in) end mill will require trial machining.

You follow the approach procedure in lesson 7.1, and the center drill approaches just fine. You let this tool machine the workpiece and all goes well. When this tool is finished, the machine moves back to its tool change position and stops. You check to be sure that all of the holes are in the right location.

Tool number three (the 6.0-mm [or 1/4-in] drill) is placed in the spindle. You use the approach procedure and all goes well. It machines the workpiece perfectly. On to tool number four.

When the 12.0-mm (or 1/2-in) end mill is placed in the spindle, you consider what it will be doing. It is machining a tight tolerance for the counter-bore depths – so you decide to trial machine. You increase the value of offset number four by 0.25-mm (or 0.010-in) and run the tool, using the approach procedure. It approaches properly and machines the counter-bored holes. The machine stops when the tool is finished (again, since optional stop is on). You measure the depth of the counter-bored holes and find them to be 9.778-mm (or 0.3652-in). You reduce offset number four by 0.222-mm (or 0.0098-in) and re-run the tool. When the machine stops at the end of the tool, you measure again and find the depth of the counter-bored holes to be 10.003-mm (or 0.3751-in). On to the next tool.

You use the approach procedure for tool number five (the 5.5 mm [or 15/64-in] drill) and it approaches just fine. It also machines the workpiece properly.

Tool number six (the 6.0-mm [or 0.250-in] reamer) is placed in the spindle. When using the approach procedure, the tool gets within 2.0-mm (or 0.1-in) of the work surface and is still moving down in Z. You press the feed hold button and look at the distance-to-go. It says this tool is still going to move 4.733-mm (or 0.187-in) in the Z minus direction. You can tell there is not enough room between the tool tip and the workpiece. Now what's wrong? Hint: The problem is in the program.

Look at block N335. The R-word is R-2.0 (or R-0.1) when it should be R2.0 (or R0.1). Another programming mistake that slipped by the person doing the off line program verification (and you, during the dry run). You must, of course, cancel the cycle, change block N335 in the program, and re-run the tool. This time the tool approaches properly. And it machines the workpiece properly.

Key Concept 10: Know How To Safely Verify Programs

You run the last two tools without incident. You bring the completed workpiece to the inspector (after cleaning and de-burring it). It passes inspection. You're now ready to begin the production run.

This example should give you a very good idea of what you are in for as a CNC setup person. While (hopefully) there shouldn't be as many mistakes to find and correct in your jobs as there are in our example job, *you must be extremely careful when verifying programs.*

Key Points for Lesson 10.1:
- The CNC cycle must be cancelled if mistakes are found.
- It is often necessary to re-run individual cutting tools.
- You must be able to spot and correct mistakes when verifying CNC programs.

Index

/

/, 98, *291*, *377*

;

;, 97

9

90 degree indexers, *314*

A

A axis, 95, *315*, *324*, *328*
A axis designator, 95
absolute, 99, 127
Absolute, *318*
absolute position display, 374
absolute positioning mode, 60, 61, 87, 88, 127, 272, 282, 283, *318*, *319*, *321*
accessory devices, *290*, 308
active program, *377*, *379*, *388*, 389
air-cutting time, 132
alarm, 57, 98, 101, 104, 105, 186, 192, 225, 242, 281, *317*, *359*, *378*, 389, *416*
alter key, 377
Alter key, *377*
approach distance, 133
Approach motions, 227
approach procedure, *360*, *361*, *362*, *363*, *407*, *413*, *417*
arc center point, 96
Arc limitations, 144
arc-in motion, 145, 146, 147, 198, 298
arc-out, 192
arc-out motion, 145, 146, 147, 192, 198, 298
auto mode, *377*, *379*, 389
automatic tool changer, 21, 23, 31, 32, 33, 37, 38, 41, 97, 119, 166, 189, 227, 228, 232, 234, 236, 244, 313, 342, 345, 356, 368, 378, 381, 382, 386, 387, 413
Automatic tool changer, 31, 37, 228
Automatic tool changer control, *381*
automatic tool changer magazine, *356*
axes, 22, 339
axis display, 73, 74
axis origin light, *386*
axis origin lights, 68, 382

B

B, 95
B axis designator, 95
background edit, 389
backward-search, 388
basic machining practice, 19, 20, 73, 77, 105, 106, 108, 117, 306, 346, 349, 364, 366
block delete, 98, 291, 292, 293, 294, 295, *296*, *377*, 380
Block delete, *291*
Boring bar, 129
boring cycle, *256*
Bridge-style vertical machining center, 24

C

C, 95
C axis designator, 95
CAM system, 183, 205
Cancel, 184
Cancel canned cycle, *252*
cancel key, 389
Canceling the CNC cycle, *407*
canned, *251*
canned cycle, 95, 96, 98, 99, 138, 148, 234, 236, 237, 244, 249, 250, 251, 252, 253, 254, 256, 258, 259, 260, 261, 262, 263, 264, 265, 266, 271, 272, 273, 274, 275, 284, 285, 289, 291, 299, 304, 309, 324
Canned cycle, *324*
canned cycles, 138
Canned cycles, *251*, *263*
Center drill, 128
centerline tool path, 116, 117, 129, 139, 148, 198
C-frame style vertical machining center, 59
circular interpolation, 124
Circular interpolation, 139
circular motion, 49, 96, 124, 125, 139, 140, 141, 142, 143, 145, 148, 189, 192, 198, 237, 297, 305, 323
Clamping the rotary axis, *323*
Clean and de-burr, *366*
climb milling, 188, 190, 306
Clockwise, 139
clockwise motion, 139, 141, 305
common fixture offset, 86, 89, 90, 218, 219, 220, 221, 338
common offset, 86
compatible G codes, 98, 234
Complex motions, *290*
conditional switches, 389, 390
Conditional switches, *380*, 389, 390
contour milling, 129, 139, 158, 179, 182, 183, 194, 205, *287*, 300, 306, 312, 355
control panel, *371*, 372, 380, 388, 389
Control programs, *288*
conventional milling, 188
coolant, 291
Coolant, 31, 37, *255*, *387*
coordinate rotation, 237, 309, 310, 311
Coordinate rotation, *309*
coordinate sheet, 115, 116
coordinates, 56, 59, 61
Counter-boring cycle, *255*
counter-clockwise, 139
crash, *264*
cursor, *377*, *388*, 389
cursor control, 388
Cursor control keys, *377*, 388
custom macro B, *289*

Index

cutter radius compensation, 97, 98, 99, 116, 129, 139, 140, 148, 158, 159, 160, 179, 180, 181, 182, 183, 184, 185, 186, 187, 188, 189, 190, 191, 192, 193, 194, 195, 196, 197, 198, 199, 200, 201, 202, 203, 204, 205, 219, 237, 249, 279, 287, 288, 298, 299, 301, 302, 304, 338, 343, 345, 355, 356, 358, 359, 361, 362, 367, 368, 413
cutter radius compensation offset number, 97
cutter-radius-compensation values, *355*
cutting conditions, 109
cutting tool measurements, *355*
Cycle indicator lights, *382*
cycle start, 378, 379, 380, 388
Cycle start button, 251, 358, 359, 360, 362, 365, 366, *379*, 380, 382, 388, 389, 390, 407, 408, 409

D

D, 97
decimal point, 62, 94
Decimal point, *377*
decimal point programming, 28, 30, 36, 49, 50, 51, 62, 94, 95, 96, 97, 104, 105, 315, 316, *377*
Deep-hole drilling cycle, *253*
<u>degrees per minute</u>, *323*, 324
Delete key, *377*
directional vectors, 96, 143, 144, 145
Directions of motion, 22, 26
display screen, *342*
Display screen control keys, *373*
distance-to-go, 343, 358, 359, 360, 374, 380, 416, 417
distributive numerical control (DNC) system, 389
DNC systems, 302, 357, 389
documentation, 41, 119, 238, *364*
dog-leg motion, 132
door close, *387*
Door open, *387*
double-arm tool changer, 228
Drill, 128
drilling cycle, 251, 252, 253, 258, 259, 260, 271
dry run, 343, *358*, 359, 360, 361, 362, 363, 389, 407, 409, 413, *415*, 416, 417
Dry run on/off switch, *380*
Dwell command, 95, 96, 99, *299*, 300, 312
dwell time, 95, 96

E

Ease-of-use, 225
edge finder, 73, 74, 75, 76, 81, 83, 84
edit mode, 377, 388
edit mode switch position, *388*
Efficiency, 225
emergency stop, 380
Emergency stop button, 379, *380*
end point, 127, 139
end-of-block, 97, 386, 387, 389
enter a new program, *389*
EOB, 97, 386, 387, 389
Exact stop check, 237, *300*
executions, 96

F

F, 96, *258*
F word, 30, 31, 36, 37, 96, 134, 139, 343, 388
fast-feed approach, 136, 196
feed hold, *358*, 380, 389
feed hold button, 343, 358, 359, 360, 366, 379, 380, 382, 389, 390, 416, 417
Feed hold button, *379*
feedrate, 96, 134, *323*
Feedrate, 30, 36, *254*
feedrate override, 380, 389
feedrate override switch, 134
Feedrate override switch, 134, 324, 343, 358, 359, 360, *379*, 389, 409, 416
Fine boring cycle, 96, *256*, 265, 271
Five degree indexers, *315*
fixture offset, 88
fixture offsets, 85, 86, 89, 90, 99, 158, 159, 161, 162, 166, 211, 212, 213, 214, 217, 218, 220, 221, 236, 301, 303, 318, 326, 328, 330, 338, 343, 352, 377, 413, 415
format, 239

G

G, 94
G codes, 28, 31, 94, 97, 98, 99, 127, 134, 139, 142, 187, 214, 234, 250, 252, 274, 291, 297, 300
G00, 99, 127, 130, 189, 251, *316*
G01, 96, 99, 127, 134, 189, 238, 251, *323*
G02, 96, 99, 127, 139, 189, *297*
G03, 96, 99, 127, 139, 189, *297*
G04, 99, *299*
G09, *300*
G10, 220, *301*
G15, *303*, *309*
G16, *303*, *309*
G17, 142, 234, 237, *304*
G18, 142, 237, *304*
G19, 142, 237, *304*
G20, 28, 38, 96, 99, 234, 237, *305*
G21, 28, 38, 96, 99, 237, *305*
G28, 88, 99, 233, 238, *251*, *386*
G30, *306*
G40, 99, 193, 237
G41, 99, 187, 190, 191, 194
G42, 99, 187
G43, 99, 131, 162
G50, *306*
G50.1, *306*
G51, *306*
G51.1, *306*
G52, *283*
G53, 233, *308*
G54, 86, 87, 99, 131, 214
G55, 99, 214
G56, 214
G57, 214
G58, 214
G59, 99, 214
G60, *308*
G61, *300*

G64, 234, 237, *300*
G68, *309*
G69, 237, *309*
G73, 96, 99, *252*
G74, *252, 254, 255*
G76, *252, 256, 257*
G80, 234, 237, *252*
G81, *252*
G82, *252, 255, 299*
G83, 96, *252, 253*
G84, *252, 254*
G85, *252, 257*
G86, *252, 256*
G87, *258*
G88, *258*
G89, 99, *252, 256, 299*
G90, 60, 87, 99, 131
G91, 61, 88, 99, *272*
G94, 31, 37, 97, 134
G95, 31, 37, 97, 134
G98, 99, *258, 260*
G99, 99, *258, 260*
Gather, *350*
gauge block, 79
geometry offsets, 377
graph, 53

H

H, 97
H word, 97, 162
hands-on experience, *338*
handwheel, 73, 75, 385, 394
Handwheel controls, *381*
handwheel mode, 341, 381, *385*
helical interpolation, 290
helical motion, *297*
hole-bottom position, 95
hole-machining, 128
home position, 68
Horizontal machining center, 21, 25, 26, 27, 33, 68, 95, 170, 171, 227, 230, 235, 244, 245, 246, 303, 313, 314, 315, 316, 318, 327, 330, 335, 340
horizontal machining centers, 244, *313*
horsepower meters, *382*

I

I, 96, 143, 146, *256, 258, 306*
imperfect threads, 129
inch mode, 28, 38, 49, 50, 54, 62, 96, 124, 134, 234, 236, 237, 244, 305, 316
Inch mode, *305*
inches per minute feedrate, 109
inches per revolution, 109
inches per tooth, 109
incremental, 99, 127, *272*
Incremental, *321*
incremental positioning mode, 60, 61, 88, 99, 127, 272, 283, 293, 301, 321, 322
indexer, 95
Indexers, 227, *314*, 315, 316, 324

Indicator lights, *381*
initial plane, *260*, 261, 262, 263, 265, 266, 267, 268, 269, 270, 271, 273, 274, 275, 416
initial setting, 156
initialized, 98, 234
initialized words, 49, 51
input key, 353, *377*
insert key, 377
Insert key, 377
inspector. *See*
Instate, 184
interference, 162
intermediate position, 88, 233, 234, 251, 317
interpolation, 96, 123, 124, 134, 139, 148, 290, 323
ipm, 109
ipr, 109
ipt, 109

J

J, 96, 143, 144, 256, *258, 306*
jog, 379, 385, 394
jog feedrate, 341, 342, 379, 381, 385
jog mode, 341, 342, 385
jogging controls, *381*

K

K, 96, 143, *306*
keyboard, *376*, 386, 387, 389
keyway, 230
Knee style vertical CNC milling machines, 23

L

L, 96, 220, *258, 272, 281, 285, 301*
L word, 220
lead, 128
leading zeros, 127
Letter Keys, *377*
linear axis, 123, 305, 315, 317, 321, 323, 324, 335
linear interpolation, 124
Linear interpolation, 134
Loading programs, *343*
Loading the program, *414*
look-ahead buffer, *300*
look-ahead-buffer, 300, 377

M

M, 97
M code, 227
M words, 97
M00, 101, 228, *255, 292*
M01, 101, 228, 238, 242, *292*
M02, 101
M03, 29, 36, 97, 101, 131, 228, *387*
M04, 29, 36, 97, 101, 228, *387*
M05, 29, 36, 97, 101, 228, *387*
M06, 31, 37, 97, 101, 228, 238, *387*
M07, 101

Index

M08, 31, 37, 101, 228, 237, *387*
M09, 31, 37, 101, 228, *387*
M19, 228, 230, 238, *256*
M30, 101, 135, 228, 238
M98, 101, *279*, *281*
M99, 101, *279*
machine function, 97
machine panel, *371*, 372, *378*, 379, 380, 381, 382, 386, 387
Machine power-up, *341*
machine zero, 68
machining process, 107
main program, *279*, 280, 281, 282, 283, 284, 285, 289, 294, 295, 296, 303, 310
manual data input, 98, 342, 356, 379, 381, 386, 387
manual data input (MDI) mode, 386
Manual data input (MDI) procedures, *393*
manual data input mode, *386*
manual functions, *380*
manual mode, *385*, 386
Manual procedures, *393*
mark-up the print, 112
math, 112, 114
MDI mode, 386, 387
mean value, 156, 157, 174, 182, 205, 348, 367
mean values, 174
measurements, *366*
memory, 42
memory protect, 388
meters, *381*
metric mode, 28, 33, 38, 49, 54, 96, 98, 99, 124, 134, 234, 237, *305*, 387
minimum length, 118
Mirror image, 113, *306*, 307, 308
miscellaneous function, 97
mistakes of omission, 50, 51, 105, 109, 120
Mistakes of omission, 105
modal, 51, 98, 127, 131, 134, 226, *252*
modal words, 49
mode switch, 358, 377, 378, 381, 383, *385*, 386, 388, 389, 407, 408
Mode switch, *378*
Monitor the cycle, *366*
Motion commonalties, 127
motion mistakes, 104, 105, 120
Motion mistakes, 105
Motion relative to zero return position, *308*
multiple program zero points, 212

N

N, 48, 94, *258*, 295
N word, *295*
nesting, *289*
Nesting, *282*
normal cutting mode, 234, 236, 237, 244, *300*, 309
Number keys, *377*
number of holes, 96

O

O, 94, 388
O word, 94

offset, 188
offset adjustments, *367*
offset display, 85, 86, 159, 343, 356, 375, 417
Offset display, *375*
offset display screen, *343*, 375
Offset setting by programmed command, *301*
offsets, 41, 85, 86, 89, 90, 99, 156, 157, 158, 159, 160, 161, 162, 163, 166, 168, 170, 172, 173, 174, 200, 205, 211, 212, 213, 214, 217, 218, 220, 221, 236, 301, 302, 303, 305, 318, 326, 328, 330, 338, 343, 346, 352, 355, 356, 367, 377, 413, 414, 415, 416
Offsets, 85
One degree indexer, *315*
operation panel, 371
operator, 40
optimize, *363*
Option G codes, 98
optional block skip, 291, 292, 295, 380
optional stop, 238, 242, 246, *292*, 361, 362, 382, 389, 417
Optional stop indicator light, *382*
Optional stop on/off switch, *380*
orient position, *256*

P

P, 96, 220, 255, *258*, 279, 280, 281, 290, 296, 301, 302, *306*
P word, 220, *281*
Pallet changer, 25, 32, 220, 227, 244, 288, 303, 308, 314, 382
pallet changers, 220
panic button, 379
parallels, *351*
parameter, 253, 309
parametric programming, *289*, 290
parentheses, 47, 238
Part families, *289*
pause time, 255, *299*
peck depth, 96, 253
peck drilling cycle, 251, 252, 253, 260
pitch, 129, *298*
plane selection, 142, 304, 305
Plane selection, *304*
polar coordinates, 304, 309, 310
Polar coordinates, *303*
polarity, 22, 26, 55, 56, 58, 86, *339*
Polarity, *317*
position display, 342, 343, 374
Position display, *374*
position display screen, *342*, 374
positioning, 130
power curve, 29
preparatory function, 94, 98
Preventive maintenance, 40, *368*
prior position, 180, 182, 185, 186, 187, 189, 190, 191, 192, 193, 194, 195, 196, 197, 198, 200, 201, 287
procedures, 377, 388, 391, 393, 394, 403
Procedures, *341*
process, 19, 40, 41, 103, 104, 105, 106, 107, 108, 109, 110, 114, 117, 118, 120, 155, 204, 230, 239, 253, 266, 279, 283, 289, 290, 293, 311, 314, 337, 346, 348, 357, 360, 363, 366, 410, 415
Process engineer, 40
Process mistakes, 105, 120

Index

Product producing companies, 39, 42, 337
Production, **42**
Production run documentation, 19, 40, 41, 119, 120, 157, 338, *363*, 364, 365, 366, 367, 368
program check display, 342, 343, 358, 359, 360, 373, 376
Program check display, *376*
program check display screen, *343*
program display, 343, 374, 407, 408
Program display, *374*
program display screen, *343*, 374
Program Editing Keys, *377*
program editing procedures, 388
Program editing procedures, *393*
Program ending format, 235
Program listing, 120
program operation mode, *389*
Program operation procedures, *393*
Program startup format, 235
program verification, 104, 105, 118, 221, 228, 290, 292, 343, 357, 358, 363, 379, 380, *405*, 407, 415, 416, 417
program zero, 19, 42, 56, 57, 58, 59, 60, 61, 62, 66, 67, 68, 72, 77, 79, 81, 82, 83, 84, 85, 86, 87, 88, 89, 90, 91, 114, 115, 118, 121, 122, 127, 131, 157, 158, 159, 162, 167, 170, 171, 173, 174, 185, 211, 212, 213, 214, 216, 217, 219, 221, 233, 236, 258, 259, 263, 282, 283, 284, 290, 301, 307, 308, 310, 311, 312, 316, *317*, 318, 321, 324, 325, 326, 327, 328, 330, 335, 337, 338, 339, 341, 342, 345, *352*, 353, 355, 358, 359, 374, 381, 385, 394, 405, 413, 415
    choosing, 58
program zero assignment values, 19, 67, 68, 69, 72, 73, 77, 79, 81, 82, 83, 84, 85, 86, 87, 88, 89, 90, 131, 135, 157, 161, 166, 167, 211, 212, 213, 217, 219, 220, 221, 234, 301, 307, 326, 328, 330, 341, 342, 352, 353, 358, 381, 385, 394, 413, 415
program zero point, 19, 42, 56, 57, 58, 59, 60, 61, 62, 66, 67, 68, 72, 77, 79, 81, 82, 83, 84, 85, 86, 87, 88, 89, 90, 91, 114, 115, 118, 121, 122, 127, 131, 157, 158, 159, 162, 167, 170, 171, 173, 174, 185, 211, 212, 213, 214, 216, 217, 219, 221, 233, 236, 258, 259, 263, 282, 283, 284, 290, 301, 307, 308, 310, 311, 312, 316, 317, 318, 321, 324, 325, 326, 327, 328, 330, 335, 337, 338, 339, 341, 342, 345, 352, 353, 355, 358, 359, 374, 381, 385, 394, 405, 413, 415
Program zero point, *324*
Programmable functions, 27, 30, 32, 36, 227
programmed point, 128
programmer, 40
Prototype producing companies, 39

Q

Q, 96, 253, *258*
quadrant, 55
qualified workholding setups, 68, 217, 220, 301, 352

R

R, 96, 139, 142, 143, 144, 146, 148, 192, *258*, 301, 417
R plane, 252, 258, 259, 260, 261, 262, *263*, 264, 266, 267, 268, 269, 270, 271, 272, 275, 285
R word, 139
radius, 96, 140
Range of cutter sizes, 181, 182, 205
Rapid approach distance, 133, 227, 360
rapid motion, 49, 124, 130, 132, 133, 134, 136, 148, 186, 265, 323, 358, 360, 379, 380, 409
Rapid motion, 130
rapid plane, 96, 99, 252, 254, 260, 264, 272
Rapid plane, *254*
rapid rate, 49, 104, 130, 131, 132, 133, 323, 379, 380
rapid traverse override, 130, 358
Rapid traverse override switch, *379*
ready position, 31, 32, 37, 97, 228
ready position tool station, 97
Reamer, 128
Reaming cycle, *257*
rectangular coordinate system, 53, 54, 56, 60, 62, 303, 304, 337, 339
relative position display, 374
Replace worn tools, *367*
re-run a tool, *408*
Re-running tools, 226
reset key, 359, 377, 378, 407, 408
Reset key, *377*
re-sharpened cutter, 182
restart command, 242, 408, 416
revolutions per minute, 109
rigid tapping, 94, *254*, 272
Rotary axes, *315*
rotary axis, 25, 26, 95, 96, 124, 132, 159, 313, 315, 316, 317, 318, 319, 320, 321, 322, 323, 324, 328, 335
rpm, 109
Running Production, 40, 168, 281, 301, 338, 345, 364, 368, 389, 407
running the first workpiece, *417*
Run-out, 182

S

S, 97
S word, 29
safety, 104
Safety, 2, 104, 133, 220, 225, 227, 234, 237, 242, 295, 304, 305, *405*, 406
safety commands, 234, 242
save the program, *363*
Scaling commands, *306*
Secondary reference position, *306*
sequence number, 48, 49, 291, 295, 296, 408
Sequence number, *295*
sequence numbers, 48
setting page, *307*
setup, 85, 88
Setup documentation, *349*
setup person, 40
Setup procedures, *393*
sfm, 109
Shifting the point of reference, 217
single block, *251*, 343, *358*, 360, 361, 362, 380, 389, 407, 409, 417
Single block on/off switch, *380*
Single direction positioning mode, *308*
single-arm tool changer, 230

Index

sizing, 157, 182, *367*
Sizing, 162, 173, 204
slash code, 98, *291*, 377, 380, 389
slash key, *377*
soft jaws, 385
soft keys, 373, 377
solid-model, 357
spindle, 28, 36, *342*
    activation, 29, 36
    direction, 29, 36
    range, 29, 36
    speed, 28, 36
Spindle control, *381*
spindle gap, 220
spindle orient position, 271
spindle probe, 81
spindle speed, 97
Spot drill, 128
statement labels, *296*
static nature, 355
straight-line motion, 134
structure, 235
sub-plate, 217
subprogram, 96, 105, 279, 280, 281, 282, 283, 284, 285, 287, 288, 289, 290, 293, 294, 295, 296, 307, 310, 311
subprogram number, 96
Syntax mistakes, 105, 120

T

T, 31, 37, 97, 163, 172, 228, 230, 231, 242, 295, 387, 415
T word, 31, 37, 97, 228, 231
table vise, 351
tapping compound, 255
tapping cycle, 252, *254*, 255, 272
target value, 156, 157, 160, 173, 204, 205, 362, 367
Tear down, *350*
temporary shift of program zero, *283*
tension/compression tap holder, *254*
Thread milling, 124, 250, 289, 290, *297*, 298, 299, 312
thread milling cutter, 297
time, 103
tolerances, 156
tool change position, 132, 164, 165, 194, 228, 229, 230, 231, 232, 233, 236, 238, 244, 313, 342, 358, 359, 360, 361, 366, 386, 387, 390, 408, 409, 416, 417
tool crib attendant, 40
Tool ending format, 235
tool length compensation, 48, 67, 87, 97, 99, 105, 130, 131, 132, 135, 136, 137, 138, 141, 142, 144, 145, 146, 147, 157, 158, 159, 160, 161, 162, 163, 164, 165, 166, 167, 168, 170, 171, 172, 173, 179, 181, 182, 184, 187, 189, 190, 191, 194, 195, 196, 197, 198, 200, 201, 202, 205, 214, 215, 217, 219, 220, 227, 229, 231, 232, 236, 237, 239, 244, 259, 260, 261, 262, 287, 290, 294, 301, 310, 338, 341, 342, 345, 355, 356, 358, 359, 360, 362, 367, 381, 413, 414, 416, 417
tool length compensation offset number, 97
tool length compensation value, 166
tool length compensation values, *355*
tool path, 117, 190
tool path verification system, 357

tool paths, 148
tool pressure, 182
Tool pressure, 155, 156, 172, 173, 174, 182, 203, 255, 256, 299, 312, 355
Tool startup format, 235
tool tip, 161
tool touch off probe, 394
Tooling engineer, 40
Tooling producing companies, 39, 337
Trial boring, *293*
trial machining, 156, 162, *361*
Trial machining, 156, 157, 160, 162, 172, 173, 182, 183, 186, 202, 203, 204, 205, 226, 292, 293, 294, 295, 312, 355, 361, 362, 363, 368, 369, 370, 406, 408, 417
trig chart, 113
trigonometry, 112
Typical mistakes, 171

U

User defined canned cycles, *289*
Utilities, *289*
utility applications, *288*

V

Vertical machining center, 21, 35
vertical machining centers, 236

W

waiting position, 31, 97
wear offsets, 377
witness mark, *256*
word order, 49
work holding setup, 41, 118, 171, 352
work holding set-up, 118
work surface tool path, 148, 180, 184, 191, 192, 195, 197
workholding setup, 88, *351*
Workpiece producing companies, 39, 41, 337

X

X, 50, 95, 96, *258*, 299, 304
X axis designator, 95

Y

Y, 95, *258*, 304
Y axis designator, 95

Z

Z, 95, 162, 164, 174, 220, *258*, 263, 275
Z axis designator, 95
Z hole bottom position, *263*
zero return, 233, *386*
zero return command, 99, *251*, 317
zero return mode, 341, 342, *385*, 386
zero return position, 68, 88, *341*, 374, 386, 387
Zero return position, 68, 77, 78, 79, 81, 84, 86, 88, 89, 130, 131, 132, 135, 136, 137, 138, 141, 142, 144, 145, 146,

Index

147, 161, 166, 170, 196, 197, 198, 201, 211, 212, 213, 215, 216, 217, 218, 219, 220, 221, 230, 233, 234, 236, 238, 242, 244, 251, 259, 260, 262, 287, 306, 308, *316*, 317, 318, 335, 341, 342, 343, 353, 358, 360, 374, 381, 382, 386, 387, 408, 415

Zero return position indicator lights, *382*

Machining Center Programming Quick Reference Card

G Codes

Code	Description	Status	Initialized	Modal	Code	Description	Status	Initialized	Modal
G00	Rapid motion	Std	Yes	Yes	G56	Instate fixture offset #3	Std	No	Yes
G01	Straight line cutting motion	Std	No	Yes	G57	Instate fixture offset #4	Std	No	Yes
G02	CW circular cutting motion	Std	No	Yes	G58	Instate fixture offset #5	Std	No	Yes
G03	CCW circular cutting motion	Std	No	Yes	G59	Instate fixture offset #6	Std	No	Yes
G04	Dwell	Std	No	No	G60	Single direction positioning	Opt	No	Yes
G09	Exact stop check, one shot	Std	No	No	G61	Exact stop check mode	Std	No	Yes
G10	Offset input by program	Opt	No	No	G64	Normal cutting (cancel G60&G64)	Opt	No	Yes
G17	XY plane selection	Std	Yes	Yes	G65	Custom macro call	Opt	No	No
G18	XZ plane selection	Std	No	Yes	G66	Custom macro modal call	Opt	No	Yes
G19	YZ plane selection	Std	No	Yes	G67	Cancel custom macro call	Opt	Yes	No
G20	Inch mode	Std	Yes	Yes	G68	Coordinate system rotation	Opt	No	Yes
G21	Metric mode	Std	No	Yes	G69	Cancel rotation	Opt	Yes	Yes
G22	Stored stroke limit instating	Opt	No	Yes	G73	Chip breaking peck drilling	Std	No	Yes
G23	Stored stroke limit cancel	Opt	Yes	Yes	G74	Left hand tapping cycle	Std	No	Yes
G27	Zero return check	Std	No	No	G76	Fine boring with no drag line	Std	No	Yes
G28	Zero return command	Std	No	No	G80	Cancel canned cycle	Std	Yes	Yes
G29	Return from zero return	Std	No	No	G81	Drilling cycle	Std	No	Yes
G30	Second reference point return	Opt	No	No	G82	Counterboring cycle	Std	No	Yes
G31	Skip cutting for probe	Opt	No	No	G83	Deep hole peck drilling cycle	Std	No	Yes
G40	Cancel cutter radius comp.	Std	Yes	Yes	G84	Right hand tapping cycle	Std	No	Yes
G41	Cutter radius comp. left	No	No	Yes	G85	Reaming cycle	Std	No	Yes
G42	Cutter radius comp. right	No	No	Yes	G86	Boring cycle	Std	No	Yes
G43	Instate tool length comp. (+)	No	No	Yes	G87	Back boring cycle	Std	No	Yes
G44	Instate tool length comp. (-)	No	No	Yes	G88	Boring cycle	Std	No	Yes
G45	Tool offset expansion	No	Yes	Yes	G89	Boring cycle with dwell	Std	No	Yes
G46	Tool offset reduction	Std	No	Yes	G90	Absolute mode	Std	No	Yes
G47	Tool offset double expansion	Std	No	Yes	G91	Incremental mode	Std	Yes	Yes
G48	Tool offset double reduction	Std	No	Yes	G92	Program zero designator	Std	No	Yes
G49	Cancel tool length comp.	Std	Yes	Yes	G98	Return to initial plane (G73-G89)	Std	Yes	Yes
G50	Cancel scaling	Opt	Yes	Yes	G99	Return to rapid plane (G73-G89)	Std	No	Yes
G51	Scaling on	Opt	No	Yes					
G52	Return to base coord. system	Opt	Yes	Yes					
G53	Shift to mach. coord. system	Std	No	No					
G54	Instate fixture offset #1	Std	No	Yes					
G55	Instate fixture offset #2	Std	No	Yes					

Notes about G codes: 1) Machine tool builders vary dramatically with regard to which G codes they make standard. 2) Parameters control the initialized state of certain G code groups (like G90-G91). 3) Not all control models include all G codes shown in this list.

Common M codes

Code	Description	Status	Initialized	Modal
M00	Program stop	Std	No	No
M01	Optional stop	Std	No	No
M02	End of program (no rewind)	Std	No	No
M03	Spindle on forward (CW)	Std	No	Yes
M04	Spindle on reverse (CCW)	Std	No	Yes
M05	Spindle off	Std	No	Yes
M06	Tool change command	Std	No	No
M07	Mist coolant	Opt	No	Yes
M08	Flood coolant	Std	No	Yes
M09	Coolant off	Std	Yes	Yes
M19	Spindle orient	Std	No	Yes
M30	End of program (rewinds)	Std	No	No
M98	Subprogram call	Std	No	No
M99	Subprogram return	Std	No	No

Other M codes you may have

Code	Description	Status	Initialized	Modal
____	Pallet change	___	___	___
____	Chip conveyor on	___	___	___
____	Chip conveyor off	___	___	___
____	Hydraulic clamp on	___	___	___
____	Hydraulic clamp off	___	___	___
____	Indexer rotation	___	___	___

As with G codes, M code numbers vary dramatically from one machine tool builder to another. Be sure to check the M codes list that comes with your machine to see what other M codes you may have.

Machining Center Programming Quick Reference Card

Letter Addresses Used With Programming

O	This word is used to designate programs in the control's memory. O0001 through O9999 can be used. The program number is always the very first word of the CNC program. No decimal point is allowed with this word.
N	This word designates a sequence number (also called *block* number). It is used for line identification only, and does not have to be in the CNC program at all. To keep programs organized, beginners should include sequence numbers and place them in a logical order. N0001 (or N1) through N9999 can be used. No decimal point is allowed with this word.
G	This word specifies a preparatory function. Preparatory functions prepare the control for what is coming in the current command or future commands (many G codes are modal). Though there are a few exceptions, G codes commonly range from G00 through G99 and normally do not include a decimal point. For a full list of G codes, see the reverse side of this quick reference card.
X	This word designates a positioning movement along the X axis. In the inch mode, a the smallest increment of programming is 0.0001 inch. Though a fixed format will be used if a decimal point is not specified in this word (X10000 will be taken as 1 inch in the inch mode or 10 mm in the metric mode), beginners should get into the habit of specifying this word with a decimal point. Example: X10.375 Secondary use: X is used with a dwell command to specify the time of dwell. G04 X0.5 is a 5 second dwell. Secondary use: X is used with hole machining canned cycles to specify the hole center position in X.
Y	This word designates a positioning movement along the Y axis. In the inch mode, a the smallest increment of programming is 0.0001 inch. Though a fixed format will be used if a decimal point is not specified in this word (Y10000 will be taken as 1 inch in the inch mode or 10 mm in the metric mode), beginners should get into the habit of specifying this word with a decimal point. Example: Y0.25 Secondary use: Y is used with hole machining canned cycles to specify the hole center position in Y.
Z	This word designates a positioning movement along the Z axis. In the inch mode, a the smallest increment of programming is 0.0001 inch. Though a fixed format will be used if a decimal point is not specified in this word (Y10000 will be taken as 1 inch in the inch mode or 10 mm in the metric mode), beginners should get into the habit of specifying this word with a decimal point. Example: Z-0.437 Secondary use: Z is used with hole machining canned cycles to specify the bottom position of the hole.
A B C	For machining centers that have rotary axes, these words are used to specify positioning along the rotary axis. The smallest increment of programming is 0.001 degree. As with X, Y, and Z, beginners should program these words with a decimal point. A position of 45 degrees is specified as A45. (or B45.) Secondary use: For machines with one degree indexers, these words are also used. Note that with a one degree indexer, no decimal point is allowed (A30 specifies a 30 degree index).
R	Used to designate the radius of a circular movement. Secondary use: Used with hole machining canned cycles to designate the rapid plane.
I, J, K	Can be used to designate the center of radius being formed with circular commands, but beginners should concentrate on using the R word. These words specify the distance and direction from the start point to the center of the arc in X, Y, and Z respectively.
F	Used to designate feedrate. Most machining centers are programmed exclusively in per minute designation, meaning a feedrate of 10 inches per minute (or 10 mm per minute) is specified as F10.0.
S	Used to designate spindle speed in rpm. No decimal point is allowed with this word. S500 specifies 500 rpm. Note that if the machine has more than one spindle range, this word also has the machine select the range.
T	For machines having double arm tool changers, T specifies the tool to be placed in the waiting position (not in the spindle). It is commonly a two digit word, not allowing a decimal point. T04 places tool station number four in the waiting position. Note that for some single arm tool changers, this word does actually cause the tool change.
M	This word specifies one of a series of two digit miscellaneous functions. No decimal point is allowed with this word. For a full list of M codes, see your machine tool builders programming manual.
D	A two digit integer word, D specifies the offset number to be used with cutter radius compensation..
H	A two digit integer word, H specifies the offset number to be used with tool length compensation..
P	P specifies the program number to call with subprogramming techniques. No decimal point is allowed with P. Secondary use: P can also be used to specify the length of time in a dwell command.
Q	Specifies the depth of peck for peck drilling cycles G73 and G83.
/	Optional block skip word (also called block delete). When programmed at the beginning of a command, this word causes the control to look at an on/off switch on the control panel. If the switch is on, the command is skipped.

Made in the USA
Middletown, DE
29 October 2024